Mathematics, Informatics, Science and Education International Conference (MISEIC 2017)

Journal of Physics: Conference Series Volume 947

Surabaya, Indonesia
9 September 2017

Editors:

Yuni Sri Rahayu
Rooselyna Ekawati
Asep Bayu Dani Nandiyanto
Agung Lukito
Setya Chendra Wibawa

ISBN: 978-1-5108-5565-6
ISSN: 1742-6588

Printed from e-media with permission by:

Curran Associates, Inc.
57 Morehouse Lane
Red Hook, NY 12571

Some format issues inherent in the e-media version may also appear in this print version.

Printed by Curran Associates, Inc. (2018)

For permission requests, please contact the Institute of Physics
at the address below.

Institute of Physics
Dirac House, Temple Back
Bristol BS1 6BE UK

Phone: 44 1 17 929 7481
Fax: 44 1 17 920 0979

techtracking@iop.org

Additional copies of this publication are available from:

Curran Associates, Inc.
57 Morehouse Lane
Red Hook, NY 12571 USA
Phone: 845-758-0400
Fax: 845-758-2633
Email: curran@proceedings.com
Web: www.proceedings.com

Table of contents

Volume 947

Mathematics, Informatics, Science and Education International Conference (MISEIC)

9 September 2017, Surabaya, Indonesia

Accepted papers received: 18 December 2017
Published online: 10 January 2018

Preface

Papers

IOP Conf. Series: Journal of Physics: Conf. Series **947** (2017) 011001 doi:10.1088/1742-6596/947/1/011001

PREFACE

The Mathematics, Informatics, Science and Education International Conference (MISEIC) 2017 is organized by Universitas Negeri Surabaya is motivated to bridge the scientists, education experts and practicioners, and students in the scientific forum through sharing ideas and issues about theoretical and practical knowledge in mathematics, informatics, science and education. This seminar will be held on September 9, 2017 and will take place in Surabaya, Indonesia. The theme for this conference is "Building Togetherness for The Upcoming Challenges in Mathematics, Informatics, Science, and Education". Thus, this conference has aims: (1) to bring together the scientists, engineers, researchers and practitioners, students, and civil society organization representatives in the scientific forum. (2) To share and to discuss theoretical and practical knowledge about innovation in applied science and engineering.

This conference hosted by the Universitas Negeri Surabaya is considered to accommodate discussion among researchers in the field of mathematics, informatics, science and education as a scientific forum. Therefore, the invited speakers in this conference are expert in the field of mathematics, informatics, science and education.

The MISEIC 2017 took place in Surabaya, Indonesia, on 9 September 2017 involving more than 300 papers are participated from various topics including pure and applied mathematics, science and technology innovation, computer science, and innovation in mathematics, science, and computer science education. There are 75 selected papers that go through a strict peer reviewed process, and these papers will be published in the present conference proceedings.

We would like to thank the organizing committee and the members of reviewers for their kind assistance in reviewing the papers. We would also extend our best gratitude to keynote speakers for their invaluable contribution and worthwhile ideas shared in the conference.

The Editors

Yuni Sri Rahayu (Universitas Negeri Surabaya, Indonesia)
Rooselyna Ekawati (Universitas Negeri Surabaya)
Asep Bayu Dani Nandiyanto (Universitas Pendidikan Indonesia)
Agung Lukito (Universitas Negeri Surabaya)
Setya Chendra Wibawa (Universitas Negeri Surabaya)

IOP Conf. Series: Journal of Physics: Conf. Series **947** (2017) 011001 doi:10.1088/1742-6596/947/1/011001

CONFERENCE PHOTOGRAPHS

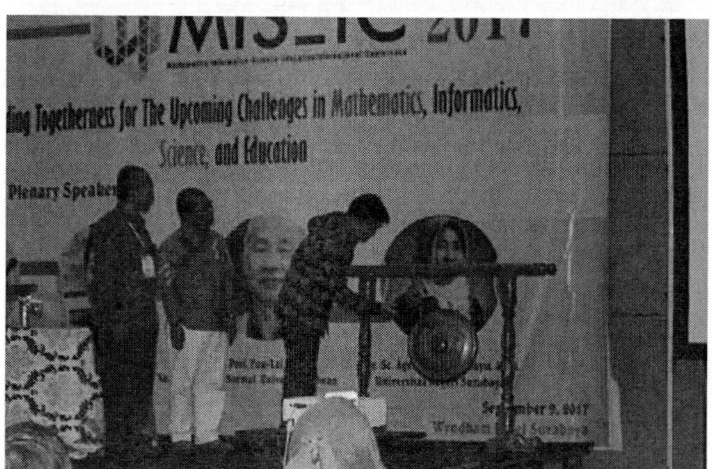

Organized by

Science, and Education

Informatics,

IOP Conf. Series: Journal of Physics: Conf. Series **947** (2017) 011001 doi:10.1088/1742-6596/947/1/011001

IOP Conf. Series: Journal of Physics: Conf. Series **947** (2017) 011001 doi:10.1088/1742-6596/947/1/011001

LIST OF COMMITTEE Chairman:
Dr. Agung Lukito, M.S. (Universitas Negeri Surabaya)

Technical Chairman:
Rooselyna Ekawati, S.Si., M.Sc., Ph.D. (Universitas Negeri Surabaya)

Scientific Commitee:
Dr. Yuni Sri Rahayu, M.Si. (Universitas Negeri Surabaya, Indonesia)
Dr.ir. W.T. van Horssen* (TU Delft, The Netherlands)
Prof. Pitoyo Hartono* (Chukyo University, Japan)
Prof. Fou-Lai Lin* (National Taiwan Normal University, Taiwan)
Dr. Ade Gafar Abdullah (Universitas Pendidikan Indonesia, Indonesia)
Dr. Eng. Asep Bayu Dani Nandiyanto (Universitas Pendidikan Indonesia, Indonesia)
Dr. Dieky Adzkiya, M.Si (Institut Teknologi 10 Nopember, Indonesia)
Prof. Dr. Sari Edi Cahyaningrum, M.Si (Universitas Negeri Surabaya, Indonesia)
Prof. Dr. Dwi Juniati, M.Si (Universitas Negeri Surabaya, Indonesia)
Dr. Sven Schulte* (University of Bremen, Germany).
Prof. Dr. Muslimin Ibrahim (Universitas Negeri Surabaya, Indonesia)
Dr. Yusuf Fuad, M.App.Sc. (Universitas Negeri Surabaya, Indonesia)
Dr. Elly Matul Imah, S.Si, M.Kom. (Universitas Negeri Surabaya, Indonesia)
Prof. Dr. Asep Kadarohman* (Universitas Pendidikan Indonesia, Indonesia)
Dr. I.G.P. Asto Buditjahjanto, S.T., M.T. (Universitas Negeri Surabaya, Indonesia)
Arie Wardhono, Ph.D. (Universitas Negeri Surabaya, Indonesia)
Dr. Abadi, M.Sc (Universitas Negeri Surabaya, Indonesia)
Dr. Manuharawati, M.Si (Universitas Negeri Surabaya, Indonesia)
Prof. Dr. Endang Susantini, M.Pd (Universitas Negeri Surabaya, Indonesia)

Peer review statement

All papers published in this volume of *Journal of Physics: Conference Series* have been peer reviewed through processes administered by the proceedings Editors. Reviews were conducted by expert referees to the professional and scientific standards expected of a proceedings journal published by IOP Publishing.

IOP Conf. Series: Journal of Physics: Conf. Series **947** (2017) 012001 doi:10.1088/1742-6596/947/1/012001

The Effect of Sodium Hydroxide Molarity on Strength Development of Non-Cement Class C Fly Ash Geopolymer Mortar

A Wardhono[1]

[1]Department of Civil Engineering, Universitas Negeri Surabaya, Indonesia
*Corresponding author: ariewardhono@unesa.ac.id

Abstract. The use of fly ash as cement replacement material can overcome the environmental issues, especially the global warming problem caused by the greenhouse effect. This is attributed to the CO2 gas produced during the cement manufacturing process, which 1 ton of cement is equivalent to 1 ton CO2. However, the major problem of fly ash is the requirement of activators to activate the polymer reactions. The most common activator used in non-cement or geopolymer material is the combination of sodium hydroxide (NaOH) and sodium silicate. This study aims to identify the effect of NaOH molarity as activator on strength development of non-cement class C fly ash geopolymer mortar. The molarity variations of NaOH were 6 Molar (M), 8M, 10M, 12M, 14M and 15M. The compressive strength test was performed at the age of 3, 7 and 28 days in accordance with ASTM standard, and the specimens were cured at room temperature. The results show that the highest compressive strength was achieved by geopolymer mortar with a molarity of 12M. It exhibits a higher strength to that normal mortar at 28 days. However, the use of NaOH molarity more than 12M tends to decrease the strength of non-cement geopolymer mortar specimens.

1. Introduction

Cement-based concrete is the most commonly used construction material. Concrete is conventionally manufactured by using Ordinary Portland Cement (OPC) as the primary binder, sand and gravel materials. The ratio of OPC in conventional concrete is approximately 10% to 15% by the total mass of concrete [1,2]. However, the use of OPC in concrete has an adverse problem for the environment. According to Davidovits [3], the production of 1 ton of OPC also released about 0.7-1 ton CO_2. The contribution of cement production alone is approximately 6% anthropogenic CO_2 gas emissions which cause the greenhouse effect and leads to the occurrence of global warming problems. Furthermore, this condition corroborated with the impact on the environment due to the depletion of quarries resources of limestone as OPC raw materials has led to the search for more environmentally friendly cement replacement materials.

Fly ash, a waste product from burning coal power plant, has been commonly used as an additive or cement replacement material due to the pozzolanic characteristic to improve mechanical and durability properties of concretes [4-6]. Recent researches also show that fly ash can be used as fully cement replacement materials with the addition of alkaline activator. This non-cement fly ash based concrete material is known as geopolymer [3, 4, 8, 9]. According to Davidovits [7], the mechanism of geopolymer reaction involves the polycondensation reaction of alumino-silicate-oxide with alkali polysialates resulting polymeric Si-O-Al bonds and forms a geopolymer matrix. This hydration product is more polymer than C-S-H gel on conventional concrete [3].

The strength properties of geopolymer are influenced by the molarity of alkaline activator, particularly in the molarity of sodium hydroxide (NaOH). Increasing the NaOH molarity will significantly increase the strength development of geopolymer specimen [10, 11]. According to Ryu et al. [11], it attributes to the activation of the reaction of the internal silicate (Si) and aluminate (Al) components caused by the increased breakage of the glassy chain of fly ash precursor, in which it is provoked by the high alkalinity resulting from the increase of the molarity of NaOH.

This study presents the effect of sodium hydroxide molarity on strength development of non-cement class C fly geopolymer mortar. The strength properties were identified by the compressive strength test at the age of 3, 7 and 28 days in accordance with ASTM standard [12].

2. Method

2.1. Materials

Class C fly ash from Paiton Power Plant Indonesia was used as primary material to develop class C fly ash geopolymer mortar specimens. It has a high ferrite (Fe) and a low silicate (Si) contents. The chemical breakdown for class C fly ash material was identified by the X-Ray Fluorescence (XRF) test using PANalytical type Minipal 4 equipment. The chemical content of class C fly ash are shown in Table 1.

Table 1. Chemical composition of class C fly ash (mass %).

Materials	SiO_2	Al_2O_3	Fe_2O_3	CaO	MgO	K_2O	TiO_2	Mn_2O_3	SO_3
Fly ash class C	13.47	4.92	53.48	18.36	0.00	1.54	1.35	0.51	0.32

Fly ash in this study was class C fly ash. It had CaO of 18.36% (> 10%), the total SiO_2+Al_2O_3+ Fe_2O_3 of 71.87% (> 50%), and SO_3 of 0.32% (< 5%) in accordance with ASTM C618 [13]. A mix of sodium silicate with SiO_2/Na_2O ratio of 3.30 and sodium hydroxide (NaOH) were used as alkaline activators.

2.2. Mix design and proportion

The details of class C fly ash geopolymer (FACG) mortars mix proportion are shown in Table 2.

Table 2. Mix proportions of class C fly ash geopolymer mortars (FACG)

Mixture	Portland Cement	Class C fly ash	Fine sand	Sodium silicate	NaOH	Water	Molarity (M)
OPC	1	-	2.75	-	-	0.40	-
FACG-1	-	1	2.75	0.32	0.21	-	6
FACG-2	-	1	2.75	0.32	0.21	-	8
FACG-3	-	1	2.75	0.32	0.21	-	10
FACG-4	-	1	2.75	0.32	0.21	-	12
FACG-5	-	1	2.75	0.32	0.21	-	14
FACG-6	-	1	2.75	0.32	0.21	-	15

Mix proportions of FACG mortars were developed from previous research [14, 15], while OPC mortar specimen as control was developed in accordance with ASTM C109 [12] with a sand to binder ratio of 2.75. A sodium silicate to NaOH (SS/SH) ratio of 1.5 was applied in accordance with previous research [14]. A water to solid ratio of 0.35 was used for FACG mortars rather than a water binder ratio due to a solid content in NaOH and sodium silicate. The quantity of water in FACG mortars was calculated as the total sum of water in sodium silicate and NaOH solutions, while the quantity of solid was determined from the mass of the solid content in activator solution and class C fly ash raw material [15]. A normal curing temperature at room temperature used for OPC mortar was applied for FACG mortars. This is attributed to the high calcium (Ca) content in the raw class C fly ash material.

2.3. Testing mortar specimens
The strength characteristics of class C fly ash geopolymer mortars in 5 x 5 x 5 cm^3 cube specimens were carried out by compressive strength test using Universal Testing Machine (UTM) in accordance with ASTM C109 [12]. Three mortar geopolymer cubes were tested at the age of 3, 7 and 28 days after casting.

3. Results and Discussions

3.1. Strength development of FACG mortars
Table 3 and Figure 1 give the strength development reported for the FACG mortar specimens for all mixes curing at room temperature.

Table 3. Compressive strength of FACG mortars curing at room temperature

Mixture	Compressive strength (MPa)		
	3 days	7 days	28 days
OPC	10.28 $\pm$ 0.17	12.93 $\pm$ 0.17	18.31 $\pm$ 0.22
FACG-1	4.92 $\pm$ 0.41	7.46 $\pm$ 0.31	12.95 $\pm$ 1.36
FACG-2	5.22 $\pm$ 0.54	8.26 $\pm$ 0.15	15.85 $\pm$ 0.69
FACG-3	5.69 $\pm$ 0.11	8.91 $\pm$ 0.16	18.58 $\pm$ 1.31
FACG-4	6.21 $\pm$ 0.32	9.16 $\pm$ 0.18	22.47 $\pm$ 0.46
FACG-5	5.36 $\pm$ 0.17	8.01 $\pm$ 0.15	15.56 $\pm$ 0.12
FACG-6	3.92 $\pm$ 0.25	6.51 $\pm$ 0.23	15.05 $\pm$ 0.41

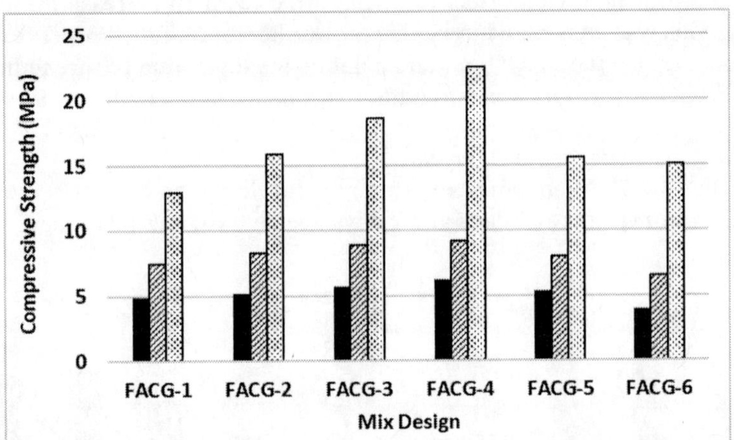

Figure 1. Strength development of FACG mortars throughout 28 days age

The result shows that FACG mortars exhibited a lower strength at early age compared to that OPC mortar. The highest early strength of FACG mortar specimens had merely achieved a compressive strength of 9.16 MPa (FACG-4), lower than OPC mortar with the strength of 12.93 MPa. However, all mixes demonstrated a significant increase in strength with time.

The highest compressive strength at early age was achieved by FACG-4 with 6.21 MPa (27.64% of final strength at 28 days) and 9.16 MPa (40.77%) at 3 days and 7 days, respectively, followed by FACG-3 with the strength of 5.69 MPa (30.62% at 3 days) and 8.91 MPa (47.95% at 7days). FACG-4 mortar was also demonstrated the highest final strength to all FACG mortars at 28 days with the strength of 22.47 MPa. The final strength of FACG-4 even exceeded the compressive strength of OPC mortar with 18.31 MPa at 28 days.

Table 4. Strength development ratio of FACG mortars from 3 to 28 days age

Mixture	Compressive strength ratio		
	3 to 28 days	7 to 28 days	28 days (Final)
OPC	0.56	0.70	1
FACG-1	0.38	0.58	1
FACG-2	0.33	0.52	1
FACG-3	0.31	0.48	1
FACG-4	0.27	0.41	1
FACG-5	0.34	0.51	1
FACG-6	0.27	0.43	1

Table 4 shows the strength development ratio of FACG mortar specimens throughout 28 days age. Ratio of strength development was determined based on the final strength at 28 days. The standard ratio of strength development of OPC (normal) mortar is 0.56 and 0.70 at 3 and 7 days, respectively as shown in Table 3. This finding was in accordance with ASTM Standard. According to ASTM C1074 [16], the maturity ratio of concrete strength development to the final strength at 28 days is 0.54 and 0.72 at the 3 days and 7 days, respectively. In contrary, all FAGC mortars demonstrates a lower strength development ratio at early age compared to OPC concrete. The highest ratio of FACG mortar was achieved by FACG-1 with the ratio of 0.38 and 0.58 at 3 days and 7 days, respectively. This ratio was significantly lower than the standard ratio of OPC mortar and ASTM C1074 standard. This indicated that the reaction rate of geopolymeric reaction are slower compared to that C-S-H reaction rate at OPC concrete. According to Silva et al. [17], the amount of aluminate (Al) available for the geopolymer reaction during synthesis process appears to have a significant effect of setting time, with an increase in molarity leading to a longer setting time. This was considered to be due to the alteration of the properties of the geopolymer during condensation. Despite the FAGC mortars exhibits the low value on strength development ratio, these mortars, i.e. FACG-4, show a better strength performance at later age.

3.2. Effect of molarity on strength development

Figure 2 displays the effect of molarity on strength development of non-cement class C fly ash geopolymer mortar specimens (a) at 3 days and 7 days age, and (b) at 28 days age.

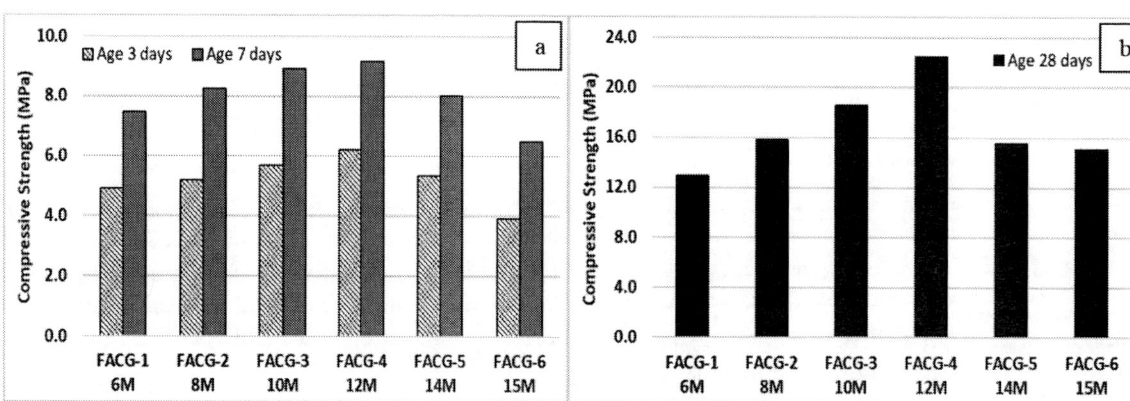

Figure 2. The effect of molarity on strength development of FACG mortar specimens at (a) 3 and 7 days, (b) 28 days

Figure 2 shows that sodium hydroxide (NaOH) molarity significantly affects the strength development of non-cement class C fly ash geopolymer (FACG) mortar specimens. The increase in molarity significantly increase the strength performance of FACG mortars. An increase of 7.55% strength showed in FACG-1 (strength of 4.92 MPa) to FACG-2 (strength of 5.22 MPa) with molarity of

6 Molar to 8 Molar at 3 days age. A further increase with an average strength increase of 9.07% has also identified in FACG-2 (8 Molar, 5.22 MPa) to FACG-3 (10 Molar, 5.69 MPa) and FACG-3 to FACG-4 (12 Molar, 6.21 MPa); with the highest strength performance was achieved by FACG-4 with 12 Molar at the age of 3 days. Similar strength performance was also identified on FACG mortars at the age of 7 days with the total increase of 22.79% from FACG-1 (6 Molar) to FACG-4 (12 Molar) with the strength increase from 7.14 MPa and 9.16 MPa. The highest increase of strength was demonstrated at the final ages (28 days) with the total strength increase of 73.51%. The final strength of FACG mortars increased from 12.95 MPa to 22.47 MPa along with the increase of molarity from 6 Molar (FACG-1) to 12 Molar (FACG-4). This finding was in accordance with previous researcher. According to Hardjito [18], the molarity of the NaOH as activator solution has a significant effect on the strength performance of the fly ash-based geopolymer specimens. In the geopolymer reaction, increasing the molarity will promote an acceleration in the reaction rate due to an increase in the soluble silicate and the higher concentration of reactants [18].

Figure 2 displays that an increase of molarity above 12 Molar did not significantly affect the strength performance of FACG mortar specimens. It tended to reduce the compressive strength at higher molarity. The increase of molarity from 12 Molar (FACG-4) to 14 Molar showed a strength reduction throughout 28 days. A decrease of 13.69% and 12.55% were shown in 3 days and 7 days with a strength reduction from 6.21 MPa to 5.36 MPa and from 9.16 MPa to 8.01 MPa, respectively. Further strength reduction was also observed at high molarity (15 Molar) with total reduction of 26.87% (at 3 days) and 18.73% (at 7 days) from 5.36 MPa to 3.92 MPa and from 8.01 MPa to 6.51 MPa, respectively. Similar behavior was also observed at 28 days age. A total strength reduction of 33.02% from 22.47 MPa to 15.05 MPa has identified at the increase of molarity from 12 Molar (FACG-4) to 15 Molar (FACG-6). Similar finding was found by previous researcher. According to Fernandez-Jimenez et al. [19], higher alkali modulus did not affect the strength development of geopolymer specimens. Increasing the alkali modulus causes a NaOH content reduction which is used to dissolve the Al_2O_3 and the SiO_2 monomer from the fly ash grain precursor material. Reducing the NaOH content will cause a reduction on the Na cations and affects the electric neutrality of Si-O-Al-O monomer in the geopolymeric matrix. This leads to a lower strength at higher molarity [19].

4. Conclusions

The effect of NaOH molarity on strength development of class C fly ash geopolymer mortars were studied experimentally from 3 to 28 days age. The primary conclusions may be drawn based on this study are: (1) Molarity of NaOH significantly affected the compressive strength development of class C fly ash geopolymer mortar specimens. (2) The optimum compressive strength was achieved by FACG-4 mortar with the molarity of 12 Molar throughout 28 days age. (3) Increasing molarity above 12 Molar did not significantly affect the strength development of class C fly ash geopolymer mortar but tend to decrease the mortar strength at higher value. (4) Despite the class C fly ash geopolymer mortar demonstrated a lower strength at early age, however, the 12 Molar geopolymer mortar showed a better strength performance compared to OPC mortar at later age.

Acknowledgment

The authors would like to express their sincere thanks to Universitas Negeri Surabaya for providing the support for this research project.

References

[1] Meyer C 2009 *Cement and Concrete Composites* **31** pp601-605
[2] Peng J Xia, Huang L, Zhao Y Bo, Chen P, Zeng L, and Zheng W 2013 *Advanced Materials Research* **610-613** pp2120-2128
[3] Davidovits J 1994 *World Resource Review* **6(2)** pp263-278
[4] Ahmaruzzaman M 2010 *Jurnal of Progress in Energy and Combustion Science* **36(3)** pp327-363
[5] Papadakis G Vagelis 1999 *Cement and Concrete Research* **29(11)** pp1727-1736

[6] Papadakis G Vagelis 2002 *Cement and Concrete Research* **32(10)** pp1525-1532
[7] Davidovits J 1994 *Proc. 1st Int. Conf. on Alkaline Cements and Concretes (Kiev) (Kiev: Scientific Research Institute on Binders and Materials)* pp 131-149
[8] Law D, Adam A, Molyneaux T, Patnaikuni I, and Wardhono A 2015 *Materials and Structures* **48(3)** pp721-731
[9] Gunasekara C, Setunge S, and Law D 2017 *ACI Structural Journal* **144(3)** pp743-752
[10] Budh C D and Warhade N R 2014 *International Journal of Civil Engineering Research* 5(1) pp83-86
[11] Ryu G Sung, Lee Y Bok, Kok K Taek, and Chung Y Soo 2014 *Construction and Building Materials* **47** pp409-418
[12] ASTM Standard **ASTM C109**:2004
[13] ASTM Standard **ASTM C618**:2004
[14] Deb P Sarathi, Nath P and Sarker P Kumar 2014 *Materials and Design* **62 p**32
[15] Wardhono A 2015 *The Durability of Fly Ash Geopolymer and Alkali-Activated Slag Concretes* (Melbourne: Royal Melbourne Institute of Technology University)
[16] ASTM Standard **ASTM C1074**:2004
[17] Silva P D, Sagoe-Crentsil K, and Sirivivatnanon V 2007 *Cement and Concrete Research* **37** pp512-518
[18] Hardjito D 2005 *Studies on Fly Ash-Based Geopolymer Concrete* (Melbourne: Curtin University of Technology)
[19] Fernandez-Jimenez A, Palomo A, and Criado M 2005 *Cement and Concrete Research* **35 pp**1204-1209

IOP Conf. Series: Journal of Physics: Conf. Series **947** (2017) 012002 doi:10.1088/1742-6596/947/1/012002

Kinetic Parameters of Thermal Decomposition Process Analyzed using a Mathematical Model

A B D Nandiyanto[1,*], R Ekawati[2], S C Wibawa[3]

[1]Departemen Kimia, Universitas Pendidikan Indonesia, Indonesia
[2]Departemen Matematika, Universitas Negeri Surabaya, Indonesia
[3]Departemen Pendidikan Informatika, Universitas Negeri Surabaya, Indonesia

Email : nandiyanto@upi.edu

Abstract. The purpose of this study was to show a mathematical analysis model for understanding kinetic parameters of thermal decomposition process. The mathematical model was derived based on phenomena happen during the thermal-related reaction. To get the kinetic parameters (i.e. reaction order, activation energy, and Arrhenius constant), the model was combined with the thermal characteristics of material gained from the thermal gravity (TG) and differential thermal analysis (DTA) curves. As an example, the model was used for analyzing the kinetic properties of trinitrotoluene. Interestingly, identical results gained from the present model with current literatures were obtained; in which these were because the present model was derived directly from the analysis of stoichiometrical and thermal analysis of the ideal chemical reaction. Since the present model confirmed to have a good agreement with current theories, further derivation from the present mathematical model can be useful for further development.

1. Introduction

Analysis of kinetic parameters has been reported as one of the important factors for understanding type of reaction.[1,2] These parameters are required for optimizing the process condition to get the best product.[3-5]

Many studies have reported how to identify the kinetic parameters, as shown by Kissinger[6], Huang et al.[7], Huang and Wu[8], and Lou[9]. Although their models have been referred by many reports, their methods have still limitations, especially for recognizing in detail what values for the reaction order, the activation energy, and the Arrhenius constant.

Based on our previous studies on the material properties [10-13], here, the purpose of this study was to show a mathematical analysis model for understanding the kinetic parameters based on thermal gravity (TG) and differential thermal analysis (DTA) curves. To confirm that the model is effective, the results gained from the present mathematical approach was compared with current literatures. As an example, kinetic parameters of thermal decomposition of trinitrotoluene were analyzed. Trinitrotoluene is well-known as a basic material that is used in wide range of applications, especially for mining uses.[14] Since the present model is in a good agreement with the current theories and literatures in the thermal decomposition process, further studies gained from this model can be useful for further development.

IOP Conf. Series: Journal of Physics: Conf. Series **947** (2017) 012002 doi:10.1088/1742-6596/947/1/012002

2. Experimental method

Mathematical model was derived based on the specific condition of the thermal decomposition process. The model was then applied for analyzing thermal-related reaction parameters of trinitrotoluene and compared with literatures, such as Kissinger[6], Huang et al.[7], Huang and Wu[8], and Lou[9]. In short, the calculation was obtained by adopting thermal characteristics (i.e. the ending temperature (T_{end}), the inflection temperature (T_{i1}), and the maximum temperature (T_m)) and heating condition (i.e. flow rate).

3. Results and Discussion

3.1. Derivation model for kinetic parameters gained from TG- DTA curves

The expression of kinetic reaction induced by thermal decomposition is described as

$$\frac{dx}{dt} = k(T).f(x) \quad (1)$$

where *x*, *t*, *k(T)*, and *f(x)* is the fraction of reactive material, the reaction time, the reaction constant, and the type of reaction model, respectively. The f(x) was assumed as $f(x) = (1-x)^n$, where *n* is the reaction order. *k(T)* was obtained from $k(T) = A.\exp(-E/RT)$, where *A* and *E* are the Arrhenius constant and the activation energy, respectively. *R* is the Boltzmann constant (8.314 J/mol.K). *T* is the process temperature (in K) that depends on the heating rate (ϕ; in K/s) and is approximated as a function of initial temperature (*To*) (expressed as $T = To + \phi.t$). In this model, we also assumed that the temperature deviation is proportional to the decomposition rate of material ($\Delta T = \beta \frac{dx}{dt}$; where ΔT and β are the temperature deviation and the proportional constant, respectively).

To solve equation (1), three boundary conditions were used:

(1) **Boundary 1**: In the end time, shown as $t = t_{end}$ and $T = T_{end}$, the result of ΔT_{end} will be $\Delta T_{end} = 0$.

(2) **Boundary 2**: In the time when reaching maximum temperature, shown as $t = t_m$ and $T = T_m$, the first derivative of ΔT is zero ($\frac{d}{dt}\Delta T = 0$).

(3) **Boundary 3**: When the process reaches inflection time ($t = t_i$), the second derivative of ΔT is zero ($\frac{d^2}{dt^2}\Delta T = 0$).

Finally, by integrating equation (1) with additional boundaries, the result is

$$\frac{T_m - T_{i1}}{T_{end} - T_m} = \frac{\ln\left[\frac{1}{2}\left(2 + n - \sqrt{4n + n^2}\right)\right]}{\ln[1-n]} \quad (2)$$

Further, by substituting the value of *n*, *E* and *A* can be obtained as

$$E = \frac{RT_O^2}{T_m - T_{end}} \ln(1-n) \quad (3)$$

$$A = \frac{E\phi}{RT_o^2 \exp\left(-\frac{E}{RTo}\right)\left[\exp\left(\frac{E}{RT_O^2}[T_m - T_O]\right) - (1-n)\right]} \quad (4)$$

Detailed information on derivation of the above mathematical equations is reported in our previous report [15].

3.2. Simplification of theoretical model from TG and DTA curves

The reaction order n is a function of the measurable characteristic process temperatures, *i.e.*, T_{end}, T_{i1}, and T_m. However, the above equations are inconvenient since solving this equation needs a trial and error approach.

To simplify the above approximation, **Figure 1** shows n versus characteristic temperatures curve based on equation (2). The curve showed that the increases in $\frac{T_m - T_{i1}}{T_{end} - T_m}$ results in the decrease in the value of n. The regression result from these data is a polynomial correlation and can be expressed as

$$n = 0.000932\left(\frac{T_m - T_{i1}}{T_{end} - T_m}\right)^4 - 0.023807\left(\frac{T_m - T_{i1}}{T_{end} - T_m}\right)^3 + 0.213463\left(\frac{T_m - T_{i1}}{T_{end} - T_m}\right)^2 - 0.796363\left(\frac{T_m - T_{i1}}{T_{end} - T_m}\right) + 1.11849 \quad (5)$$

The regression result in equation (5) can be used for approximating the value of n. However, the correlation is effective for n of less than 1 and $\frac{T_m - T_{i1}}{T_{end} - T_m}$ of between 0.10 and 10.

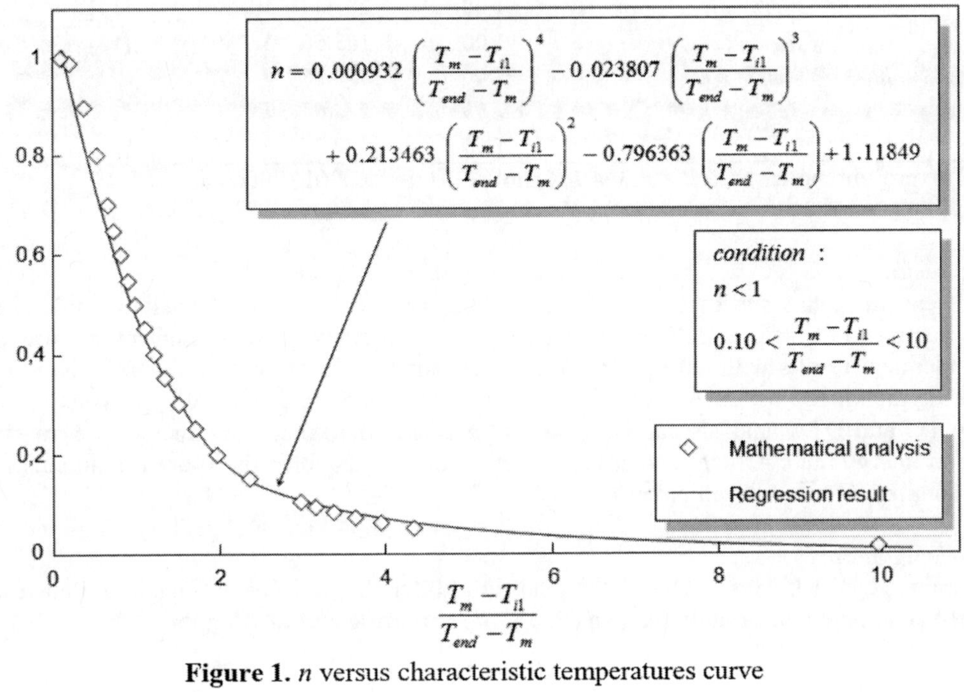

Figure 1. n versus characteristic temperatures curve

3.3. Verification and application of the mathematical model

Table 1 shows kinetic parameters of trinitrotoluene results based on above analysis compared with previous studies. Since the present model is effective in the specific condition, the calculation was limited to the approximation of material with n of less than 1 and $\frac{T_m - T_{i1}}{T_{end} - T_m}$ of between 0.10 and 10. The present model analyzed the value of n using above correlation shown in **Figure 1**, whereas the values of E and A were calculated based on equations (3) and (4), respectively. The result showed that the present model successfully predicted the kinetic parameters in detail including n, E, and A, while other reports[6-9] have some limitations. For instance, one literature can predict the n value, while the other can not estimate E or A. In addition, our present study for the approximation of n, E, and A values is better than our previous study [15], in which more detailed values can be obtained.

Table 1. Kinetic parameters of trinitrotoluene obtained by various methods.

Method	ϕ (°C/min)	n	E (kcal/mol)	A	Ref.
Values in literature	6	1.00	24.20	-	[7]
	10	-	14.00	-	[7]
	15	0.00	29.40	-	[7]
Kissinger		-	19.30	-	[6]
Huang and Wu	6	1.65	22.00	$7.1\ 10^{5}$	[7]
	10	1.75	21.00	$2.9\ 10^{5}$	[7]
	15	2.66	-	-	[7]
Luo	6	1.00	251.00	$4.1\ 10^{93}$	[9]
	10	1.00	193.70	$1.8\ 10^{79}$	[9]
	15	1.00	126.40	$3.7\ 10^{44}$	[9]
Our previous study	6	1.00	364.69	$6.0\ 10^{93}$	[15]
	10	1.00	281.44	$1.8\ 10^{79}$	[15]
	15	1.00	183.66	$3.7\ 10^{44}$	[15]
Present study	*6*	*1.00*	*356.15*	$4.2\ 10^{93}$	
	10	*1.00*	*274.85*	$1.8\ 10^{79}$	
	15	*1.00*	*179.35*	$3.7\ 10^{44}$	

Note: The present model used R = 8.314 J/K.mol and 1 J = 0,000239006 cal.

4. Conclusion

The present study has successfully derived the mathematical analysis model for understanding the kinetic parameters based on TG dan DTA curves. The accuracy of the present model was confirmed by the identical results with current literatures. The analysis of the present model was also done for calculating the kinetic parameters of trinitrotoluene. Since the mathematical approximation confirmed that the TG and DTA analysis can be used for analyzing the kinetic parameters (i.e. reaction order, activation energy, and Arrhenius constant), further derivation from the present mathematical model can be useful for further development.

Acknowledgements

We acknowledged RISTEK – DIKTI for grant in aid Penelitian Terapan Unggulan Penelitian Tinggi (PTUPT) and Penelitian Produk Terapan (PPT) for supporting and funding this study.

References

[1] Nandiyanto A B D, Munawaroh H S H, Kurniawan T and Mudzakir A 2016 Influences of temperature on the conversion of ammonium tungstate pentahydrate to tungsten oxide particles with controllable sizes, crystallinities, and physical properties *Indonesian Journal of Chemistry* **16**(2) pp. 124-129

[2] Anshar A M, Taba P and Raya I 2016 Kinetic and Thermodynamics Studies the Adsorption of Phenol on Activated Carbon from Rice Husk Activated by ZnCl2 *Indonesian Journal of Science and Technology* **1**(1) pp. 47-60.

[3] Zhao J and Yang X 2003 Photocatalytic oxidation for indoor air purification: a literature review *Building and Environment* **38**(5) pp. 645-654.

[4] Putra Z A 2016 Early phase process evaluation: Industrial practices *Indonesian Journal of Science and Technology* **1**(2) pp. 238-248

[5] Nandiyanto A B D, Sofiani D, Permatasari N, Sucahya T N, Wiryani A S, Purnamasari A, Rusli A and Prima E C 2016 Photodecomposition profile of organic material during the partial solar eclipse of 9 march 2016 and its correlation with organic material concentration and photocatalyst amount *Indonesian Journal of Science and Technology* **1**(2) pp. 132-155.

[6] Kissinger H E 1957 Reaction kinetics in differential thermal analysis *Analytical chemistry* **29**(11) pp. 1702-1706

[7] Huang C C, Wu T S and Leu A L 1991 Determination of kinetic parameters for decomposition reaction from a single DTA curve *Thermochimica acta* **188**(1) pp. 119-128

[8] Huang C C and Wu T S 1992 Obtaining kinetic data of energetic material decomposition by differential thermal analysis *Thermochimica acta* **204**(2) pp. 239-250

[9] Luo K M 1995 Calculation of kinetic parameters from DTA curves using the characteristic temperature *Thermochimica acta* **255** pp. 241-254

[10] Nandiyanto A B D, Permatasari N, Sucahya T N, Purwanti S T, Munawaroh H S H, Abdullah A G and Hasanah L 2017 Preparation of Potassium-Posphate-embedded Amorphous Silicate Material from Rice Straw Waste *IOP Conference Series: Materials Science and Engineering* **180**(1) pp. 012138

[11] Nandiyanto A B D, Permatasari N, Sucahya T N, Abdullah A G and Hasanah L 2017 Synthesis of Potassium Silicate Nanoparticles from Rice Straw Ash Using a Flame-assisted Spray-pyrolysis Method *IOP Conference Series: Materials Science and Engineering* **180**(1) pp. 012133

[12] Nandiyanto A B D, Putra Z A, Andika R, Bilad M R, Kurniawan T, Zulhijah R and Hamidah I 2017 Porous Activated Carbon Particles from Rice Straw Waste and Their Adsorption Properties *Journal of Engineering Science and Technology* **12** pp. 1-11

[13] Nandiyanto A B D, Sofiani D, Permatasari N, Sucahya T N, Wiryani A S, Purnamasari A, Rusli A and Prima E C 2016 Photodecomposition profile of organic material during the partial solar eclipse of 9 march 2016 and its correlation with organic material concentration and photocatalyst amount *Indonesian Journal of Science and Technology* **1**(2) pp. 132-155.

[14] Zlateva V and Pavlova S 1998 The impact of trinitrotoluene on eyes in miners. *Meditsina truda i promyshlennaia ekologiia* **1998**(2) pp. 26-29

[15] Nandiyanto A B D 2017 Mathematical Approximation Based on Thermal Analysis Curves for Calculating Kinetic Parameters of Thermal Decomposition of Material *Journal of Engineering Science and Technology* **12** pp. 76-90

The Classification of Diabetes Mellitus Using Kernel k-means

M Alamsyah[1], Z Nafisah[1], E Prayitno[1], A M Afida[1], and E M Imah[1,*]

[1]Mathematics Department, Universitas Negeri Surabaya, Ketintang, Surabaya 60231, East Java, Indonesia
*Corresponding Author. Email: ellymatul@unesa.ac.id

Abstract. Diabetes Mellitus is a metabolic disorder which is characterized by chronicle hypertensive glucose. Automatics detection of diabetes mellitus is still challenging. This study detected diabetes mellitus by using kernel k-Means algorithm. Kernel k-means is an algorithm which was developed from k-means algorithm. Kernel k-means used kernel learning that is able to handle non linear separable data; where it differs with a common k-means. The performance of kernel k-means in detecting diabetes mellitus is also compared with SOM algorithms. The experiment result shows that kernel k-means has good performance and a way much better than SOM.

1. Introduction

Diabetes Mellitus is a chronic metabolic disease which is caused by excessive insulin secretion in irregular glucose. Diabetes Mellitus will increase blood glucose in the body causing the occurrence of some risks such as heart disease, blindness, kidney failure and death [1]. According to International Diabetes Federation, as many as 382 million people live with diabetes in the worldwide. By 2035, this will increase to be twice as 592 million. Diabetes is one of the major health problems of all over the world. All forms of diabetes increase the risk of long-term complications. These typically will develop after many years (10–20), but may be the first symptom in those who have not received a diagnosis before that time [8]. Diabetes mellitus is divided into 2 types, Diabetes mellitus type 1 which is usually occurs in children where the damage from the body produces insulin and the body cannot use insulin properly. Diabetes mellitus type 2 commonly occurs in adults where the body does not produce enough insulin or cells [2].

Medical diagnosis is the process of classification. The classification process is performed on data containing 8 components. Components are composed of components that can indicate the presence of diabetes in a person. This data will be processed to classify whether the patient has diabetes or not. However, the limitation of medical personnel to manually detect large amounts of data is still a constraint. Recently, there are many methods and algorithms used to classify the diabetes disease in the literature. Anuja et al used supporting vector machine for diabetes disease classification, her study presented that the classification accuracy was 78% [3]. Durairaj et al applied back propagation algorithm and the classification accuracy was 91% [4]. Saxena et al used k nearest algorithm with K value of 3 which obtained a classification accuracy of 70% and with K value 5 obtained a classification accuracy of 75% [5]. Although the accuracy produced by some studies have exceeded 50% even closer to 100%, but the accuracy is still less satisfactory.Therefore it is necessary to develop or re-research with algorithms that can produce greater accuracy.

Clustering is a means of data classification which points to homogeneous groups that arise in a number of fields such as pattern recognition, machine learning, data mining and image processing. One of the most popular clustering algorithms is k-means, where homogeneous groups are identified by minimizing the clustering error defined as the sum of the squared Euclidean distances between each data point and the corresponding cluster center [9]. However, k-means has an obvious limitation that it is sensitive to the initial seeds selecting. Random initial centers always lead k-means to trapped in the local optimum easily [11].The k-means clustering algorithm can be enhanced by a kernel function; by using an appropriate nonlinear mapping from the original (input) space to a higher dimensional feature space, one can extract clusters that are non-linearly separable in input space [7]. Kernel k-means is an extension of the standard k-means algorithm that maps data points from input space to a higher dimensional feature space through a nonlinear transformation and minimizes the clustering error in feature space [9].Kernel k-means have been used to identify clusters that are non-linearly separable in the input space [7].

This study used from of data that have been taken to diagnose patients with and without diabetes. The use of k-means algorithm can actually classify the data, but unfortunately the k-means algorithm has very small distance difference and has no effect, so with the use of Kernel k-means algorithm set of data will be transformed to a higher dimension. As a result, data that has a very small distance is considered to be affected. This will certainly alter the degree of accuracy of the classification process. In addition, in this study the researcher compared the level of classification accuracy with Kernel k-means with SOM (Self Organizing Map) algorithm. SOM (Self Organizing Map) is a commonly tool used for classification to diagnose a particular disease of patients.

This paper is organized as follows: Section II consists of research method, and describes some basic concepts of Kernel k-means, and also SOM. Section III describes the data-set used. Section V describes design system of this experiment. The experimental results are given in section VI, and section VII which presents the conclusion of the study.

2. Kernel k-Means

Kernel k-Means [9] is the development of the k-Means Algorithm that uses the methods of kernel for high-dimensional of mapping data in the new space so that it can be separated linearly. This is done to improve the accuracy of cluster results. In Kernel k-means, the data are expected to be well separated because the overlapped data or outlier data can be linear in the new dimension space. The grouping distance also used in the Euclidean distance. This method is better than the k-means cluster when there is an outlier (nonlinear) data. By using the Kernel cluster method, the distance statement is noticed in the Kernel function. The kernel function that used in this study is Gaussian kernel.

Suppose that $u_i = \emptyset(x_i)$ is the transformation ofx_i, so the Euclidean distance formula can be seen on Eq. 1.

$$\mathrm{d}(x_i, x_j) = \left\|\emptyset(x_i) - \emptyset(x_j)\right\|^2 \tag{1}$$

Kernel K-Means Algorithms:
Step 1: Choose δ (x_i, C_k) (1≤ i ≤ N, 1 ≤ k ≤ K) with the initial value,
And form the initial cluster K C_1, C_2, ..., C_K
Step 2: For each clusterC_k, calculate | C_k | and g (C_k)
Step 3: For each sample of exercise x_i and C_k cluster, compute f (x_i, C_k) and then
Findx_i at Cluster nearby.
δ (x_i, C_k)

$1, f\ (x_i, C_k) + g\ (C_k) < f(x_i, C_j) + g(C_j)$

$\boldsymbol{for all j \neq k}$

$0, otherwise$

Step 4: Repeat steps 2 and 3 to find.
Step 5: For each clusterC_k., select the closest sample to the center as a representative of
C_k.,$mk = Arg\ min$ D (Φ (x_i),z_k). X_i, where δ $(X_i, C_k) = 1$

3. SOM (Self Organizing Map)

Self Organizing Map (SOM) is a grouping method in the form of two-dimensional topography like a map to facilitate observation of the grouping results distribution. SOM requires the determination of learning rate, the function of learning, the number of iterations desired in the process of grouping. To provide the results of the Self Organizing Map method grouping, it doesn't need require objective functions such as K-means and Fuzzy C-means for optimal conditions in iteration, in addition, SOM will not stop its iteration as long as the number of iterations determined has not been reached.

The SOM algorithm is based on a competitive algorithm which is found on the vector quantification principle: at each cycle of life in the network, the unit from Kohonen's layer codebook is most similar to the input win. This unit is called Winner Unit (WU). After that, the WU codebook is modified to get it even closer to the input. The codebooks belonging to the units that are physically near the WU (which are part of the neighborhood) are also put closer to the input of a given data. The algorithm calculates a first stage during the parameters of neighborhood and corrections of weights are set and the codebook initialization is carried out; this stage is followed by the cyclic stage of codebook adjustment. In this stage, the codebooks are modified for the network to classify the input records [6,8].

Initially the weights and learning rate are initialized. The input data to be clustered are presented to the network. Once the input vectors are given, based on the initial weights, the winner unit is calculated either by Euclidean distance method or sum of products method. Based on the winner unit selection, the weights are updated for particular winner unit using competitive learning rule. The SOM training process can be summarized in following steps [6, 7, and 8]:

Step 0: initialization weights (Wij), Determine the distance parameter (R), Determine the learning rate (α).
Step 1: if the stop condition is wrong, take steps 2-8.
Step 2: for all input vector x do step 3-5.
Step 3: for every j,

$$D_j = \sum_{i}^{n} (w_{ij} - X_i)^2$$

Step 4: get j so D (j) is minimal
Step 5: for all j that is within range and for all i,

$$w_{ij}new = w_{ij} + \alpha(X_i - w_{ij})$$

Step 6: update the learning rate
Step 7: update the distance
Step 8: test the stop condition

$$(\text{If } | \text{Wij (new)} - \text{Wij} | \le \epsilon \text{ then Stop})$$

4. Experimental Setup

In this experiment, the data-sets are collected from machine learning databases [10]. The data-sets used for the experiments were 768 data consist of 8 attributes. The attributes include the number of pregnancy, plasma glucose concentration a 2 hours in an oral glucose tolerance test, diastolic blood pressure (mm Hg), triceps skin fold thickness (mm), 2-hour serum insulin (mu U/ml), body mass index (kg/m^2), diabetes pedigree function, age (years).

The data-set originally donated by Vincent Sigillito from the Applied Physics Laboratory at the Johns Hopkins University, one of the most well-known data-sets for testing classification algorithms [10]. This data-set consists of records describing 768 female patients of Pima Indian heritage which are at least 21 years old living near Phoenix, Arizona, USA. From the 768 patients in the PID data-set, classification algorithms used a training set with 576 patients and a testing dataset with 192 patients.

The study is separated in several stages. The first stage is input data from database. Then input data is implemented using kernel k-means and SOM algorithm. Input data should be divided into training

IOP Conf. Series: Journal of Physics: Conf. Series **947** (2017) 012003 doi:10.1088/1742-6596/947/1/012003

and testing data. The ratio of training and testing data is 1:1.The results of classification were compared, in order to get the conclusion of a more efficient method. The detail of the research process can be seen on Figure 1.

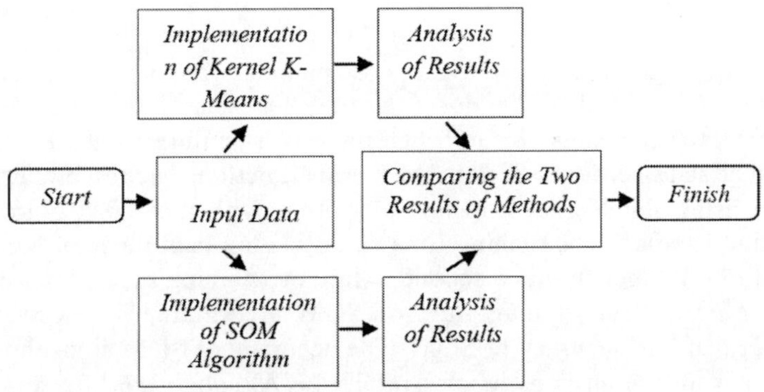

Figure 1. System design

5. Results and Discussion

In Kernel k-means from diabetes data-set that the researcher used, it was chosen two centroid randomly as the first centroid. It become the differences class of the people that infected by diabetes and the class of the people thatare not infected by diabetes. In this program, we choose Gaussian kernel. After running the program with optimal iteration, that is the fifty-th iteration, result of iteration will be achieved. From the result, it can be seen that the sum of right data and the sum of wrong data.

Based on experiments, every experiment has difference accuration, because of the choosing centroid process from the set of data; randomly chosen. Actually, the number of optimal iteration get the right accuration.

In SOM algorithm the diabetes data-set divided into 2 parts, 384 for training data and 384 for testing data. The experiment with SOM algorithm was done 10 times with different learning rate. Based on the experiments, changes in learning rate affect the number of iterations or epoch occurred and the test time. When the learning rates are smaller, the epoch occurs less and the time required for testing is also faster, the learning rates are greater, the epoch occurs more and the time required for testing is also longer. The change of learning rate does not affect the accuracy of the test. The experimental results of Kernel k-Means and SOM algorithms can be seen on Figure 2.

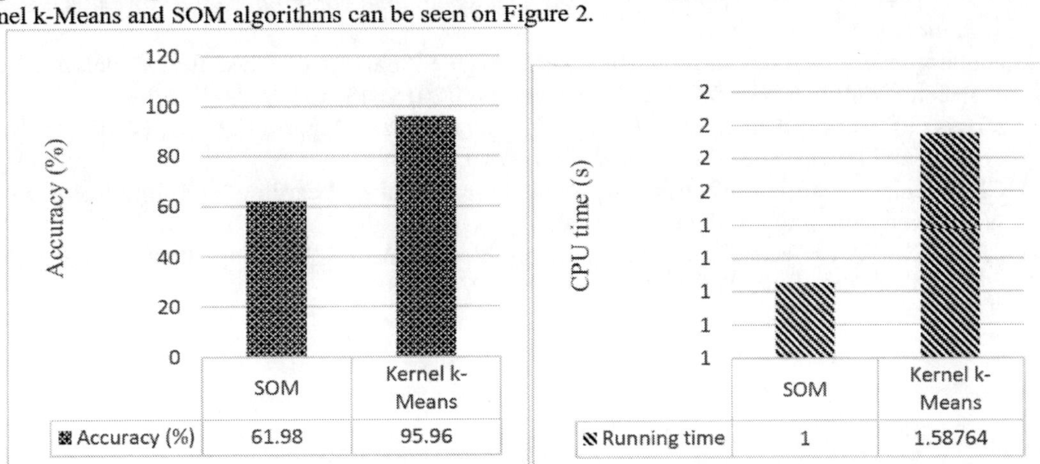

(a) (b)

Figure 2. a) Accuracy Kernel k-Means and SOM algorithms. b) CPU time for building the program

Figure 2.a shows that the accuracy of Kernel k-means is 33.9 point better than the accuracy of SOM. CPU time of Kernel k-means is longer than SOM, but the differentiation between them is very small, not more than 6 seconds.

6. Conclusion

In this study, classification is performed to diagnose patients from data that has been obtained. From the experiment the researcher tried to diagnosis the diabetes with Kernel k-means algorithm. Kernel k-means is better than the K-means cluster when there is a nonlinear data. By using the Kernel cluster method, the distance statement is noticed in the Kernel function. The researcher compared the results of the classification using the Kernel k-means algorithm with the SOM algorithm. The result SOM algorithm proved to have a faster running time classification than those of Kernel k-means algorithm. Running Time of SOM Algorithm is 1 second, while the running time of Kernel k-Means algorithmis 1.58674 second, slower than running time of SOM algorithm. The accuracy of Kernel k-means clustering is better than the accuracy of SOM. The accuracy of SOM algorithm is as much as 61, 98% and Kernel k-Means algorithm is as much as 95.96%. Although the k-means kernel algorithm has a slower in classification, the result is higher. Based on the level of accuracy, method Kernel k-means algorithm has better performance than SOM Algorithm.

References

[1] V VV&Ravikumar A 2014 Study of Data Mining Algorithms for Prediction and Diagnosis of Diabetes Mellitus*International Journal of Computer Application***95**(17) 12–16
[2] Karthikeyan T &Vembandadsamy K 2015An Analytical Study on Early Diagnosis and Classification of Diabetes Mellitus*International Journal of Computer Application*Vol**5**
[3] AnujaKumari V & Citra R 2013 Classification of Diabetes using Support Vector Machine*International Journal of Engineering Research and Applications (IJERA)* 1797-1801
[4] Durairaj M &Kalaiselvi G 2015 Prediction of Diabetes using Back propagation Algorithm *International Journal of Emerging Technology and Innovative Engineering (IJETIE)* 21-25
[5] Sexana K, Khan Z, & Singh S 2014 Diagnosis of Diabetes Mellitus using K Nearest Neighbor Algorithm*International Journal of Computer Science Trends and Technology (IJCST)* 36-43
[6] Massini G 2010*Applications of Mathematics in Models, Artificial Neural Networks and Arts, Chapter 13* (Netherland: Springer)
[7] Dhillon I Set al 2004Kernel k-means, Spectral Clustering and Normalized Cuts*KDD '04 Proceedings of the tenth ACM SIGKDD international conference on Knowledge discovery and data mining* 551-56
[8] Kandhasamy J P &Balamurali S 2015 Performance Analysis of Classifier Models to Predict Diabetes Mellitus*Procedia Computer Science***47**(2015) 45 – 51
[9] Tzortzis G F &Likas A C 2009 The Global Kernel K-Means Algorithm for Clustering in Feature Space*IEEE Transactions on Neural Networks* 1181-94
[10] UCI Machine Learning Repository- Center for Machine Learning and Intelligent System (retrieved from http://archive.ics.uci.edu)
[11] Qi Jet al 2017 An effective and efficient hierarchical K-means clustering algorithm*International Journal of Distributed Sensor Networks* 2017 Vol**13**(8)

Iris Recognition Using Feature Extraction of Box Counting Fractal Dimension

C Khotimah[1,a)] and D Juniati[1,b)]

[1]Departement of Mathematics, Universitas Negeri Surabaya , Surabaya, Indonesia

Email: [a)]chusnulkhotimah@mhs.unesa.ac.id , [b)]dwijuniati@unesa.ac.id

Abstract. Biometrics is a science that is now growing rapidly. Iris recognition is a biometric modality which captures a photo of the eye pattern. The markings of the iris are distinctive that it has been proposed to use as a means of identification, instead of fingerprints. Iris recognition was chosen for identification in this research because every human has a special feature that each individual is different and the iris is protected by the cornea so that it will have a fixed shape. This iris recognition consists of three step: pre-processing of data, feature extraction, and feature matching. Hough transformation is used in the process of pre-processing to locate the iris area and Daugman's rubber sheet model to normalize the iris data set into rectangular blocks. To find the characteristics of the iris, it was used box counting method to get the fractal dimension value of the iris. Tests carried out by used *k*-fold cross method with k=5. In each test used 10 different grade K of K-Nearest Neighbor (KNN). The result of iris recognition was obtained with the best accuracy was 92,63 % for $K = 3$ value on K-Nearest Neighbor (KNN) method.
Keywords: Biometrics, Iris Recognition , Fractal Dimension.

1. Introduction

Biometric is a science that is now growing rapidly. The unique and distinctive characteristic of human beings makes biometrics of many researchers and academics as an effective and efficient identification system. Some biometric systems such as fingerprint recognition, face recognition and iris recognition. Iris recognition is selected by researcher because it has unique and relatively stable patterns. Several methods have been used for iris recognition systems [1]. The main differences of some iris recognition systems are the methods used that is how to extract and analyze the iris features. Iris recognition usually three main steps: iris preprocessing which consist of localization iris and normalization iris, feature extraction, and feature matching. Many experiments about iris recognition. Daugman who the first indroduce iris recognition with title "High Confidence Visual Recognition of Persons by A Test of Statistical Independent"[2]. Daugman used 2-D Gabor wavelet for feature extraction and Hamming distance. In 2013, Chen et al published journal with title "Iris Recognition on Bidimensional Empirical Mode Decomposition and Fractal Dimension". Chen used Bidimensional Empirical Mode Decomposition Localization (BEMD) and Fractal Dimension for feature extraction, the result describe that fractal dimension can use for feature extraction of iris recognition [3]. Segmentation Iris consist of iris localize and iris normalize. iris localization used Hough transform. Hough transform is the method used to find the regular shape, e.g. line or circle in shape [4]. Furthermore, for normalization iris annular into a rectangular block ~~used~~ Daugman's rubber sheet model. Daugman's rubber sheet model is mapping every point in annular iris to pseudo-polar coordinates. The second process, feature extraction using box counting to get the fractal dimension value of the iris. Box counting method effectively is used to calculate the dimensions of an object with an irregular shape. Iris texture is irregular, so box counting suitable for feature extraction. Feature extraction for iris recognition is finding the characteristics of the iris, which used box counting method to get the fractal dimension value of the iris. The last process, feature matching used k-fold cross-validation and K-Nearest Neighbor (KNN) method [5]. The fractal dimension value of the iris devides into two parts training data, and testing data. Then, K-Nearest Neighbor (KNN) with Euclidean distance is used to classification.

2. Method for Iris Recognition

2.1. *Canny edge detection*

Edge detection processes find edge from the object. Edge of the image on the border between two-pixel regions that have a contrasting color. The purpose of edge detection is to extract important feature on an image, such as lines, circles, edges. In essence, edge detection is useful to know the boundaries between an object against the background object picture [1].
The following is a Canny edge detection algorithm:

a. Reduce noise image using Gaussian filter using standard devices (σ). This process aim to make thin lines of image undetect as edge.
b. Calculate gradient intensity of image with determining gradient direction to particular direction.
c. Non maximum suppression process to muffle undesired pixel into point edge.
d. Thresholding Hysteresis Process or process of minimizing the emerging edge to produce accurate edge line. In other words, this process converts the pixel value to zero if it is deemed unfit to be an edge.

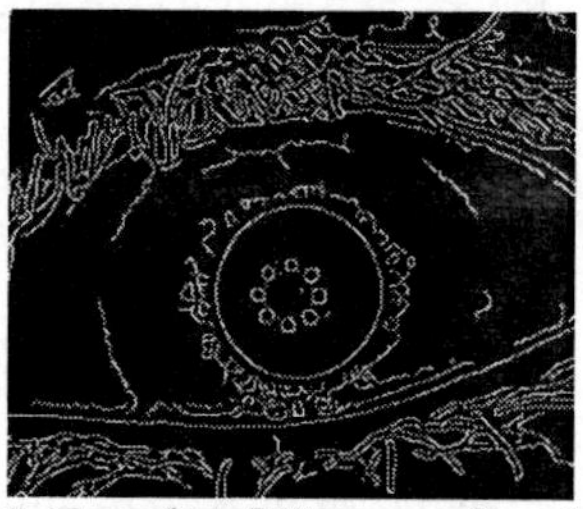

Figure 1. Result of Canny edge detection

2.2. *Iris localization using Hough Transform*

Hough transform is a method for finding regular shapes from the image as line, circle image, etc. The circular Hough transform can be employed to deduce the radius and center coordinates of the pupil and iris regions [4]. Hough transform process consists of 3 step. The first step is edge detection using Canny edge detection. The purpose of edge detection is to decrease points in the image. When edge point has been found, votes are cast in Hough space for the parameters of circles passing through each edge point. These parameters are the center coordinates x_c and y_c, and the radius r, (x_j, y_j)edge point with $j = 1,2, \ldots$ .which are able to define any circle according to the equation

$$(x_j - x_c)^2 + (y_j - y_c)^2 = r^2 \quad (1)$$

Here result of Hough transform

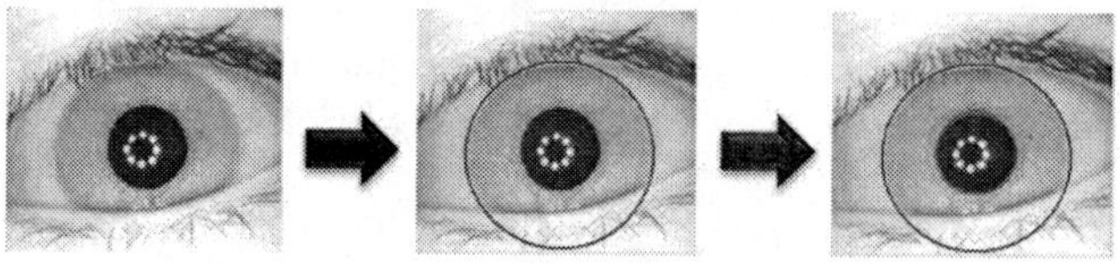

Figure 2. Process of Hough transform

2.3. *Iris Normalization*

Normalization of the iris dataset is the stage where the annular iris obtained from the normalized segmentation results into a two-dimensional rectangular block. The pupils of each individual are of different sizes, therefore the conversion of Cartesian coordinates to the normalization of pseudo-polar coordinates. The iris normalization function is to produce iris regions of equal size which make it easier to calculate the texture value. Daugman's model rubber sheet model or so-called Daugman's algorithm reprints every point in the iris area (between the two pupil areas and the limbus boundary) annular to the apparent polar coordinate (r, θ) where r∈ [0,1] and θ∈ [0, 2π] [4]. Reconstruction of iris regions I from Cartesian coordinates (x, y) to normalized polar coordinates (r, θ) is expressed as follows:

$$I(x(r,\theta), y(r,\theta)) \rightarrow I(r,\theta) \quad (2)$$

with

$$x(r,\theta) = (1-r)x_p(\theta) + rx_l(\theta) \quad (3)$$

and

$$y(r,\theta) = (1-r)y_p(\theta) + ry_l(\theta) \quad (4)$$

where

$$x_p(\theta) = x_{p0}(\theta) + r_p * cos(\theta) \quad (5)$$
$$y_p(\theta) = y_{p0}(\theta) + r_p * sin(\theta) \quad (6)$$
$$x_l(\theta) = x_{l0}(\theta) + r_l * cos(\theta) \quad (7)$$
$$y_l(\theta) = y_{l0}(\theta) + r_l * sin(\theta) \quad (8)$$

where, x_p, y_p are the coordinates of the sample point of the pupil and x_l, y_l are the coordinates of the sample point of the limbus along direction θ. x_{p0}, y_{p0} are the pupil center coordinates and x_{l0} , y_{l0} are the iris center coordinates[4].

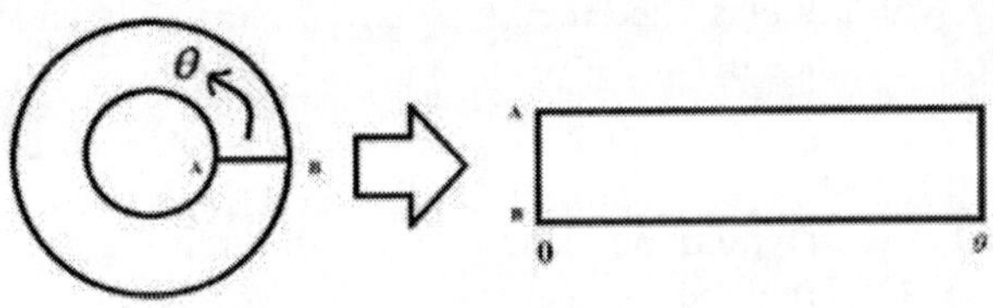

Figure 3. Result of iris normalization

Iris annular became into the rectangular block with size 64×512. Despite all normalized iris templates being the same size, eyelashes and eyelids will still be visible on the templates, reducing iris recognition performance. Therefore, an ROI (Region of Interest) is selected to remove the influence of eyelashes and eyelids, by extracting features only from the upper half of the region $32 \times 512$the section closest to the pupil, which provides the most discriminating information [3].

2.4. *Box Counting*

Fractal dimension is not like Euclide dimension, the fractal dimension has a non-integer number of dimension. Fractal dimension is very important because it can measure the real data[6]. For example, a measure of the cloud to predict rain, a measure of soil to determine soil fertility, a measure of dust to determine air cleanliness, and so on. Many methods are used to find fractal dimensions such as Hausdorff method, Richardson method, Hurst exponent method and box-counting method. In this research used the box-counting method to determine iris image dimension. Calculating the dimensions of an image using the box-counting method is done by covering the object area with a square box with varying sizes [7].

This is an equation for measure fractal dimension using the box-counting method:

$$Dim = \frac{\log(N(r))}{\log\left(\frac{1}{r}\right)} \quad (9)$$

where $N(r)$ is value of boxes containing the image, r is ratio of the length segment or in the word r is scale. The following is an algorithm for finding fractal dimension using box counting [6] :

1. Covering the object area with square boxes with size r bagian. r value change from 1 to 2^n, with $n = 0,1,2,\ldots$
2. Calculate value of $N(r)$, the number of boxes containing the object.
3. Calculate value of $\log\left(\frac{1}{r}\right)$ and $\log(N(r))$
4. Creating a straight line from value of $\log\left(\frac{1}{r}\right)$ and $\log(N(r))$
5. Calculate the slope of a straight line with the following equation:

$$\propto = \frac{\left(\sum_{k=1}^{n} xy\right) \frac{\left(\sum_{k=1}^{n} x\right)\left(\sum_{k=1}^{n} y\right)}{n}}{\left(\sum_{k=1}^{n} x^2\right) \frac{\left(\sum_{k=1}^{n} x\right)^2}{n}} \quad (10)$$

Where $x = \log\left(\frac{1}{r}\right)$, $y = \log(N(r))$, n= amount of data used, and $\propto$ is slope which be called the fractal dimension

2.5. *k-fold cross-validation*
Cross-validation is a method to divide data into two parts, training data, and testing data. In computing, cross-validation aims to avoid overlapping on data testing. *k*-fold cross-validation is a part of cross-validation. In *k*-fold cross-validation, the data is partitioned into k equally (or nearly equally). Process of testing and training repeat k times [8].

2.6. *K*-Nearest Neighbor
K-Nearest Neighbor (KNN) is a classified method based on learning data that has a close distance to the object. To determine the distance between two points are used Euclidean distance, defined as follows:

$$D(x, y) = \sqrt{\sum_{i=1}^{n} (x_i - y_i)^2} \quad (11)$$

where D is a distance between a point on testing data x and training data y that will be classifiying.
Here is the algorithm of KNN method:
a. Start
b. Determine the number of K
c. Calculate distance of testing data to training data using distance formula
d. Sorting the data based on small to large from Euclidean distance
e. Specifies the group of testing data to majority label of *K*
f. Stop

3. Methodology
3.1. *Data*
In this study, the data of iris was chosen from *Chinese Academy of Science* (CASIA) interval-V4 database and randomly selected. In this experiment was used 60 images from 10 classes. Each image has resolution of 280 × 320 pixel in JPG format.
3.2. *Procedur of iris recognition*

IOP Conf. Series: Journal of Physics: Conf. Series **947** (2017) 012004 doi:10.1088/1742-6596/947/1/012004

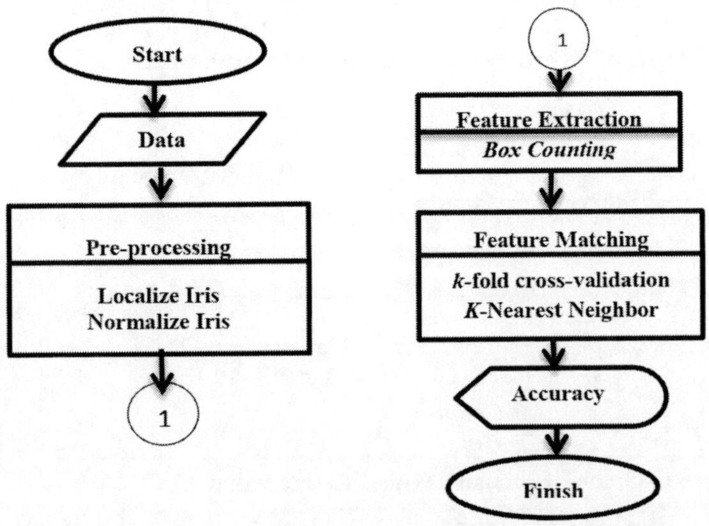

Figure 4. Flowchart of experiment

The iris recognition process begins with segmentation iris which consist of localize iris using Hough transform and normalize iris using Daugman's rubber sheet model. After segmentation process, then feature extraction with calculate the value of fractal dimension using box-counting method. Then, the next step is feature matching. The feature matching consist of two steps, that is *k*-fold cross validation method with *k*=5 was devide randomly into 5 sections with equal size and classification method was used in this study is *K*-Nearest Neighbor (KNN) with Euclidean distance.

4. Experimental result

The iris recognition process begins segmentation process consists of localizing iris using Hough transform and normalize iris using Daugman's rubber sheet model. In this experiment used 60 image *Chinese Academy of Science* (CASIA) interval-V4. This is example of segmentation process of "S1002L01.jpg".

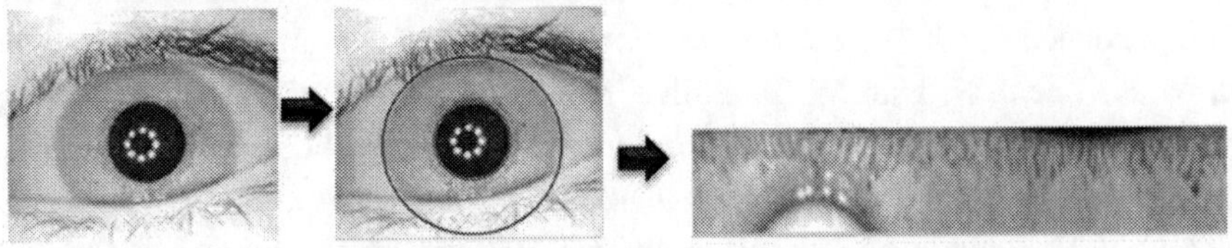

Figure 5. Example of segmentation process of "S1002L01.jpg"

After segmentation process, the next process calculated the value of fractal dimension in each image using the box-counting method. Example, iris of "S1002L01.jpg" has the value of fractal dimension equal 1,2827. In this experiment, all data have a value of fractal dimension in range 1,25523 – 1,38377. Furthermore, feature matching consists of *k*-fold cross-validation and *K*-Nearest Neighbor. It is begins that the value of fractal dimension of all data divide using 5-fold cross-validation, 4 partitions were used for training data and the other one for testing data. Testing data is classified using K-Nearest Neighbor to know which class of data and then to knows accuracy.

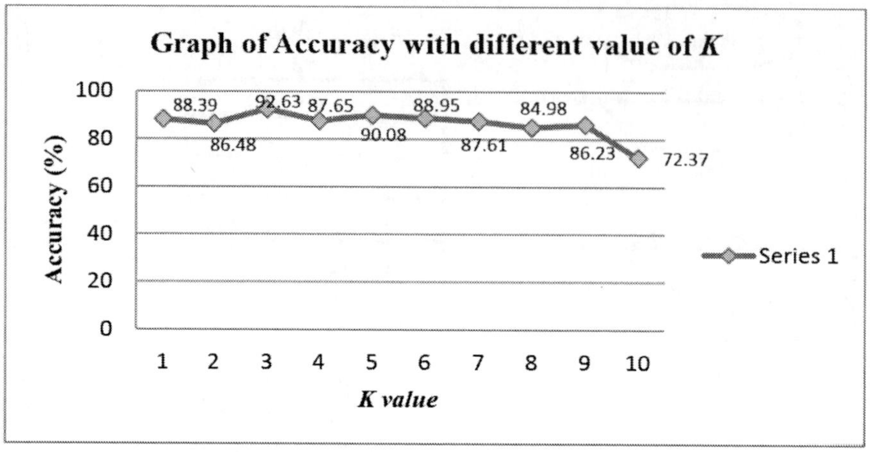

Figure 6. Graph of accuracy with different value of K

The best accuracy value is 92.63 % with K=3

5. Conclusion

Based on the result, it was known that fractal dimension of box counting method can be used to iris recognition. The process of iris recognition was three step: pre-processing iris consist of localize iris using Hough transform and normalize iris using Daugman's rubber sheet model, feature extraction using box counting, and feature matching using KNN and *k-fold cross-validation*. Iris recognition using feature extraction of box counting fractal dimension has accuracy value 92.63 % with 3-NN and 5-*fold cross-validation.*

Reference

[1] Jain Ramesh, Kasturi Rangachar, Schunck Brian G 1995 *Machine Vision* (New York: McGraw-Hill, Inc)

[2] Daugman J G 1993 High Confidence Visual Recognition of Persons by a Test of Statical Independence *IEEE Transaction on Pattern Analysis and Machine Intelegence* **15** p 11

[3] Chen W K, Lee J N, Han W Y, Shih C K, and Chang K C 2013 Iris Recognition based on bidimensional empirical mode decomposition and fractal dimension *Information Science* **221** pp 439-451

[4] Masek L 2003 *Recognition for Human Iris Pattern for Biometrics Identification* (Australia: University of Western Australia)

[5] Safitri D W and Juniati D 2017 Classification of diabetic retinopathy using fractal dimension analysis of eye fundus image *AIP Conference Proceedings.* 020011(2017)

[6] Backes A R and Bruno O M 2008 *A new Approach to Estimate Fractal Dimension of Texture Image Image and Signal Processing* ed A Elmoataz and O Lezoray (Heidelberg: Springer) chapter 2 pp 136-143

[7] Juniati D and Budayasa I K 2016 *Geometri Fractal & Aplikasinya* (Surabaya: Unesa University Press)

[8] Refaeilzadeh P, Tang L, Liu H 2008 Cross-Validation *Encyclopedia of Database Systems* ed L Liu and M T Ozsu (Arizona: Arizona State University) pp 532-538

[9] Institute of Automation, Chinese Academy of Science. *CASIA Iris Image Database*, http://biometrics.idealtest.org/

IOP Conf. Series: Journal of Physics: Conf. Series **947** (2017) 012005 doi:10.1088/1742-6596/947/1/012005

Gaussian Process Regression Model in Spatial Logistic Regression

A Sofro and A Oktaviarina
Mathematics Department, Universitas Negeri Surabaya, East Java, Indonesia
E-mail: ayuninsofro@unesa.ac.id

Abstract. Spatial analysis has developed very quickly in the last decade. One of the favorite approaches is based on the neighbourhood of the region. Unfortunately, there are some limitations such as difficulty in prediction. Therefore, we offer Gaussian process regression (GPR) to accommodate the issue. In this paper, we will focus on spatial modeling with GPR for binomial data with logit link function. The performance of the model will be investigated. We will discuss the inference of how to estimate the parameters and hyper-parameters and to predict as well. Furthermore, simulation studies will be explained in the last section.

1. Introduction

In recent years, modeling for Binomial data has been developed massively. The conventional way to handle this situation is using a logistic regression model by assuming the independence among the data. However, in practice this is a strong assumption since the majority health data are dependent data. One of popular approach to solve the problem is to use conditional autoregressive model to model the correlation structure, which is first proposed in [1].

Some other methods have also been developed in the last decade, including simple disease mapping method [1]. This method has been extended into spatial time generalized linear mixed model ([3]; [4]; [5]). Generalized linear mixed model using prior distribution for spatially structured random effect is a common alternative way, see [6]. But they limited the correlation structure using intrinsic conditional autoregressive model, [7]. Some authors pointed out that there are misleading regarding the spatial dependence covariance matrix, [10]. Moreover, it seems that practitioners need to understand how to determine the precision of the covariance matrix rather than puzzling effects, [11]. Thus more study needs to be done on modeling spatial correlation structure since this is a very common problem in practice([8]; [9]).

The aim of this paper is to propose a model which can describe the spatial covariance structure flexibly. Kriging is also one of approach in geostatistics, [14] but it seems offer limitation, such as difficulty to accomodate a large number of covariates. Therefore, to achieve this we used a Gaussian process regression (GPR) model. Recently GPR and the related methods has been developed fast and has a wide application in machine learning and other areas [12]. Some recent development can be found in [13] and [15]. Covariance structure of GPR is defined by a covariance kernel which depends on a set of covariates. Consequently it provides a very flexible covariance structure coping with a variety of the variables such as geographic position, distance among the spatial areas and other variables related to culture and life style etc.

In this paper we will explain the details how to use a GPR model to model dependency in Binomial data and provide the technical details. Simulation studies will be presented to demonstrate the performance of the proposed method.

2. Gaussian Process Regression

Let y_i be a response variable and $\mathbf{x}_i$ be P-dimensional covariates. A Gaussian process regression model can be specified as follows :

$$y_i = f(\mathbf{x}_i) + \epsilon_i, \quad i = 1, ..., N \tag{1}$$

where $\epsilon_i \sim i.i.d \;\; \mathcal{N}(0, \sigma^2)$, σ^2 unknown and

$$f(.) \sim \mathcal{GP}(\mu(\cdot), k(\cdot, \cdot)),$$

if

$$\mathbf{f} = (f_1, f_2, .., f_N) \sim \mathcal{N}(\boldsymbol{\mu}, \mathbf{K}),$$

where the i-th element of $\boldsymbol{\mu}$ is $\mu(\mathbf{x}_i)$, the (i, j)-th element of $\mathbf{K}$ is $k(\mathbf{x}_i, \mathbf{x}_j)$ and $k(\cdot, \cdot)$ is a covariance function.

It is common to assume a zero mean function in the Gaussian process prior, i.e $\mu(.) = 0$. The most popular choice for the covariance function is the following squared exponential kernel,

$$k(\mathbf{x}_i, \mathbf{x}_j) = v_0 \exp\left\{ -\frac{1}{2} \sum_{p=1}^{P} w_p (x_{ip} - x_{jp})^2 \right\} \tag{2}$$

where v_0, w_p, $p = 1, ...P$ denotes the set of hyper-parameters and are defined as $\boldsymbol{\theta}$.

3. Spatial Logistic Regression with GPR

For the ith observation, let z_i be the response variable, τ_i be a spatial related random effects and U_i be the covariates involved in the fixed effects, where $i = 1, ...n$ and n is the sample size. Typically in this model, the data are assumed to be conditionally independent and follow Binomial distribution using logit as link funtion. Hence, a spatial logistic regression with GPR can be expressed

$$\begin{aligned} z_i|\pi_i &\sim Bin(1, \pi_i) \text{ independently}, \\ \log\left(\frac{\pi_i}{1-\pi_i}\right) &= \mathbf{U}_i^T \boldsymbol{\beta} + \tau_i. \end{aligned} \tag{3}$$

The correlation structure is defined via the random effect item $\tau(.) \sim \mathcal{GP}(0, k(\cdot, \cdot))$ which has been discussed in the previous section. The logit link function in equation (3) can be derived as follow

$$\frac{\pi_i}{1-\pi_i} = \exp(\mathbf{U}_i^T \boldsymbol{\beta} + \tau_i) \tag{4}$$

$$\pi_i + \pi_i \exp(\mathbf{U}_i^T \boldsymbol{\beta} + \tau_i) = \exp(\mathbf{U}_i^T \boldsymbol{\beta} + \tau_i)$$

$$\pi_i = \frac{\exp(\mathbf{U}_i^T \boldsymbol{\beta} + \tau_i)}{1 + \exp(\mathbf{U}_i^T \boldsymbol{\beta} + \tau_i)}. \tag{5}$$

A suitable choice of a kernel covariance function and its hyper-parameters can improve the prediction accuracy, [13]. Rather than making assumption regarding prior of hyper-parameters, we choose using all observed data. One of the popular methods is using empirical Bayesian estimation to select the hyper-parameters. In practice, $\boldsymbol{\theta}$ and $\boldsymbol{\beta}$ can be estimated at the same time.

3.1. Emprical Bayesian Estimates

Now, we focus on the empirical Bayesian approach. Let us assume that we have observed a set of data $\mathcal{D} = (z_i, U_i, \mathbf{x}_i), i = 1, ..., N, \mathbf{x}_i \in T \subset \mathcal{R}^P$, where z_i is observation of variable response, U_i is covariates variable. $\mathbf{x}_i$ are covariates modelled covariance structure τ_i and P is the dimension of the input vector $\mathbf{x}_i$.

The random effects $\boldsymbol{\tau}$ are unknown, the marginal density of $\mathbf{z}$ does not have a convenient closed-form representation. To estimate parameters, we can do so by maximizing the following marginal density $\mathbf{z} = (z_1, ..., z_N)^T$

$$\begin{aligned} p(\mathbf{z}|\boldsymbol{\beta}, \theta) &= \int p(\mathbf{z}|\boldsymbol{\beta}, \boldsymbol{\tau})p(\boldsymbol{\tau})d\boldsymbol{\tau} \\ &= \int p(\mathbf{y}, \boldsymbol{\tau})d\boldsymbol{\tau} \end{aligned} \tag{6}$$

where $\boldsymbol{\tau} \sim \mathcal{N}(\mathbf{0}, \mathbf{K})$.

However, the above marginal density function is analytically intractable. One of the methods used to address this issue is to use a Laplace approximation. The log likelihood of equation (6) can be written as

$$l(\boldsymbol{\beta}, \boldsymbol{\theta}) = \sum_{i=1}^{N} \log \int \exp(\Phi(\boldsymbol{\tau}))d\boldsymbol{\tau}. \tag{7}$$

Note that

$$\Phi(\boldsymbol{\tau}) = \log p(\mathbf{z}|\boldsymbol{\beta}, \boldsymbol{\tau}) - \frac{1}{2}\log|\mathbf{K}(\boldsymbol{\theta})| - \frac{1}{2}\boldsymbol{\tau}^T\mathbf{K}^{-1}(\boldsymbol{\theta})\boldsymbol{\tau} - \frac{N}{2}\log 2\pi. \tag{8}$$

and the likelihood of $p(\mathbf{z}|\boldsymbol{\beta}, \boldsymbol{\tau})$ is following

$$\begin{aligned} L(p(\mathbf{z}|\boldsymbol{\beta}, \boldsymbol{\tau})) &= \prod_{i=1}^{n} \left(\pi^{z_i}(1-\pi_i)^{1-z_i}\right) \\ &= \prod_{i=1}^{n} \left(\frac{\pi_i}{1-\pi_i}\right)^{z_i} (1-\pi_i). \end{aligned} \tag{9}$$

From equation (4) , we can substitute to equation (5). Hence, we get

$$\prod_{i=1}^{n} \left(\exp(\mathbf{U}_i^T\boldsymbol{\beta} + \tau_i)\right)^{z_i} \left(1 - \frac{\exp(\mathbf{U}_i^T\boldsymbol{\beta} + \tau_i)}{1 + \exp(\mathbf{U}_i^T\boldsymbol{\beta} + \tau_i)}\right).$$

Then the log likelihood of $p(\mathbf{z}|\boldsymbol{\beta}, \boldsymbol{\tau})$ can be written as following

$$\sum_{i=1}^{n} \left(z_i(\mathbf{U}_i^T\boldsymbol{\beta} + \tau_i) - \log(1 + \exp(\mathbf{U}_i^T\boldsymbol{\beta} + \tau_i))\right). \tag{10}$$

We put $\boldsymbol{\tau}_0$ be the maximiser of $\Phi(\boldsymbol{\tau})$. Thus a Laplace approximation is

$$\int \exp(\Phi(\boldsymbol{\tau}))d\boldsymbol{\tau} = \exp\left\{\Phi(\boldsymbol{\tau}_0) + \frac{N}{2}\log(2\pi) - \frac{1}{2}\log|\mathbf{H}|\right\} \tag{11}$$

where $\mathbf{H}$ is the negative of the second derivative of $\Phi(\boldsymbol{\tau})$ respect to $\boldsymbol{\tau}$ and evaluated at $\boldsymbol{\tau}_0$. We have $\mathbf{H} = \mathbf{C} - \mathbf{K}$ and $\mathbf{C}$ is a Hessian matrix. Hence, we can get $\mathbf{C}$ is a diagonal matrix,

$$\mathbf{C} = Diag\left(\frac{\exp(\mathbf{U}_1^T\hat{\boldsymbol{\beta}} + \tau_{01})}{(1 + \exp(\mathbf{U}_1^T\hat{\boldsymbol{\beta}} + \tau_{01})^2}, ..., \frac{\exp(\mathbf{U}_1^T\hat{\boldsymbol{\beta}} + \tau_{0n})}{(1 + \exp(\mathbf{U}_n^T\hat{\boldsymbol{\beta}} + \tau_{0n})^2}\right).$$

In order to estimate the parameters, we maximize the likelihood function with Laplace approximation in equation (7). To apply Laplace approximation, we need to find $\boldsymbol{\tau}_0$.

3.2. Predictions

It is of interest to predict z^* at a new test input $\mathbf{x}^*$. The main purpose in this section is to calculate $E(z^*|\mathcal{D})$ and $Var(z^*|\mathcal{D})$.

Let $\tau(\mathbf{x}^*) = \tau^*$ be the underlying latent variable at $\mathbf{x}^*$. The expectation of z^*, conditional on τ^* is given by

$$E(z^*|\tau^*, \mathcal{D}) = \frac{\exp(\mathbf{U}^{*T}\hat{\boldsymbol{\beta}} + \tau^*)}{1 + \exp(\mathbf{U}^{*T}\hat{\boldsymbol{\beta}} + \tau^*)}. \tag{12}$$

It follows that

$$E(z^*|\mathcal{D}) = E[E(z^*|\tau^*, \mathcal{D})] = \int \left(\frac{\exp(\mathbf{U}^{*T}\hat{\boldsymbol{\beta}} + \tau^*)}{1 + \exp(\mathbf{U}^{*T}\hat{\boldsymbol{\beta}} + \tau^*)} \right) p(\tau^*|\mathcal{D}) d\tau^*. \tag{13}$$

One method to calculate the above expectation is to approximate the integration using a Laplace approximation. We can rewrite it as

$$\begin{aligned} p(\tau^*|D) &= \int p(\tau^*|\boldsymbol{\tau}, D) p(\boldsymbol{\tau}|D) d\boldsymbol{\tau} \\ &= \int p(\tau^*, \boldsymbol{\tau}|D) d\boldsymbol{\tau} \\ &= \frac{1}{p(\mathbf{z})} \int p(\mathbf{y}|\boldsymbol{\tau}) p(\tau^*, \boldsymbol{\tau}) d\boldsymbol{\tau}. \end{aligned} \tag{14}$$

For convenience, we denote $(\boldsymbol{\tau}, \tau^*)^T$ and its covariance matrix $K_{N+1,N+1}$ by $\boldsymbol{\tau}_+$ and $\mathbf{K}_+$ respectively. Thus, the equation (13) can be written as

$$\frac{1}{p(\mathbf{y})} \int \left[\left(\frac{\exp(\mathbf{U}^{*T}\hat{\boldsymbol{\beta}} + \tau^*)}{1 + \exp(\mathbf{U}^{*T}\hat{\boldsymbol{\beta}} + \tau^*)} \right) \prod_{i=1}^{N} p(z_i|\hat{\boldsymbol{\beta}}, \tau_i) \right] \left[(2\pi)^{-\frac{(N+1)}{2}} |\mathbf{K}_+|^{-\frac{1}{2}} \exp(-\frac{1}{2} \boldsymbol{\tau}^T \mathbf{K}_+^{-1} \boldsymbol{\tau}_+) \right] d\boldsymbol{\tau}_+ \tag{15}$$

The calculation of the integral is not tractable, since the dimension of $\boldsymbol{\tau}_+$ is usually very large. We still use a Laplace approximation. Note that

$$\tilde{\boldsymbol{\Phi}}(\boldsymbol{\tau}_+) = \log \left(\frac{\exp(\mathbf{U}^{*T}\hat{\boldsymbol{\beta}} + \tau^*)}{1 + \exp(\mathbf{U}^{*T}\hat{\boldsymbol{\beta}} + \tau^*)} \right) + \sum_{i=1}^{N} \log p(z_i|\hat{\boldsymbol{\beta}}, \tau_i) - \frac{N+1}{2} \log(2\pi) - \frac{1}{2} \log |\mathbf{K}_+| - \frac{1}{2} \boldsymbol{\tau}_+^T \mathbf{K}_+^{-1} \boldsymbol{\tau}_+$$

where $\log p(\mathbf{z}|\boldsymbol{\beta}, \boldsymbol{\tau}) = \sum_{i=1}^{n} \left(z_i(\mathbf{U}_i^T\hat{\boldsymbol{\beta}} + \tau_i) - \log(1 + \exp(\mathbf{U}_i^T\hat{\boldsymbol{\beta}} + \tau_i)) \right)$. Equation (14) can be expressed as

$$p(z^*|\mathcal{D}) = \frac{1}{p(\mathbf{z})} \int \exp(\tilde{\boldsymbol{\Phi}}(\boldsymbol{\tau}_+)) d\boldsymbol{\tau}_+.$$

Let $\hat{\boldsymbol{\tau}_+}$ be the maximiser of $\tilde{\boldsymbol{\Phi}}(\boldsymbol{\tau}_+)$, then by using Laplace approximation we have

$$\int \exp(\tilde{\boldsymbol{\Phi}}(\boldsymbol{\tau}_+)) d\boldsymbol{\tau}_+ = \exp(\tilde{\boldsymbol{\Phi}}(\hat{\boldsymbol{\tau}_+}) + \frac{N+1}{2} \log(2\pi) - \frac{1}{2} \log |\mathbf{K}_+^{-1} + \mathbf{C}_+|) \tag{16}$$

where $\mathbf{C}_+$ is the negative of the second derivative of

$$(\mathbf{U}^{*T}\hat{\boldsymbol{\beta}} + \tau^*) - \log(1 + \exp(\mathbf{U}^{*T}\hat{\boldsymbol{\beta}} + \tau^*)) + \sum_{i=1}^{n} \left(z_i(\mathbf{U}_i^T\hat{\boldsymbol{\beta}} + \tau_i) - \log(1 + \exp(\mathbf{U}_i^T\hat{\boldsymbol{\beta}} + \tau_i)) \right)$$

with respect to $\boldsymbol{\tau}_+$ and evaluated at $\hat{\boldsymbol{\tau}}_+$. Here $\mathbf{C}_+$ is a diagonal matrix, i.e

$$\mathbf{C}_+ = Diag\left(\frac{\exp(\mathbf{U}_1^T\hat{\boldsymbol{\beta}}+\tau_{01})}{(1+\exp(\mathbf{U}_1^T\hat{\boldsymbol{\beta}}+\tau_{01})^2},\ldots,\frac{\exp(\mathbf{U}_1^T\hat{\boldsymbol{\beta}}+\tau_{0n})}{(1+\exp(\mathbf{U}_n^T\hat{\boldsymbol{\beta}}+\tau_{0n})^2},0\right).$$

Since $p(\mathbf{z})$ has already been investigated, the predictive mean (13) can be calculated.

To calculate $Var(z^*|\mathcal{D})$, we use the formula :

$$Var(z^*|\mathcal{D}) = E[Var(z^*|\tau^*,\mathcal{D})] + Var[E(z^*|\tau^*,\mathcal{D})]. \tag{17}$$

Because $Var(z^*|\tau^*,\mathcal{D}) = E(z^*|\tau^*,\mathcal{D})(1-E(z^*|\tau^*,\mathcal{D}))$, therefore

$$E[Var(z^*|\tau^*,\mathcal{D})] = E(z^*|\mathcal{D}) - (E(z^*|\mathcal{D}))^2. \tag{18}$$

From the model definition, we have

$$Var(E[z^*|\tau^*,D]) = E[E(z^*|\tau^*,D)]^2 - [E[E(z^*|\tau^*,\mathcal{D})]]^2$$

$$= \int\left(\frac{\exp(\mathbf{U}^{*T}\hat{\boldsymbol{\beta}}+\tau^*)}{1+\exp(\mathbf{U}^{*T}\hat{\boldsymbol{\beta}}+\tau^*)}\right)^2 p(\tau^*|\mathcal{D})d\tau^* - [E(z^*|\tau^*,\mathcal{D})]^2. \tag{19}$$

The first item in (19) can be obtained by Laplace approximation similar to $E(z^*|\mathcal{D})$ in (13) and the second item is the square of (14).

4. Numerical Examples

In this section we demonstrate a numerical example to explore the performance of the proposed model. We will present a scenario comparing some models. Here, the two models considered are logistik regression (Model 1) and spatial logistik regression with GPR (Model 2).

In the simulation study, we first generate random data from a Binomial distribution with true values $\beta_0 = 1$ and $\beta_1 = 2$. Data are generated from model (3) which assumes τ_i follows a Gaussian process regression model as defined in equations (1) and (2). The true hyper-parameters are $v_0 = 0.04, w_1 = 1$ and $w_2 = 1$.

Here, the covariance kernel is squared exponential which $\mathbf{x}$ are the location of each area measured by its longitude and latitude from Dengue fever data, see [16]. We select 30 of them as training data and the remaining as test data. To measure performance, we calculate the sample mean of the estimated parameters and the root mean squared error (RMSE) between the estimated and the true values of the parameters.

Table 1 shows the sample mean from different methods and also the value of RMSE (RMSE) based on one hunded replications. The value of average RMSE (average RMSE) between π and

	Sample Mean		RMSE	
Method	β_0	β_1	β_0	β_1
True value	1	2		
Model 1	1.00134	1.99499	0.01507	0.02880
Model 2	1.00020	2.00010	0.00018	0.00055

Table 1. Mean of the estimated parameters and RMSE between the estimated and true values of the parameters based on one hundred replications for logistic regression (Model 1) and spatial logistic regression with GPR (Model 2).

$\hat{\pi}$ also can be seen in Table 1. It shows clearly that the performance of model 2, i.e spatial logistik regression with GPR is the best in term of estimating parameters.

Meanwhile, Table 2 shows prediction performance of the proposed model. We investigate a different number of training data (m) from similar generated data in the previous scenario. The aim is to analyze the accuracy of prediction of the model 2. We calculate the average of RMSE between the predicted values and the actual observed data based on one hundred replications. From the Table 2, It can be said that the more increased number of data, the more improve the performance of prediction as it is expected.

m	Average RMSE
20	0.00976
30	0.00815

Table 2. The average of RMSE between $\boldsymbol{\pi}$ and $\hat{\boldsymbol{\pi}}$ based on one hundred replications for different number of training data

5. Conclusion

In this paper, we proposed a Gaussian process regression model for the covariance structure. It allows the use of geographic position and other variables to define the spatial correlation structure, and thus it provides a very flexible model. The simulation study shows its good performance. The computation of the method is quite efficient.

References

[1] Besag J and Kooperberg C 1995 On Conditional and Intrinsic Autoregressions *Biometrika* vol. **82** pp 733-746
[2] Knorr L and Besag J 1998 Modelling risk from a disease in time and space *Statistics in Medicine* vol. **17** pp 2045-2060
[3] Sun T, Kim H and He Z Q 2000 Spatio-temporal interaction with disease mapping, *Statistics in Medicine* vol. **19** pp 2015-2035
[4] McNab Y C and Dean C B 2001 Autoregressive spatial smoothing and temporal spline smoothing for mapping rate *Biometrics* vol. **57** pp 73-85
[5] Silva G L, Dean C B, Niyonsenga T and Vanessa A 2008 Hierarchical Bayesian spatio-temporal analysis of revacularization odds using smoothing splines *Statistics in Medicine* vol. **27** pp 2381-2401
[6] Banerjee S, Carlin B P and Gelfand A E 2004 Hierarchical Modeling and Analysis for Spatial Data Boca Rotan Chapman and Hall
[7] Rue H and Held L 2005 Gaussian Markov Random Field Theory and ApplicationsBoca Rotan Chapman and Hall
[8] Haining R 1990 Spatial Analysis in the Social and Environmental Sciences Cambridge Cambridge University Press
[9] Bernadinelli L, Clayton D and Montomoli C 1995 Bayesian estimates of disease maps: how important are priors?*Statistics in Medicine* vol. **14** pp. 2411-2431
[10] Wall M M 2004 A close look at the spatial structure implied by the CAR and SAR models *Journal of Statistical Planning and Inference* vol. **121** pp. 311 324
[11] Renato and Krainski E 2009 Neighborhood Dependence in Bayesian Spatial Models *Biometrical Journal* vol. **51** pp. 851869
[12] Rasmussen C E and Williams C K 2006 Gaussian Process for Machine Learning Cambridge England: MIT Press
[13] Shi J Q and Choi T 2011 Gaussian Process Regression Analysis for Functional Data, Boca Rotan Taylor and Francis
[14] diggle P J, Tawn J A and Moyeed A 1998 Model-based geostatistics *Royal Statistical Society* vol. **47** pp. 299-350
[15] Wang B and Shi J Q 2014 Generalized Gaussian Process Regression Model for non-Gaussian Functional Data *Journal of American Statistical Association* vol. **109** 1123-1133
[16] Sofro A 2016 Convolved Gaussian Process Regression Models for Multivariate Non-Gaussian Data PhD Thesis, Newcastle University United Kingdom

IOP Conf. Series: Journal of Physics: Conf. Series **947** (2017) 012006 doi:10.1088/1742-6596/947/1/012006

Some Basic Properties of Uniformly Symmetrically Continuous (Real Valued) Functions on Metric Spaces

Manuharawati[1*], D N Yunianti[1], M Jakfar[1]

[1]Mathematics Departement, Universitas Negeri Surabaya, Jalan Ketintang, Surabaya 60231, Indonesia.

Email*: manuharawati@unesa.ac.id

Abstract. We give the definition of uniform symmetric continuity for real valued functions defined on a metric space. Then we investigate the basic properties of uniformly symmetrically continuous functions and compare them with those of symmetrically continuous functions and uniformly continuous functions. Several examples are also given.

1. Introduction

Brown and Page in [1] have defined a metric space. Given a set $X \neq \emptyset$. The function $d: X \times X \rightarrow \mathbb{R}$ is called by metric on X if and only if for every $x, y, z \in X$ satisfy:

(i) $d(x,y) \geq 0$, and $d(x,y) = 0$ if and only if $x = y$.

(ii) $d(x,y) = d(y,x)$

(iii) $d(x,y) \leq d(x,z) + d(z,y)$.

The set X which has a metric d on X is called by a metric space, and it is denoted by (X,d). The metric space (X,d) is said to be discrete if $d(x,y) = 1$ for $x \neq y$ and $d(x,y) = 0$ for $x = y$.

The concept of continuity and any concept involving continuity plays a very important role in real analysis, functional analysis, and topology. There are various types of continuities, four of which will be discussed in this paper.

Definition 1. *Given metric spaces (X,d) and $(\mathbb{R}, d^*)$, $A \subset X$. The function $f: A \rightarrow \mathbb{R}$ is said to be* ***continuous*** *at a point $c \in A$ if for every $\varepsilon > 0$ there exists a $\delta > 0$ such that $d^*(f(x), f(c)) < \varepsilon$, whenever $x \in A$, $d(x,c) < \delta$.*
We say that f is continuous on A if f is continuous at each point $c \in A$.

Definition 2. *Given metric spaces (X,d) and $(\mathbb{R}, d^*)$, $A \subset X$. The function $f: A \rightarrow \mathbb{R}$ is said to be* ***uniformly continuous*** *on A if for every $\varepsilon > 0$ there exists $\delta > 0$ such that $d^*(f(x), f(y)) < \varepsilon$, where $x, y \in A$, and $d(x,y) < \delta$.*

IOP Conf. Series: Journal of Physics: Conf. Series **947** (2017) 012006 doi:10.1088/1742-6596/947/1/012006

Definition 3. *Given metric spaces* (X, d) *and* $(\mathbb{R}, d^*)$, $A \subset X$. *The function* $f: A \to \mathbb{R}$ *is said to be* ***symmetrically continuous*** *at* $c \in A$ *if for every* $\varepsilon > 0$ *there exists* $\delta > 0$ *such that for all* $h \in \mathbb{R}$, $|h| < \delta$ *and for every* $x, y \in A$, $d(x, c) = d(y, c) = |h|$ *then*

$$d^*(f(x), f(y)) < \varepsilon.$$

We say f *is symmetrically continuous on* A *if* f *is simmetrically continuous at every point* $c \in A$.

Like continuous functions, functions have been studied thoroughly by many authors, see for example in [1-5,7-13]. In particular in [6], Marcus defined a concept of uniformly symmetrically continuous function on $\mathbb{R}$.

In this paper, we would like to declare the generalization of the concept of uniformly symmetrically continuous, that is the uniformly symmetrically continuous (real valued) function, so that aplicable on metric spaces. The purpose of this study is to obtain properties of uniformly symmetrically continuous function on metric spaces and compare them with those of symmetrically continuous functions and uniformly continuous functions.

2. The Result

Definition 4. *Given two metric spaces* (X, d) *and* $(\mathbb{R}, d^*)$, $A \subset X$. *The function* $f: A \to \mathbb{R}$ *is said to be uniformly symmetrically continuous on* A *if for every* $\varepsilon > 0$ *there exists* $\delta > 0$ *such that for all* $h \in \mathbb{R}$, $|h| < \delta$ *and for every* $x, y, z \in A$, $d(x, z) = d(y, z) = |h|$ *then*

$$d^*(f(x), f(y)) < \varepsilon.$$

The next propositions are the properties of the concept of continuity. First, the relation between the concept of continuity and uniformly continuity is explained

Proposition 5. *Given two metric spaces*(X, d) *and* $(\mathbb{R}, d^*)$, $A \subset X$. *If* f *is uniformly continuous on* A, *then* f *is continuous on* A, *but the convers is not satisfied.*

Proof:

This follows immediately from the definition. The convers is not satisfied because there exists the counter example. For example, given the usual metric space $\mathbb{R}$, that is $\mathbb{R}$ with $d(x, y) = |x - y|$. The function $f: \mathbb{R} \to \mathbb{R}$ and

$$f(x) = 1/x$$

is a continuous on interval $(0,1)$, but f is not uniformly continuous on its interval. ■

Second, the relation between the concept of continuity and symmetrically continuity is described as follows

Proposition 6. *Given two metric spaces*(X, d) *and* $(\mathbb{R}, d^*)$, $A \subset X$. *If* f *is continuous at* $c \in A$, *then* f *is symmetrically continuous at* $c \in A$, *but the convers is not satisfied..*

Proof:

Let f is continuous at $c \in A$. Let $\varepsilon > 0$ be given, there exists a $\delta > 0$ such that $d^*(f(x), f(c)) < \frac{\varepsilon}{3}$ whenever $x \in A$ and $d(x, c) < \delta$. Let $h \in \mathbb{R}$, $|h| < \delta$ and $x, y \in A$, $d(x, c) = d(y, c) = |h|$, we obtain

$$d^*(f(x), f(y)) \leq d^*(f(x), f(c)) + d^*(f(c), f(y)) = \frac{\varepsilon}{3} + \frac{\varepsilon}{3} < \varepsilon.$$

It show that f is symmetrically continuous at $c \in A$.

The convers is not satisfied because there exists the counter example. For example (look at [6]), given the usual metric space $\mathbb{R}$, that is $\mathbb{R}$ with $d(x, y) = |x - y|$. The function $f: \mathbb{R} \to \mathbb{R}$ with

$$f(x) = \begin{cases} 0, & x \neq 3 \\ 2, & x = 3 \end{cases}$$

ss symmetrically continuous at $c = 3$, but f is not continuous at $c = 3$. ■

Corolarry 7. *Given two metric spaces* (X,d) *and* $(\mathbb{R},d^*)$, $A \subset X$. *If* f *is continuous on* A, *then* f *is symmetrically continuous on* A.

It is obvious that every uniformly symmetrically continuous function is symmetrically continuous. Let us record this as a proposition.

Proposition 8. *Given two metric spaces* (X,d) *and* $(\mathbb{R},d^*)$, $A \subset X$. *If* f *is uniformly symmetrically continuous on* A, *then* f *is symmetrically continuous* A, *but the convers is not satisfied..*
Proof:
This follows immediately from the definition. ■

The next, we want to explain the relation between the concept of uniformly continuity and uniformly symmetrically continuity.

Proposition 9. *Given two metric spaces* (X,d) *and* $(\mathbb{R},d^*)$, $A \subset X$. *If* f *is uniformly continuous on* A, *then* f *is uniformly symmetrically continuous on* A, *but the convers is not satisfied.*
Proof:
Let f is uniformly continuous on A. Let $\varepsilon > 0$ be given, there exists a $\delta > 0$ such that $d^*(f(x),f(y)) < \frac{\varepsilon}{3}$ whenever $x,y \in A$ and $d(x,y) < \delta$. Let $h \in \mathbb{R}$, $|h| < \delta$ and $x,y,z \in A$, $d(x,z) = d(y,z) = |h|$, then we get

$$d^*(f(x),f(y)) \leq d^*(f(x),f(z)) + d^*(f(z),f(y)) = \frac{\varepsilon}{3} + \frac{\varepsilon}{3} < \varepsilon.$$

It show that f is uniformly symmetrically continuous on A. ■

Corolarry 10. *Given two metric spaces* (X,d) *and* $(\mathbb{R},d^*)$, $A \subset X$. *If* f *is uniformly continuous on* A, *then* f *is symmerically continuous on* A, *but the convers is not satisfied..*

In connection with four types of the continuities mentioned above, the problem arises to investigate their relation. For a given two metric spaces (X,d) and $(\mathbb{R},d^*)$, let $A \subset X$ and we define

$$USC = \{f: A \to \mathbb{R} \mid f \text{ is uniformly symmetrically continuous on } A\}$$
$$UC = \{f: A \to \mathbb{R} \mid f \text{ is uniformly continuous on } A\}$$
$$C = \{f: A \to \mathbb{R} \mid f \text{ is continuous on } A\},$$
$$SC = \{f: A \to \mathbb{R} \mid f \text{ is symmetrically continuous on } A\}.$$

Then we have the following result

$$\begin{array}{rl} UC \subset C & but\ \ C \not\subset UC \\ C \subset SC & but\ \ SC \not\subset C \\ USC \subset SC & but\ \ SC \not\subset USC \\ UC \subset USC & but\ \ USC \not\subset UC \end{array}$$

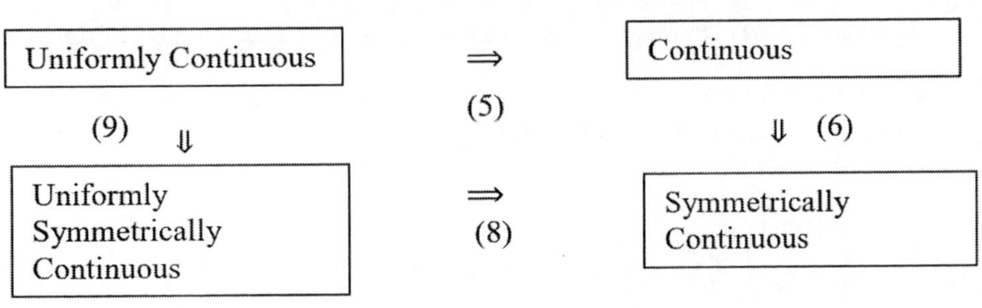

Figure 1. The connection between four types of the continuities

IOP Conf. Series: Journal of Physics: Conf. Series **947** (2017) 012006 doi:10.1088/1742-6596/947/1/012006

3. The Basic Properties

Although uniformly symmetrically continuous functions and uniformly continuous functions are different, they share some basic properties as will be shown in this section.

Theorem 11: Given two metric spaces (X,d) and $(\mathbb{R},d*)$; $A\subset X$, The functions $f,g:A\longrightarrow\mathbb{R}$. If f and g is uniformly continuous on A and $d*$ is usual metric ($d(x,y) = |x-y|$), Then
(a) $f+g$ is uniformly symmetrically continuous on A.
(b) kf is uniformly symmetrically continuous on A whenever $k \in \mathbb{R}$.

Theorem 11 above can be proved by Definition 4.

4. Conclusion

By the given two metric spaces (X,d) and $(\mathbb{R},d*)$ with $A\subset X$, we have the following conclusion: (1) We success in defining the concept of uniformly symmetrically continuous of the real valued function on a metric space. The concept is an abstraction to domain of the concept of uniformly symmetrically continuous of the real valued function on $\mathbb{R}$, (2) The concept of uniformly symmetrically continuity of the real valued function on a metric space is a generalization of the concept of symmetrically continuity of the real valued function on a metric.

References

[1] Brown A L and Page A 1970 *Elements of Functional Analysis* (New York: Van Nostrand Reinhold Company).
[2] Belna C L 1983 Symmetric continuity of real functions *Proc. Amer.Math. Soc.* **87** pp 99–102
[3] Darji U 1993 Symmetric behavior in functions *Proc. Amer. Math. Soc.* **119** pp 915– 923
[4] Dobos J 1989-1987 On quasi-uniform convergence and symmetrically continuous functions *Acta Math. Univ. Comenian.* **48/49** pp 149–152
[5] Humke P D and Laczkovich M 2005 Symmetrically approximately continuous functions, consistent density theorems, and Fubini type inequalities *Trans. Amer. Math. Soc.* **357** 31–44
[6] Manuharawati 2014 Kekontinuan simetris fungsi bernilai real pada ruang metrik *Prosiding Konferensi Nasional Matematika* XVII - 2014
[7] Marcus S 1989 Symmetry in the simplest case: The real line *Comput. Math. Appl.* **17** pp 103–115
[8] Okropiridze M 1998, Symmetrically continuous functions of two variables *Proc. A. Razmadze Math. Inst.* **118** pp 127–134
[9] Pongsriiam P, Khemaratchatakumthorn T, Termwuttipong I and Triphop N 2004 Symmetrically continuous functions *East–West J. Math.* **Special Volume** pp 89–100
[10] Pongsriiam P, Khemaratchatakumthorn T, Termwuttipong I and Triphop N 2004 Some remarks on symmetric continuity of functions *Thai J. Math.* **Special** Issue pp 31–36
[11] Ponomarev P 1967 Symmetrically continuous functions *Mat. Zametki.* **1** pp 385– 390
[12] Rosen H 2002-2003 Darboux symmetrically continuous functions *Real Anal. Exchange* **28** pp 471–475
[13] Szyszkowski M 2002 *Symmetrically Continuous Functions* Ph D Thesis West Virginia University pp 56
[14] Szyszkowski M 1999-2000 Symmetrically continuous functions on various subsets of the real line, *Real Anal. Exchange* **25** pp 547–564
[15] Thomson B S 1994 *Symmetric Properties of Real Functions* (CRC Press)

IOP Conf. Series: Journal of Physics: Conf. Series **947** (2017) 012007 doi:10.1088/1742-6596/947/1/012007

Student's Critical Thinking in Solving Open-Ended Problems Based on Their Personality Type

L D Fitriana[1,*], Y Fuad[1], and R Ekawati[1]

[1]Mathematics Department, Universitas Negeri Surabaya, Ketintang, Surabaya 60231, East Java, Indonesia
*Corresponding Author. Email: lindafitriana@mhs.unesa.ac.id

Abstract. Critical thinking plays an important role for students in solving open-ended problems. This research aims at describing student's critical thinking in solving open-ended problems based on Keirsey's personality types, namely rational, idealist, guardian, and artisan. Four students, with the higher rank in the mathematics' test and representing each type of Keirsey personality, were selected as the research subjects. The data were collected from the geometry problem and interviews. The student's critical thinking is described based on the FRISCO criteria. The result underlines that rational and idealist students fulfilled all FRISCO criteria, and but not for guardian and artisan students. Related to the inference criteria, guardian and artisan students could not make reasonable conclusions and connect the concepts. Related to the reason of criteria, rational student performed critical thinking by providing logical reason that supported his strategy to solve the problem. In contrast, the idealist student provided subjective reason. This results suggest that teachers should frequently train the students' logical thinkingin every lesson and activity to develop student's critical thinking and take the student's personality character into account, especially for guardian and artisan students.

1. Introduction

It has been recognized that one of primary aims of education is to foster student's ability to think critically. Critical thinking is to think logically and accurately when evaluating reasons as the basis for action [3, 4, 8, 10]. Critical thinking involves challenges in problem-solving because it is stated as one factor that affects a person's ability to solve problems [2, 8]. Critical thinking relates to the judgments as the basis for decision making when solving the problem. A student who thinks critically tends to use his or her knowledge and intelligence effectively to achieve the correct and logical decisions. On the contrary, when a student does not think critically then he tend to make illogical and unjustified decisions, unless he is lucky or be able to make the right decisions with, perhaps, some wrong reasons. Ennis [6] states six critical thinking criterias which can be used as a mental checklist for critical thinking and guarantees in which the final decision are concluded as the most important things. These criterias include focus, reason, inference, situation, clarity, and overview which are shown on Table 1.

Table 1. FRISCO Criterias for Critical Thinking

Criteria	Indicators
Focus	Figuring out the problem.
Reason	- Looking for reasons for and against the decision in a certain way (pro and contra reasons). - Looking for evidence.

	- Identifying and judging the acceptability of the reasons.
Inference	Making a reasonable conclusion that suits the reason.
Situation	Identifying the thing which is relevant to the significance of the thinking activity and some of the rules that guide it.
Clarity	Clarifying the meaning of terms and the way that is used.
Overview	Checking what have decided and inferred.

Critical thinking needs to be integrated in the curriculum so that students can learn and can apply them to improve their performance and reasoning ability. Some schemes that can develop critical thinking involves more than one good answer or explanation, comparing various answers to a question and judgement which performs the best strategy or answer. In line with it [1], another kind of critical thinking activity is involving students' presentations with dilemmas or situations from which two solutions which are equally valid can be proposed.

An open-ended problem is a structured problem in which it has more than one correct solution or way to elicit the solution [11, 15]. Through the provision of open-ended problems, students are given the opportunity to view the problems from several perspectives and are expected to provide evidence or logical arguments which supports their judgement or choice [5, 10]. In this study, students are expected to demonstrate their critical thinking ability in choosing strategy by providing supportive reasons.

Every student tends to have differences in thinking and solving problems. This is reasonable because every student has a different personality. Based on the condition that each student is different from each other, employing the Keirsey's criterias can be used to classify students' personality into four types of personality types: rational, idealist, guardian, and artisan [9]. The classification can be observed through how students derive their energy (introvert versus extrovert), receive and process the information (sensing versus intuitive), make decisions (thinking versus feeling), and organize their lives (judging versus perceiving) [13]. All the rationals are intuitive and thinking, idealists are intuitive and feeling, guardians are sensing and judging, and artisans are sensing and perceiving.

Moreover, the previous research showed that thinking has a significant correlation with the personality types [7]. However, this study tries to describe student's critical thinking in solving open-ended problems based on the Keirsey personality types. Students with different personality types may indicate differences in critical thinking when solving the given problem.

2. Method

This study used qualitative approach to investigate student's critical thinking in solving the given problems which formed two geometry questions. A purposive sampling technique were applied to select the 8th grade students from the Unesa's Labschool of Surabaya. Twenty students in the chosen class were given the Keirsey personality and mathematical tests. Based on the test results, four students with the highest rank who represent each type of Keirsey personality were selected as the research subject. Then, the problems were given to the students in order to promote as much student's critical thinking as possible. All subjects were given the first problem and were individually interviewed based on their answers. The second one was conducted two weeks later to ensure the validity of the data. The obtained data was qualitatively analyzed based on the FRISCO criterias through three steps including data condensation, data presentation, and data interpretation.

3. Results And Discussion

The given open-ended problems are described as follows.

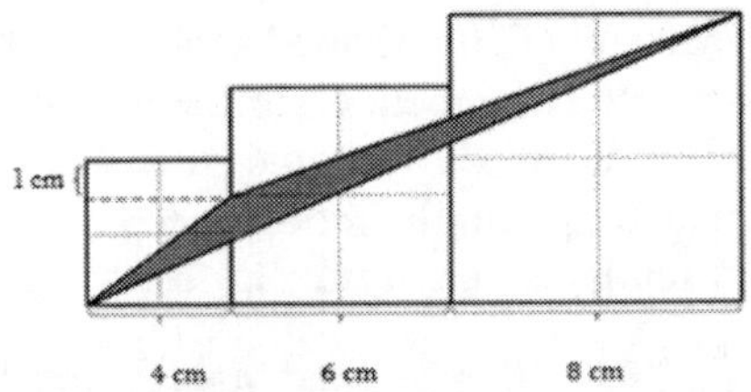

a. Determine the shaded area!
b. If the length sides of each square is a half of the original, is the shaded area changed to be a half of the original area?
c. If the length sides of each square is k times of the original length, k ≠ 1, is the shaded is also changed to be k times of the original area?

3.1 Critical Thinking of Rational Student

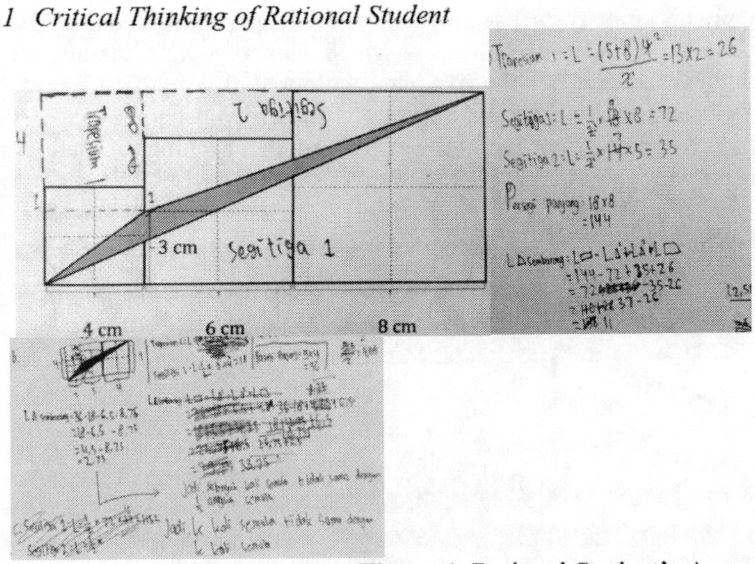

Figure 1. Rational Student's Answer

Focusing on the criteria, the rational student didn't fully write about what is known and what is asked from the problem. He simply stated there are three squares that has one side of 4 cm, 6 cm, and 8 cm. In the interview, he is, however, able to explain what is being asked and what is known. He mentioned the information which is implicitly deducted from the problem, namely the length sides of the square that can be directly identified. He associated the known information with the square properties, that has four equally length sides. According to [16, 17], an intuitive student prefers to focus on the connections and relationships among the facts, informations or concepts. He is also able to find two strategies which are used to determine the shaded area. The first one is to draw auxiliary lines so that a rectangle can be obtained. In addition, the subtracting the area of the rectangle with the first triangle area, second triangle area, and the btrapezoid area can also be obtained. The second one is done by using the Heron's formula, i.e. $L = \sqrt{S(S-a)(S-b)(S-c)}$, S which involves the half of the triangle perimeter; and a, b, c as length sides of the triangle.

Relating to reason criteria, the rational student chose the first strategy because it is simpler than the second one. In the interview, he could explain the supporting reasons for the selection of his strategy. The rectangle can be composed by triangles 1, triangle 2, trapezoid, and shaded area. Hence, the shaded area can be determined by his chosen strategy. On the other hand, he did not use the small planes to determine the shaded area. It is because he did not know the length sides of the triangle and it will be difficult in the calculation. It concludes that he can show the pro and contra reasons in choosing a certain strategy. According to [12], an intuitive student solved the given problems according to "right versus wrong" principle.

Based on the calculation, the rational student can fulfill the inference criteria by making a correct conclusion, namely the shaded area which is 11 cm^2. If the length sides of the squares are changed to be a half, then the shaded area is 2.75 cm^2, which is not a half but a quarter of the original area. If the length sides of the square are changed to be k $(k > 1)$ times of the original, the shaded area will not be k times but k^2 times of the original area. Moreover, he underlined that the argument is similar to a square and a right triangle in which the length sides are converted to k times of the original, then the area becomes k^2 times of the original. The conclusion is taken by associating with the ratio concept and the concept of a plane area. According to [17], an intuition student tends to be careful in observing the patterns and in continually looking for new possibilities.

Furthermore, the rational student maintained the context of the problem by observing the length sides of the changing square to answer the problem. He was also able to identify an appropriate formula which supports his chosen strategy. To apply his strategy, he used formula of rectangle, triangle, and trapezoid area. It shows that the rational student can identify such situations which relevant to his thinking process skills. In addition, he explained that the shaded area is the area that is coloured differently. It shows that the rational student fulfills the clarity criteria because he understood the terms "shaded area". After finding the answer, the rational student demonstrates the overview criteria by checking the truth of his decision, conclusion, and all of his answer.

3.2 Critical Thinking of Idealis Student

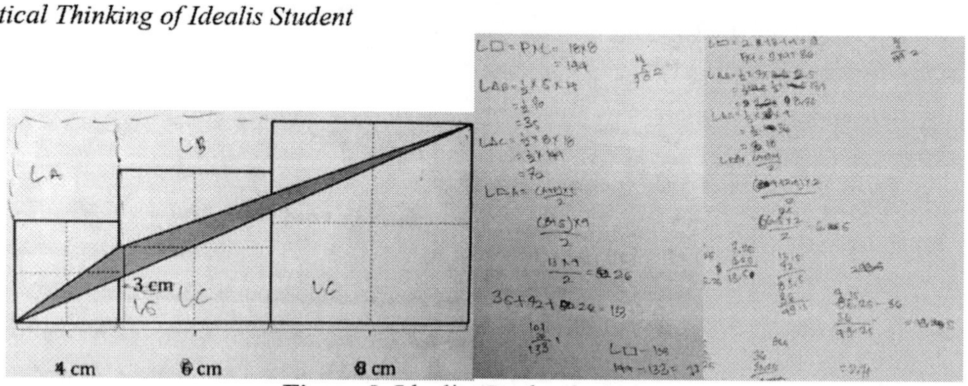

Figure 2. Idealist Student's Answer

Focusing on the criteria, the idealist students did not write about what is known and what is asked from the question. In the interview, he can fully explain what is being asked and what is known. He employ only one strategy that can be used to determine the shaded area. His strategy is by drawing an auxiliary line so that a rectangle can be formed as well as the subtracting the area of the rectangle with the trapezoid area (L_A), first triangle area (L_B), and the second triangle area (L_C). He said that his strategy is easy to implement. The condition shows that the idealist student gave a good reason about his selection strategy subjectively. According to [16], one of the terms that can be used to describe feelings is subjective. Besides, a feeling student frequently takes criticism very personally [12].

In terms of reason criteria, the idealist student explained the same reason about his proposed strategy as well as the rational student. As well as the rational student, for the inference criteria, the idealist student concluded correctly that the shaded area is $11\ \text{cm}^2$. When the length sides of the squares are changed to be a half of the original lengths, the shaded area is $2.75\ \text{cm}^2$, which is not a half but a quarter of the original area. However, if the length sides of the squares become k times of the original, the shaded area will not be k times of the original area. The idealist student insists that it is because the k is substituted by ½ then the shaded area is not a half of the original one. It shows such intuitive natures of the rational type. This condition supports the statement of [14] that someone with intuitive nature has a tendency to see patterns and connect some known information when taking and processing information. Relating to the clarity and overview criteria, the idealist student showed all these criteria as well as the rational student. However, he defined that the shaded area is the area that is black coloured.

3.3 Critical Thinking of Guardian Student

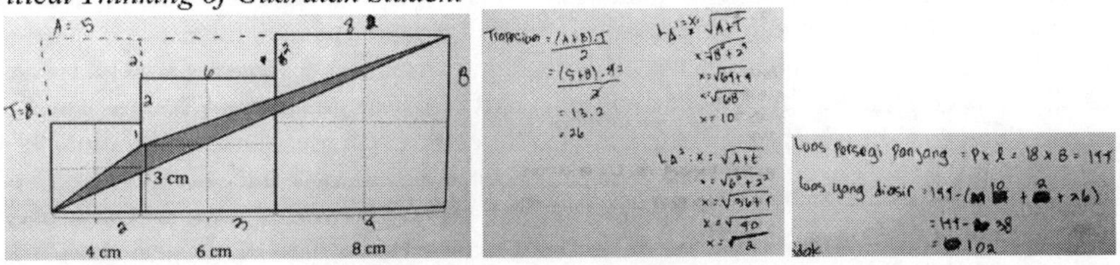

Figure 3. Guardian Student's Answer

Focusing on the criteria, the guardian student did not write about what is known and what is asked by the question on his answer sheets. In the interview, he was not fully understood the problem. He found only one strategy to determine the shaded area, which is similar to the strategy used by rational and idealist students. In solving the problem, he was consistently used his strategy. It is because a judging student likes everything which is established from the beginning [9].

Relating to the reason criteria, the guardian student could not explain the reasons beyond his chosen strategy as well as rational and idealist students. The guardian student cannot fulfill the inference criteria. It is because he concluded the shaded area incorrectly. He explained that the shaded area will be a half of the original when the length sides of squares are a half of the original. He incorrectly connected the related information or concepts. According to [17], a sensing student prefers to think about thing that is presented clearly.

After choosing the strategy, the guardian student could not identify the formula that appropriates his strategy. To find the area of the right triangle, he used the Pythagoras formula so that he can determine the hypotenuse of the triangle, which is not the triangle area (Figure 3). It shows that the guardian student does not fulfill the situation criteria. The guardian student explained his understanding about the term "shaded area". As well as the rational student, the guardian student underlined that the shaded area is the area that is coloured differently. After determining the answer, he said that he did not check the truth of his answer or his conclusion consistently. It underlines that the guardian student does not fulfill the overview criteria.

3.4 Critical Thinking of Artisan Student

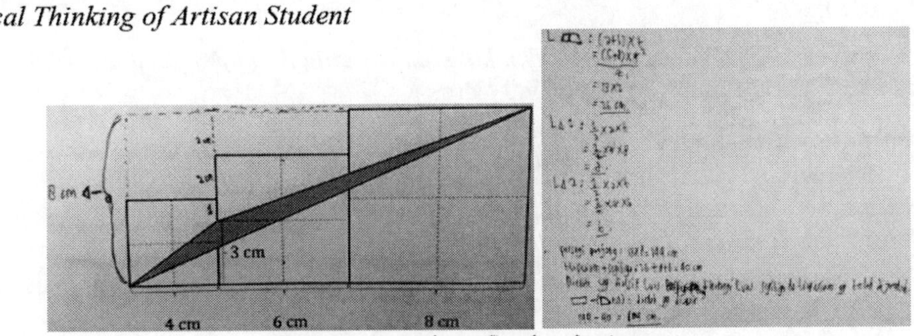

Figure 4. Artisan Student's Answer

On the answer sheet, the artisan students also did not write about what is known and what is asked from the question. However, in the interview, he showed that he can fulfill the focused criteria by explaining what is being asked and what is known from the problem. His explanation is based on the information that there are three squares with one side of 4 cm, 6 cm, and 8 cm, respectively. According to [16], a sensing student prefers to see things by relying on their senses. He found only one strategy to determine the shaded area, which is similar to the strategy used by rational, idealist, and guardian students.

Related to the reason criteria, the artisan student explained the similar reasons about his proposed strategy as well as rational and idealist students. He underlined that the rectangle can be constructed by the triangle 1, triangle 2, trapezoid, and shaded area. Hence, the shaded area can be determined by subtracting the area of rectangle with the triangle 1, triangle 2, and trapezoid areas.

Although his idea is correct, the artisan student concluded the shaded area incorrectly. He explained, as the guardian student, that the shaded area will be a half of the original area when the length sides of the square are changed to be a half of the original. He also could not understand the notation "k" because k performs an unusual number of his knowledge. It is because sensing student who prefers something that is presented clearly [17].

The artisan student was able to identify the formula that is appropriate to his strategy. It shows that the artisan student fulfills the situation criteria. However, he determined the length side of the triangle incorrectly. In addition, the artisan student fulfills the clarity criteria by explaining the terms "shaded area" correctly. He underlined that the shaded area is the area that is coloured black. He similarly defined the shaded area as the idealist student did. After finding his answer, he said that he didn't checked his answer and conclusion consistently.

4. Conclusion

Rational and idealist students could fulfill all FRISCO criterias which were not fullfilled by guardian and artisan students. In solving the given problem, all subjects were able to focus in explaining the what is known and what is asked although the guardian student was not considered well enough. On the reason criteria, only rational, idealist, and artisan students could clarify their reasons and can make a reasonable conclusion. On inference criteria, only rational and idealist students could make conclusion correctly. On situation and clarity criterias, all students had demonstrated their ability to solve the problem by involving all situations which are presented on their question sheets, and understand all terms. However, guardian student could not choose the appropiate formula. On the overview criteria, the guardian and artisan students unchecked the answer consistently. In this study, not all students could think critically based on FRISCO criterias. Therefore, the teachers should frequently train students on how to develop critical thinking skills, especially for guardian and artisan students.

References

[1] Aizikovitsh E and Cheng D 2015 Developing Critical Thinking Skills from Dispositions to Abilities: Mathematics Education from Early Childhood to High School *Creative Education* pp 455-62

[2] Butterworth J and Thwaites G 2013 *Thinking Skills Critical Thinking and Problem Solving*. 2nd ed (New York: Cambridge University Press)

[3] Carroll R T 2004 Becoming a Critical Thinker *Scepdic*

[4] Chukwuyenum A N 2013 Impact of Critical thinking on Performance in Mathematics among Senior Secondary School Students in Lagos State *Journal of Research & Method in Education (IOSR-JRME)* **3**(5)

[5] Emry K, Lewis L, and Morfet C 2008 Open-Ended Maths Tasks Space, Chance, and Data *Hawker Brownlow Education*

[6] Ennis R H 1996 Critical Thinking Dispositions: Their Nature and Assessability *Informal Logic* **2004** (3)

[7] Foong L M, Shariffudin R S, and Mislan N 2012 Pattern and Relationship Between Multiple Intelligences, Personality Traits and Critical Thinking Skills Among High Achievers in Malaysia *IPEDR* **1**(3) p 27

[8] Jacob S M and Sam H K 2008 Measuring Critical Thinking in Problem Solving through Online Discussion Forums P*roc. Int. Conf. of Engineers and Computer Scientists*

[9] Keirsey D 1998 *Please Understand Me II* (Del Mar: Promotheus Nemesis Book Company)

[10] Lai E R 2011 Critical Thinking: A Literature Review *Pearson's Research Reports*

[11] Munroe L 2015 The Open-Ended Approach Framework *European Journal of Educational Research* **4**(3)

[12] Myers IB 1980 *Introduction to type* (Washington D C: Consulting Psychologists Press)

[13] Myers I B and Briggs K 1962 *MBTI Manual: A Guide to the Development and Use of the Myers-Briggs Type Indicator* (Washington D C: Consulting Psychologists Press)

[14] Russo C J 2012 The Impact of Personality Type and Gender on Students' Performance in A Business Capstone Course *Academy of Educational Leadership Journal* **16**(1)

[15] Shimada S 1997 *Lesson Study for Effective Use of Open-Ended Problems* (Reston VA: The National Council of Teachers of Mathematics) Inc

[16] Tyagi A 2008 Personality Profiles Identification Using MBTI Test for Management Students: An Empirical Study *Journal of the Indian Academy of Applied Psychology* **34**(1)

[17] Woosley E R 2001 A Look at The Relationships Between MBTI Psychological Types and Mode Conflict Styles *Guidance and Counseling – Mental Health journal*

Designing Tasks to Examine Mathematical Knowledge for Teaching Statistics for Primary Teachers

T Y E Siswono[1], A W Kohar[2], S Hartono[3]

[1,2,3] Faculty of Mathematics and Natural Science, Universitas Negeri Surabaya, Kampus UNESA Ketintang, Surabaya, Indonesia, 60231

Email: [1]tatagsiswono@unesa.ac.id

Abstract. Mathematical knowledge for teaching (MKT) is viewed as fuel resources for conducting an orchestra in a teaching and learning process. By understanding MKT, especially for primary teachers, it can predict the success of a goal of an instruction and analyze the weaknesses and improvements of it. To explore what teachers think about subject matters, pedagogical terms, and appropriate curriculum, it needs a task which can be identified the teachers' MKT including the subject matter knowledge (SMK) and pedagogical content knowledge (PCK). This study aims to design an appropriate task for exploring primary teachers' MKT for statistics in primary school. We designed six tasks to examine 40 primary teachers' MKT, of which each respectively represents the categories of SMK (common content knowledge (CCK) and specialised content knowledge (SCK)) and PCK (knowledge of content and students (KCS), knowledge of content and teaching (KCT), and knowledge of content and curriculum (KCC)). While MKT has much attention of numbers of scholars, we consider knowledge of content and culture (KCCl) to be hypothesized in the domains of MKT. Thus, we added one more task examining how the primary teachers used their knowledge of content (KC) regarding to MKT in statistics. Some examples of the teachers' responses on the tasks are discussed and some refinements of MKT task in statistics for primary teachers are suggested.

1. Introduction

In this recent decade, there has been a growing research interest in the notion of knowledge that mathematics teachers need to master in order to make their teaching effective. This is primarily originated from the work of Shulman [1] which was then modified by other researchers such as Ball et al [2] for their mathematical knowledge for teaching (MKT). In particular, the research in the investigation of statistical knowledge for teaching has also obtained a greater attention by scholars. The attention is based on common reports which revealed that better teaching of statistics requires better training for the teachers. As with no specific training, they are likely to fall back on what are often erroneous intuitions and beliefs [3].

It is important, therefore, to get some prior studies by examining primary teachers' competencies regarding basic statistical knowledge, particularly when it comes to basic concepts such as a central tendency covering mean, median, modus, which is included in typical mathematics curriculum for primary students. To that, some have concerned on exploring teachers' knowledge of subject matter on statistics [4, 5,6], while others concerned on pedagogical content knowledge on statistics (PCK) [7, 8].

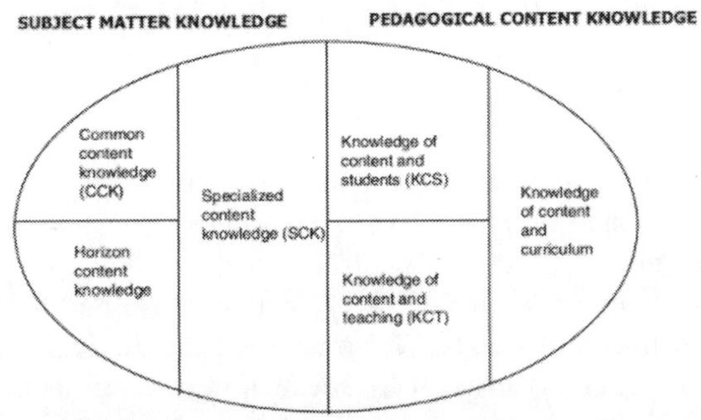

Figure 1. Domains of Mathematical Knowledge for Teaching [2]

MKT domain proposed by Ball et al [2] as shown in figure 1 encompasses the commonly expected mathematical knowledge of adults as well as the specialised mathematical knowledge of the classroom needed by teachers to carry out the "work of teaching mathematics" [2].They categorized the domain into two types namely subject matter knowledge (SMK) and pedagogical content knowledge (PCK). SMK consists of common content knowledge (CCK), specialized content knowledge (SCK) and horizon content knowledge (HCK), while PCK consists of knowledge of content and student (KCS), knowledge of content and teaching (KCT), and knowledge of content and curriculum (KCC).

In SMK, Ball *et al* [2] defines CCK as the mathematical knowledge and skill used in settings other than teaching. Teachers need to know the topic they teach such as they must recognize an inaccurate definition [2]. In short, SCK is defined as a mathematical knowledge and a unique skill in teaching. For example, it requires understanding certain interpretations of the mean in ways that students should figure out, such as inferencing mean from a fixed total [9], appreciating the difference between models of mean [2], and distinguishing interpretation of mean as data reduction, fair share, typical value, and signal in noise [10].

With regard to PCK, Ball *et al* [2] defines KCS, as the knowledge combining students and mathematics knowledge. Teachers must anticipate what students are likely to think and what they will find confusing. In coping with the concept of mean, for example, teachers need to understand the common errors that students make when solving a mean-related problem, or figuring out the students' difficulties and misconceptions in the concept of mean. KCT, on the other hand, is defined as the knowledge that combines knowledge in teaching and mathematics. For instance, teachers need to understand which representation they should use to introduce the concept of mean, what sequence of problems they should use to begin the work on the concept of mean, or how to chronologically arrange the solution methods that should be presented in a whole class discussion. In knowledge of content and curriculum (KCC), Ball *et al* [2] actually is not yet sure whether this may be a part of their category of knowledge of content and teaching or whether it may run across several categories. However, they refer KCC to Shulman's [1] initial argument which defines knowledge of curriculum as a set of programs designed for teaching certain subjects and topics at particular level, the diversity of available teaching materials related to such programs and a number of characteristics that are both indications and contra indications use in a specific curriculum [1]. To cope with the topic of mean at primary level, for example, teachers need to understand how to determine learning goals about the topic of mean for a particular activity or what prior knowledge related to mean that primary students should know.

However, we found limited resources that examined teachers' MKT both in SMK and in PCK in a statistics concept, particularly in the topic of mean for primary teachers. Whereas previous studies, as we found, only examined SMK [4] and PCK [8] separately. In addition, there has been only a limited research on PCK for statistical education. Sosa [11] finds that the majority of studies addresses the knowledge of the content (KC) to be taught (about 88%), while less than a half study specified PCK or knowledge regarding to student learning process. Thus, in this present study, we designed a set of tasks that examine MKT in a such topic and discussed how primary teachers worked on the MKT tasks. Here, our concern is MKT in central tendency topic, particularly, 'mean' taught at primary school level. In addition, MKT encompasses CCK, SCK, KCS, KCT, and KCC. We also add one more task about how culture is introduced in teachers' teaching knowledge and practice. This task discusses a potential teachers' knowledge we hypothesized as one of the teachers' MKT namely knowledge of content and culture (KCCl).

2. Method

In designing MKT tasks for the statistic subject matter for primary teachers, we were guided by some theoretical pieces of literature, mainly by the works conducted by Shulman [1] and Ball et al. [2] and the findings which are relevant to MKT for central tendency. In coping with CCK, we considered referring the relevant theory from Ortiz and Font [4] which examined teachers' understanding on the estimation of an unknown quantity. For SCK, we considered the model of task developed by Zaskis [9] which examined teachers' understanding of a mean interpretation inferred from a fixed total. Furthermore, KCS concerned on examining teachers' knowledge of students' common errors and difficulties in a mean-related problem, while KCT and KCC respectively concerned on teachers' knowledge about how to teach relevant representations under the topic of mean and the order of prior knowledge taught before learning mean. One additional task regarding to KCCl was added as our hypothesis (see Table 1).

To confirm the validity of the MKT tasks, we asked an expert to validate the questions by giving constructive suggestions regarding content and language use. To discuss teachers' MKT on the concept of mean, we analyzed forty primary teachers' responses on the MKT tasks which exemplify the representatives of various responses found in all MKT types.

Table 1. MKT Tasks for Statistics at Primary Level

No	MKT types	Questions	Focus concern	MKT Test definitions [2]
1	Common content knowledge	Eight students from a class weigh a bag of sugar using the same instrument, with the following values in grams being obtained: 6.2, 6.0, 6.0, 6.3, 6.1, 6.23, 6.15, and 6.12. What would be the best estimate of the sugar's real weight?	Estimation of an unknown quantity	Knowing the facts and concepts of the discipline of mathematics, i.e. mean
2	Specialized content knowledge	The weight of an object made from clay is 91.1 grams. The object was split into twelve smaller objects with different weight. What was the value of the average weight? Explain your strategy	Focus on inference from a fixed total	Presenting explanations for certain procedures
3	Knowledge of content and students	Imagine your students are asked to solve this problem. "A set of data from eight students' math scores give mean, median, and mode respectively of 7, 8, and 7.2. Find three possible sets of those eight scores which fit the requirement. " What *errors* and *difficulties* which may you probably find from your students' work? Explain.	Common errors and difficulties regarding the concept of mean, median, and mode in a problem-solving task	Anticipating student diificulties and errors when presented with a mathematical task
4	Knowledge of content and teaching	Which representations (figures, learning media, visual display, etc) would you use to introduce the concept of mean? How would you use such representations in your teaching?	Models of representation and its teaching sequence of mean	Selecting models, representations and procedures that support the development of mathematical understandings
5	Knowledge of content and curriculum	a. What topics before 'mean, median, and mode' that primary students should understand? b. Do you think all primary graders (from 1 to 6) are able to learn about mean, median, and mode? If Yes, How? If No, Why?	Learning trajectory on mean, median, and mode for primary students	Knowing what should be taught regarding central tendency in order

6	Knowledge of Content and Culture*	a. Which local wisdom in Indonesia would you think is appropriate with the topic of mean, mean, and mode? b. How would you bring that local wisdom into your teaching practice of mean, median, and mode?	Local wisdom on mean, median, and mode	-

*hypothesized

3. Results

3.1. Teachers' SMK on the concept of mean

We obtained four categories of CCK regarding to the estimation of an unknown quantity. Category 1 corresponds the responses that indicate the use of the arithmetic mean and the term 'mean' explicitly. Category 2 corresponds the indication of the use of arithmetic mean but does not indicate the term 'mean' explicitly. At last, category 3 portrays those which do not indicate the use of arithmetic mean and the term 'mean' explicitly. To grab overall pictures, Figure 2 shows clarifies the four categories mentioned.

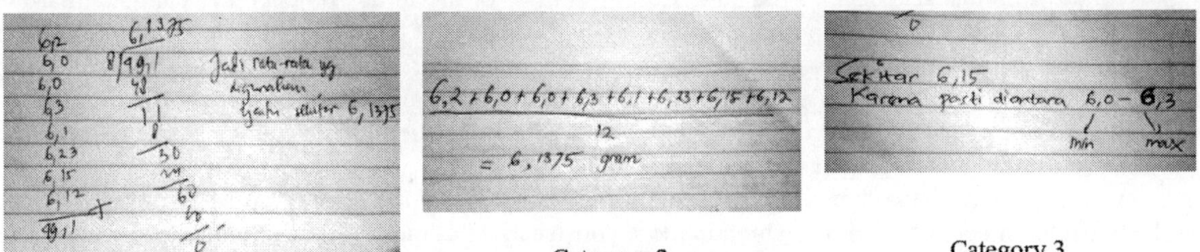

Category 1 Category 2 Category 3

Figure 2. Teachers' CCK responses

Regarding to the teachers' SCK, we found three categories of responses namely (1) correct responses which indicate the use of inference from a fixed total, (2) correct responses which indicate the use of a strictly applied algorithm, and (3) incorrect responses. The incorrect ones were due to the miscalculation when applying the mean algorithm. Figure 3 illustrates the examples of the first two categories.

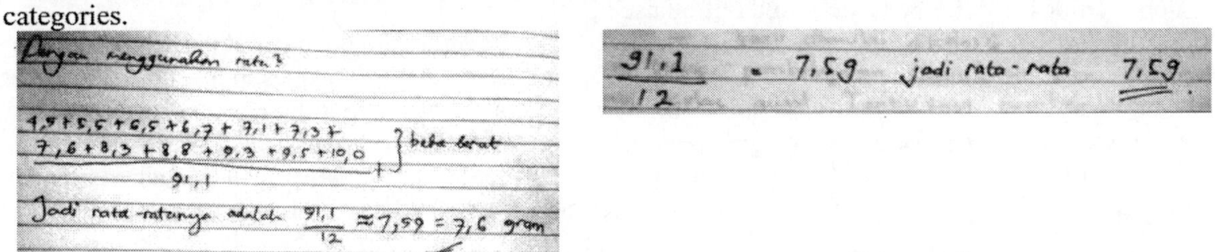

Category 1 Category 2

Figure 3. Examples of SCK in Category 1 and 2

Category 1 shows that the teacher likely did not really understand the concept of mean from the perspective of the inference from a fixed total. The teacher added up all the data and divided the sum by numbers of data, which means the teacher tried to first list a set of possible weight of clay which is different each other and applied the standard arithmetical mean. In contrast, category 2 indicates a teacher's understanding in a fixed total of mean concept, indicated by the teachers' direct calculations; dividing the total of data, instead of listing a possible set of different data, by the number of objects.

3.2. Teachers' PCK on the concept of mean

Discussing KCS, we divide the teachers' responses into two types namely difficulties and errors. There are three errors that students experience, such as they (1) do not know that median should be in between the fourth and the fifth datum when the data has been in order, (2) do not think that mode is the only one that can be found from the set, and (3) know how to make the sets of data but create incorrect calculation. Meanwhile, we found two categories of student difficulties namely (1) finding more than a set of possible data and (2) determining which one of mean, median, or mode that should be considered as the easiest way to use a mathematical formula. One of the teachers' responses based on the latter difficulty is given as follows.

> "When selecting mean first, students will likely think of the total data, obtaining $8 \times 7 = 56$, and make eight data whose sum, median, and mode respectively is 56, 8, 7,2. When selecting mode, they have already understood that the mode, which is 7.2, must be contained in the set of data. Thus, 7.2 must be most frequently found from the set. I think the best way is by starting from the mode, the median, and finally, the mean, because mode is always found in a set of data"

The response above shows that the teacher knows that the errors regarding the misunderstanding of the mode and median could lead to the student difficulties in selecting what should be firstly considered for any calculation.

Regarding KCT, we obtained two types of representations the teachers proposed for teaching mean concept: figural and physical activities. The figural representations, as shown in figure 4, were selected to be introduced in the teachers' teaching sequences. The idea of teaching using such representations is that teachers would start from giving a set of similar objects, i.e. candies, marbles, small stones, balls or other familiar objects. The students would get the different number of objects. The students' task is to think of how to share their objects each other so that every student has the same number of the objects. Meanwhile, the physical activities were suggested to be used for learning mean concept by having students to do some measurement activities, such as measuring students' height or weight, then finding its average using a mean formula.

Comparing those two types of representations, we argue that the figural representations are closer to the expected model of teaching which is more constructive, attractive, and interactive since students should discuss with their peers how to share the objects and to think about 'a number' which represents the same number of objects for each student. However, either the figural representations or the physical activities the teachers offered seemed to have students learned only about the 'meaning of mean', instead of the 'concept of mean', whereas many educators suggest to set learning sequences which aim students to, for examples, make sense with other central tendency topics such as median and mode, identify that average does not necessarily equal with one of the values summed [12] or various types of mean interpretations as described by Konold & Pollatsek [10].

With regard to KCC, the teachers responded that students should understand about ordering numbers, a basic arithmetical operation involving decimals, and extreme values in a set of data. When asked to argue whether it is possible for all primary graders to learn about the concept mean, median, mode, we found there were two categories namely agree and disagree statements. Those who agreed stated that actually, students at lower grades have learned about basic statistics such as counting objects based on certain representations, determining which object has the largest, the smallest, and the most middle number (median) and pointing the most common height out from the students (mode). However, those who disagreed argued that mean, median, and mode can only be taught at upper primary grade, since their complexities when using formula and involving non-integer number as well as interpreting word problems related to those three topics.

3.3. Teachers' KCCl on the concept of mean

Our proposal regarding KCCl was originated from an idea proposing that differences between students' and teachers' cultural backgrounds and teachers' limited knowledge of the students' cultures present challenges for teachers to plan and to teach in ways that harness on their students' background experiences [13]. In MKT, therefore, teachers are encouraged to map the cultural knowledge around students' environment, such as local wisdom, cultural and natural heritages, which potentially affect on students' understanding of a certain topic in mathematics and, then, finally apply the mappings into their instructional practices. For example, when teaching 'mean' at grade six, teachers can bring the folklore found at the place where students are schooling. See [14].

The teachers in this study answered the KCCl task with various responses, ranging from giving examples of using local *batik* (traditional cloth), traditional song, dances, and games, to the increase of visitors in a temple. The idea of using those examples is by asking students to obtain data showing the trend of a certain data which is related to, for examples, the numbers of batik home industries or the traditional games preferred by students. However, we found no teachers responded by showing how MKT are explicitly integrated into the local wisdom they mentioned. The only responses we found is that there was a response explained about how to use a traditional game, called *congklak*, to introduce mean, median, and mode. But, it seems limited to explore why and how to bring the games into teaching practices.

4. Discussion and Conclusion

In our MKT tasks, we designed all the MKT types from the teacher participants particularly on the topic of mean-related concepts. However, we suggest that the MKT tasks can still be improved regarding the content of focus concern included in the set of MKT tasks. For example, further research can give additional CCK referred to Ortiz and Font used Batanero's statistical problem to examine teachers' CCK [4] such as about obtaining equal shares in order to achieve a uniformed distribution, finding an element that represents a set of given values whose distribution is approximately symmetrically framed by a context being one of data comparison, concerning a situation on which the task is to determine the value that is most likely to be obtained when selecting a random element from a population; and about the weighted arithmetic mean. Moreover in PCK, further designers can also consider the PCK on statistics of Watson et al.'s which consists of four aspects: recognizing big idea, anticipating student answers, employing content-specific strategies, and constructing shifts to general [8]

To conclude, we would highlight that the design of MKT task for statistics in primary school resulted. We suggest completing the MKT tasks with the characteristics of MKT component as well as central tendency concepts. This is to develop a more comprehensive task designed to examine teachers' MKT on central tendency, instead of mean, median, mode, separately. In addition, further studies may also use an experimental method to examine the validity and the reliability of the MKT tasks.

References

[1] Shulman L S 1986 *Educational researcher* **15** 4-14
[2] Ball D, Thames M H, & Phelps G 2008 *Journal of teacher education* **59** 389-407
[3] Stohl H 2005 Probability in teacher education and development In: G A Jones (Ed) Exploring probability in schools Challenges for teaching and learning (pp 345-366) (New York: Springer)
[4] Ortiz JJ & Font V 2014 *REDIMAT* **3** 192-219
[5] Leavy A & O'Loughlin N 2006 *Journal of Mathematics Teacher Education* **9** 53-90
[6] Lampen E 2015 Teacher narratives in making sense of the statistical mean algorithm Pythagoras **36** 1-12
[7] Estrella S, Olfos R, & Mena-Lorca A 2015 *Educação e Pesquisa* **41** 477-493
[8] Watson J, Callingham R, & Nathan E 2009 *Crossing divides* **2** 563-570
[9] Zazkis D 2013 *International Journal of Mathematical Education in Science and Technology* **44** 204-213
[10] Konold C & Pollatsek A 2004 Conceptualizing an average as a stable feature of a noisy process In *The challenge of developing statistical literacy, reasoning and thinking* (pp 169-199) (Springer Netherlands)
[11] Sosa JEP and Astudillo MTG 2010 *Conocimiento didáctico del contenido sobre la representación de datos estadísticos: Estudios de casos con profesores de estadística en carreras de psicología y educación* (Doctoral dissertation, Dissertação de Doutoramento apresentada à Universidade de Salamanca)
[12] Strauss S & Bichler E 1988 *Journal for Research in Mathematics Education* **19** 64-80
[13] Averill R 2012 *Journal of Urban Mathematics Education* **5** 157-181
[14] Lestariningsih L Putri R I I & Darmawijoyo D 2015 *Journal on Mathematics Education* **3** 165-174

IOP Conf. Series: Journal of Physics: Conf. Series **947** (2017) 012009 doi:10.1088/1742-6596/947/1/012009

The improvement of autism spectrum disorders on children communication ability with PECS method Multimedia Augmented Reality-Based

Taryadi[1] and I Kurniawan[2]

[1,2]Dept. of Information Technology, STMIK Widya Pratama Pekalongan, Indonesia,

Email: tari_ball@stmik-wp.ac.id, ichwan.kurniawan@gmail.com

Abstract. This research was done to examine the usage of PECS method (Picture Exchange Communication System) multimedia augmented reality based as a learning alternative in training the communication of autism childen. The aim of this research were developing an approach to improve the communication ability before and after an intervension with PECS multimedia augmented reality method. The subject of this research were 12 autism children in Inclusion school in Pekalongan region. The experiment method was used with the single subject research approach. The research resulted that the average ability level in communication before and after the treatment has shown 47% while during the treatment the average level is 65%. Whereas there is an improvement after intervension stage with the average of 76%..

1. Introduction

Communication is an activity of transferring message to other people by using (a) vocal gesture, and eye movement combine with the attention to the object or people; or (b) by conventional gesture or symbol (such as, saying words or language symbol). [1]. the three main functions used in communication in pre-linguistic period are attention, collective attention, and body movement. [2]. These three matters become the basic composer of communication ability, on account of these there process, children start to understand information of their world, also understanding that other people perspective may be different from them. Furthermore, children will develop more complex communication ability such as: asking help, negotiation, commenting, as well as social communication. [3].

One of the characteristics of child development with autism spectrum disorders (ASD) is lack of communication development. And in fact, the often complain raised by parents when they take their children to autism therapist is a communication barrier. Communication barrier causes a difficulty in children learning and their behaviour seems to be uncontrolled. The stage of communication is not running smoothly in ASD child. Furthermore the communication environment is an important relation to get the outcome in gaining the bonding with ASD child.[4][5].

In addition to the communication barrier, social mutualism has also become one of the weaknesses of autism child. Some researchers have shown that the improvement of the stage of communication is by improving collective attention and moving will release the behaviour of ASD child communication. [6].

Unfortunately, it is a difficult matter to gain and train the collective attention communication in ASD children. The motivation to start the collective attention is a skill from social attention and social interaction. [3]. The collective attention activity occurred by the response from the children and a part of interest or positive influence, not by verbal response. Despite of that teacher cannot ask children to pay attention if they do not start the attention, thus treatment is needed to the ASD child teacher to start the attention. [7].

Method of communication intervention is Picture Exchange Communication System. Since PECS is not involving the social interaction or an understanding by the support of an adult, thus it will be useful for the learning of ASD children. The general treatment to the children gives a significant effect to them. The research on ASD children showed that the accuracy of PECS method in session of communication is requested. [8].

Most of the autism children have better visual memory than auditory memory. Autism children are able to understand and remember more through visual learning or visual thinking, by visual learning they tend to be easily concentrate and understand such learning. The use of media with visual strategy can possibly be implemented in teaching the communication ability.[9]

Based on the concept of PECS method, an augmented reality multimedia will be developed for autism child in communication learning. The application is expected to be able to support an autism child in improving the communication skill and gives an effective practical value. Herein, the augmented reality multimedia application with PECS method is such way of intervention in improving the autism children in communication skill.

The aim of this research was to see how far the usage of PECS method that is developed by augmented reality multimedia technology may support the improvement of the ASD child communication.

2. PECS for ASD child

PECS (Picture Exchange Communication System) is an approach in communication training using verbal symbols. [10]. PECS is designed by Andrew Bondy and Lori Frost in 1985 and started to publish in 1994 in the United States. At the early times PECS was used for pre-school students with autism syndrome and other unusual condition related to the communication problem. PECS was used by the student whose language development is not good and the one who is not having a willingness to communicate to others. In subsequent developments the use of PECS has expanded to the age stage and so on. [11].

In regard to PECS, it is not automatically children will be talking, but by the support of pictures or symbols hence the understanding of the verbal transferring language can be clearly understood. At the early stage, children are introduced with non-verbal symbols, whereas at the final stage of PECS usage, children are motivated to be able to speak. Despite PECS is not a program for autism children in teaching about how to speak, it is believed that it will support them to speak. [12].

Yoder and Stone made comparison between children that used PECS and other system. It is resulted that autism children that were trained by PECS become more verbal than others. PECS will become much more effective in supporting the autism child become more verbally if it is implemented under six years old. [9].

3. Multimedia Augmented Reality

Augmented Reality (AR) is either direct real-time interface or indirect of a physical appearance from the real object by adding an online object which resulting an additional information on the existing object. This AR is gathered the real things and the existing virtual object, this virtual object is just adding the real object and not replace it, while the purpose of this AR is to simplify the real object by inserting the online object thus the information is not only for the user interface but also for user who are not connected directly to the user interface from the real object, such as live streaming video. The main tools for AR are displaying, input tool, tracking, and computer. [13].

Recently AR is growing rapidly and there has been many application or library used in developing AR. AR needs video streaming with camera used as the source of input picture, then to track and detect the marker. The 3D model will appear after the marker has been detected. The 3D model is made by the software for 3D design. [14].

AR has been implemented in many fields, such as education, medical, entertainment, military, design, robotics, and many others. AR has also been applied on tools used by people like mobile phone. [15].

AR has been used in Education as modelling in learning media out of the class. Any material that needs modelling may be built by AR with the expectation gives more understanding to students. As one of the examples is AR may be used in human body modelling to make the sense more real. [16]. In addition, AR can also be used in modelling the solar system in science subjects. There are many AR applications that have been developed as a supporting media in student learning activity.

4. Research Model

This qualitative research, the research focusing on the understanding based on the certain research methodology tradition to explore the human problem or social problem in natural base. The design used were A---B----C which consists of treatment—treatment----effect treatment. The further step is to record the behaviour measurement before treatment, during treatment and after treatment thus the effect of the treatment will be yielded. The effect of the treatment is shown in behaviour recording graph.

The research design is to observe one target of behaviour change in a person or a group of people with its focus. The first stage is an observation or baseline stage. Recording the subject behaviour before any treatment given was done in this stage. After the baseline composed, the treatment was given to the subject and recorded the subject behaviour during the intervention in order to give way or may reach certain criteria. The next stage was to see the effect of the treatment comparing to the baseline performance. The easiest way to see the effect of the treatment is by behaviour recording graph. The further stage is to record the behaviour measurement before the treatment, during the treatment, and after the treatment thus the effect of the treatment will be shown. The effect of the treatment will be displayed in behaviour recording graph.

After the PECS has done, the next stage is data analysis with descriptive qualitative method. The analysis was done to determine the description of PECS treatment in improving the stage of communication in ASD child. The collected data was analysed by visual analysis, which interpreting the result of the graph data accurately and meaningfully by seeing the behaviour change happened based on the displayed graph.

Qualitative analysis was done by visual inspection analysis to see the change time by time and yielded the M score from daily observation of the behaviour. Visual analysis is done to know the success frequency of behaviour in asking subject either in baseline stage or during intervention PECS stage I up to IV as well as to know the duration needed by the subject in requesting the wanted item by making sentence strip and give to the communication interlocutor independently. Qualitative analysis was done to the data obtained from the observation, interview, sharing and it also covers discussion with parents or therapist.

5. Result And Discussion

Multimedia application as the treatment for ASD child has built augmented reality technology. The developed application may run in android based gadget. The outlook from augmented reality multimedia for ASD child learning can be seen in Figure and Figure 2.

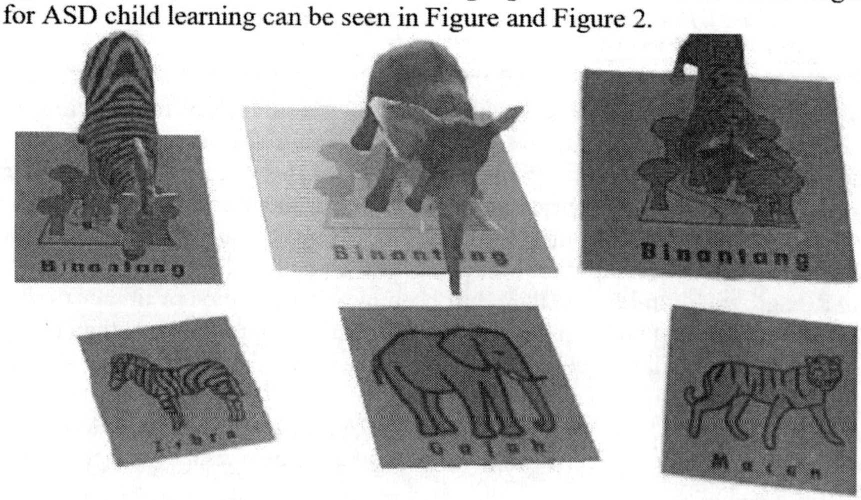

Figure 1. The multimedia outlook of augmented reality for animals

Figure 2. Multimedia outlook of augmented reality for toys

The examining process was done by teacher to ASD child, where teacher is willing to do the intervention to ASD child using android based gadget media. The Multimedia will be shown to the children to show the media and help in communication. The examining process may be seen in figure 3.

Figure 3. The intervention process of ASD child with augmented reality multimedia

Based on the previous result, during and after the intervention done to ASD child with augmented reality multimedia may be seen on figure 4. The result of inter-observer agreement calculation during the PECS implementation on children shows the average score is 76%. Thus it can be concluded that the observation done by observer is reliable.

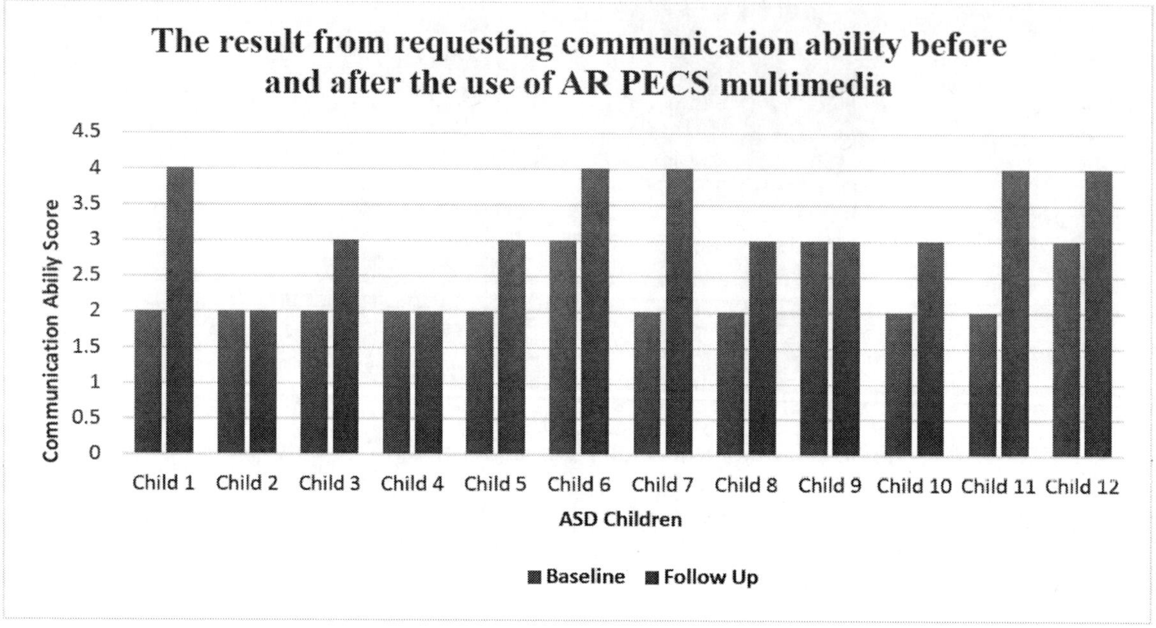

Figure 4. The difference between communication ability before and after treatment

Figure 4 shows the results of tests that have been done on 12 children with autism. Test results showed that there was a significant improvement in the autistic children's communication ability after being treated with multimedia augmented reality. The increase achieved an average of 76% of the communication skills of children before treatment.

The improvement of communication ability happened in all research subject from requester stage to early communication stage, but one subject did not improve and requested to stay at requester stage level. The result of participation ability in asking communication through media of augmented reality multimedia with PECS can be seen on figure 5.

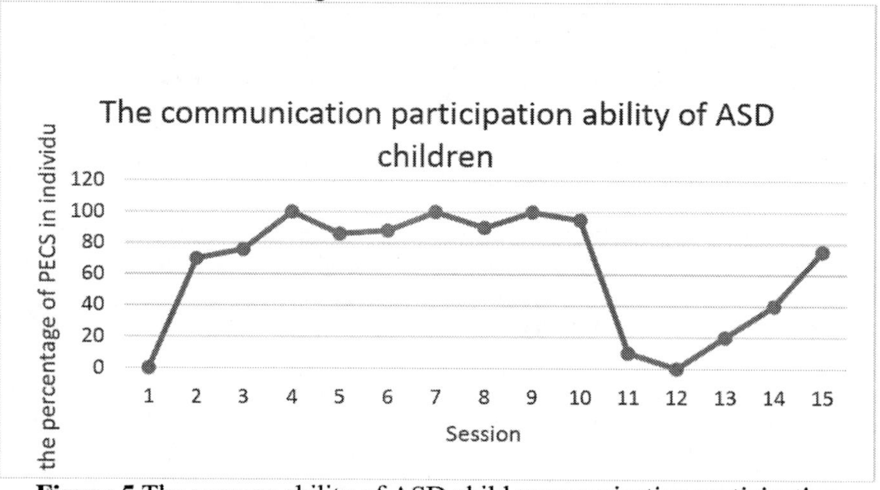

Figure 5. The average ability of ASD child communication participation

The conclusion of the above explanation may be seen on Figure 5 that the ability of requesting through picture exchange need 15 session with 230 trial. The result of requesting communication ability through picture exchange show improvement in advance despite there are decreasing exchange ability at the beginning of stage 4 due to participant needs to make an adjustment at the early stage related to the numbers of shaping process at phase IV and raise continuously at session 15 until it reach the independent exchange of almost 70%. The most assistance used is complete assistance that is therapist helping physically and a bit verbal help like calling name.

There are three factors influencing speed in learning PECS, they are kinds of items that motivate each individual at initiating communication, the use of the concrete symbols as the real ones, requesting centred learning that emphasized more to motivation and reward given to the autism child to communicate functionally.

By the different condition of each subject such as spectrum level, cognitive, behaviour, and different emotion, the subjects have their own ability in improving the communication ability and PECS phase. The reinforcement of children has also influence on their willingness in requesting or communicating to other people. However it's different with the subject that has an interest to the reinforcement, it will be more easily in stimulating to do the requesting communication. The fluctuate condition is also bothering the child in the process. Building the routinely in autism children is such important thing in learning something. [17].

In general, autism children learn or finish any task more easily by concrete and structured ways than the abstract ones. PECS has structured and levelled stages. The usages of PECS picture that exchange with the items are more easily understood than verbal and auditory stimulus.

The education given to autism child needs to be done in stages from the smallest one continue to the bigger and more complex and so forth until the children are able to socialize with the non-disabled children. [9].

6. Conclusion

Based on the result comparison of communication ability before, during and after the treatment process to ASD child by using PECS method augmented reality multimedia based, it is yielded that there is an improvement of communication ability from requester stage to early communication stage.

Some influencing factors in improving the communication ability of the participant are the interest to the teacher, physical condition such as hungry, full, sleepy, and their autism spectrum level. In addition, it best gives the children understanding ability and children personality characteristics such as cheerful, active or introvert and hypoactive or hyperactive. The surrounding factor has also strongly influence on improving the communication ability of autism child such as supporting condition and huge motivation from the family will support a lot, but ignoring condition to the child will obstruct the improvement of communication ability.

Acknowledgement

The research is supported and funded by Research Directorate and Society Dedication of General Directorate of Research Reinforcement and The Development of Research Ministry, Technology, and Higher Education in year 2017.

Researcher also sends best gratitude to the teachers of the inclusion school in Pekalongan region, Pekalongan Regency, Pemalang Regency that has supported a lot in this research. The acknowledgement also give a special gratitude to the parents of ASD child that permit the researcher in doing the treatment of PECS method with augmented reality multimedia based to improve the communication ability. To all parties that helped in collecting data, data calculation, researcher team that support in all process of the research.

References

[1] Yoder P J, Mc Cathren R B, Warren S F and Watson A L 2010 Important distinctions in measuring maternal responses to communication in prelinguistic children with disabilities *Communication Disorders Quarterly* pp 135–147

[2] Wetherby A M and Prutting C A 2014 Profiles of communicative and cognitive-social abilities in autistic children *Journal of Speech and Hearing Research* pp 364–377

[3] Mc Evoy R E, Rogers S J and Pennington B F 2013Executive function and social communication deficits in young autistic children *Journal of Child Psychology and Psychiatry and Allied Disciplines* pp 563–578
[4] Mundy P, Sigman M, Ungerer J and Sherman T 2013 Defining the social deficits in autism: The contribution of non-verbal communication measures *Journal of Child Psychology and Psychiatry* pp 657–669
[5] Stone W L, Ousley O Y, Yoder P J, Hogan K L and Hepburn S L 2007 Nonverbal communication in two- and three-year-old children with autism *Journal of Autism and Developmental Disorders* pp 677–696
[6] Mc Duffie A 2004 The relationship of four prelingustic skills and rapid word-object mapping to vocabulary outcomes in young children with autism *Unpublished doctoral dissertation* Vanderbilt University
[7] Bondy A and Frost L 2012 PECS and Other Visual Communication Strategi in Autism *Woodbine* (House: Library of Congress Cataloging in Publication Data)
[8] Yoder P and Stone W L 2006 RandomieComparion of Two Communication Interventions for Preschoolers with Autism Spectrum Disorders *Journal of Consulting and Clinical Psyhology* Vol.**74** No. 3 pp 426-435
[9] Aldred C Green J and Adams C 2014 A new social communication intervention for children with autism: Pilot randomized controlled treatment study suggesting effectiveness *Journal of Child Psychology and Psychiatry* pp 1–11
[10] Whalen C and Schreibman L 2013 Joint attention training for children with autism using behaviour modification procedures *Journal of Child Psychology and Psychiatry* pp 456–468
[11] Bondy A S and Frost L A 2014 PECS: The Picture Exchange Communication System training manual *Cherry Hill* (NY: Pyramid Educational Consultants)
[12] Ganz J B, Simpson R L and Corbin J N 2008 The Impact of the Picture Exchange Communication System on Requesting and Speech Development in Preschoolers WithAutism Spectrum Disorder and Similar Characteristics *Research in Autism Spectrum Disorder* pp 157-169
[13] Hodgdon L Q 2012 Solving Social Behavioral Problem Through The Use of Visually Supported Communication. In KA Quill (Ed). Teacher Children With Autism Strategies to Enhance Communication and Socialization *International Journal on Smart Sensing and Iintelligent Systems* pp 134-14
[14] Jacko J A 2013 H*andbook of Research on Ubiquitous computing Technology for Real Time Enterprises* CRC Press
[15] Billinghurst H M and Bruce T 2007 *Emerging Technologies of Augmented Reality: Interfaces and Design* Idea Group Publishing Idea Group Inc United Statesof America
[16] EricK 2008 *Augmented Learning Research and Design of Mobile Educational Games* Massachusetts Institute of Technology
[17] Tincani M 2014 Comparing The Picture Echange Communication System And Sign Language Training For Children With Autism *Focus on autism and development Dissabilities* pp 152-163

IOP Conf. Series: Journal of Physics: Conf. Series **947** (2017) 012010 doi:10.1088/1742-6596/947/1/012010

The Role of Self-Assessment in Foundation of Mathematics Learning

Masriyah[1]

[1]Mathematics Department of Universitas Negeri Surabaya

E-mail: [1]masriyah@unesa.ac.id

Abstract. This research is motivated by the low performance of students who took Foundations of Mathematics course. This study was aimed to describe (1) the learning outcomes of students who learned Mathematics Foundation after learning axiomatic applying self-assessment; (2) the difficulty of students and the alternative solutions; and (3) the response of students toward Foundation of Mathematics learning taught by applying self-assessment. This research was a descriptive research. The subjects were 25 mathematics students who studied Foundation of Mathematics in odd semester of the 2015/2016 academic year. Data collection was done using questionnaires, and testing methods. Based on the results of data analysis, it can be concluded that the learning outcomes of students were categorized as "good."; Student responses were positive; the difficulties lied in the sub material: Classification of Axiom Systems and the requirements, Theorem and how the formation, and finite geometry. The alternatives deal with these difficulties are to give emphasis and explanation as needed on these materials, as well as provide some more exercises to reinforce their understanding.

1. Introduction

This research was motivated by low performance of students who took Foundations of Mathematics course. Foundations of Mathematics course is basis of all Mathematics courses at the Mathematics Department. The foundation of Mathematics course consists of three topics, namely: (1) Logic, (2) Axiomatic, and (3) Set. The logic and set materials had been studied by students in senior high school, while the axiomatic material is a relatively new for them, and they often have difficulty to learn it. It can be seen from the learning outcomes of the midterm test of axiomatic material is always much lower than the other two topics. As an illustration, in the sub summative test ranged from 0 to 100, the three topics in last year three classes can be presented in the Table 1.

Table 1. The average results for Sub Summative Test of Students

Classes	Average of test score		
	Logic	*Axiomatic*	*Set*
C	70,3	44,8	68,2
A	73,4	46,9	71,6
U	71,1	48,7	70,0

The low scores in axiomatic material test showed that this material was difficult to understand by students. The cause of difficulties experienced by students has not yet known by the lecturers, so the lecturers could not respond optimally. For that, the researcher needs a certain technique to discover more about the causes of the difficulties of students, so that lecturers can determine solution alternatives. Meanwhile, self-assessment is an assessment technique that asks students to assess themselves with regard to the status, process and achievement levels of competency studies. Self-assessment techniques can be used to measure the competence of cognitive, affective and psychomotor [1]. Application of self-assessment can also help lecturers detect students' difficulties, so they can plan an appropriate technique to overcome the students' difficulties. Thus they can design a learning process that is expected to overcome the difficulties of students. In the end, the application of self-assessment in learning process is expected to improve leaning outcomes' students.

Self-assessment is arguably the most powerful means for a tertiary education organisation to understand and improve its educational performance [2]. Besides he also explained that self-assessment is arguably the most powerful means for a tertiary education organization to understand and improve its educational performance. Sitzmann et al also found that many researchers continue to interpret self-assessments of knowledge as a valid indicator of learning [3]. The improvements in the mathematics results indicate that the use by young children of self-assessment techniques appears to be effective in fostering improvement in one area of academic achievemen [4]. Moreover, self-assessment is an important skill to develop for lifelong learning [5]. Many researchers are turning to self-assessment as a means of harnessing this self-reflection phase to improve student learning. Researchers generally agree that self-assessment, or self-evaluation, can be defined as students judging their own work, based on evidence and explicit criteria, for the purpose of improving future performance[6,7].

The use of self-assessment as a type of formative assessment for learning is effective [8]. Researchers claim that self-assessment enhances motivation of beginners in a new area of study [9], focuses student attention on learning objectives and the assessments used to ensure them [7], and promotes a mastery goal orientation in which the focus is on improving knowledge, understanding, and skill [6]. Several studies on self-assessment and rubric use document positive effects on student self-efficacy [10, 11]. Challenges to the effective implementation of self-assessment include changing students' views of their work to that of a set of goals [12], ensuring students have a clear understanding of the learning objectives [13], and allowing enough time for students to adequately develop self-assessment abilities [12]. In order to effectively implement self-assessment in their classrooms, teachers must develop self-assessment procedures that provide students with explicit criteria for evaluating their work, that involve students in the decision-making process, that encourage students to think about the quality of their performance, that manifest realistic goal setting, and that are integrated with the instruction [7].

Meanwhile, the axioms can play many roles in mathematics [14]. In addition, axiomatic is indispensable knowledge in mathematics, it is needed to understand all branches of mathematics, because mathematics is a knowledge using axiomatic deductive systems. Therefore, an understanding of axiomatic is indispensable for students. Thus all students should understand the axiomatic material well. If students have difficulty in understanding the axiomatic material, then most likely they will have difficulty in understanding almost all of mathematics materials..

The objectives of this research are to describe: (a) the learning outcomes' students at the course of foundations of mathematics after learning axiomatic applying self-assessment; (b) the difficulties' students and the alternative solutions in the mathematics foundation course after learning axiomatic applying self-assessment; and (c) responses of students who studied foundation of mathematics taught by applying self-assessment.

2. Self Assessment

Self-assessment is an assessment technique that asks students to assess themselves with regard to the status, process and achievement levels of competency studies [1]. Self-assessment techniques can be used to measure the competence of cognitive, affective and psychomotor: (a) Assessment of cognitive competence in the classroom, for example: students were asked to assess the mastery of knowledge and skills of thinking; (b) Assessment of affective competency assessment, for example, students can be asked to create text that contains the outpouring of feelings towards a particular object. Furthermore, students are asked to make an assessment based on criteria or reference were prepared; (c) In connection with psychomotor competency assessment, students can be asked to assess the skills that they have learned based on criteria or reference were prepared.

The use of self-assessment can provide a positive impact on the personality development of students. The use of self-assessment also has some advantages, among others: (a) can foster confidence students, because they were trusted to assess themselves; (b) students realize their strengths and weaknesses, because when they were given the confidence, they should introspect on their strengths and weaknesses;

(c) can encourage, familiarize and train students to be honest, because they are required to be honest and objective in making judgments.

Self-assessment conducted based on clear and objective criteria. Therefore, self-assessment in the classroom should be done through the following steps:(a) Determine the competence or the capability to be assessed, (b) Determine the assessment criteria that will be used, (c) Formulate the assessment format, can be a scoring guidelines, a list of the check, or the scale, (d) Ask students to assess themselves, (e) Examines random sample assessment results, to encourage the students in order to continually assess ourselves accurately and objectively, (f) Give feedback to the students based on results of the study on the samples of assessment results taken randomly.

3. Foundation of Mathematics Material

Foundation of Mathematics is a basic course in the Department of Mathematics, which consists of three main topics, namely: (1) Logic, (2) axiomatic, and (3) Set. In this study, it is limited to the axiomatic material, comprising: (a) Facts, concepts, relations, operations and principles, (b) Inductive and Deductive Mindsets, (c) Undefined term and Initial Statement, (d) Axiom system and the requirements and classification, (e) The types and elements of definition, the intension and extension, (f) Theorem, the elements and formation, and (g) Finite geometry.

An axiomatic system is a list of undefined terms together with a list of statements (called "axioms") that are presupposed to be "true." A theorem is any statement that can be proven using logical deduction from the axiom [15]. In order for a set of axioms can constitute a system, the necessary requirements are important. They are: (1) Consistent, (2) Independent, (3) Complete, and (4) Economical.

An axiom system is said to be"**consistent**" when the statements in a collection of the axioms is not contradictory. Non-contradictory statement is not only in the sense only, but also in terms of terms and symbols used. It is said to be "**independent**" if each of these statements in a collection of axioms that are not mutually dependent. It means that one of axiom of the system must not be derived or obtained from the other axioms. It is said to be "**complete**" if each statement which is derived from the system can be proven whether it is true or false. (Of course, it is in the scope of the dichotomous logic). When the axioms in an axiomatic system is not complete, then we can not obtain the theorems. It is said to be "**economic**" when the symbols or terminology used is not redundant, and all of the statements in the system of axioms have no same meaning.

In every science, there is a way to clasify the materials, and each classification has a certain basis. It was held not to be intended to complicate to someone who study it, instead it will be easier for those who study it. In mathematics, it is known some classification of axiom. There are two ways to classify it, namely: (1) axiom of "self evident truth" and "non-self-evident truth", (2) axiom of "material", "formal" and "formalized".

Consider this statement: "The highest judge in mathematics that can be used to determine whether a statement is true or false is the structure. The highest judge in the SCIENCE is Reality".

3.1. The concept and its formation

A concept is an abstract idea that can be used to perform classification of objects. It can be established through an abstraction. As a simple example in daily life we can say that the bicycle, train, car, rickshaw is a vehicle. But houses, trees, stones is not vehicle. It means "vehicle" is a concept. The concept of vehicle can be viewed as an abstraction of a particular vehicle. In mathematics, it is known many concepts. Eg: "triangle", "square" and so on. It is also known the concept of "metric space", "group", and there are many others.

How is the formation of a concept? It can be conducted through: (1) ***an abstraction***, for example: the establishment of concept number though two times of abstraction. (2) ***an idealization***, for example: "flatness" of a plane and "straightness" of a line. (3) ***abstraction*** and ***idealization***, for example: the formation of "cube" or "cone" concepts, (4) ***an additional terms in the previous concept***, for example: the formation of "rhombus" from the "parallelogram" concept.

3.2. Definition

Definition of a concept is "an expression that can be used to restrict the concept". There is a freedom to make a definition of a concept, but the most important is the consistency of the definition. There are many types of definitions, namely: (a) **Analytical Definition.** A definition is said to be analytical definition if it is mentioned proximum genus and specific differentiation, (b) **Ginetical Definition.** A definition is said to be ginetical definition if it shows or reveals the occurrence of a defined concept, (c) **Definition using the formula.** A definition is not always expressed in the form of regular expressions. It can also be expressed in mathematical sentences.

3.3 Theorem

A theorem is statement that must be proved the correctness. It can be stated as an implication (if . . . then . . .). The specific properties of the theorem is not always obtained by deductive reasoning, but also may be found through field experience or empirical data. However, finally the truth must be proven by deductive thought patterns in the structure.

Thus, a theorem or a specific property can be obtained through inductive steps, and then verified by way of deductive reasoning. The properties of a sequence can be "found" by trial and error, and then can be verified by using mathematical induction. Likewise, some properties or theorems in the theory of network or graph. It has been argued that the form of a theorem is an implication. But there is also in a bi-implication form. In contrast, the definition is the sentence must always be interpreted as a biimplication. It is included the theorem: the "lemma" and "corrolary".

4. Method

This is a descriptive research. It was conducted at Mathematics Department of Surabaya State University. The subjects were 25 mathematics students who studied Mathematics Foundation in odd semester of the 2015/2016 academic year.

Data collection methods of this research were test and questionnaire. Test was used to obtain data on how was the students' ability after the learning process of Foundation of Mathematics. Besides, the students' work also can be used to know the students' difficulties by seeing which material of the test has been answered incorrectly. There are two questionnaires, namely: (1) Self-assessment questionnaire (2) Responses questionnaire.

The test results were analyzed and determined average score, then it was categorized as follows.

$60 \leq \bar{x} \leq 100$	good	
$40 \leq \bar{x} < 60$	good enough	$\bar{x}$: average of test
$0 \leq \bar{x} < 40$	not good	

The results of questionnaire were analyzed by grouping the types of students' difficulties, and then determined the percentage of each types of students' difficulties. These difficulties were also cross checked with the students' work in midterm test. The results of responses questionnaire of students were analyzed by calculating percentage of students who gave positive responses, then determined the category of students' response. The students' response was said to be positive if at least 70% of the students responded positively to at least 80% aspect in question.

5. RESULT AND DISCUSSION

The research was conducted during the course of learning Foundation of Mathematics for material axiomatic that apply self-assessment in Mathematics Education. Learning axiomatic materials was carried out within 3 (three) meetings, on Wednesday, 9, 16 and 23 September 2015. Having carried out three times of learning process, the researcher held sub summative test on September 30, 2015.

The results of the midterm test of Foundations of Mathematics course with the axiomatic material including the category of "Good" with an average score of 75.68. At the end of the second and third meetings of the Foundation of Mathematics learning course, the students were given self-assessment questionnaire of cognitive aspect, and at the end of the third meeting, they were given self-assessment questionnaire of affective aspects. The results of the self-assessment questionnaire of cognitive and affective aspects that are given to all students can presented as follows. All of students have understood the materials: Characteristics of Mathematics, Types of base objects of Mathematics, Inductive Mindset,

Undefined Term and Initial Statement. About 88% students have understood "Structure of Axiomatic Deductive", 68% students have understood System of axioms and the requirements, and 75 % have understood Classification of axioms.

The most difficult of the eight sub-topics are "Classification of Axiom", "System of axioms and the requirement", and "Theorems and the formation".The reasons given were: (1) Because it is a new material, do not understand the material, (2) Because the material is difficult to understand, confusing material, (3) Because if the problem was changed, it often made be confused.

Almost all the students can understand the material being taught. Even for sub topics 1 to 4 and sub topic 6, all students (100%) can understand the material being taught. However, there are some students (72%) stated that among the eight topics covered, the material "Classification Axiom" is the most difficult material, (32%) stated that System of axioms and the requirement is the most difficult material, while as many as 20% of the students stated that " theorems and the formation " material is the most difficult material. Based on these results, the researcher tried to give emphasis and explanations as necessary on three materials, as well as provide some more exercises to strengthen their understanding. Furthermore, based on the results of the self-assessment questionnaire of affective, it can be stated that all of students do the tasks given, pay attention to the direction given faculty, try to follow well the learning process of Mathematics Foundation for axiomatic material. Most of them (80,95%) said that during the discussion groups, they listen and pay attention if their friends or lecturer speak

It also be stated that the attitudes of students towards learning course Foundations of Mathematics for axiomatic materials are positive, because 100% aspects of their self-assessment items have been responded positively by the students. Based on data analysis presented in Table 6, it can be stated that the students' response is "positive," because every item of questionnaire was responded positively at least 71.43% of students.

Based on all data analysis, it can be stated that the students learning outcomes at the Foundations of Mathematics course after learning that applied self-assessment was categorized as "good". Difficulties of students are in sub material: Axiom, Classification of Axiom Systems and the requirements; theorem, the components and formation, as well as finite geometry. Many students confuse to differentiate between definition and theorem. It seemed when they were asked to provide an example of a theorem, they provide example of definition and vice versa. The alternative solutions are to give emphasis and explanations as necessary on both the materials, as well as provide some more exercises to strengthen their understanding. The response of students toward Mathematics Foundation for axiomatic material that is performed by applying self-assessment is positive

6. Conclusion and Recommendation

Based on the research result, it can be concluded that self assessment is effective used to improve students'performance. Self assessment can help lecturers detect students' difficulties, so they can plan an appropriate technique to overcome the students' difficulties.

Based on the conclusin, there are two recommendationa to the lecturers: (1) to apply self assessment in their lessons; (2) to overcome the difficulties experienced by the students as soom as possible,.

References

[1] Depdiknas 2010 Model Penilaian Kelas SMP/MTs (Jakarta: BPPPN Pusat Kurikulum)

[2] NZQA, Kia N T, 2015 http://www.nzqa.govt.nz/providers-partners/self-assessment/what-is-self-assessment/

[3] Sitzmann et al. 2010 Self Assessment of Knowledge: Cognitive Learning or Affective Measure? *Academy of Management Learning & Education* 9 (2) pp 169–191

[4] Fontana D & Fernandes M 1994 Improvements in math performance as a consequence of self-assessment in Portuguese primary school pupils *British Journal of Educational Psychology* **64** pp 407-417

[5] Race P 2001 A briefing on self, peer and group assessment *Assessment Series* **9** LTSN Generic Center. Retrieved 12-12-2008, from

IOP Conf. Series: Journal of Physics: Conf. Series **947** (2017) 012010 doi:10.1088/1742-6596/947/1/012010

http://internt.iha.dk/paedagogik/seminarer/Chris%20Rust/ASS009PhilRace.pdf
[6] McMillan J H and Hearn J F 2008 Student self-assessment: The key to stronger student motivation and higher achievement *Educational HORIZONS* **87** 40-49.
[7] Rolheiser C & Ross J A 2001 Student Self-Evaluation: What Research Says and What Practice Shows.‖ Retrieved December 28, 2009, *//www.cdl.org/resource-library/articles/self_eval.php*
[8] Hotard 2010 *The Effects Of Self-Assessment On Student Learning Of Mathematics* (Louisiana State University: Faculty of Natural Sciences)
[9] Zimmerman B 2002 Becoming a Self-Regulated Learner: An Overview *Theory Into Practice* **41** (2) 64-70.
[10] Olina Z & Sullivan H J 2002 Effects of teacher and self-assessment on student performance *Paper presented at the Annual Convention of the American Educational Research Association* April 1-5 2002 New Orleans
[11] Andrade H L, Du Y, & Mycek K 2010 Rubric-referenced self assessment and middle school students' writing *Assessment in Education: Principles, Policy & Practice* **17**(2) 199-214.
[12] Black P J, Harrison C, Lee C, Marshall B & Wiliam D 2002 *Working inside the black box: Assessment for learning in the classroom* (London: Nelson)
[13] Black P & Wiliam D1998b Assessment and Classroom Learning *Assessment in Education* **5** (1) 7-74
[14] Schilmm D 2014 Axioms in Mathematical Practice *Philosophia Mathematica* **21** (1) Issue 1 February 2013 pp 37–92
[15] Hamblin 2007 *Eucliean and Non-Euclidean Geometry* Fall 2007 Presentation: Cyclic Quadrilaterals

Multiple Sensing Application on Wireless Sensor Network Simulation using NS3

I F Kurniawan[1], **R Bisma**[2]

Informatics Department, Universitas Negeri Surabaya, Surabaya, ID

E-mail: ibnufebry@unesa.ac.id[1], rahadianbisma@unesa.ac.id[2]

Abstract. Hardware enhancement provides opportunity to install various sensor device on single monitoring node which then enables users to acquire multiple data simultaneously. Constructing multiple sensing application in NS3 is a challenging task since numbers of aspects such as wireless communication, packet transmission pattern, and energy model must be taken into account. Despite of numerous types of monitoring data available, this study only considers two types such as periodic, and event-based data. Periodical data will generate monitoring data follows configured interval, while event-based transmit data when certain determined condition is met. Therefore, this study attempts to cover mentioned aspects in NS3. Several simulations are performed with different number of nodes on arbitrary communication scheme.

1. Introduction

Wireless Sensor Network (WSN) consists of a finite set of devices that have capability to sense surrounding condition and communicate to each other using wireless medium. These devices collaboratively exchange sensor data, and forward to sink node. According to [1], the device is known as mote which is autonomous, compact and have communication and computation capability. By installing these devices, user is able to get actual condition by looking into capture data. Several works have shown important roles of this technology such as garbage monitoring [2,3], and agriculture [4,5]. Those applications involve distributed nodes in large area. While, more simplistic and locale monitoring scheme is presented in [6] where nodes are installed to monitor temperature and humidity of office building via ZigBee networking protocols.

Latest hardware technology have enabled users to perform multiple monitoring task by providing several sensor instruments on single device. These sensors have various capture abilities depending on users' requirement. This breakthrough is achieved by providing plugand-play support of different sensors on board. Each sensor is installed on a board, and can be stacked each other. This is typical implementation on Arduino, and Raspberry Pi.

Regardless of mentioned contribution, this study aims to simulate multiple-sensing application in wide area using NS3. Simulation is still important stage to model nodes behavior prior to real deployment. Therefore, this study attempts to provide better view on NS3 classes to perform the simulation. A number of sensing devices are installed on each node in order to capture different reporting schemes, such as periodic and event-based.

Following section in this paper is organized as follows: section 2 discusses previous works conducted on Wireless Sensor Network, and NS3. Sequentially, section 3 provides system assumption that is used on simulation process. Implementation of sensing node, and conclusion are given in section 4, and 5, correspondingly.

IOP Conf. Series: Journal of Physics: Conf. Series **947** (2017) 012011 doi:10.1088/1742-6596/947/1/012011

2. Related Studies

Numerous WSN implementation have been studied by researchers in last decade. WSN has been widely applied on different area of application, such as military, environmental, logistics, humancentric, and robotics application [1]. The wide variety of implementation primarily depends on sensor usage on different context. For instance, WSN usage on military extents from information collection to enemy detection, surveillance on battleground, or target classification [7,8]. The classification method relies on seismic, and acoustic signal sensing as input data. Another example as described in [9] where sensor network aids the user by informing relative position and movements of the enemy units.

In environmental monitoring, WSN has played several roles not only in indoor and outdoor monitoring, but also emergency services. One example of indoor monitoring have been provided by [10] in which researchers and graduate students have utilized motes to monitor light and temperature condition. The sensors have capability to capture temprature, light, frames status (doors, windows), air flows, and indoor pollution. Consequently, indoor environment control can be optimized. More sophisticated application is written in [1] where WSN may support nature disaster mitigation effort. Author of this study can integrate fire and smoke detection system with light signals. This composition produces safety guidance for residents during fire accident. The collaborative sensor network can aid people to find safest route to escape the bulding.

Outdoor monitoring has shown common implementation area for WSN besides indoor monitoring. WSN implementation in [11] has enabled people to observe animal habitat remotely. Compared to indoor, outdoor monitoring has wider region and intended to sense temperature, barometric, pressure, as well as humidity condition. Furthermore, passive infrared sensors and photoresistors are used. This combination is intended to natural environment of one bird species. To achieve this, sensor nodes are distributed near birds' burrows, and its surroundings. Bird's spawn data are aggregated by nodes and passed through to a gateway. The sink node has more powerful ability such as data compilation and forwarding to local base station as database. Finally, people may access the surveillance information from database via internet. This shows WSN monitors in heterogeneous way, and multi-level network.

Prior to real life deployment, WSN demands good preparation in order to obtain efficient effort. Therefore, some studies make use of network simulator such as NS-3 [12] to substitute TinyOS application [13]. Initially entitled as Yet Another Network Simulator, NS-3 has gained attention of researchers due to its extensible and robustness property. This simulator has motivated [14] to do benchmarking on a number of open source network simulator frameworks. According to the study, NS-3 requires much less memory usage, and computational time than 4 other frameworks. Since then, many works have been accomplished with the help of this framework. Author of [15] proposes enhancement on physical layer to maximize WSN utility. The study modified typical NS-3 implementation of physical layer to give more accurate and detailed representation of signal. It further claims that completed work is able to conduct all signal processing steps which occur during transceiver frame decoding.

3. System Model

This study environment uses a number of sensing nodes with equal attributes such as computation ability, limited energy capacity, and wireless transmission range. Additionally, these nodes are equipped with environment sensor device, and able to directly transmit data to single base station (BS) node. Contrast to sensing nodes, BS acts as sink node that accepts any monitoring data from sensing nodes. These data will be aggregated by application type and stored in local storage for future user's needs.

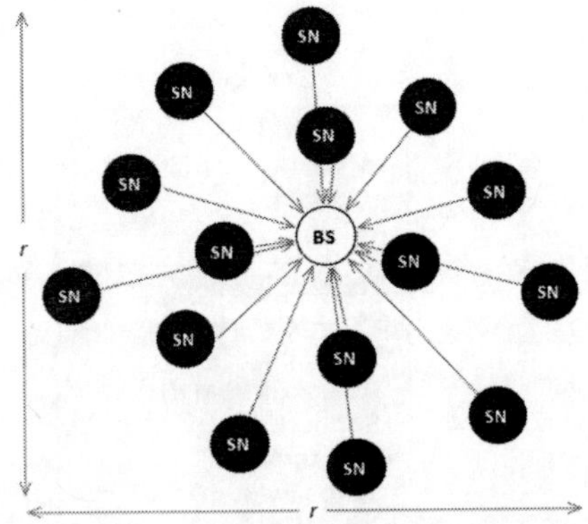

Figure 1. Network Topology used

Figure1 depicts network topology used in this study. Sensing nodes (SN) are deployed randomly throughout area (r×r) and able to deliver packet to one BS node which is located in simulation area. These nodes are configured to deliver monitoring data according to its availability. While periodic data are produced in regular fashion, event-based are obtained occassionally. Therefore, each sensing nodes are expected to transmit periodic data, at least. If there is event occurs, then the data will be transmitted together with periodic ones. To achieve this, packet transmission rate in simulation needs to be configured to follow certain distribution. This study uses uniform random distribution. Periodic data have shorter delay transmission, in contrast event-based ones have longer. These intervals are determined in initial phase of simulation.

Communication among these nodes are established using WiFi standards (802.11). These nodes have wireless transmission coverage with r distance. Furthermore, wireless propagation delay and loss are set to constant and fixed RSS, correspondingly.

4. NS3 Implementation

A set of operation in NS3 must be followed to begin new application development. Main application SensorNodeApplication serves as model application, while SensorNodeHelper is provided to ease SensorNodeApplication instantiation. These files are located in model and helper folder, respectively. To complete this new application setup in NS3, wafscript file needs reconfiguration to accomodate proper .waf compilation process. Brief sensing node class diagram is provided on fig. 2.

Sensing node implementation in NS3 can be achieved by adopting existing OnOffApplication model due to its basic functionality of sending and receiving packets based on certain distribution. Each sensing node are deployed randomly, initially configured its packet interval and technical capabilities from main class. Since this study deals two types of sensing application, two uniform random generators are used during sensing node's lifetime. Method AssignStreams is written to handle random number generation. During its initial operation, within StartApplication to be exact, some of event handler such as accepted (HandleAccept) , closed (HandlePeerClose), falied (HandlePeerError) and incoming connection (HandleRead) are registered and packet transmission is scheduled.

NS3 provides scheduling mechanism in which programmer may instruct specific upcoming task/event occurrence. Interestingly, NS3 also enables programmer to cancel this schedule. By utilizing this features, different sensing application are preconfigured its transmission time according to random generator. This task is handled by SchedulePeriodicData and ScheduleEventData function for periodic, and event-based data, respectively. Additionally, schedule cancelation needs to be carried out to accomodate either unexpected or scheduled

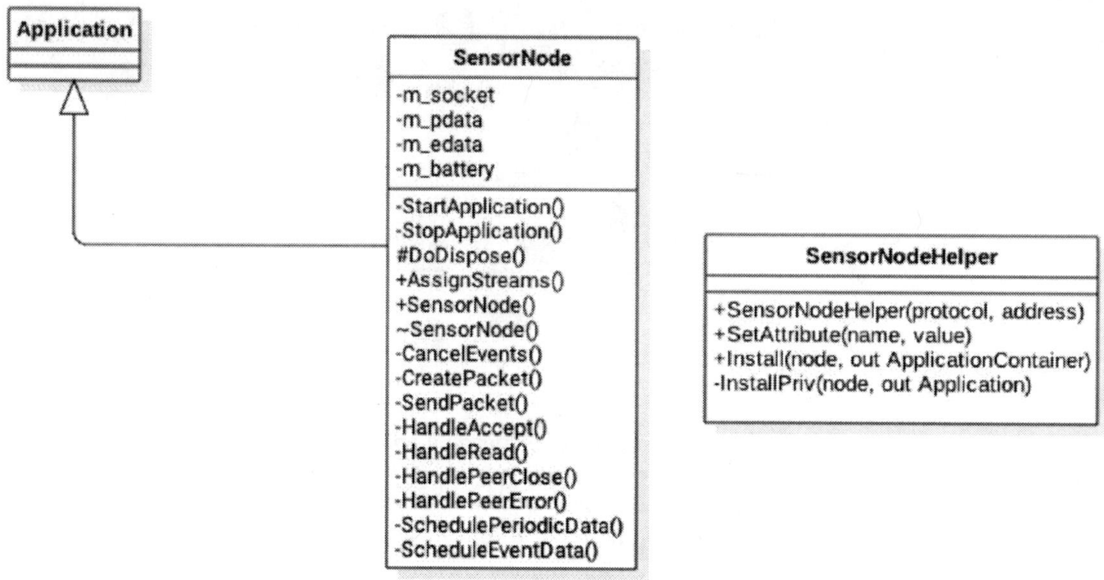

Figure 2. Class Diagram

stop, such as power outage, and Simulation::Stop call. In our simulation, this is handled by CancelEvents.

NS3 simulation is began with constructing main application which acts as entry execution poin. In this file, set of operation is conducted which extent from random number seed generation, nodes creation, wireless physical, and medium layer configuration, speed and range propagation loss setup, to IP address assignments. After this setup completes, SensorNodeApplication is created and embedded to each nodes and schedule simulation start and end time. Finally, any packet generated from these sensing nodes are consumed by BS which is implemented using PacketSink application.

Finally, after successfully constructing the sensing nodes, main application for running experiments is prepared. In this last step, several steps are done such as wireless environment, wireless device setup, simulation timing, and trace file configuration. In wireless environment and device setup, loss and delay propagation as well as channel properties are set. Meanwhile, nodes setup includes NetDevices, Internet suite protocol, location, and energy installation step. At last, to capture any actions done in simulation, trace file and NetAnim instance are prepared.

5. Conclusion

In this article, sensing node with multiple sensor instruments are developed by using NS3 framework. The provided class diagram is constructed based on existing OnOffApplication with several modifications, such as event-based and periodical scheduling, and incoming packet handling. Whereas, constant mobility, and energy source, and consumption model are configured in main simulation class.

Due to NS3's modularity, future simulation with different strategies and attributes are possible to achieved. WSN nodes with movement capabilities and various energy consumption model provide many research possibility. These area requires good strategy to keep entire system last longer.

References

[1] Arampatzis T, Lygeros J and Manesis S 2005 A Survey of Applications of Wireless Sensors and Wireless Sensor Networks

[2] Joshi J, Akshay, Bagga A, Aggarwal R, Deka M, Urjit and Abhinav 2016 *Proceedings of the 7th International Conference on Computing Communication and Networking Technologies* ICCCNT '16 (New York, NY, USA: ACM) pp 7:1–7:5 ISBN 978-1-4503-4179-0 URL http://doi.acm.org/10.1145/2967878.2967886

[3] Mamun M A A, Hannan M A, Hussain A and Basri H 2015 *IEEE Sensors Journal* **15** 561–567 ISSN 1530437 X

[4] Mainwaring A, Culler D, Polastre J, Szewczyk R and Anderson J 2002 *Proceedings of the 1st ACM International Workshop on Wireless Sensor Networks and Applications* WSNA '02 (New York, NY, USA:
ACM) pp 88–97 ISBN 1-58113-589-0 URL http://doi.acm.org/10.1145/570738.570751

[5] Storey M, Blair G and Friday A 2002 *Mobile Networks and Applications* **7** 377–387 ISSN 1572-8153 URL http://dx.doi.org/10.1023/A:1016543700801

[6] Ferdoush S and Li X 2014 *Procedia Computer Science* ISBN 9781479998555 ISSN 18770509

[7] Li D, Wong K D, Hu Y H and Sayeed A M 2002 *IEEE signal processing magazine* **19** 17 – 29

[8] Meesookho C, Narayanan S and Raghavendra C 2002 *Sensor Array and Multichannel Signal Processing Workshop Proceedings, 2002* (IEEE) pp 370–374

[9] Sinopoli B, Sharp C, Schenato L, Schaffert S and Sastry S S 2003 *Proceedings of the IEEE* **91** 1235–1246

[10] 2001 Brainy Buildings Conserve Energy URL http://www.coe.berkeley.edu/labnotes/0701brainybuildings.html

[11] Mainwaring A, Culler D, Polastre J, Szewczyk R and Anderson J 2002 *Proceedings of the 1st ACM international workshop on Wireless sensor networks and applications* (Acm) pp 88 – 97

[12] Lacage M and Henderson T R 2006 *Proceeding from the 2006 workshop on ns-2: the IP network simulator WNS2 '06* 12 ISSN 1595935088

[13] Levis P, Madden S, Polastre J, Szewczyk R, Whitehouse K, Woo A, Gay D, Hill J, Welsh M, Brewer E and
Culler D URL http://users.eecs.northwestern.edu/ peters/references/TinyOSSpringer05.pdf

[14] Khan M A, Hasbullah H and Nazir B 2014 *I4CT 2014 - 1st International Conference on Computer, Communications, and Control Technology, Proceedings* 324–328

[15] Kumar S, Ranjan P, Radhakrishnan R and Tripathy M R 2015 *Computational Intelligence and Communication Networks (CICN), 2015 International Conference on* (IEEE) pp 79 – 84

IOP Conf. Series: Journal of Physics: Conf. Series **947** (2017) 012012 doi:10.1088/1742-6596/947/1/012012

Students' Creativity: Problem Posing in Structured Situation

I K Amalina[1,*], M Amirudin[1], and M T Budiarto[1]

[1]Department of Mathematics, Universitas Negeri Surabaya, Surabaya, East Java, Indonesia

*Corresponding author: ijtihadiamalina16070785040@mhs.unesa.ac.id

Abstract. This is a qualitative research concerning on students' creativity on problem posing task. The study aimed at describing the students' creative thinking ability to pose the mathematics problem in structured situations with varied condition of given problems. In order to find out the students' creative thinking ability, an analysis of mathematics problem posing test based on fluency, novelty, and flexibility and interview was applied for categorizing students' responses on that task. The data analysis used the quality of problem posing and categorized in 4 level of creativity. The results revealed from 29 secondary students grade 8, a student in CTL (Creative Thinking Level) 1 met the fluency. A student in CTL 2 met the novelty, while a student in CTL 3 met both fluency and novelty and no one in CTL 4. These results are affected by students' mathematical experience. The findings of this study highlight that student's problem posing creativity are dependent on their experience in mathematics learning and from the point of view of which students start to pose problem.

1. Introduction

Creativity is the important component for nation advancement. Nation advancement is influenced by education, one of them is mathematics. Then creative thinking is important for mathematics [1, 2]. Creativity is processed to create new ideas that are appear which is the process or the result and involve originality [3].

Problem posing task is very important because it can improve students' abilities to pose questions, and one that is considered important in Indonesian curriculum. However, based on observation, students abilities in posing questions were considered as poor, so problem posing task is needed. Problem posing is defined as tasks designed by teachers that require students to create one or more questions [1, 4]. Most students in Canada are only given an amount of task to solve mathematical problems posed by a teacher or those that appear in the book [5]. Therefore, students are only capable to solve mathematical problems without posing them. The students' mathematical experience is incomplete if it has never had given a chance to solve a problem made by themselves [7]. By using the problem posing, students are exposed to complex situations that make them feel responsible [7]. In addition, posing a problem gives students a sense of belonging to math. Education in America, China, and Italy emphasize the problem posing as necessity in a learning process of mathematics [8, 9, 10].

There are three situations of problem posing such as free, semi-structured, and structured situations [11]. In free situation students pose problems without limitation, semi-structured situation students are given an open situation and invited to explore the structures in the situation and facilitate them using the knowledge, skills, concepts, and relationships of their previous mathematical experiences, while in structured situation students pose problem by varying the existing situation therefore requires high creativity. Problem posing plays a role in determining student creativity. Creativity is understood as the cognitive ability to create and discover. While problem posing refers to generate something new or revealing something new from a set of data that requires creativity [2]. The ability to pose a problem is a sign of creativity therefore there is a close relationship between problem posing and creativity [2, 3, 12]. Hence, problem posing can be in the form of testing a creative thinking.

The structured situation is a problem posing situation that is rarely analyzed [7]. Researchers analyzed teachers' mathematical problem posing in a structured situation and their difficulties [7]. However, they just analyzed in ratio and proportion topics. Another study about assesses students' creativity in mathematics [13]. In this study, students posed mathematical problems than assessed based on fluency, flexibility, and novelty. The situation used is semi-structured problem posing. However, this study directly analyzes using three components without analyzing the quality of students' problem posing. Study about structured situation on students of grade 8 and 9 to modify it conducted by Stoyanova & Ellerton [11]. However, they just wrote about the framework.

Therefore, this study aimed to investigate and describe students' creativity in problem posing. The situation used is structured situation. Problems posed by students analyzed based on quality of problem posing [9, 14]. Then they analyzed based on fluency, flexibility, and novelty. This qualitative research used the test of problem posing and interview.

2. Method

This qualitative study was conducted on 29 junior high school students in Surabaya Indonesia, in which 3 students with different creative thinking level at level 1,2, and 3 are taken. The data used in this study were collected from February to March 2016. The instrument used is the test of problem posing that used to analyze creative thinking level. Data collection used tests and interviews. Test done during 60 minutes with a situation and students posed at least 3 problems. Each student was interviewed during 20 minutes. The aim of this interview is to know deeply about the student consistency in that level by using questions which concern to fluency, flexibility, and novelty.

Problems posed by students analyzed based on mathematics problem, non-mathematical problem, irrelevant mathematical problem; plausible or implausible mathematical problem; problem with sufficient or insufficient information [9, 14]. Then they analyzed based on fluency, flexibility, and novelty [12] and grouped in CTL. The indicators of CTL that is used in this study explain in Table 1.

Table 1. Indicators of CTL

Level	Characteristic of Creative Thinking Level
Level 4 (Very Creative)	Students satisfied all components of creative thinking or only flexibility and novelty in posing problems.
Level 3 (Creative)	Students were fluent and then they were flexible or demonstrate novelty.
Level 2 (Quite Creative)	Students were able to show flexibility or novelty in posing problems without fluency.
Level 1 (Almost Not Creative)	Students were able to show fluency without novelty and flexibility in posing problems.
Level 0 (Not Creative)	Students were not able to show any components of creativity.

Adapted from reference [15]

3. Results

Students posed at least 3 problems. The results obtained by all the mathematical problems posed are plausible questions of the 69 mathematics problems posed. Students posed 64 plausible mathematical problems with sufficient information of a total of 69 plausible mathematical problems. Of the 64

IOP Conf. Series: Journal of Physics: Conf. Series **947** (2017) 012012 doi:10.1088/1742-6596/947/1/012012

problems posed by students are analyzed based on fluency, novelty, and flexibility then grouped based on CTL. The results were no student in CTL 4, there were 2 students in CTL 3, a student in CTL 2, there were 14 students in CTL 1, and 12 students in CTL 0. Because there is no student was on CTL 4, so the researcher only took a student from each of CTL 1, 2, and 3. Here are the problems, answers, and interviews of selected students.

> Two sailors travel from Jakarta to America. They carry 2 ships, the first ship contains 250kg of meat, 600kg of flour, and 1200kg of potatoes. In the second ship, the number of potatoes is increased which is 300kg from the first ship, while the amount of meat and flour equal to the first ship. Due to the long sailing time, the total decaying potatoes on both ships are 245kg. Pose mathematics problems by varying the given situation and define the solutions. Varying the problem can be done by adding information or modifying a given situation

Figure 1. Structured Situation Used in Research

3.1. Student in CTL 3

A Student in CTL 3 posed three problems and solution shows in Figure 2.

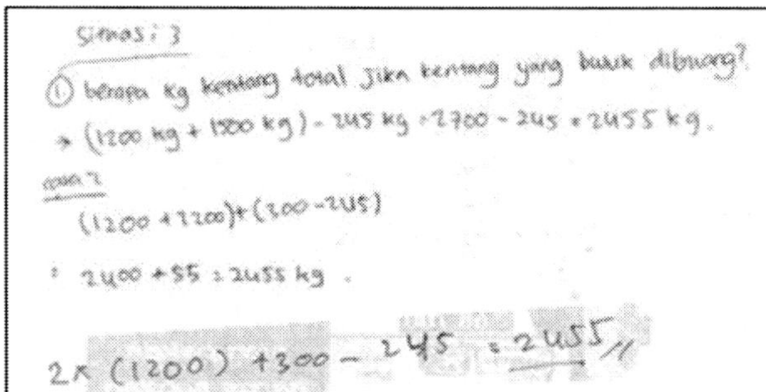

What is the total of potatoes if the rotten potatoes is removed?

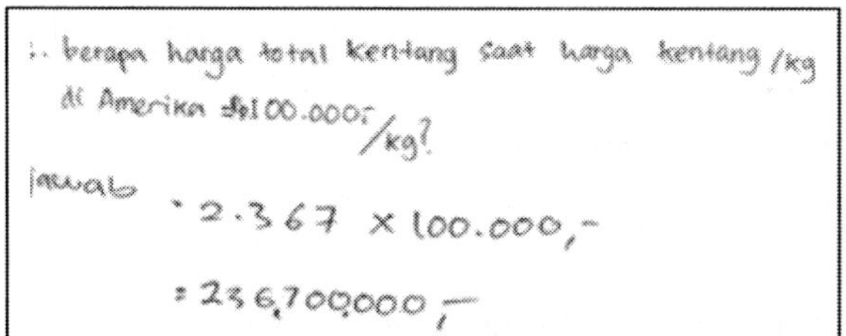

What is the total price of potatoes when the price of a kilogram potato in America Rp 100.000

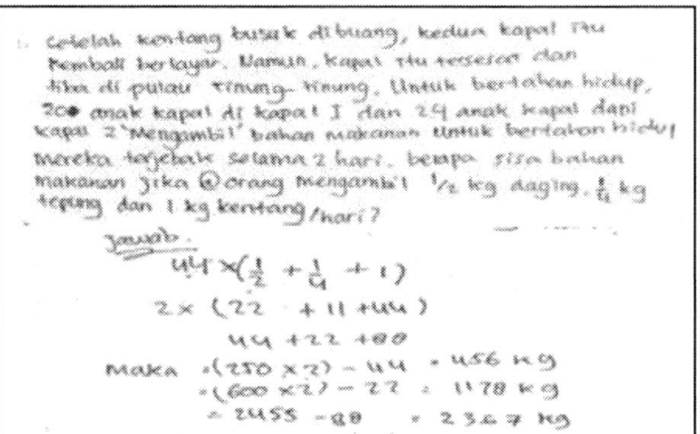

After the rotten potatoes dumped, both of ships are back sailing. But they lost and arrive in Tinung-Tinung island. 20 people in the first ship and 24 people in the second ship take food to survive. The lost for about 2 days. What is the reminder of food if every person takes ½ kg meat, ¼ kg flour, and 1kg potatoes each day.

Figure 2. Problems and Solutions Posed by Student in CTL 3

The student posed the first problem of the given situation about the total potatoes if the rotten potatoes removed. It is about addition and subtraction. There were 2 ways but the same ways, just differ in the order of working step. The researcher interviewed the student to make the answer in another way, but she did actually the same way by applying the concept of integer operations which is $(1200 \times 2) + 300 - 245$. Therefore, student did not meet flexibility.

The second problem posed by the student was unclear information in the problem, based on interview results obtained that the problem number 1 happened is like a given situation but the rotten potatoes are thrown away. After the potatoes are removed, the number 2 ship gets lost and the food is taken by the crew, after being taken, the remainder is brought to America.

The student posed problems by adding information from the given situation. The answer was the correct answer. The concept used was different concept from problem number 1. The concept on problem number 1 is the concept of integer operation while the concept used in problem number 2 included the concept of fractions. The context is a different context because of the addition of different information included. So student met the novelty. A student could not solve in different ways. Student only solved 1 correct answer. So she did not meet the flexibility.

The third problem required an explanation through interviews. Based on the interview results from problem number 3, she can explain it as follow

Researcher : What do you mean by what is the total price of potatoes, do you mean that the potatoes at the beginning?

Student : After being eaten by the crew, what is the total price of potatoes when the price of potato in America is Rp 100.000,00 / kg?.

After explained through the interview, the question posed had the correct answer. The same concept used is the integer operation but the context used is different, by adding information. So, the student met the novelty.

Of three questions posed by the student have correct answers and based on interview therefore she met fluency. So it can be concluded the student met fluency and novelty so she is in CTL 3.

3.2. Student in CTL 2

Student in CTL 2 posed three problems and solution shows in Figure 3.

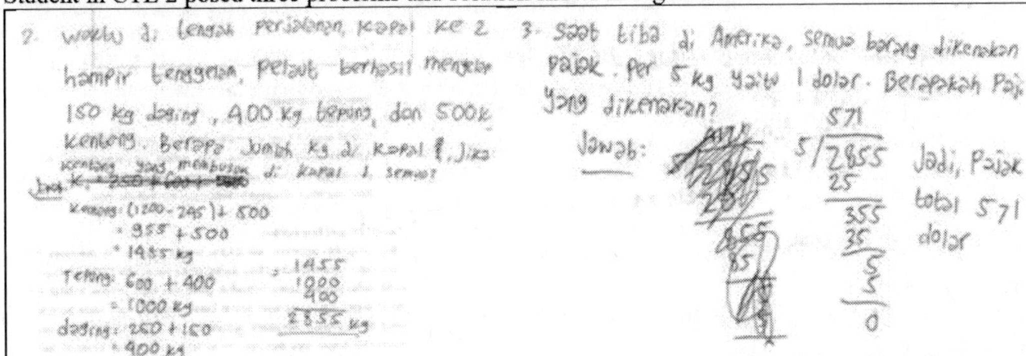

In the middle of the journey, the second ship almost disappear. The sailor is success to save 150kg meats, 400kg flours, and 500kg potatoes. What is the number of food if all the rotten potatoes in the first ship?

In the America, All of food is taxed a dollar every 5kg. What is the tax must be paid?

Figure 3. Problems and Solutions Posed by Student in CTL 2

Student poses the first problem with the wrong answer so it is not included in the analysis to determine the creativity. The second problem needs to be clarified through interviews. Based on the interview, the second problem means

Researcher : Can you explain more about your problem?

Student : The second ship is almost drowned by the halfway travel time. Sailors managed to save 150kg of meat, 400kg of flour, and 500kg of potatoes. What is the amount of food (in kg) on the ship if all the potatoes are decomposed on the first ship?

Potatoes are decaying 245kg on first ship. It is the continue of story number 1 so the second ship was only 150kg of meat, 400kg flour, and 500kg of potatoes.

The second question is a problem with addition of information from the given situation and correct answer.

The third problem is a different context from the first and second problems that have been posed. It is about tax, when arriving in America, all food is taxed. Per 5 kg is 1 dollar. How much tax is charged?. On the third problem, the student posed the problem with the correct answer with the addition of information. Problem number 3 is the continuation of the problem number 2 therefore the tax is the remaining items in question number 2. This showed that the student could clarify the information in question number 3.

Based on the information above, the student posed 2 problems of different context with the correct answer. Then it can be concluded that the student met the novelty. But the student only posed 2 questions with the correct answers, the student could not answer in different ways he did not meet the flexibility and fluency thereof. So it can be concluded the student is in CTL 2.

3.3. Student in CTL 1

Student in CTL 1 posed problem and solution shows in Figure 4

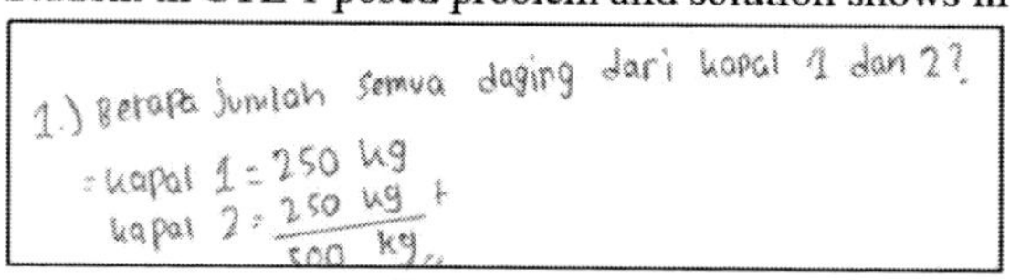

How many all the meat in the first ship and second ship? First ship = 250kg
Second ship= 250kg +
500kg

Figure 4. Problem and Solution Posed by Student in CTL 1

The student posed three problems with the same concept about the addition of integers. The context used is also the same context. The answers of three problems are the correct answer but the student could not solve in a different way. Based on the information above, the student posed 3 different problems that have the same concept and context. Then it can be concluded that the student met the fluency. So, it can be concluded the student is in CTL 1.

Based on the interview, students in CTL 1, 2, and 3 use the previous mathematical experience. It can be seen in this interview

A Student in CTL 1 : My teacher has taught me the problems like these.
A Student in CTL 2 : I pose them based on the books which I have read before.
A Student in CTL 3 : I have read some questions in magazine about mathematics quiz so I use that experience

Students also pose the problem by think through the answers first. It can be seen one of student's interview script

A Student in CTL 3 : I think the answer first because I am afraid if I pose the problem first then there is no answer that satisfy this problem

4. Discussion

Based on the result of this research, no student met the flexibility component because students solve the problem in one way or in two ways but by using the same ways. Therefore, no students satisfied CTL 4 (very creative). The characteristic of CTL 4 is students satisfied all components of creative thinking or only flexibility and novelty in posing problems [15]. So flexibility is emphasized on CTL 4. Students did not satisfy flexibility because they are unusual to use more than one way to solve a problem or they did not have experience about that. This result is the same as the result of the research that no students in CTL 4 because did not have the experience to solve in more than a way [16]. It shows that mathematical experiences are important in posing problems [11].

The highest level that reached by the student is in CTL 3, she met the novelty and fluency. She posed 3 plausible mathematical problems with sufficient information and correct answers. The concept used in one problem is different from another. There is additional information on the problem so it has a different context with the previous problem. It is a little bit different with one of the research result. The result is students pose similar problems, but in this result of research student pose problems from the given situation with additional information [7]. This shows that she met fluency and novelty. It appropriate with the theory that students in CTL 3 were fluent and then they were flexible or demonstrate novelty [15].

Student in CTL 2 posed 2 problems, both of these have additional information so the context used is different. So he met novelty, in accordance with the theory if the students satisfy the novelty or flexibility component then the student in CTL 2 (quite creative) [15]. Student who met fluency is in CTL 1 [15]. The concept used in the three problems is the same concept given. The context used is the same context as the given situation without any additional information.

The way for posing the problem is by thinking through the answers first and using the previous mathematical experience. That result same as the research result that student uses the previous mathematical experience for posed problems but there are some different points which are students posed problems by thinking outside of the box, start with mathematical concepts and relate to real life, and come up with relevant concepts and connect them [17].

5. Conclusion

Based on the results of the study, we conclude that the creativity of students in posing the problem depends on the experience of students in learning mathematics. The more students experience in learning mathematics the more creative the students are also in posing the problem. Beside the learning experience, the students' creativity in posing the problem also depends on the students' way in posing the problem. Some are guided by the answer first, and also directly pose the problem by looking at different points of view from a known problem.

Acknowledgements

This research reported herein was supported by Dr. Tatag Yuli Eko Siswono as an instrument validator and all the help that very meaningful, and Anjani Nur Panjalin as language validator. The opinions expressed herein do not necessarily reflect the position, policy, or endorsement of the supporting agency.

References

[1] Siswono TYE & Budayasa I K 2006 Implementasi Teori tentang Tingkat Berpikir Kreatif dalam Matematika. *Makalah disampaikan pada Seminar Konferensi Nasional Matematika XIII dan Kongres Himpunan Mahasiswa Indonesia*, Semarang, July 24th-27th 2006

[2] Singer F M & Voica C 2015 Is Problem Posing a Tool for Identifying and Developing Mathematical Creativity? In Florence Mihaela Singer, Nerida F. Ellerton and Jinfa Cai (Ed). *Mathematical Problem Posing: From Research to Effective Practice* pp 141—174 (New York: Springer)

[3] Leung S S 1997 *On the Role of Creative Thinking in Problem Posing*, (Online), (http://www.emis.de/journals/ZDM/zdm973a4.pdf, accessed on December 29th 2015)

[4] Kwek M Lin 2015 Using Problem Posing as a Formative Assessment Tool. In Florence Mihaela Singer Nerida F Ellerton and Jinfa Cai (Ed). *Mathematical Problem Posing: From Research to Effective Practice* pp 273—292 (New York: Springer)

[5] Silver, Edward A 1994 *On Mathematical Problem Posing For the Learning of Mathematics*, **14**(1) 19—28

[6] Polya G 1957 *How To Solve It* (Princeton: Princeton University Press)

[7] Sengul S & Katranci Y 2014 Structured problem posing cases of prospective mathematics teachers: experiences and suggestions *International journal on new trends in education and their implications* **5** issue 4 article 17 ISSN 1309-6249

[8] Barlow A T & Cates J M 2006 The Impact of Problem Posing on Elementary Teachers' Beliefs about Mathematics and Mathematics Teaching *School Science and Mathematics* **106** (2) pp 64—73.

[9] Bonotto Cinzia & Santo Lisa Dal 2015 On the Relationship Between Problem Posing, Problem Solving, and Creativity in the Primary School. In Florence Mihaela Singer, Nerida F. Ellerton and Jinfa Cai (Ed) *Mathematical Problem Posing: From Research to Effective Practice* pp 103—123 (New York: Springer)

[10] National Council of Teachers of Mathematics 2000 *Principles and Standards for School Mathematics* (Reston VA: NCTM) Inc

[11] Stoyanova E & Ellerton N F 1996 *A Framework for Research into Students' Problem Posing in School Mathematics*, (Online) (http://www.merga.net.au/documents/RP_Stoyanova.pdf, accessed on October 26th 2015)

[12] Silver E A 1997 *Fostering Creativity Through Instruction Rich in Mathematical Problem Solving and Problem Posing*, (Online), (http://www.emis.de/journals/ZDM/zdm973a3.pdf, accessed on October 26th 2015).

[13] Siswono TYE & Rosyidi A H 2005 Menilai Kreativitas Siswa dalam Matematika *Prosiding Seminar Nasional Matematika dan Pendidikan Matematika Peran Matematika dan Terapannya dalam Meningkatkan Mutu Sumber Daya Manusia Indonesia* Unesa February 24th 2005

[14] Leung S S & Silver E A 1997 The Role of Task Format, Mathematics Knowledge, and Creative Thinking on Arithmatic Problem Posing of Prospective Elementary School Teachers. *Mathematics Education Research Journal* **9**(1) pp 5—24.

[15] Siswono TYE 2011 Level of Student's Creative Thinking in Classroom Mathematics. *Educational Research and Review* **6**(7) pp 548—553.

[16] Siswono TYE & Kurniawati 2005 Identifikasi Proses Berpikir Kreatif Siswa dalam Pengajuan Masalah Matematika dengan Informasi Berupa Gambar: Penerapan Model Wallas MATEMATIKA *Jurnal Matematika atau Pembelajarannya* **11**(1) april 2005 ISSN: 0852-7792

[17] Harpen Van Xianwei & Presmeg N 2015 An Investigation of High School Students' Mathematical Problem Posing in the United States and China In Florence Mihaela Singer, Nerida F. Ellerton and Jinfa Cai (Ed). *Mathematical Problem Posing: From Research to Effective Practice* pp 293—308 (New York: Springer)

Critical-Inquiry-Based-Learning: Model of Learning to Promote Critical Thinking Ability of Pre-service Teachers

S Prayogi[1], L Yuanita[2], and Wasis[2]
[1]Institut Keguruan dan Ilmu Pendidikan (IKIP) Mataram, Jl. Pemuda No. 59A, Mataram 83126, Indonesia.
[2]Universitas Negeri Surabaya, Jl. Ketintang, Surabaya 60231, Indonesia.

Email: [1]saifulprayogi@ikipmataram.ac.id

Abstract. This study aimed to develop *Critical-Inquiry-Based-Learning* (CIBL) learning model to promote critical thinking (CT) ability of preservice teachers. The CIBL learning model was developed by meeting the criteria of validity, practicality, and effectiveness. Validation of the model involves 4 expert validators through the mechanism of the *focus group discussion* (FGD). CIBL learning model declared valid to promote CT ability, with the validity level (Va) of 4.20 and reliability (r) of 90,1% (very reliable). The practicality of the model was evaluated when it was implemented that involving 17 of preservice teachers. The CIBL learning model had been declared practice, its measuring from learning feasibility (LF) with very good criteria (*LF*-score = 4.75). The effectiveness of the model was evaluated from the improvement CT ability after the implementation of the model. CT ability were evaluated using the scoring technique adapted from *Ennis-Weir Critical Thinking Essay Test*. The average score of CT ability on pretest is -1.53 (uncritical criteria), whereas on posttest is 8.76 (critical criteria), with N-gain score of 0.76 (high criteria). Based on the results of this study, it can be concluded that developed CIBL learning model is feasible to promote CT ability of preservice teachers.

1. Introduction

Critical thinking (CT) is one of the essential skills that must learner possess in 21st century [1], and become trend and center of attention in learning, even curriculum authority in several advanced country has mentioned CT skills in their curriculum as learning objectives [2], moreover its has become a major focus and competency in learning at all levels of their education [3]. CT is reasonable and reflective thinking focused on deciding what to believe or do [4], its detailed description of some characteristic including the process of interpretation, analysis, evaluation, inference, explanation, and self-regulation [5]. CT is one of the higher-order thinking skills element, that can be taught [6]. The educators believe that improving learner's critical thinking is very important [7], however few of them have an idea how to teach it [8]. Teaching critical thinking requires holistic approach and should involve a set of appropriate learning models oriented on purpose that can to make learner manipulated their cognitive skills [9].

Several previous study recommended inquiry activity as teaching foundation to promote 21st century needs, including critical thinking [10]. Through inquiry activities, learners construct their knowledge actively so that desired learning outcomes can be achieved [11]. Inquiry process cannot be separated from critical thinking [12], thinking development and inquiry-teaching process can improve critical thinking skills [13]. When learner trains to investigate, it could help them to develop their critical thinking ability and scientific reasoning [14].

Teaching critical thinking skills to the prospective teacher has become attention for a while. Teacher educators have to teach a cognitive skills to prospective teacher before they trained it to their students in the classroom [15]. When the teachers and prospective teachers are trained and able to think critically, they will be able to improve student's analyzing ability [16]. Education before becoming a teacher is proper time to intervention activities which promote their critical thinking [17]. Therefore, school or educational institution has to improve critical thinking ability to the prospective teachers [18].

This study aims to develop a specific learning model that promotes critical thinking ability of preservice teachers. The learning model in question is the Critical-Inquiry-Based-Learning (CIBL) learning model. The CIBL learning model were developed based on study about inquiry learning which is then attributed to the aspects of critical thinking into it. Inquiry has become a learning model which aim to make learner thinking [19]. It can be foundation and opportunities to develop critical thinking based on inquiry activities. Aspects of prior knowledge, advance organizer and motivation in learning reinforce the CIBL learning model. The learning steps in the CIBL model are orientation, exploration, analysis, inference, evaluation, and reflection. The results of model development and implementation in the class are described in this article.

2. Methods

This research is a development research that paired theory of Borg and Gall [20], and theory of Nieveen and Plomp [21]. The process of validating the CIBL learning model is done through a focus group discussion (FGD) mechanism involving 5 experts. After the CIBL model is declared valid, it will validate the learning tools as a supported of CIBL model, and instrument of critical thinking ability test (CTAT) by 2 validators. The reliability of the learning model is calculated using the percentage of agreement equation by Emmer and Millett in [22], they are reliable if the percentage of agreement is $\geq$ 75%. The subjects implementation of model are 17 preservice teachers (PT) of physics in Institute Teacher Training and Education Mataram, West Nusa Tenggara, Indonesia. Practicallity of the model evaluated from the learning feasibility (LF). The effectiveness of the model was evaluated from the improvement CT ability after the implementation of the model. CT ability are evaluated using the scoring technique adapted from *Ennis-Weir Critical Thinking Essay Test,* where the highest score is +3 and the lowest score is -1. The indicators of CT ability in this study are analysis, inference, evaluation, and decision making. While to know the score change of CT ability, its analyzed using *n-gain* equation [23]. Criteria of validity, learning feasibility and critical thinking ability shown as Table 1.

Table 1. Criteria of validity, learning feasibility and critical thinking ability

Validity		**Learning Feasibility**		**Critical Thinking Ability**	
Interval	*Criteria*	*Interval*	*Criteria*	*Interval*	*Criteria*
Va > 4,21	Very valid	*LF* > 4,21	Very good	CT > 8,8	Very critical
3,40 < Va ≤ 4,21	Valid	3,40 < *LF* ≤ 4,21	Good	5,6 < CT ≤ 8,8	Critical
2,60 < Va ≤ 3,40	Quite valid	2,60 < *LF* ≤ 3,40	Adequate	3,6 < CT ≤ 5,6	Critical enough
1,79 < Va ≤ 2,60	Less valid	1,79 < *LF* ≤ 2,60	Less	0,8 < CT ≤ 3,6	Less critical
Va ≤ 1,79	Invalid	*LF* ≤ 1,79	Poor	CT ≤ 0,8	Not critical

3. Results and Discussion

The validity results show that CIBL learning model which is developed was declared valid. The validity level (Va) of CIBL model is 4.20 (its valid if; 3.40 < Va < 4.20) with the percentage of reliability is 90.1% (reliable). The recommendation from validators in FGD are are CIBL model is valid with some revision. The content validity in this study has two main components as the basic development of CIBL model, which are needs and state-of-the-art knowledge.

Strengthening in need aspects is very important because it is the main basis of the development of a model, also if there is need aspect that cause the model has to be developed. Need aspects of the CIBL model development got an average validity score 4.25 from four validators with very valid criteria.

Besides need aspects, state-of-the-art knowledge aspect become very important of content validity to measure as a basic development of model. Strengthening in state-of-the-art of knowledge aspect as a support of development of a model got an average validity score 4.31 with very valid criteria.

Construct validity as an aspect of CIBL model development were focused on theoretical and empirical review that support the model. The syntax of the CIBL model as validation material that model phase shows logical, related, and mutual order of learning activity. The score of this aspect is 4.13 from four validators with valid criteria.

The average score of validation of learning tools is 4.24 with very valid criteria and the percentage of reliability is 90.2% (reliable). Then, the instrument of CTAT in this study is developed to assessing preservice teachers CT ability. The average validity result of instrument is 4.75 for each component of content validity and 3.13 for language and question-writing aspect. The final validity result of the instrument is 3.94 with valid criteria, while the reliability of the instrument is 79.37% with reliable criteria. Norris argued that the facts about the uniqueness level of CT is not finished yet because of many theories in different view, so making measurement and assessment of CT is difficult. Assessing CT distractor with another subject, because transferring to other context may be different with uniqueness of knowledge in CT [24]. In this study, we made simplification with developing the instrument of CT test based on 4 of CT ability indicators that are, analysis, inference, evaluation, and decision making.

In the implementation step, the practicality of the model was measured from learning feasibility. Observation result of learning feasibility using CIBL learning model were done very good ($LF = 4.75$). Its cause of the supports, especially the availability of the learning tools, including handsbook (module) and worksheet. When learning tools are designed well, it can give information which help learner more effective to accomplish learning objectives [25]. Learning tools that are good designed are functioned as communication tool, tool of learning plan, learning plan for students, learning resources, and tool for learning evaluation [26]. The support from module is also very important in this study. The material in the book is arranged systematically so it can condition students to learn [27]. The worksheet in this study were designed as the guidelines of learning feasibility in inquiry activity according to CIBL model to train student's CT ability. Scientific process skill is able to be tools that can develop thinking skills including critical thinking skill, efficient learning feasibility, and correct problem solving [12]. Scientific process skill has a great effect in learning because it helps learner to improve higher mental skill, such as critical thinking, decision making, and problem solving ([28], [29]).

The assessment result of preservice teachers CT ability showed that average score of critical thinking of 17 PT in the pretest was -1.53 with not critical criteria (not critical, if $X \leq 0.8$). The assessment of CT ability after the implementation of CIBL model (posttest) showed that average score of 17 PT was 8.76 with critical criteria (critical, if $5.6 < X \leq 8.8$), also the N-gain was 0.76 with high criteria (results are shown in Table 2). The study results showed preservice teachers CT ability tend to increase from not critical to critical criteria, so it can be state that CIBL model development is effective to improve CT ability. This result is also inseparable from the validity of CIBL model that aims to promote CT ability. CIBL model has accommodated several recommendations in learning which is the main idea in developing CIBL model, such as training critical thinking by presenting contextual or real life case in learning, motivating to openly discuss, and encouraging experimental activity oriented by inquiry activity ([30], [31]). CIBL model based on inquiry activity was integrated with worksheet oriented scientific process skill.

Table 2. The results of PT critical thinking ability

CTs interval	*Criteria*	*Pre test*			*Post test*			*N-gain*	*N-gain criteria*
		Mean	*F*	*%*	*Mean*	*F*	*%*		
$X > 8,8$	Very critical	-1.53	0	0	72.34	12	70.59	0.76	High
$5,6 < X \leq 8,8$	Critical		0	0		3	17.65		
$3,6 < X \leq 5,6$	Critical enough		0	0		1	5.88		
$0,8 < X \leq 3,6$	Less critical		2	11.76		1	5.88		
$X \leq 0,8$	Not critical		15	88.24		0	0		
Amount			17	100		17	100		

Learning feasibility has become effectiveness factor of CIBL model. Learning steps (syntax) in CIBL model, that are orientation, exploration, analysis, inference, evaluation, and reflection are designed consistently to train intact PT critical thinking throughout the learning process. Learning activity based on inquiry can improve critical thinking skills of the student [32]. Learning orientation was done by facing learner with contradictive information (anomalous data) and then by presenting advance organizer. Presenting contradictive information or anomalous data is viewed as a solution that can attract interest, ensuring the prior knowledge emerges ([33], [34]), and helping learners to reflect their ideas to give explanation about learned phenomena [35]. Giving explanation based on ideas or phenomena is one of the study and become main indicator of critical thinking.

Exploration is the second phase in CIBL model after orientation phase. When learners train to investigate, it will help them to improve their critical thinking and scientific reasoning. Through investigating process, learner will collaborate to create new knowledge and learn to think critically. After exploration phase, then learner do analyze, inference, and evaluate. Analysis, inference, and evaluation processes are main indicators of critical thinking. Those indicators were adopted into syntax or phases of CIBL model. This is appropriate with researcher's previous recommendations or findings that several indicators of critical thinking are in low category [36]. According to [30], integrating critical thinking indicators into learning can train them more to critical thinking.

The result of this study is also supported by valid learning tools of CIBL model, such as syllabus, lesson plan, module and worksheets, that help learner to train their critical thinking ability. The learning tools is supporting factor which help the students to organize CIBL model that used in learning, so that they can train their critical thinking ability. Constructivism learning theory of Vygotsky said that well-organized learning can produces the mental development. Critical thinking is mental processes [37]. The good designed learning tools can give information that help learners are more effective to achieve learning objectives. The module also supported the improvement of student's CT ability. Module is equipped with attributions to achieve critical thinking objectives. The module as a part of learning tools contains information, tools, and text that are required by teacher in presenting materials and skills that have to be learned by students, those skills are critical thinking in this context.

The test of CT ability showed that the N-gain score is 0.76 (high criteria), but there are five of PT with medium N-gain score (the range score, 0.30-0.70), and one of five of PT is less critical at the posttest. These results are affected by motivation of the PT in learning. When the contradictive information is presented in the beginning of the learning, there are several PT that shows less attention and do not give responses, even though this learning activities mean to motivate PT in learning besides to think critically. This is shown from observation result of PT activity in learning. Motivation factors toward content and context learning material are very important in learning. Interest and motivation factors can affect the process of believe establishment that occur when learners get new knowledge or are faced on new situation in learning even when they are presented with new information that is contradictive with their prior conception [38]. Someone's motivation to engage in cognitive activity can determine impact of delivered information [39]. High motivation in learning showed high level of cognitive necessity, this become learner's predictor to engage in the intellectual challenging activity [40], which is critical thinking. Students with high cognitive activity and motivation will be better in elaborating information, showing performance which are better in cognitive assignments, and more effective on complex problem solving [24].

4. Conclusion

The CIBL learning model was developed based on inquiry processes in learning by accommodating the aspects of prior knowledge, motivation, and thinking ability in learning. The results show that the CIBL learning model has been valid, practice, and effective to promote the critical thinking ability of preservice teachers, including aspects of analysis, inference, evaluation and decision making. These results are very important as empirical evidence of CIBL model development to promote the critical thinking ability of preservice teachers. The wide-scale implementation and dissemination of CIBL model are needed to evaluate the implementation of models in different places.

Acknowledgement
This research is part of the dissertation of doctoral study at the Surabaya State University (UNESA).

References

[1] Partnership for 21st Century Skills 2011 *Frame Work for 21st Century Learning* www.p21.org
[2] Bailin S 2002 Critical Thinking and Science Education. *Sci. Edu* **11(4)** 361–375
[3] Schmaltz R M, Jansen E and Wenckowski N 2017 Redefining Critical Thinking: Teaching Students to Think like Scientists *Fronti. Psych* **8(459)** 1-4
[4] Ennis R H 1996 *Critical Thinking* (New York: Prentice-Hall)
[5] Facione P 2011 *Critical Thinking. What It Is and Why Its Counts* (California: Measured Reason and The California Academic Press)
[6] Woolfolk A 2009 *Educational Psychology* (New York: Pearson)
[7] Albrecht W S and Sack R J 2000 *Accounting education: Charting The Course Through A Perilous Future.* Accounting Education Series vol 16 Sarasota, FL: American Accounting Association
[8] Duron H, Limbach B and Wough W 2006 Critical Thinking Framework to Any Dicipline. *Int. J. Teac. Lear. Higher Edu* **17 (2)** 160-166
[9] Thompson C 2011 Critical Thinking Across The Curriculum: Process Over Output *Int. J. Human. Soc. Sci* **1(9)** 1-7
[10] Wasis 2016 Higher Order Thinking Skills (HOTS): Konsep dan Implementasinya *Sem. Nas. PKPSM IKIP Mataram* (Mataram: West Nusa Tenggara-Indonesia)
[11] Samarapungavan A, Mantzicopoulos P and Patrick H 2008 Learning Science through Inquiry in Kindergarten *Sci. Edu* **92(5)** 868-908
[12] Sriarunrasmee J, Suwannatthachote P and Dachakupt P 2015 Virtual Field Trips with Inquiry learning and Critical Thinking Process: A Learning Model to Enhance Students' Science Learning Outcomes *Proced. Soc. Behav. Sci* **197** 1721-1726
[13] Llewellyn D 2002 *Inquiry Within Implementing Inquiry-base Science Standard* (New York: Corwin Press)
[14] Barrow L 2006 A Brief History of Inquiry-From Dewey to Standards. *J. Sci. Teac. Edu* **17** 265-278
[15] Ashton P 1988 *Teaching Higher-order Thinking and Content: An Essential Ingredient in Teacher Preparation* (Gainesville: University of Florida Press)
[16] Mei Y L, Swee E L, Jung C and Leah A 2003 What Hong Kong Teachers and Parents Think About Thinking *Early Child Develop. Care* **173(1)** 147-158
[17] Warburton E C 2008 Changes in Dance Teachers' Beliefs About Critical-Thinking Activities *J. Edu. Hum. Develop* **2(1)** 1-16
[18] Innabi H and ElSheikh O 2007 The Change in Mathematics Teachers' Perceptions of Critical Thinking after 15 Years of Educational Reform in Jordan *Edu. Stud. in Math* **64 (1)** 45-68
[19] Arends R 2012 *Learning to Teach.* Ninth Ed (New York: McGraw-Hill)
[20] Borg W R and Gall M D 1983 *Educational Research: An Introduction* (London: Longman Inc)
[21] Nieveen N and Plomp T 1999 *Prototyping to Reach Product Quality* (Dordrecht: Kluwer Academic Publisher)
[22] Borich G D 1994 *Observation Skills for Effective Teaching* (Columbus: OH Merrill)
[23] Hake R R 1999 Analyzing Change/Gain Scores Retrieved from <http://lists.asu.edu/cgi-bin/wa?A2=ind9903&L=aera-d&P=R6855
[24] Norris S 1989 Can We Test Validly for Critical Thinking? *Edu. Res* **18(9)** 21-26
[25] Parkes J and Harris M B 2002 The Purposes of a Syllabus *College Teaching* **50(2)** 55-61
[26] Fink S B 2012 The Many Purposes of Course Syllabi: Which are Essential and Useful? Available on http://www.syllabusjournal.org/syllabus/article.pdf.

[27] Levin B 2008 *How to Change 5000 Schools: A Practical and Positive Approach for Leading Change at Every Level* (Cambridge: Cambridge University Press)
[28] Lee A T, Hairston R V, Thames R, Lawrence T and Herron S S 2002 Using a Computer Simulation to Teach Science Process Skills to College Biology and Elementary education majors *Comp. Simul. Bioscene* **28(4)** 35- 42
[29] Koray Ö, Köksal M S, Özdemir M and Presley A I 2007 The Effect of Creative and Critical Thinking based Laboratory Applications on Academic Achievement and Science Process Skills *Elem. Edu. Online* **6(3)** 377-389
[30] Miri B, Ben-Chaim D and Zoller U 2007 Purposely Teaching for the Promotion of Higher-order Thinking Skills: A Case of Critical Thinking *Res. Sci. Edu* **37(4)** 353-369
[31] Fine M and Desmond L 2015 Inquiry-Based Learning: Preparing Young Learners for the Demands of the 21st Century. *Edu. Voice* **VIII** 2-11
[32] Thaiposri P and Wannapiroon P 2015 Enhancing Students' Critical Thinking Skills Through Teaching and Learning by Inquiry-based Learning Activities Using Social Network and Cloud Computing *Proced. Soc. and Behav. Sci* **174** 2137-2144
[33] Ambrose S and Lovett M 2014 Prior Knowledge is More Than Content: Skills and Beliefs Also Impact Learning *Appl. Sci. Learn. Edu* **1(2)** 7-19
[34] Chinn C A and Brewer W F 1998 An Empirical Text of A Taxonomy of Responses to Anomalous Data in Science. *J. Res. Sci. Teac* **35(6)** 623-654
[35] Limon M and Carretero M 1997 Conceptual Change and Anomalous Data: A Case Study in The Domain of Natural Sciences. *Eur. J. Psyc. Edu* **12 (2)** 213-230
[36] Qing Z, Jing G and Yan W 2010 Promoting Preservice Teachers' Critical Thinking Skills by Inquiry-based Chemical Experiment *Proced. Soc. Behav. Sci* **2** 4597-4603
[37] Sternberg R 1986 Critical Thinking: Its Nature, Measurement, and Improvement http://eric.ed.gov/PDFS/ED272882.pdf
[38] Pintrich P R 1999 *Motivational Beliefs as Resources for and Constraints on Conceptual Change.* In W. Schnotz, S. Vosniadou, dan M. Carretero (Eds), New perspectives on conceptual change (pp. 33-50) (Amsterdam: Pergamon)
[39] Cacioppo J T, Petty R E, Feinstein J A and Jarvis W B G 1996 Dispositional Differences in Cognitive Motivation: The Life and Times of Individuals Varying in Need for Cognition. *Psycho. Bull* **119(2)** 197-253
[40] Steinhart Y and Wyer R S 2009 Motivational Correlates of Need for Cognition *Europ. J. Soc. Psyc* **39** 608-621
[41] Luong C, Strobel A, Wollschlager R, Greiff S, Vainikainen M P and Preckel F 2017 Need for Cognition in Children and Adolescents: Behavioral Correlates and Relations to Academic Achievement and Potential *Learn. Indiv. Differ* **53** 103-113

Students' Spatial Performance: Cognitive Style and Sex Differences

U Hanifah[1], D Juniati[2], T Y E Siswono[3]

[1] Postgraduate Program in Mathematics Education, Universitas Negeri Surabaya
[2,3] Department of Mathematics, Universitas Negeri Surabaya

Email: [1]umihanifah10@mhs.unesa.ac.id; [2]dwi_juniati@yahoo.com [3]tatagsiswono@unesa.ac.id

Abstract. This study aims at describing the students' spatial abilities based on cognitive styles and sex differences. Spatial abilities in this study include 5 components, namely spatial perception, spatial visualization, mental rotation, spatial relations, and spatial orientation. This research is descriptive research with qualitative approach. The subjects in this research were 4 students of junior high school, there were 1 male FI, 1 male FD, 1 female FI, and 1 female FI. The results showed that there are differences in spatial abilities of the four subjects that are on the components of spatial visualization, mental rotation, and spatial relations. The differences in spatial abilities were found in methods / strategies used by each subject to solve each component problem. The differences in cognitive styles and sex suggested different choice of strategies used to solve problems. The male students imagined the figures but female students needed the media to solve the problem. Besides sex, the cognitive style differences also have an effect on solving a problem. In addition, FI students were not affected by distracting information but FD students could be affected by distracting information. This research was expected to contribute knowledge and insight to the readers, especially for math teachers in terms of the spatial ability of the students so that they can optimize their students' spatial ability.

1. Introduction

Spatial ability is a capability to visualize the image. According to Gardner [1], spatial ability is a skill to capture the world of space appropriately or, in other words, it is a skill to visualize the image. Everyone's spatial ability was definitely different because of his or her different characteristics. In accordance with Eddy [2], there are five factors that distinguish the characteristics of an individual: age, motivation/attitude, personality, learning/cognitive style and aptitude/intelligence. In this study, the characteristics of this research subjects used are the cognitive styles. Cognitive style is a typical way of learning, both in terms of how to receive and manage information, attitudes toward information, and habits related to the learning environment. It is in accordance with the opinion of Zhang and Sternberg [3] who state that the cognitive style is an individual characteristic in thinking, feeling, remembering, solving problems, and making decisions. Witkin and Goodenough [4] categorize cognitive styles in several types, one of which is the field-independent (FI) and field-dependent (FD) cognitive style. According to Witkin and Goodenough, the individuals who have a field-independent cognitive style are able to separate elements from the background so that it is easy for them to rebuild new information. Meanwhile, the individuals who have a field dependent cognitive style tend to be difficult to separate an item from an intact part that makes it difficult to rebuild new information. Each individual can be divided into two sexes, namely male and female. In addition to cognitive style factors, from the differences in sex, there are differences in problem-solving skills between women and men. Zhu [5] states that the gender differences in using strategy during mathematical problems solving fall into two classes: (a) on one hand, gender difference within groups with high-level spatial skills arise through the ability of integrating many problem-solving strategies, in which females do better than males; (b) on the other hand, gender difference within groups with low-level spatial skill arise from the ability in using other skills to compensate, in which males outperformed females. Naurzalina et al.'s study [6] also found that males tend to spend less time and made less mistakes compared to females. In this study, the researcher took the junior high school students because according to psychological theory Piaget [7], they should have been at the stage of formal operations so that students were expected to be able to think abstractly. Nevertheless, there are still many students who still have difficulty in thinking abstractly about the geometry that is widely used in math. Thus, the spatial ability of junior high school students is low.

In fact, the spatial ability of the students is still underperformed. The fact was revealed through a study conducted by Guven & Kosa [8], which states that the students' spatial skills are quite low. In particular, the average of 12 questions in Views section is 3.8 and in general the average of 36 questions is 15.7. The finding shows the insufficiency of students' spatial skills. It is surprising that although the students have learnt three-dimensional objects and their features in the early stages of elementary school as well as studied the three-dimensional objects in various lessons in Turkey, the averages value are still low.

Based on the research background, the purpose of this study is to describe the spatial ability of male and female of junior high school students who have cognitive field independent and field dependent. Spatial abilities in this study include five components, namely spatial perception, spatial visualization, mental rotation, spatial relations, and spatial orientation. This research is a descriptive research with qualitative approach which began by determining the subject using GEFT and math ability test, and then it continued by giving spatial ability test and interview. The validity of the data was checked through the use of time triangulation. Data reduction, display data, and conclusion are parts of data analysis used in this study. The results showed that there are differences in spatial abilities of the four subjects that are on the components of spatial visualization, mental rotation, and spatial relations. The differences in spatial abilities are found in the methods / strategies used by each subject to solve each component problem. The differences in cognitive styles and sex difference suggest different choice of strategies used to solve problems. This research is expected to contribute knowledge and insight to readers, especially for math teachers about the students' spatial ability so that can optimize it in their own classroom context.

2. Spatial Ability

According to Maier [9], spatial ability is divided into 5 components. *Spatial perception* is the ability to view the objects from vertical or horizontal viewpoint. *Spatial visualization* comprises ability to visualize a configuration in which there is movement or displacement among (internal) parts of the configuration.

Mental rotation involves the ability to rapidly and to accurately rotates 2D- or 3D- figures. *Spatial relation* refers to the ability to comprehend the spatial configuration of objects or parts of an object and their relation to each other. *Spatial orientation* is the ability to orient oneself physically or mentally in space.

3. Method

This research is a qualitative descriptive research. This study was conducted to describe the spatial characteristics of junior high school students in terms of cognitive style and sex differences in detail and systematic way. In determining the research subject in this research, the researcher involved 30 male and 30 female junior high school students.

The subject selection process of each category was done in the following way. Students were given mathematics test to measure their mathematical ability. It was done to choose the subjects who have equal mathematical ability, i.e. maximal score difference is 5. After the researcher gave mathematic ability test, students were given cognitive style test i.e. Group Embedded Figure Test (GEFT) developed by Witkin et al [10] and defined subject groups FI and FD according to the scores obtained

IOP Conf. Series: Journal of Physics: Conf. Series **947** (2017) 012014 doi:10.1088/1742-6596/947/1/012014

by the subjects. The students who got score between 0-9 were categorized as groups of students who had FD cognitive style and between of 10-18 were categorized as groups of students who had FI cognitive styles.

Then, it is chosen that the subjects are one male with FI cognitive style, one female with FI cognitive style, one male with FD cognitive style, and one female with FD cognitive style. The selection was based on the test results and the teacher's consideration of the students' ability in mathematics and communication.

There were four subjects selected and given spatial ability tests. The spatial ability test that was given to the subjects consists of 6 questions and should be done within 60 minutes. The spatial ability test was adopted from Maier. After a spatial ability test was performed, the four subjects were interviewed to complete the data that could not be revealed through a written test. Then, the researcher checked the result of the test and interviewed the subjects in different time to check the validity of data. In this study, the researcher made use of time triangulation. The analysis of data was carried out by firstly reducing data, displaying data, and drawing conclusion (Miles & Huberman [11]). The conclusion was sought to understand the characteristics of described students' spatial abilities based on spatial ability tests and interview results.

4. Spatial Ability Test

Spatial ability test was used to know student's spatial performance. In this test, the students are given 60 minutes to answer 6 questions related to spatial abilities. The spatial ability test appears in Figure 1. *Problem 1* was about the water level. The left picture shows about half a glass of water. In the right picture, the students are asked to draw the correct water level in equal to the leftmost image. In *Problem 2,* the students are asked to identify the pyramid-shaped webs shown on the left side. In *Problem 3,* the students are asked to select one of the four selected drawings that are identical to the standard image on the left side. In *Problem 4,* the students are asked to select one of the four selected images representing the standard image on the left side. In *Problem 5 and 6,* the students are asked to draw the pictures showing the images as the results of the other side.

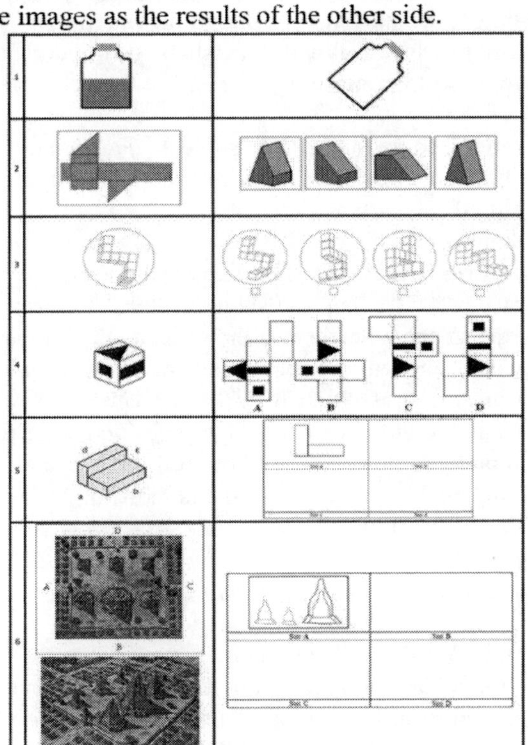

Figure 1. Spatial Ability Test

5. Results and Discussion

This section will carry out the results and the discussion of research results based on the data obtained related to the students'spatial abilities based on differences in cognitive style and sex. This section will examine the similarities and differences in students' spatial performance.

5.1. Spatial Abilities of Male with Field Independent Cognitive Style

In spatial perception component, the subject determined the position of water surface by remembering and knowing that water was a liquid that fill the entire room so that the surface is always straight and horizontal. Although the container altered in any way, the surface of water will always remain straight.

In spatial visualization component, the subject determined the outcome of change or displacement part of the solid figures by looking at one by one all the parts of the webs and solid figures. Then, the subject matched the parts of the webs that same with the part of solid figures.

In mental rotation component, the subject determined the rotation result by imagining the rotation of solid figures in the appropriate direction. The subject imagined it without counting the number of cubes on the solid figures at first. The subject must ensure that the solid figure has the same direction

In spatial relation component, the subject determined which webs that can form a known solid figure by arranging the webs into solid figures. After that, the subject looked at the relation among the parts by matching the position or direction of each part of the webs and solid figures.

In spatial orientation component, the subject determined the shape of the figures when he viewed from different angles by considering the size of the drawing and the things drawn as seen from the specified sides.

5.2. Spatial Abilities of Male with Field Dependent Cognitive Style

In spatial perception component, the subject determined the position of water surface by remembering and knowing that water has flat surface following the shape of the container. However, the position of water depicted by the subject is still not quite right. Because the subject thought that if the place tilted, then the water level also tilted.

In the spatial visualization component, the subject determined the outcome of change or displacement part of the solid figures by looking at the shape and position of some parts of the webs and solid figures and matching those parts.

In mental rotation component, the subject determined the rotation result by imagining the rotation of solid figures in the appropriate direction. The subject imagined it with counting the number of cubes on the solid figures at first. After that, the subject rotated the image. Once rotated, he looked at the direction of the parts of the image whether it is the same as the known solid figures or not.

In spatial relation component, the subject determined which webs could form a known solid figure by arranging the webs into solid figures. After that, the subject looked at the relation among the parts by matching the position or direction of each part of the webs and solid figures.

In spatial orientation component, the subject determined the shape of the figures when viewed from different angles by imagining and drawing the visible portion of the specified side. The subject observed the position and size of the figures.

5.3. Spatial Abilities of Female with Field Independent Cognitive Style

In spatial perception component, the subject determined the position of water surface by remembering and knowing that water is a liquid that fill the entire room so that the surface is always straight and horizontal. Although the container altered in any way, the surface of water will always remain straight.

In spatial visualization component, the subject determined the outcome of change or displacement part of the solid figures by looking at one by one all the parts of the webs and solid figures. Then, the subject matched the parts of the webs that same with the part of solid figures.

IOP Conf. Series: Journal of Physics: Conf. Series **947** (2017) 012014 doi:10.1088/1742-6596/947/1/012014

In mental rotation component, the subject determined the rotation result by imagining the rotation of solid figures in the appropriate direction. The subject imagined it with counting the number of cubes on the solid figures at first. The subject must ensure that the solid figure has the same direction. She looked at the bend / curve whether is it the same as the known solid figures.

In spatial relation component, the subject determined which webs could form a known solid figure by looked at the relation among the parts by matching the position or direction of each part of the webs and solid figures. After that, she arranged the webs into solid figures.

In spatial orientation component, the subject determined the shape of the figures when viewed from different angles by imagining the figure if it is on that side and drawing it.

5.4. Spatial Abilities of Female with Field Dependent Cognitive Style

In spatial perception component, the subject determined the position of water surface by remembering and knowing water has flat surface following the shape of the container. However, the position of water depicted by the subject is still not quite right. Because the subject thought that if the place tilted then the water level also tilted.

In spatial visualization component, the subject determined the outcome of change or displacement part of the solid figures by making the same webs and arranging it into solid figures.

In mental rotation component, the subject determined the rotation result by counting the number of cubes on the solid figures at first. After that, the subject imagined it with standing on the adjusted side to form solid figures. The subject observed the direction per section of the solid figures that it had same direction with known solid figures.

In spatial relation component, the subject determined which webs could form a known solid figure by drawing the webs, after that she arranging the webs into solid figures. When she drew the webs, she noticed the shape, position and direction of the solid figures.

In spatial orientation component, the subject determined the shape of the figures when viewed from different angles by imagining to standing on the specified side, looking at the shape of the visible figures with respect to the size and position of the figures.

5.5. Similarities and Differences in Students' Spatial Abilities

5.5.1. Similarities in Students' Spatial Abilities. Based on the results of spatial ability tests and interviews on male FI & FD and female FI & FD obtained some things about the similarities between the male FI & FD and female FI & FD. Here are the similarities of spatial abilities of male FI & FD and female FI & FD.

- In the spatial perception component, the four subjects remembering and knowing that water is a liquid that filled the entire room so that the surface is straight and horizontal.
- In spatial orientation component, the subject determined the shape of the figures when viewed from different angles by considering the size of the drawing and drawn as seen from the specified sides.

5.5.2. Differences in Students' Spatial Abilities. Based on the results of spatial ability tests and interviews, the researcher obtained some information that showed the differences in spatial ability between male subjects FI & FD and female FI & FD. Differences in sex indicate that the methods of problem solving done by the four subjects are also different. In accordance with the results of Zhu' study [12], it was found that that men and women showed different strategy choices used to obtain solutions. Besides sex, cognitive style differences also have an effect on solving a problem. This is in line with the opinions of Witkin, Moore, Goodenough, & Cox [13] that suggest that cognitive style differences affect a person's strategy of accepting, thinking, solving problems, learning, and connecting to others. Here are the differences in spatial abilities of male subjects FI & FD and female FI & FD.

Table 1. Differences in Students' Spatial Abilities

	FI MALE	FD MALE	FI FEMALE	FD FEMALE
Spatial Perception	The position of the water surface is flat, straight, horizontal	The position of the water surface is skewed	The position of the water surface is flat, straight, horizontal	The position of the water surface is skewed
Spatial Visualization	The subject matches all parts of the webs that same with the part of solid figures.	The subject matches some parts of the webs that same with the part of solid figures	The subject matches all parts of the webs that same with the part of solid figures	The subject creates a web and then folds it
Mental Rotation	The subject imagines the figure without counting the number of cubes on the solid figures at first.	The subject imagines it with counting the number of cubes on the solid figures at first.	The subject counts the number of cubes on the solid figures at first. After that, she matches bends or turns per section	The subject imagines counting the number of cubes on the solid figures at first. After that, she looks at objects from the other side
Spasial Relation	The subject arranges the webs into solid figures. After that, the subject looks at the relation among the parts by matching the position or direction of each part of the webs and solid figures.	The subject arranges the webs into solid figures. After that, the subject looks at the relation among the parts by matching the position or direction of each part of the webs and solid figures.	The subject subject looks at the relation among the parts by matching the position or direction of each part of the webs and solid figures. After that, the subject arranges the webs into solid figures.	The subject draws, cuts, arranges the webs into solid figures

6. Conclusion

The results showed that there are differences in the spatial abilities of the four subjects that are on the components of spatial visualization, mental rotation, and spatial relations. Meanwhile, in spatial perception and spatial orientation, the four subjects have the same spatial ability. The results showed that there are some differences done by the four subjects about the methods used to solve the problem. In terms of the sex differences, male students imagined the figures yet female students needed the media to solve the problem. Besides sex, cognitive style differences also have an effect on solving a problem. FI students were not affected by distracting information, while FD students can be affected by distracting information.

7. Acknowledgments

We would like to thank all those who have helped the researcher until the completion of this paper.

References

[1] Gardner H 1983 *Frames of mind: The theory of multiple intelligences* (New York: Basic Books)
[2] Eddy E 2012 *On The Involvement Of Cognitive Processes In The Acquisition Of English Grammar By Slovak Learners* (Slovakia: University of Prešov)
[3] Zhang L and Sternberg R 2006 *The nature of intellectual styles* (Mahwah, NJ: Erlbaum)
[4] Witkin H A and Goodenough D R 1981 *Cognitive styles: Essence and origins* (NY: International University Press)
[5] Zhu Z 2007 *International Education Journal* **8** 187-203
[6] Naurzalina D, Karimova A, Sarkulov M, Tolegenova A, Zholamanova A and Almurzayeva B 2015 *The European Proceedings of Social & Behavioural Sciences* 96-105
[7] Piaget J 1972 *The Psychology of Intelligence* (Totowa, NJ: Littlefield)

[8] Guven B and Kosa T 2008 *The Turkish Online Journal of Educational Technology* **7** 100-1007
[9] Maier P H 1996 *Spatial Geometry and Spatial Ability- How to Make Solid Geometry Solid?* In E. Cohors-Fresenborg H Reiss K Toerner G and Weigand H G Selected Papers from the Annual Conference of Didactics of Mathematics 63-75.
[10] Witkin H A, Oltman P, Raskin E, and Karp S 1971 *A Manual for the Embedded Figures Test.* (Palo Alto, CA: Consulting Psychologists Press)
[11] Miles M B and Huberman A M 1984 *Qualitative Data Analysis: A Sourcebook of New Methods* (California: SAGE publications Inc).
[12] Zhu Z 2007 *International Education Journal* **8** 187-203
[13] Witkin H A, Moore C A, Goodenough D R, and Cox P W 1977 *Review of Educational Researh Winter* **47** 1-64.

IOP Conf. Series: Journal of Physics: Conf. Series **947** (2017) 012015 doi:10.1088/1742-6596/947/1/012015

Designing PISA-Like Mathematics Tasks In Indonesia: Experiences and Challenges

Z Zulkardi[1,3], A W Kohar[2]

[1]Department of Mathematics Education, Universitas Sriwijaya, Kampus FKIP Unsri Bukit Besar Jl. Srijayanegara Palembang, Indonesia, 30136
[2]Department of Mathematics and Natural Science, Universitas Negeri Surabaya, Kampus UNESA Ketintang, Surabaya, Indonesia, 60231

[3]Corresponding author: zulkardi@gmail.com

Abstract. The insignificant improvement of Indonesian students in PISA mathematics survey triggered researchers in Indonesia to develop PISA-like mathematics tasks. Some development studies have been conducted to produce valid and practical PISA-like problems that potentially effect on improving students' mathematical literacy. This article describes the experiences of Indonesian task designers in developing PISA-like mathematics tasks as well as the potential future studies regarding to mathematical literacy as challenges for policy makers, researchers, and practitioners to improve students' mathematical literacy in Indonesia. The results of this research indicate the task designers to consider domains of PISA like: context, mathematical content, and process as the first profiles of their missions. Our analysis shows that the designers mostly experienced difficulties regarding to the authenticity of context use and language structure. Interestingly, many of them used a variety of local wisdom in Indonesia as contexts for designing PISA-like tasks. In addition, the products developed were reported to be potentially effects on students' interest and elicit students' mathematical competencies as mentioned in PISA framework. Finally, this paper discusses future studies such as issues in bringing PISA task into an instructional practice.

1. Introduction

The issue of using real-world context in mathematics education in recent decades has gained much attention from scholars and practitioners. The attention is aimed, particularly, at developing students' ability to formulate, employ, and interpret mathematics in a variety of contexts, in which PISA (Programme International for Student Assessment) call such ability as mathematical literacy [1]. However, many evidences indicate that students around the world find difficulties when dealing with context-based problems like what is found in PISA tasks [1]. In particular, Indonesian student achievement in PISA survey shows an insignificant improvement during the period of 2000-2015 [2]. Some evidence were also reported to explain the factors influencing such insignificant achievements, such as opportunity to learn regarding to non-supportive teaching practice on the context-based problem [3], teachers' insufficient knowledge on designing context-based problem [4], and inadequacy of learning resources supporting the use of context-based mathematical problems provided by Indonesian typical mathematics textbook [5].

To minimize the latter two factors, some effort have been carried out by researchers and practitioners in Indonesia, one of which is by providing learning resources such as developing mathematics problem like in PISA with a variety of purposes. Some concerned on designing problems used in a set of learning sequence through a design research study [6], while the other concern on designing problems which can be used for assessment through a developmental research [7,8].This effort is in line with the inclusion of mathematical literacy in Indonesian school mathematics curriculum which follows the movement of *PMRI* (Indonesian version of realistic mathematics education) as a part of reformation of current curriculum in Indonesia. The movement indicated, for examples, by the suggestion of improving the instructional quality using PISA problems and introducing PISA problems within workshop for students and teacher [9]. Designing problems like in PISA (or PISA-like tasks) is also recognized as one of the main contents in the curriculum of mathematics education in Indonesian teacher education. For example, Stacey et al. [9] reported that numerous studies on developing PISA-like problem had been conducted by teachers at Sriwijaya University as their final research projects.

However, designing context-based mathematics problem like in PISA seems to be challenging for the beginners like students' teachers. It is not only about making problem which is plausible to solve but also making problem authentically in which the real-world included and modeled mathematically like the framework of PISA in term of mathematic concern. Some studies reported several obstacles regarding to create problems like in PISA. Paolucci and Wessels [10], for example, reported that while the item developers in their study were relatively successful with creating appropriate and motivating contexts, they were far less proficient with turning these contexts into an appropriate problem. Siswono et al [4] also reported that most of the teacher participants in their study put difficulties on finding real world context as the main obstacle when designing problem-solving tasks. Thus, it needs to be clarified whether these kinds of obstacles are experienced by task designers in Indonesia. In addition it needs to be confirmed how they deal with the PISA framework on developing PISA-like tasks. Accordingly, other task designers can learn from them for future attempts regarding task design on mathematical literacy.

To answer the above challenge, this paper presents the experiences and challenges of mathematics tasks designers in Indonesia to develop PISA-like tasks. The designers concerned the scholars or the students who took mathematical literacy as the main topic for their research projects either in undergraduate programs or graduate programs. The experiences tell how the designers apply their frameworks to develop PISA-like tasks, how they employ certain techniques to design PISA-like tasks, how they deal with PISA framework on designing PISA-like tasks, how they utilize contexts for their tasks, and what potential effects of PISA-like tasks they develop on students' mathematical literacy. Meanwhile, the challenges include the difficulties experienced by the designers when developing PISA-like tasks, namely creating more authentic, more accessible regarding to the use of language structure, and more cognitively-demanding tasks. It is expected that the discussion of this paper would be useful for those who are interested in further study particularly in designing PISA-like mathematics tasks and how to translate the tasks into instructional practices.

2. PISA-like tasks designers' experiences

We organize the discussion about PISA-like task designers' experiences into five sections as follows:

2.1. Applying formative evaluation as the framework for developing PISA-like tasks

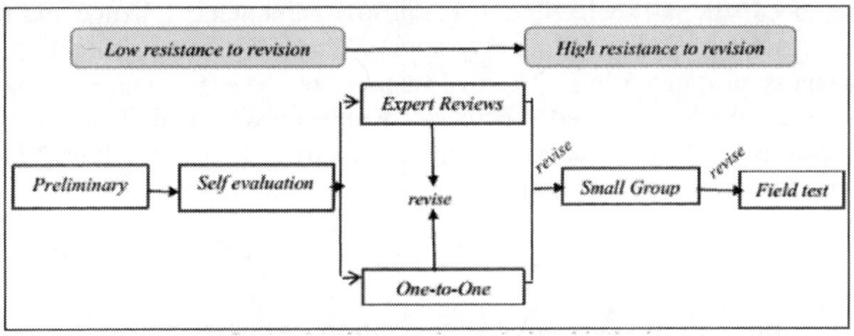

Figure 1. Formative evaluation [11]

Many of the task designers in Indonesia applied formative evaluation as the main framework for developing PISA-like tasks. Tessmer [11] defined the formative evaluation as a judgment of the strengths and weaknesses of the instruction in its development stages, for the purposes of revising the instruction to improve its effectiveness. Thus, the emphasis of this type of research design is on the development of repetitive cycles [11]. The stages of the framework are preliminary stage and prototyping stage which include self-evaluation, expert reviews and one-to-one, small group, and field test [11]. The components were involved in this evaluation are designers, experts, and learners (in this case the student).

At the preliminary stage, the task designers took several activities: studying pieces of literature (development research on PISA-like tasks, the framework of mathematical literacy, current curriculum related with PISA results, and how to design PISA-like tasks), finding experts and learners, and developing procedures for conducting research in schools with partner teachers. Kohar et al. [7], at this stage, for instance, studied some literature and found the research focus of the study which not only describes the process and the development of the tasks, but also examines the students' mathematical literacy performanced by zooming in the three processes: formulate, employ, and interpret the students experience during solving the tasks. *At the self-evaluation stage*, the task designers evaluated and examined the initial prototype, i.e. a set of PISA-like tasks equipped with item profiles, alternatives answer, and scoring rubrics. Furthermore, they also designed several instruments such as interview guidelines to find out the students' mathematical literacy performance or other expected objectives suitable with their research focus, and a set of questionnaires to see the potential effect of the tasks.

At expert reviews, the task designers gave the prototype to several experts, typically in three ways:face-to-face review, panel review, and mails review. Face-to-face reviews are performed by giving the prototype in order to confirm the experts' reviews directly. Panel review was organized by discussing the prototype with several experts at once that save time compared to interviewing the experts separately [11]. Meanwhile, mails review was carried out by sending the prototype to the experts via email and then obtaining constructive suggestions from them. The experts were selected by the task designers were typically from mathematics expert group of PISA in Australia [7,12,13], PMRI experts [7,8,12,13,], experienced PISA-like task designers [7,13], and experienced teachers [7,12,13]. *At one-to-one*, the prototype was examined to some students [4-6 students] particularly to evaluate the clarity of tasks regarding to language and figural display and study the emergence of alternative problem solutions. The results from both experts' review and one-to-one were then simultaneously used to revise the prototype. The revised prototype was then evaluated in a*small group* consisting of 9-15 students [7,8,12,13]. At this stage, the task designers firstly obtain the students' performances in detail so that it can be considered as an overview to explore the potential effect of the prototype on students' mathematical literacy at *field test*.

2.2. Adopting PISA framework as the main basis of profiling PISA-like tasks

PISA-like task designers in Indonesia mainly consider the framework of mathematics PISA, either released in 2003 or 2012 [1]. The framework, which briefly defines mathematical literacy as an individual's capacity to formulate, employ, and interpret mathematics in a variety of contexts, contains a model to guide them characterizing the PISA-like tasks they designed. The model depicts that mathematical literacy starts from real-world problems, which are categorized into the category of contexts (personal, occupational, societal, and scientific) and contents (change and relationship, space and shape, quantity, and uncertainty and data) (see [1]). Those two categories are then completed with one additional category or domain in PISA 2012, namely process (formulate, employ, and interpret), showing the most dominant process need to be performed when solving the real-world problems. While PISA 2003 only used paper-based assessment of mathematics (PBAM), while PISA 2012 used both PBAT and computer-based assessment mathematics (CBAM) for its item profiles.

We found a variety of PISA-like tasks developed by Indonesian task designers with various focus of domains. Table 1 summarizes several examples reported in some of national and international journals.

Table 1. Profile of PISA-like tasks in Indonesia: Some examples

Authors	Domain concern			Level	Mathematical competencies	PBAM	CBAM
	context	content	process				
Ahyan et al. [8]				√		√	
Anisah et al. [15]		√			√	√	

Charmila et al. [14]	√				√	
Kamaliyah et al. [16]				√	√	
Kohar et al. [7]	√	√	√		√	
Kertayasa et al. [13]						√
Lesmana et al. [17]				√		√
Oktiningrum & Hartono [12]	√				√	

Table 1 explains that while some concerned on only one domain or one mathematical competency, as exemplified by Charmila et al. [14], which focus on a local context of a province, others concerned on more than one domain. Anisah et al. [15], for example, consider the content: quantity and the mathematical competency: reasoning and argumentation for their items. Interestingly, there were also task designers concerning on the level of PISA task for their items, such as Kamaliyah et al. [16] who developed the sixth level of PISA-like tasks. In recent years, some task designers developed their tasks using computer-based technology. For instances, Kertayasa et al. [13] created an online version of their tasks which easily accessed by many people, as well as Lesmana et al. [17].

2.3. *Using problem posing as the method of designing PISA-like tasks*

With regards to the method in which the task designers were inspired to design PISA-like task, some of them primarily applied problem posing forms: problem generation and problem reformulation [18], each of which respectively refers to the process of posing a problem based on a set of given information, and the process of posing a problem related to a problem that is or was the focus of problem-solving. In addition, Bairac's [19] methods of composing mathematics problem (paraphrasing, analogizing, changing the data in the statement, generalizing, and combining), and Grundmeier's [20] problem reformulation methods (switching, changing the context, changing the given, changing the wanted, extension, adding information, and rewording) were also used as methods of designing PISA-like tasks. Kohar at al [7], for example, combined such three methods when designing their PISA-like tasks. Figure 2 below shows an example of how they used such techniques.

Body Mass Index [7]

The following chart shows comparison of world population over 20-year-old with Body Mass Index, *BMI≥30*

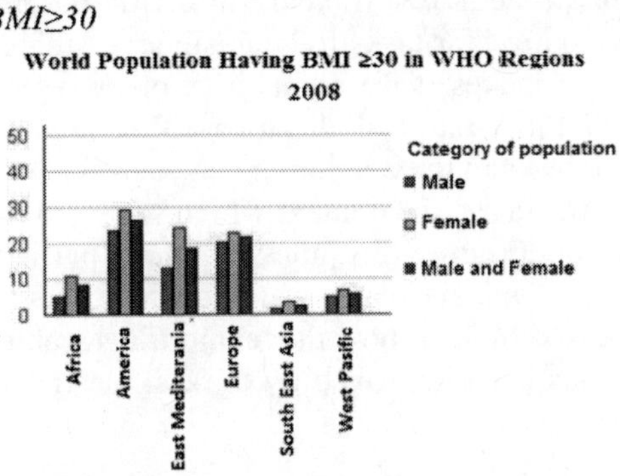

Circle "Yes" or "No" for each of the following options.

Statement	Is this correct?
Less than 5% of South Asian people are obese	Yes/No
The number of Americans with a minimum BMI of 30 is greater than that of European with a minimum BMI of 30	s/No
In all WHO regions, there are more obese women than obese men	Yes/No
In Africa, Eastern Mediterranean, and Southeast Asia, almost every man with BMI ≥ 30, there are two women with BMI ≥ 30	Yes/No

Figure 2. A task designed by using problem generation and paraphrasing particular information

2.4. *Employing local wisdom as contexts to design PISA-like tasks*

Developing mathematics problem using local wisdom has been obtaining a growing attention for PISA-like tasks designers in Indonesia in recent years. Not only because the awareness of the benefit of using local wisdom as contexts to encourage students' mathematical thinking process [14], but also the ease of accessing local context which is very diverse found by the task designers in Indonesia. Moreover, it is indicated that when working on mathematical literacy problem, students don't need to apply mathematical knowledge and rules to a new situation in a simple matter. Rather, it involves engagement with attaining the various attributes of the context of the problem in the task. Thus, developing PISA-like task which facilitates students' familiar situation like found in local wisdom becomes significant.

The local wisdom found in the PISA-like tasks include cultural heritages such as traditional craft and houses [14,16], traditional games[6], temples [7], and natural heritages such as local tourism objects [12]. Figure 3 shows an example of a PISA-like task which used *Prambanan* temple as cultural heritage.

Prambanan Temple[7]

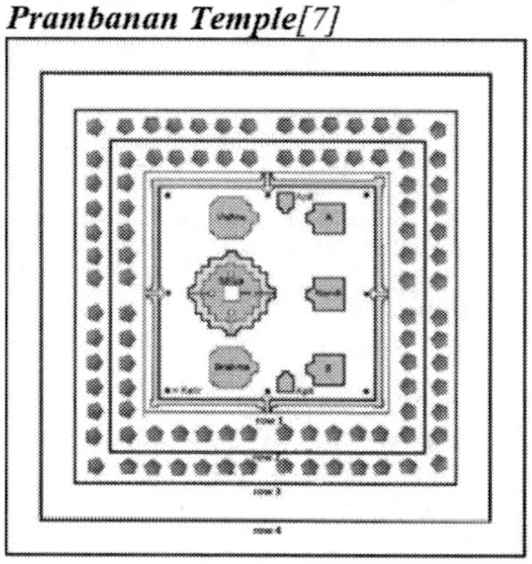

The picture below shows the arrangement of some *Perwara* temples (indicated by " ⬟ ") when already restored on row 1 and 2 at *Prambanan* temple. The arrangement of the remaining temples which will be restored on the 3rd row and the 4th row follows the pattern of previous arrangements. What is the total number of temples that still need to be restored? Please explain your answer.

Figure 3. A task designed by harnessing cultural heritage as a local wisdom

2.5. Exploring potential effect of PISA-like task on students' interest and mathematical literacy performances

To examine the quality of PISA-like tasks, the task designers explored to what extent the PISA-like tasks they developed can promote students' mathematical literacy, elicit students' mathematical competencies, and generate students' interest and seriousness in solving the tasks. The PISA-like tasks were reported to elicit students' reasoning, communication, problem-solving strategies, representation, and mathematising [7], and creativity [21]. Additionally, they were also able to motivate students to work on the context-based problems [7,12].

3. PISA-like tasks designers' challenges

3.1. Designing more authentic PISA-like tasks

Designing PISA-like tasks is more than designing plausible problem of sufficient information. Rather, it is the effort that needs to look for and translate authentic context for tasks to be designed. Authentic contexts are known as a unique feature of PISA task which makes it different from other types of problem-solving tasks [1]. Thus, it becomes a challenge for PISA-like task designers. We found that many reports reveal their difficulties on designing PISA-like tasks in the beginning stages of formative evaluation. The comments related to camouflage context of the tasks, the context which seems to have been rather artificially derived in order to provide an opportunity to practice some particular mathematical skill, rather than being a genuine problem that someone really wants or need to solve, became mostly found from experts review. Figure 4 shows an example of how the 'camouflage context' appears on the tasks before revision and how the situation has been improved by the task designer to improve the task authenticity.

Food Boxes[7]

Before revision:
Some food boxes are placed on the table as shown below. Count the total number of the food boxes!

After revision:
A committee of a seminar event forgot to count the total amount of the food boxes ordered when she put it on the table. However, she can still calculate the total amount of boxes by observing the pattern of rice boxes arrangement that he arranged on the table as shown in the figure above. Help the committee to count the number of boxes. Explain your strategy!?

Figure 4. The transformation from camouflage context to a more authentic context.

3.2. Designing PISA-like task with accessible language for students

We found that it is not easy for the beginners to arrange pieces of information in a task which is clearly understood by students. The problems experienced by the task designers include mostly about having too much information provided in the tasks, using unfamiliar terms, unspecific unit of contexts, and unacceptable use of mathematical symbols. Regarding the first problems, experts suggest to task designers to avoid some repetition of ideas or sentences, avoid introducing the problems of the task with context stories that are too general, and if it possible, convert sentences that are too long in the form of tables or figures [7].

3.3. Designing higher order thinking skill (HOTS) PISA-like task

HOTS task is known as tasks which promote higher order thinking skills (HOTS) and a concept of education reformed based on learning taxonomies (such as Bloom's taxonomy). The idea is that some types of learning require more cognitive processing than others, but also have more generalized benefits. Regarding PISA task, HOTS task is the PISA task which is considered at the level of 4,5, or 6 out of six levels of PISA-task. The challenge is that task designers are encouraged to design PISA-like task which meets the requirement of HOTS and also contains authentic context.

4. Implications and Conclusion

Regarding to the authenticity of PISA-like tasks, our analysis suggests that PISA-like task designers should aware of posing tasks of camouflage context. Thus, they need to consider levels of studying of context usage for mathematics problems. A theoretical framework was developed by Salgado [22] to help a task designer in designing context-based problem by considering the level of context usage of a problem which is being posed and analyzed and how the context is used to formulate a problem in mathematical terms and to interpret the answer about the context of the given problem. It is expected that the task designers that can pose tasks having at least level 1 or 2, instead of 0. Once PISA-like tasks are designed and evaluated, the further challenge is to bring those tasks into teaching practice. In connecting to that problem, Wijaya et al [3] suggested using consultative teaching for teaching practice regarding context-based task like PISA task from comprehension stage, transformation stage, mathematical processing stage and encoding stage. Alternatively, since the framework of PISA mathematical literacy is heavily influenced by the realistic mathematics education (RME) movement, which stresses the importance of solving mathematical problems in real-world setting, we argue that this learning approach can be employed and maintained as a promising basic theory to develop mathematical literacy-based teaching program.

To conclude, we would highlight that PISA-like task designers in Indonesia have contributed to provide PISA-like tasks as potential learning resources for improving students' mathematical literacy. However, they found some challenges mainly regarding to profiling tasks and designing tasks which have a genuine or authentic task. By accustoming to design this type of tasks, the task designers participate a mathematical literacy environment which is helpful for not only improving students' performance in PISA survey but also prepare students to solve more complex problems in current situation.

References

[1] OECD2013 PISA 2012 *Assessment and analytical framework: mathematics, reading, science, problem solving and financial literacy*(Paris: OECD Publishing)
[2] OECD2016PISA 2015*Results in focus* (Paris: OECD Publishing)
[3] WijayaA, van den Heuvel-PanhuizenM, and DoormanM2015*Math Educ Res J***27**637-662
[4] Siswono T Y E Kohar A W Kurniasari I and Astuti Y P *2016 Journal of Physics: Conference Series***693** 12015-12033
[5] Wijaya A, van den Heuvel-Panhuizen M, Doorman M *2015 Educ Studies in Math***89** 41–65
[6] Jaelani A Putri R I& Hartono Y 2014 *Journal on Math Educ***4** 29-40
[7] Kohar A W, Zulkardi, Darmawijoyo 2014 Developing PISA-Like Mathematics Tasks To Promote Students' Mathematical Literacy In: Ratu Ilma (Eds) *The Second South East Asia Design/*

Development Research (SEA-DR) International Conference April 26-272014 Unsri Palembang April 26th-27th2014 Palembang

[8] Ahyan S 2014*Journal on Math Educ***5** 47-56

[9] Stacey K, Almuna F, Caraballo RM,Chesne J,Garfunkerl S, Gooya Z,...Zulkardi Z2015 PISA's influence on thought and action in mathematics education, In K Stacey & R Turner (Eds) *Assessing mathematical literacy: The PISA experience* (pp 275-306) (Dordrecht: Springer International Publishing)

[10] Paolucci C and Wessels H 2017*Journal of Teacher Educ***68** 330-344

[11] Tessmer M 1993 *Planning and conducting formative evaluations: Improving the quality of education and training*(London: Kogan Page)

[12] Oktiningrum W & Hartono Y 2016 Developing PISA-" Like" Mathematics Task with Indonesia Natural and Cultural Heritage as Context to Assess Students Mathematical Literacy *Journal on Mathematics Education***7** 1-10

[13] Kertayasa I K & Zulkardi Z (2014) Achievement Of Indonesian Student In Assessment By Using Online Mathematical ProblemsIn: Ratu Ilma (Eds) *Proceeding of The Second South East Asia Design/ Development Research (SEA-DR) International Conference April 26-272014* (Palembang: Universitas Sriwijaya)

[14] Charmila N Zulkardi Z & Darmawijoyo D Pengembangan Soal Matematika Model PISA Menggunakan Konteks Jambi *Jurnal Penelitian dan Evaluasi Pendidikan***20** 198-207

[15] Anisah A Zulkardi Z & Darmawijoyo D 2013 Pengembangan Soal Matematika Model PISA pada Konten Quantity Untuk Mengukur Kemampuan Penalaran Matematis Siswa Sekolah Menengah Pertama *Jurnal Pendidikan Matematika***5**

[16] Kamaliyah K Zulkardi Z & Darmawijoyo D 2014 Developing the sixth level of PISA-like mathematics problems for secondary school students *Journal on Math Educaton***4**9-28

[17] Lesmana H &Ilma R I 2013 Pengembangan Soal Non Rutin Berbasis Komputer Untuk Melatih Penggunaan Kemampuan Matematika Siswa *Ekuivalen-Pendidikan Matematika***4**

[18] Silver E A 1994 On mathematical problem posing*For the learning of mathematics***14**19-28.

[19] Bairac R 2005 Some methods for composing mathematical problems International *Journal for Mathematics Teaching and Learning* Available from http://wwwcimtplymouthacuk/journal/bairacpdf

[20] Grundmeier T A 2015 Developing the problem-posing abilities of prospective elementary and middle school teachers In FMSinger & NF Ellerton (Eds*) Mathematical problem posing: From research to effective practic*e (pp 411-432) (New York NY: Springer)

[21] Novita R & Putra M 2016 Using task like PISA's problem to support student's creativity in mathematics *Journal on Math Educ***7** 31-42

[22] Salgado F A 2016 Developing a Theoretical Framework for Classifying Levels of Context Use for Mathematical Problems In White, B, Chinnappan, M & Trenholm, S (Eds) *Opening up mathematics education research (Proceedings of the 39th annual conference of the Mathematics Education Research Group of Australasia)*, pp 110–117 (Adelaide: MERGA)

IOP Conf. Series: Journal of Physics: Conf. Series **947** (2017) 012016 doi:10.1088/1742-6596/947/1/012016

High-Impact Aerobic and Zumba Fitness on Increasing VO_2MAX, Heart Rate Recovery and Skinfold Thickness

T J Suminar[1], N W Kusnanik[1*], and O Wiriawan[1]

[1]Universitas Negeri Surabaya

*Corresponding author: niningwidyah@unesa.ac.id

Abstract Purpose of this study is to determine the significant effect of high-impact aerobics exercise, and Zumba fitness on increasing VO_2Max, decreasing of heart rate recovery, and decreasing of skinfold thickness. A sample of this study is 30 members aerobics of student activity unit. Type of this study was quantitative by using a quasi-experimental design method. The design of this study used Matching-Only Design. Data were Analyzed by using the t test (paired t-test). The samples divided into three groups consisted of experimental group I, experimental group II, and control group. They were given a treatment for 8 weeks or 24 meeting. For the data, retrieval is done by MFT test, heart rate recovery test, and skinfold thickness test. Furthermore, the result was analyzed by using SPSS 21 series. In conclusion, significant effect of high-impact aerobics and Zumba fitness on increasing VO_2Max, heart rate recovery, skinfold thickness.

1. Introduction

Physical fitness requires power, endurance and flexibility in order to do any physical activity. Sport is defined as physical activity formed of a game that contains the efforts against the elements of nature, other people, ourselves and the complexity of the organization. Furthermore, pulse rate is the basis for the correct and measurable physical activity. From the pulse rate intensity of a person in physical activity will be known, and it can also show how hard the heart works. Someone who has a normal pulse rate (slower) is believe that person has good VO_2Max and will not get exhausted when they doing high-impact aerobics and Zumba exercise. VO_2Max has an influence on the body's ability to perform the recovery. Faster recovery allows athletes to reduce the interval of rest between training sessions and matches, also to improve overall training load [3]. In addition, fat in the body can be detected through anthropometric examination [4] conclude that skinfold fat can be determined by using a skinfold thickness, more than 85% body fat stored in the subcutaneous tissue, whereas subcutaneous fat thickness was significantly associated with weight. The aim of this study is to examine significant effect on high-impact aerobics and Zumba toward VO_2 Max, heart rate recovery, and skinfold thickness). In this this research, I would to bring forward the idea that my study will be reference for high-impact aerobics, heart rate recovery as well as skinfold thickness.

2. Research Methods

This research is categorized as quantitative by quasi-experimental methods (quasi-experimental design). The design of this research is using the Matching Only Design. This design does not use random as a way to incorporate the subject into the group, but by matching. The instruments used in this study are Multistage Fitness Test (MFT) to test VO_2Max, the pulse oximeter to measure pulse, and skinfold thickness to measure the thickness of subcutaneous fat.

3. Results

The Experiment Group I of high-impact aerobics was given a pre-post results on each dependent variable which is VO_2Max, heart rate recovery, and skinfold thickness. The data will be shown in the table 1.

Table 1. Initial Test Data Acquisition And Final Test Group I

	VO_2MAX (ml / kg / min)		PULSE RECOVERY (dt / min)		THICK FAT (ml)	
Average	Pre	Post	Pre	Post	Pre	Post
	30.73	32 , 82	100.7	94.7	26.37	22.78
St Dev	2.51	2.59	1.27	7.61	3.77	4.92
Increased%	6.37%		5.96%		13.61%	

The Table 1 shows significant improvement after being given a treatment for eight weeks with 24 sessions. It can be seen from the final tests average of 32.82 and 30.73 on average initial test. The increased percentage of from pretest to post test is amounted to 6.37%. So the differences of the average show an increase after given the treatment during training for eight weeks with the frequency of exercise for three times in a week.

Table 2. Preliminary Test Data Acquisition And Final Test Group II

	VO_2MAX (ml / kg / min)		PULSE RECOVERY (dt / min)		THICK FAT (ml)	
Average	Pre	Post	Pre	Post	Pre	Post
	30.1	34.74	97.2	92.1	26.89	23.43
St Dev	3.16	2.63	5.69	3.95	3.24	2.52
Increased%	13.36%		5.25%		12.87%	

The Table 2 above show a significant improvement after being given treatment for eight weeks with 24 sessions. It can be seen from the average of the final test of 34.74 and the main test of 30.1. it happens at increased percentage from pretest to post test amounted to 13.36%. So the difference of the average showed an increasing trend after a treatment was given during the training for eight weeks with a frequency of exercise for three times in a week.

Table 3. Initial Test Data Acquisition And Final Test Group III (control)

	VO_2MAX (ml / kg / min)		PULSE RECOVERY (dt / min)		THICK FAT (ml)	
Average	Pre	Post	Pre	Post	Pre	Post
	28.86	29.61	97.5	106.1	27.61	31.79
St Dev	3.37	3.41	4.76	3.24	2.71	2.90

Increased%	1.96%	4.15%	4.96%

Based on Table 3, the control group also showed an increase in the dependent variable at the third VO_2Max, decreased of pulse rate recovery, and decreased of fat thickness. It can be seen from the final of the data test, at the variable of VO_2Max that the result final test and the initial test are respectively 29.61 and 28.86 with a percentage increase of 1.96%. The pulse rate of recovery variable that the result final test and the initial test are respectively 97.5 and 106.1 with a percentage increase of 4.15%. While the reduction of the fat variable in the final test data obtained 27.61 and initial tests 31.79 with an percentage of increasing as much as 4.96%. In summary, the control group also got an increased impact at the third dependent variables although the increases are relatively small if compared to the experimental group.

4. Discussion

High-impact aerobic have benefits to help in improving the immune system, heart, and circulatory conditions [5]. High-impact aerobics are aerobic exercises that lead to movements in which feet do not touch the floor. Alex, et al [6] define high-impact aerobic exercise as an aerobic exercise which done with fast movement fast rhythm tempo. High-impact aerobics are conducted with movement and hard pounding such as: running, jumping, jogging, or movements that are done quickly and accompanied by the music with strong tempo. A physical exercise that classified in high-impact is when it reaches the maximum heart rate as much as 80-89%, where the resting rate of heart is increased. Furthermore, after the strenuous physical activities on such high-impact aerobics, if it more rapidly toward normal pulse rate, then the heart rate recovery will improve too.

Increasing the intensity of exercise can be done by adding weight training with jumping jacks movement by accelerating the movement frequency. The pulse rate is a reduction of heart rate recovery after an activity, where the heart rate returns to normal as before physical activity. Pascoal et al., [7] stated that heavy physical activities such as Zumba increase the pulse rate faster. Ljubojević [2] explains that "The Zumba fitness is a new kind of dance workout, inspired by Latin American music and Latin American dances. The exercise combines the basic of dance merengue, salsa, samba, cumbia, reggaeton and other Latin American dances. The movement uses not only basic aerobic steps, but Also enriches their composition with the other dance like hip-hop, belly dancing, Indian, African dance, etc ". Meanwhile, according to Maryann [8], Zumba is a Latin inspired aerobic dance with a strong Latin influence with a fusion of rhythms roommates are merengue, salsa, cumbia, reggaeton, mambo, chachacha, soca, Bhangra, belly dance, flamenco, hip hop, tango, and samba ". Jayanti [9] explains that Zumba is a dance that can quickly burn calories and fat in the body because the movement in Zumba dance is the characteristic of cardio movement such as jumping, spinning, moving faster, etc.

Zumba Fitness burns on average 369 calories, or about 9.5 kcal per minute. The average heart beat recovery was 154 beats per minute (bpm). In this research, exercises that given is the high-impact aerobics exercise that carried out in 24 meetings has an increase of 6.37% with average before treated as much as 30.73 ml/kg/min and after treated increased to 32.82 ml/kg/min to the ability of VO_2Max. The test results show paired sample *t* test with significance level of 0.001-5.195 ($p = 0.05$), which means there is a significant effect of high-impact aerobic exercises to VO_2Max. This is supported by research conducted by Jaywant, [10] that aerobic exercise influence on the increasing in VO_2Max and lose weight. Therefore, many women choose high-impact aerobics exercises for fitness activities. It also in line with the research of Mustedanagić, et. al.,[11] that aerobic exercise performed for 60 minutes, consisted of 10 minutes warm up, 35 minutes halves aerobics, body shape for 10 minutes and cooling for 10 minutes, can reduce the folds of fat under the skin and can reduce weight gain in women who are obese.

In some sports activities, which require speed and explosive power, exceeding fat will decrease. In detail, acceleration of fat declination is proportional increases when force are applied. Conducted research, suggest that the skinfold thickness is recorded in 7 skinfold sites including triceps, biceps, subscapular, supra spaniel, abdominal, front high medial calf [12]. The need for an increase of O_2 during exercise recovery (excess post-exercise oxygen consumption, or EPOC) is caused by several factors, but the most known is the factor of oxygen deficit or payment oxygen that occurs during exercise when the contractile activity supported by ATP from the sources of non-oxidative e.g. creatine phosphate and anaerobic glycolysis [13]. Both energy supply and energy demand will be increased because of increasing body composition [14]. Munzirini [15] suggest that high-intensity aerobic exercise influences body fat reduction in adolescent girls. When someone is doing Zumba and aerobics Kaori combustion occurs more. it can be marked on how much the sweat out. There are several other benefits of Zumba exercise, for example, can lose weight, accelerate blood flow, improve the respiratory tract and relieve stress, Sukma [4] suggest that Zumba is a sport combined from 30% fitness movement and 70% dance movement, Zumba exercise can burn 600 to 1000 calories/hour. The fast movement also generates not only burning calories and fat but at the same time also nourish the heart [16]. Gunawan, et al [17] find that exercise Zumba can also increase VO_2Max, if it is conducted over 8 weeks and exercised for 60 minutes. Haghjoo et al. [18] said that Zumba physical fitness training within a period of 8 weeks had a significant effect on the reduction of body fat percentage of women with a significant level ($p = 0.001$).

5. Conclusion

Based on the results of research about the influence of high-impact aerobics and Zumba to the VO_2Max, Pulse Recovery (Heart Rate Recovery), and the decreases of the fat thickness (skinfold Thickness), it can be concluded that: (1) there are significant effects between high-impact aerobics of the VO2Max, pulse Recovery (Heart Rate Recovery), and the decrease in thickness of the fat (skinfold thickness), (2) there are significant effects between Zumba exercises to the VO_2Max, pulse Recovery (Heart Rate Recovery) and a decrease in thickness of the fat (skinfold thickness), and (3) there is significant difference between high-impact aerobics and Zumba to the VO_2Max, pulse Recovery (Heart Rate Recovery), and the decrease in fat thickness (skinfold thickness).

6. References

[1] Brick L 1996 *Fitness Aerobics (Fitness Spectrum Series)* (New Zealand: Human Kinetics)

[2] Ljubojevic A, Jakovljevic V, and Poprzen M 2014 Effects Of Zumba Fitness Program On Body Composition Of Women *Original Scientific Paper (Online)* **10** (1) pp 29-33 (http://www.sportlogia.com/no9engl/eng4.pdf)

[3] Whyte G 2006 *Advances in sport and exercise science series "the physiology of training. Philadelphia*: Elsevier's Health Sciences Rights Department. (http://www.imd.inder.cu/adjuntos/article/613/The%20Physiology%20of%20Training.pdf)

[4] Sukma 2013 Efek Zumba Terhadap Penurunan Tebal Lemak Bawah Kulit Dan Berat Badan Member Df Fitness Dan Aerobic *Jurnal FIK UNY*

[5] Rosidah 2013 *Perbedaan Pengaruh Latihan Senam Aerobik Low Impact Dan Mix Impact Terhadap Tingkat Kesegaran Jasmani Pada Siswa Putri Smk Negeri 1 Surakarta Tahun Pelajaran 2012/2013* (Jurusan Pendidikan dan Kesehatan Universitas Sebelas Maret: Surakarta)

[6] Alex M S, Subiono H S, and Sutarji 2012 Pengaruh Latihan Senam Aerobik Low-Impact dan High-impact Terhadap Kesegaran Jasmani *Journal Of Sports Science and Fitness* pp41-45

[7] Pascoal M E, Robert D, and Poluan H 2014 Studi Senam Zumba Sesaat Terhadap Kadar Kolesterol Total Pada Mahasiswa Jurusan Gizi Politeknik Kesehatan Kementerian Kesehatan Manado *GIZIDO* **6** 1

[8] Maryann 2012 *Energy Expenditure and Physiological responses to 60 minute Zumba Aerobic Sessions (group class versus home) in Healthy Adult Female* (United Kingdom: University of Chester)

[9] Jayanti Y, Jimmy F R, and Wenny S 2013 *Pengaruh Latihan Zumba Terhadap Nilai FEV1 Forced Expratory Volume in one Second.* (Manado: Fakultas Kedokteran Universitas Sam Ratulangi)

[10] Jaywant P J 2013 Effect of Aerobic Dance on the Body Fat Distribution and Cardiovascular Endurance in Middle Aged Women (2013) *Journal of Exercise Science and Physiotherapy* **9** 1 pp 6-10

[11] Mustedanagić J, Bratić M, Milanović Z, and Pantelić S 2016 The Effect Of Aerobic Exercise Program OnThe Cardiorespiratory Fitness And Body Composition Of Female College Students *Acta Universitatis Physical Education And Sport* **14** 2 Pp 145 – 158

[12] Kusnanik N W 2001 *A comparison Of Field And Laboratory Testing Of Sport Specific Fitness For Female Field Hocky Players* (Australia: University Of Canberra)

[13] Sherwood L 2012 *Fisiologi manusia: dari sel ke sistem (6th Ed.) (Terjemahan Brahm U. Pendit).* (Singapore: Cengage Learning Asia Pte Ltd 5 Shenton Way)

[14] Kusnanik N W and Hartati H 2017 Physical and Physiological Profil Of Junior High Student In Indonesia *Sports Science* **10** 1 pp 96-99

[15] Munzirini M R 2016 *Perbedaan Pengaruh Senam Aerobik Intensitas Sedang Dengan Senam Zumba Di Air Terhadap Penurunan Lemak Tubuh Pada Remaja Putri Universitas 'Aisyiyah Yogyakarta* (http://opac.unisayogya.ac.id)

[16] Schiff T N 2014 *Dampak Aktivitas Senam Aerobik, Tari Zumba, Dan Jaipong Terhadap Perubahan Kadar Lemak Tubuh* (Bandung: Universitas Pendidikan Indonesia)

[17] Gunawan A, Polii, and Damajanti 2015 Pengaruh Senam Zumba Terhadap Kebugaran Kardiorespiratori Pada Mahasiswa Fakultas Kedokteran Universitas Sam Ratulangi Angkatan 2014 *Jurnal e-Biomedik* **3** 1

[18] Hoghjoo M, Zar A, and Hosseini A S 2016 Effect of 8-week Zumba training on overweight women's body composition *Pars Journal of Medical Sciences* **14** 2

Students' Covariational Reasoning in Solving Integrals' Problems

N V Harini[1,*], Y Fuad[1], and R Ekawati[1]

[1]Mathematics Department, Universitas Negeri Surabaya, Indonesia.
**Coresponding Author:* vindrin.an@gmail.com / yusuffuad@unesa.ac.id

Abstract. Covariational reasoning plays an important role to indicate quantities vary in learning calculus. This study investigates students' covariational reasoning during their studies concerning two covarying quantities in integral problem. Six undergraduate students were chosen to solve problems that involved interpreting and representing how quantities change in tandem. Interviews were conducted to reveal the students' reasoning while solving covariational problems. The result emphasizes that undergraduate students were able to construct the relation of dependent variables that changes in tandem with the independent variable. However, students faced difficulty in forming images of continuously changing rates and could not accurately apply the concept of integrals. These findings suggest that learning calculus should be increased emphasis on coordinating images of two quantities changing in tandem about instantaneously rate of change and to promote conceptual knowledge in integral techniques.

1. Introduction

Several studies [1,2,3,4] have revealed that the thinking of calculus students that are interpreting and representing the varying rate of change for intervals of a function's domain is slow to develop, with specific problem reported in students' ability to interpret graphical function information. Studies by [1,2,5] have underlined that calculus students do not appear to view a graph of a function as means of defining a covarying relationship between two variables. Whereas this covariation view of function has also been essential for understanding concepts of calculus [1,6].

In [7] had showed that not only covariational reasoning abilities of teachers were weak and lack depth, but also their predictions about students' reasoning abilities bounded by their own thoughts related to the problem. However, its result only focused on the level of covariational reasoning and did not investigate deeply about teachers' mental action. Moreover, many researchers [8,9,1,10] had investigated about covariational reasoning, but none of them exposed covariational reasoning that related with definite integral concepts. The definite integral concept is one of the main concepts in studying calculus. Many researchers [11,12,13,14] indicated that students have consistently difficulties in understanding and implementing the concept of definite integral, with specific problem, namely computational error and misconception in evaluate the proposed integrals. Therefore this study focuses the investigation on how students' reason during their studies concerning two covarying quantities in integral problems.

IOP Conf. Series: Journal of Physics: Conf. Series **947** (2017) 012017 doi:10.1088/1742-6596/947/1/012017

2. Method

This study was conducted on the international class of the mathematics' education study programme of Universitas Negeri Surabaya (Unesa), in academic year 2016 which was chosen by purposively sampling. The international class consists of 4 male and 21 female. From the test of represented 2nd semester calculus, we obtained that 6 students (score >80), 12 students (80≥ score ≥70), and 7 students (score <70). Six students had sign as volunteer subjects of the research. Six volunteerly students who had studied definite integrals topic in calculus class were set up to answer three items of the proposed problems relating to an analysis of covarying two quantities.

The instrument consists of three items, and had validated by two senior lecturers of the mathematics' department of Unesa. The six students invited to complete three written problems that involve interpreting and representing how quantities vary in tandem, which were identically designed in context but the problems have different representations, namely algebraic representation, graphical representation, and numerical representation. Three of these students (labeling with A, B, and C) were volunteerly invited to participate in the interview sessions that concentrated on how students reason during solving covariational problems. The selection of the interview subjects was also based on diversity responses on their written test. The analysis focused on students' use of reasoning covariationally while responses to the written test, which related with five mental actions [1] of covariational reasoning (see Table 1).

Table 1. Mental action of covariational reasoning

Mental Action	Description of Mental Action	Behaviors
Mental Action 1 (MA1)	Coordinating the value of one variable with changes in the other.	Labeling the axes with verbal indications of coordinating the two variables (e.g., y changes with changes in x).
Mental Action 2 (MA2)	Coordinating the direction of change of one variable with changes in the other variable.	Constructing an increasing straight line. Verbalizing an awareness of the direction of the change in the output while considering changes in the input.
Mental Action 3 (MA3)	Coordinating the amount of change of one variable with changes in the other.	Plotting points/constructing secant lines. Verbalizing an awareness of the variable amount of change of the output while considering changes in the input.
Mental Action 4 (MA4)	Coordinating the average rate-of-change of the function with uniform increments of change in the input variable.	Constructing contiguous secant lines for the domain. Verbalizing an awareness of the rate of change of the output (with respect input) while considering uniform increments of the input.
Mental Action 5 (MA5)	Coordinating the instantaneous rate of change of the function with continuous changes in the independent variable for the entire domain of function.	Constructing a smooth curve with clear indications of concavity changes. Verbalizing an awareness of the instantaneous changes in the rate of change for the entire domain of the function (direction of concavities and inflection points are correct).

Adopted from [1]

3. Results and Discussion

The given problem (see Figure 1) prompted students to construct a graph of a dynamic situation with a continuously changing rate. Table 2 shows the kinds of responses that the six students provided on the written test. Only one student of these high performing 2nd term calculus students given an acceptable solution, while four students constructed acceptable graph except discontinuous part and one other constructed collected of line segments. During the follow up interview to describe the graph's shape, the three interview subjects provided varied responses.

Student A provided responses that indicated all five mental actions that linear with [1]. When prompted to explain the rational for his acceptable graph (see figure 2), student A stated the following statements:

> *I remember the relationship of volume, debit, and time. Volume is area of debit and time and I used definite integral to find data of changing volume in each hour. Then I constructed v(t) graph in the Cartesian coordinate, time in horizontal axe and volume in vertical axe (MA1). Based on my graphical sketch, I consider that v(t) must contain increasing function at t=0 untill t=5 and decreasing function when t=5 untill t=7 (MA2). By using the definite integral concept, I compute the amount of change in v(t) each hour (MA3). Then, I find the exact value of v(t) during 7 hours. Oh but it still perform a set of points. No, it must be continuous curve since the input (time) is real number (MA4). Okay, let me try to make shorter interval when computing value of v(t) using definite integrals. Finally, I get that the type of graph that presents volume during 7 hours. When t=0 untill t=3 the curve is concave since it is a quadratic function that have positive a.*

Student B produced an acceptable graph except discontinuous part (see figure 3). When prompted to explain why she constructed that type of graph, she stated as follows

> *I just draw graph like what the result of my integral process. I check all initial and terminal points in each intervals of piecewise function, and I get that.*

She appeared to have difficulty in constructing curve of continuously changing rate. Further prompting revealed that she was conceptually able to construct acceptable graph, but she lack in the process of integral. She missed the constant C in integral process, therefore her computation and plotting point did not lead to acceptable graph. This condition is similarly to that of [13].

Student C constructed a set of segments (see figure 4). When prompted to explain the rationale for her graph, she stated

> *Since the function of Q(t) is linear so the graph of v(t) should be linear". It indicated that she appeared to have difficulty to construct image of continuously changing instantaneous rate".*

Table 2. Data Responses of Problem

Responses types	Number of students out of 6 providing each response type
All aspects of graph were acceptable	1
Acceptable graph, except discontinuous part	4
Collected of line segments	1

Let Q(t) represent the rate at which amount of water in pool changed during 7 hours periods. Assume that the initial volume was 800 m^3 and there were two kinds of pipes those connected with the pool, a pipe that filling water to the pool and other throwing water out of pool. Construct graph of v(t), given that $Q = \frac{dv}{dt}$.

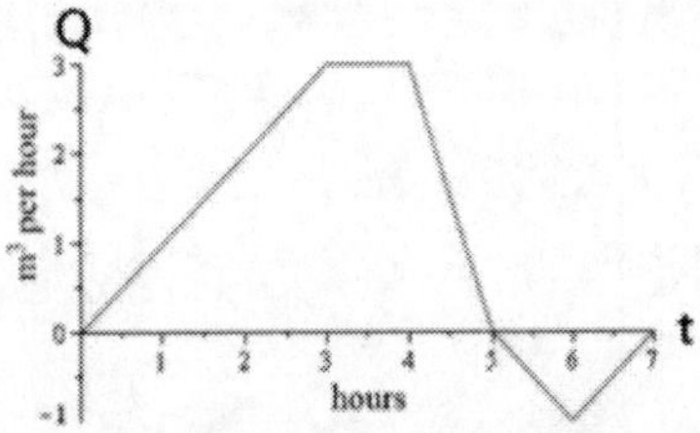

Figure 1. Covariational Problem

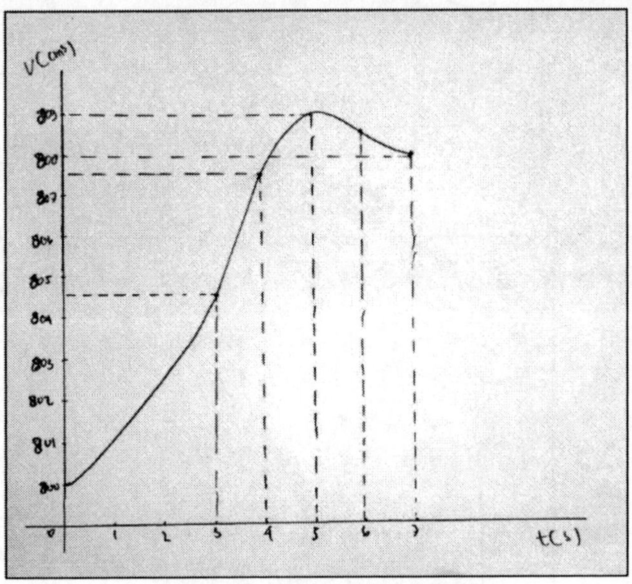

Figure 2. Graph of Student A

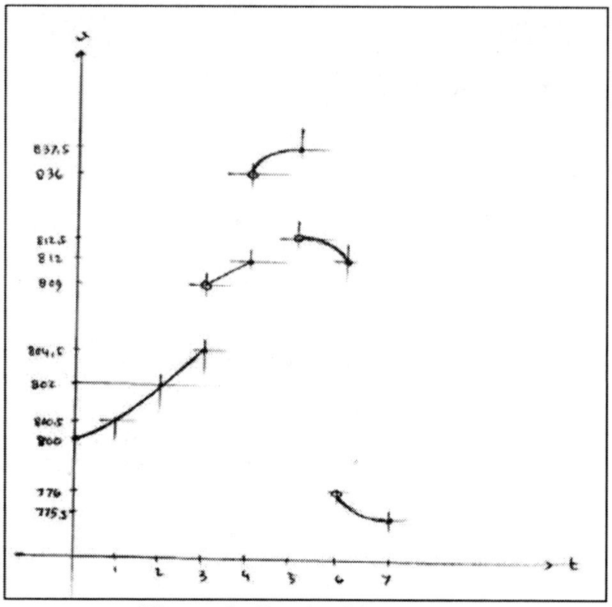

Figure 3. Graph of Student B

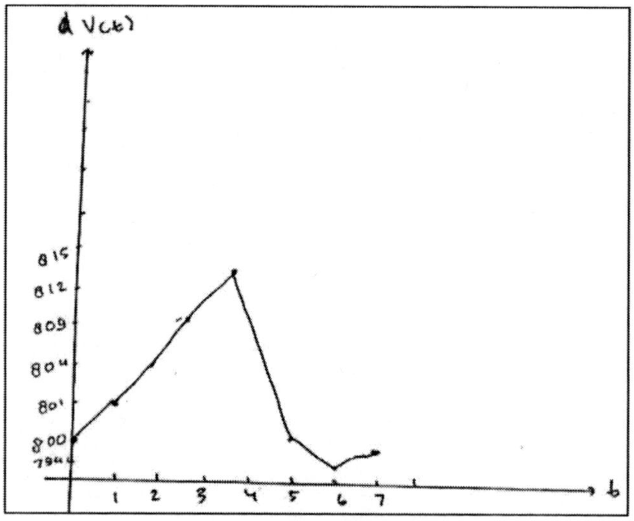

Figure 4. Graph of Student C

4. Conclusion

Based on the results, students were able to construct the relation of dependent variable that change in tandem with independent variable and worked on all five mental actions. Those mental actions are: coordinating the value of one variable with changes in the other, coordinating the direction of change of one variable with changes in the other variable, coordinating the amount of changes of one variable with changes in the other, coordinating the average rate of change of the function with uniform increments of change in the input variable, and coordinating the instantaneous rate of change of the function with continuous changes in the independent variable for the entire domain of function. However, students appeared to have difficulty in forming image of continuously changing rate and could not accurately apply integral technique. This findings suggest that learning in calculus should place increased emphasis on coordinating image of two quantities changing in tandem about instantaneous rate of change and conceptual knowledge in integral technique. Further studies are needed to investigate how to develop students' reasoning in coordinating the instantaneous rate of change of the function with continuous changes in the independent variable for the entire domain of the function.

References

[1] Carlson M, Jacobs S, Coe E, Larsen S, and Hsu E 2002 Applying Covariational Reasoning while Modeling Dynamic Events: A Framework and a Study *Journal for Research in Mathematics Education* **33** (5) 352-78
[2] Monk S 1992 Students' Understanding of a Function Given by a Physical Model in Harel G and Dubinsky E The Concept of Function: Aspects of Epistemology and Pedagogy **25** 175-93
[3] Thompson P W 1994 Images of Rate and Operational Understanding of the Fundamental Theorem of Calculus *Educational Studies in Mathematics* **26** (2-3) 229-74
[4] Subanji 2011 *Teori Berpikir Pseudo Penalaran Kovariasional* (Malang: UMPRESS)
[5] Tall D 1993 Students' Difficulties in Calculus *Proceedings of Working Group 3 on Students' Difficulties in Calculus, ICME-7 1992, Québec, Canada* 13–28
[6] Cotrill J, Dubinsky E, Nichols D, Schwingendorf K, Thomas K, and Vidakovic D 1996 Understanding the Limit Concept: Beginning with a Coordinated Process Schema 1-23
[7] SenZeytun A, Cetinkaya B, and Elbas A K 2010 Mathematics Teachers' Covatiational Reasoning Levels and Predictions about Students' Covariational Reasoning Abilities *Educational Science:*Theory and Practice **10**(3) 1601-12
[8] Moore K C, Paoletti T, Musgrave S 2013 Covariational Reasoning and Invariance among Coordinate Systems *Journal of Mathematical Behavior* **32** 461-73
[9] Strom A D 2006 The Role of Covariational Reasoning in Learning and Understanding Exponential Functio*n Proceeding PME-NA* **2** 624-30
[10] Zieffler A S and Garfield J B 2009 Modelling the Growth of Students' Covariational Reasoning during an Introductory Statistics Course *Statistics Education Research Journal* **8**(1) 7-31
[11] Grundmeier T A, Hansen J, and Sousa E 2006 An Exploration of Definition and Procedural Fluency in Integral Calculus **16**(2) 178-91
[12] Mahir N 2009 Conceptual and Procedural Performance of Undergraduate Students Integration **40**(2) 201-11
[13] Serhan D 2015 Students' Understanding of the Definite Integral Concept *International Journal of Research in Education and Science (IJRES)* **1**(1) 84-8
[14] Tall D O and Vinner S 1981 Concept Image and Concept Definition in Mathematics, with Particular Reference to Limits and Continuity **12** 151-6

Students' Misconceptions on Properties of Rectangles

R W Ningrum[1], M Yulianti [2], D D Z Helingo[3], M T Budiarto[4]

[1,2,3]Postgraduate, Universitas Negeri Surabaya
[4]Department of Mathematics, Universitas Negeri Surabaya

Email:[1]Rachmanianingrum16070785029@mhs.unesa.ac.id

Abstract. This research aims to describe students' misconception of junior high school on the concept of the properties of rectangles and the alternative to overcome them. This research was qualitative descriptive research using test with the CRI (Certainty of Response Index) method and interview method. There were three students as subjects in this study who was identified to have the most misconceptions among students in the class . Based on the data analysis, the subjects showed similar misconceptions. One of student misconception was on the properties of rectangles. Student thought that rectangles were always regularly shaped, or in other word no rectangle was irregular shaped. Students' misconception caused by picture given to the problem. The misconceptions may due to the lack of understanding of rectangle properties. Alternatives can be done such as re-explain (remediation) or by using cognitive conflict strategies. In particular, our results indicated that 8th grade students had misconceptions in understanding the properties of rectangles. By knowing the misconceptions, students can possibly know how new knowledge is interpreted as well as having an impact on student's ability to learn correct information in the future.

1. Introduction

Misconception is a concept which has no related idea with the concept of recognized experts [1].If students' misconception is not well cured immediately, and then the students' understanding of the higher topics will be hampered. Misconception is the phenomenon of different conceptions of a person with the conception of the experts [2].This misconception arises from the fact that one before knowing the correct concept they already have their own concept which is formed from reasoning, intuition, culture, or the other. The concept is maintained to explain the symptoms that exist in the vicinity but the concept is different from the actual concept. In addition, studying mathematics is not about studying the concepts and structure of mathematics [3].

Based on research conducted by Suha, who reviewed the misconception on the flat triangle matter in 4th grade of primary school can be seen that misconception may occur in the type of theoretical triangle [4]. In addition, the causes that occur in research is students' interest in learning and less given conceptual understanding. Identification of misconception is defined as a way to detect students' learning which is thought to have misconceptions of concept; in this case the conception of students is different from that of experts. The identification in this study was by using written diagnostic test with Certainty of Response Index (CRI) method. The Certainty of Response Index (CRI) that introduced by Hasan, Bagayoko, and Kelley in 1999, in their journal entitled "Misconceptions and The Certainty of Response Index (CRI)" is a way of measuring the level of confidence or the certainty of respondents in answering each question (matter) that is given [5][6]. In addition, interview diagnosis is also used in identifying misconceptions that occurred.

Most asserts that not all misconceptions can be changed [7], because it takes a fast and deep handling effort to correct and change the concept to be true in a short time. When it is known and affected students, the teacher can give alternatives to overcome the misconceptions. Simple alternatives are commonly done such as re-explaining. This re-explanation allows students to have better understanding of the concept. There are many attempts to overcome misconceptions such as those described by Swedosh and Clark (1997); by demonstrating cognitive conflict. The cognitive conflict approach can create a sense of students' dissatisfaction with the conception that they possess a change in the student's strong conceptions that conform to the scientific conception [8].The effort overcome the misconceptions or misunderstandings is used to emphasize that learning is not only related to conceptual understanding, but also to have better problem solving and have laboratory investigation skills [9][10]. The objects of this project are rectangle. This study involved three students who have most misconceptions among the students in the classroom. After this survey, an eighth grader teacher can analyze students' misconceptions and help them to improve their geometric knowledge. In this paper, it is described some of the causes of the students misconceptions. Students' misconception was caused by picture given. The misconceptions may due to lack of understanding of rectangle properties.This article give the alternative solution to identify the misconception and to overcome them. By identifying the misconceptions, students can comprehend how the new knowledge is interpreted and have an impact on student's ability to learn correct information in the future.

2. Students' misconceptions, causes of misconception , misconception's identify and alternative solution

The conceptual framework is grounded in the students' conception on the concept of rectangle.

2.1. Misconceptions

Misconceptions include understanding and thought that are not based on information that is right, or not understood by many people [11]. Misconception is the phenomena of different conceptions of a person with the conception of the experts [12].This misconception arises from the fact that before knowing the correct concept they already have their own concept which is formed from reasoning, intuition, culture, or the other. The concept is maintained to explain the symptoms that exist in the vicinity but the concept is different from the actual concept.

According to Soedjadi, misconceptions arise because of preconceptions [13]. Preconception is a beginning concepts' of a person who has an interpretation about an object. This initial concept is derived from education formal level. It is not impossible that a student has a different initial view of what he or she is learning. Under these circumstances, preconception becomes a misconception. Based on the opinion of the experts above, it can be concluded in this study that misconception is a conception that is inconsistent with the scientific notion or understanding received by scientists.

2.2. Causes of Misconception

According to Sodjadi (2000) there are 4 things that cause misconception. The first is the meaning of words, the second is practical aspects, the third is simplification, and the fourth is images [13]. Suparno (2005) explains that there are five factors that causes misconceptions in students namely students, teachers, textbooks, context, and way of teaching [12].

2.3. Misconcepton's identify

One of identification of misconception is open-ended questions in an essay test that can make it easy to detect students' misconceptions, because the students tend to answer according to their thinking [12]. Certainty of response index (CRI) is a way introduced by Hasan to measure the level of confidence or certainty of respondents in answering each given question [6]. The function of CRI is A diagnostic tool that allows teachers to modify the way they teach [5]. When a concept has been clearly articulated by the teachers, the student can accept the concept appropriately and there won't be misconception, therefore, it can be said that teaching or learning process has been successfully applied. But if the opposite happens, then the teacher must earlier identify to what is actually the cause of learning is not as expected, then can modify the way of teaching.

2.4. Rectangle

A rectangle is a special parallelogram. It is an equiangular parallelogram. The relation among the special parallelogram can be described in the Figure 1 by using a circle to represent each set. [14]

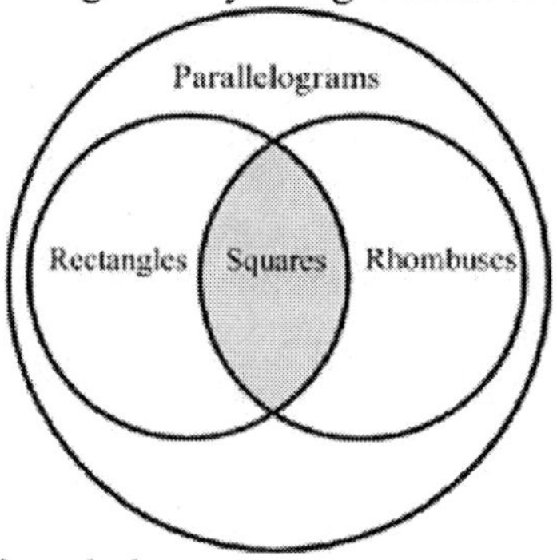

Figure 1. The relation among the special parallelogram

Since every rectangle and every rhombus must be a parallelogram, the circle for the set of rectangles and the circle for the set of rhombuses must be inside the circle for the set of parallelograms. Since every square is both a rectangle and a rhombus, the overlapping shaded section must represent the set of squares. Here are the principles involving properties of rectangle:
Principle 1: A rectangle has all the properties of parallelogram.
Principle 2: Each angle of a rectangle is a right angle.
Principle 3: The diagonals of a rectangle are congruent.
After knowing misconception and causes that occurs in students' case, it can be given an alternative to overcome misconception. Alternative is the simplest thing like describing frequent. Matter is possible be informed about the restarted in order for students to have better understanding of the concept.

2.5. Alternative solution

After knowing misconception and the cause of that which is occurs in students, it is possible to have alternative to overcome misconception. Alternative is the simplest thing like describing frequent. Matter is possible to be informed about the restarted in order for students to have better understanding of the concept. There are a lot of efforts in addressing misconception as presented by Swedosh and Clark (1997), which demonstrates to cognitive conflict. It is a cognitive approach conflict that can create a sense of dissatisfaction of the students of conception [8]. Based on the results of research conducted by Adnyani, Sadia and Natajaya where there is a difference in misconception a significant between the strategies of the cognitive conflict than the strategy of conventional sense. The strategy of the cognitive conflict is superior in providing the misconceptions to students when compared with a conventional learning strategy [15]. This strategy of the cognitive conflict will facilitate students to construct and explore their knowledge of the encountered phenomenon. The strategy of conventional rarely implemented the strategy of activation of knowledge and to be on the matter and the completion of the curriculum. By a strategy of cognitive conflict, teachers can change students to construct the initial conception of students who are not in accordance with the conception with the conception of experts.

3. Method

This research is a qualitative descriptive research using test method and interview. In this study, students were given written diagnostic test. The subject of this study was based on the results of the diagnostic tests written about misconceptions. After the test, three students who had misconceptions of 17, 16 and 16 questions from 21 questions become the research subject.

Those three research subject were given misconception test. The material of misconception test was about rectangular concept such as the definition and properties. The rectangular includes square, rectangle, rhombus, parallelogram, kite, and trapezoid. The misconception test arrange in the question of diagnostic tests consisting of 21 questions such as "Look at the following geometry. Build any kind of which is a rectangular figure! Why?"

IOP Conf. Series: Journal of Physics: Conf. Series **947** (2017) 012018 doi:10.1088/1742-6596/947/1/012018

The misconception test analyzed using certainty of response (CRI) method. Every question had column of CRI score. Student gave CRI score in 0-5 scale for each question. The CRI score was a students' belief in giving answer for every question. Student gave 0 score for lowest belief and 5 score for highest belief. The CRI score can diagnose the misconception. The information of misconception test result and the cause of student misconception known by interviewing after the misconception test were done. Analysis result and discussion used for describing student misconception, the causes, and the alternative to handle it.

4. Results and Discussion

Written diagnostic misconception test was given to second grade in junior high school. Based on the results of misconception tests, three students were selected as the subject of research. Table 1 presents the subjects in this study.

Table 1. List of Eighth Grade Student of Misconceptions Test Results

No	Research subject	Misconceptions calculation
1.	RAD	17
2.	PB	16
3.	MAA	16

The analysis of misconception tests is in Figure 2 which showed that each subject had different misconception on the concept of definition and characteristics of the rectangular figure and the student misconception causes. In addition, alternative solutions can be given to overcome the misconception that happened.

7. Which image included rectangle?

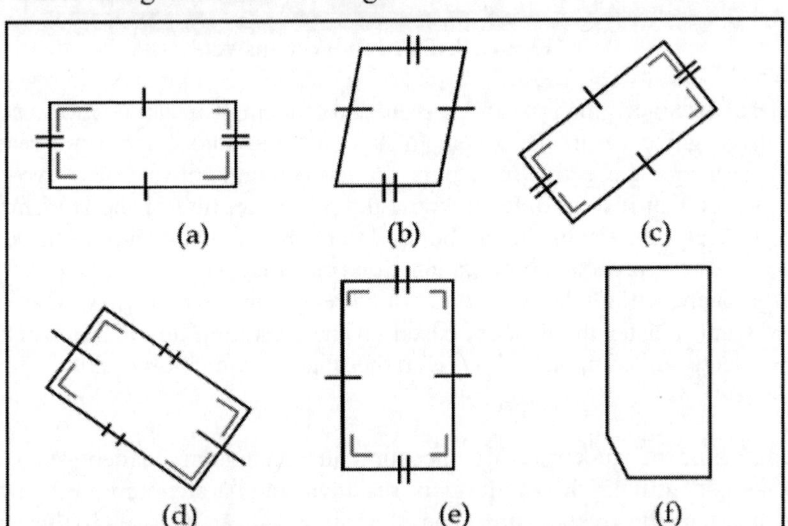

Figure 2. Question Number 7 in Misconception test

In this study, the first, second, and third subject gave 5 CRI score for their answer in seventh question. A score of 5 indicates the high level of student confidence in answering questions [5]. They answer questions with the correct knowledge or concepts without any hesitation at all. These values are true values without any notion at all. In question number seven, subjects gave a wrong answer and high CRI score; it indicates the student misconception [5].

The research subjects had misconception in the properties of rectangle. Figure 2, Figure 3, and Figure 4 showed the results of the response of the students' answers. They assume that the rectangle was always in the same position and has the same shape as any other rectangle. In this case, they were experiencing misconceptions to the concept of the properties of rectangle. This was in accordance with the theory described by Suparno, who stated that misconception is a concept that is not in accordance with scientific concepts or concepts established by experts[1].

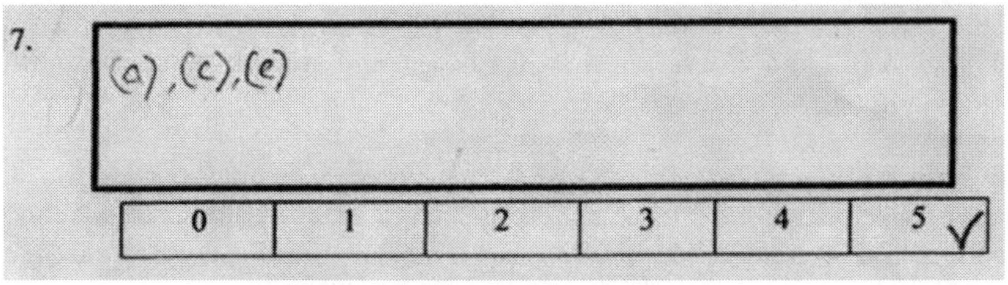

Figure 3. First Subject answers

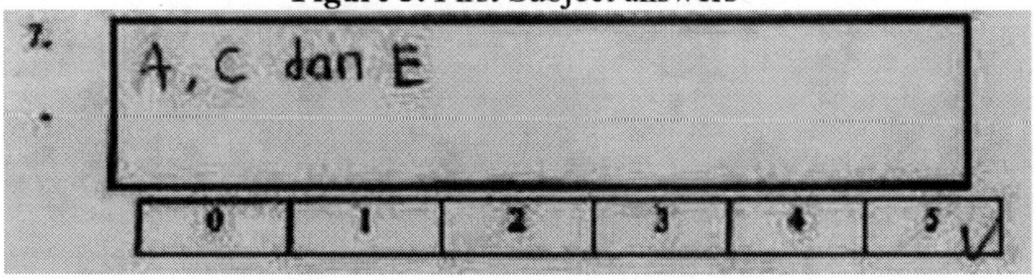

Figure 4. Second Subject answers

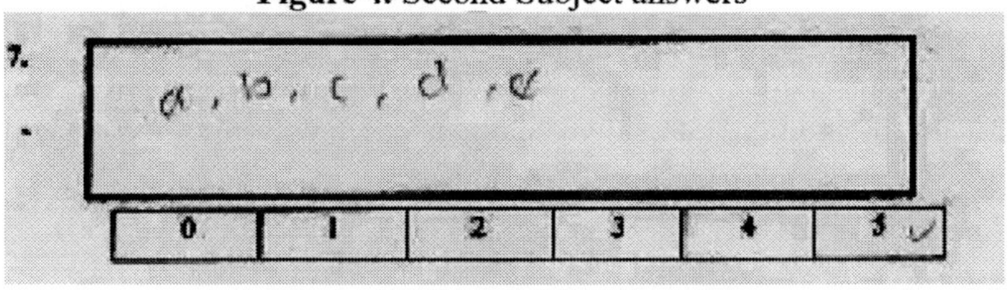

Figure 4. Third Subject answers

Based on these misconceptions it can be seen that students' misconception occurred because the students did not recognize the image given. In this case the students did not really understand the properties of the rectangle correctly. They only see the rectangle always in a two-dimensional figure presented by a picture that rose in order to know the misconception of the students to the concept of rectangles. Alternatively, the treatment can be used to overcome the causes of these misconceptions is to use cognitive conflict approach and re-explanation (remediation).

Misconception happened in whole concept which relates to the distinctive features. Hopefully, the teachers can give and transfer the concept based on the definition and distinctive features of square. Therefore, the students will comprehend it well rather than only memorizing it.

5. Conclusion

There were many kind of students' misconception in rectangular concept such as identifying the rectangle properties, identifying the similarity of rectangle and parallelogram properties. This students' misconception caused by the given picture. The alternative that can be given to this students' is by using conflict cognitive approach in learning mathematics. By knowing the misconceptions, the students can comprehend how new knowledge is interpreted and have an impact on students' ability to learn correct information in the future.

References

[1] Suparno P 2013 *Miskonsepsi dan Perubahan Konsep dalam Pendidikan Fisika* (Jakarta: PT. Gasindo)

[2] Berg E van Den 1991 *Miskonsepsi fisika dan remidiasi* (Salatiga: Universitas Kristen Satya Wacana)

[3] Baroody dkk 2007 An Alternative Reconceptualization of Procedural And Conceptual Knowledge *Journal forResearch in Mathematics Education* **15**(33) pp 35-131
[4] Suha H R N 2015 *Miskonsepsi Bangun Datar Segitiga Kelas IV SekolahDasar* (Skripsi tidak diterbitkan. Yogyakarta: Universitas Sanata Dharma)
[5] Saleem H D B and E L Kelley 1999 Miskonsepsi and The Certainty Responde Index *(CRI)*. *Journal of Phys. Educ.***5** pp 294-299
[6] Tayubi Y R 2005 *Identifikasi Miskonsepsi Pada Konsep-Konsep Fisika Menggunakan Certainty of Response Index (CRI)* (Online) (http://file.upi.edu/Direktori/JURNAL/JURNAL_MIMBAR_PENDIDIKAN/MIMBAR_NO_3_2005/Identifikasi_Miskonsepsi_Pada_KonsepKonsep_Fisika_Menggunakan_Certainty_of_Response_Index_(CRI).pdf Diakses pada tanggal 16 Oktober 2015)
[7] Moore 2006 *Mathematical Misconceptions in school Childern* (Master's Dissertation)
[8] Swedosh, Philip and Clark J 1997 *Mathematical misconceptions we have an effectivemethod for reducing incidence but will the improvement persist?* (Online). (http://www.merga.net.au/documents/RP_Swedos h_Clark_1997.pdf diakses tanggal 2 November 2015)
[9] Marek E A & Fleener M J 1992 *Testing in the LearningCycle. Science Scope* **15**(6) pp 48-49
[10] Türkmen H dan Usta E 2007 The Role Of Learning CycleApproach Overcoming Misconceptions in Science *Kastamonu Educational Journal* **15**(2)
[11] Türkmen H dan Usta E 2002 *Learner's Pocket Dictionary Sixth Edition* (New York: Oxford University Press)
[12] Suparno P 2005 *Analisis Miskonsepsi Fisika dan Remidiasi* (Salatiga: Universitas Kristen Satya Wacana)
[13] Soedjadi R 2000 *Kiat Pendidikan Matematika di Indonesia* (Jakarta:Direktorat Jenderal Pendidikan Tinggi DepartemenPendidikan Nasional)
[14] Rich B and Thomas C 2009 *Geometry Fourth Edition* (United State: McGraw-Hill Companies)
[15] Adnyani et. al. 2013 Pengaruh Strategi Pembelajaran Konflik Kognitif Terhadap Penurunan Miskonsepsi Fisika Ditinjau Dari Gaya Kognitif Siswa Kelas X di SMA Negeri 1 Bebandem *Journal Program Pascasarjana Universitas Pendidikan Ganesha Program Studi Administrasi Pendidikan* **4**

IOP Conf. Series: Journal of Physics: Conf. Series **947** (2017) 012019 doi:10.1088/1742-6596/947/1/012019

Mathematics Literacy of Secondary Students in Solving Simultanenous Linear Equations

R S I Sitompul[1], I K Budayasa[1], Masriyah[1]

[1]Department of Mathematics Education, Universitas Negeri Surabaya
Corresponding Author.Email: romastisitompul@mhs.unesa.ac.id

Abstract. This study examines the profile of secondary students' mathematical literacy in solving simultanenous linear equations problems in terms of cognitive style of visualizer and *verbalizer*. This research is a descriptive research with qualitative approach. The subjects in this research consist of one student with cognitive style of visualizer and one student with cognitive style of *verbalizer*. The main instrument in this research is the researcher herself and supporting instruments are cognitive style tests, mathematics skills tests, problem-solving tests and interview guidelines. Research was begun by determining the cognitive style test and mathematics skill test. The subjects chosen were given problem-solving test about simultaneous linear equations and continued with interview. To ensure the validity of the data, the researcher conducted data triangulation; the steps of data reduction, data presentation, data interpretation, and conclusion drawing. The results show that there is a similarity of visualizer and *verbalizer*-cognitive style in identifying and understanding the mathematical structure in the process of formulating. There are differences in how to represent problems in the process of implementing, there are differences in designing strategies and in the process of interpreting, and there are differences in explaining the logical reasons.

1. Introduction

Literacy has become an interesting debate to national and international countries. Many countries incessantly conduct literacy assessments and look for ways to improve literacy. Mathematic teachers have to realize that in order to face an increasingly developed world, it is necessary for the young generation to be responsive and critical. However, Indonesia in international level is considered unable to contribute a generation that have a good literacy skill. The fact revealed by Wardhani [1] that the learning result assessment instrument designed by secondary teachers in Indonesia in the syllabus development model published by *BSNP* (National Education Standards Agency) [2] generally presents an instrument of learning outcomes with inadequate manner to facilitate students in developing literacy skills which impact on the low achievement of the students in the event of world literacy assessment.

The literacy issues of the PISA and TIMSS (Trends International Mathematics and Science Study) study, which measure international students' math skills and science, require reasoning and problem-solving skills that emphasize issues and situations in everyday life. The ability to think mathematically is an important ability in students that is needed to be developed to deal with problems, both in mathematical problems and problems in real life. Mathematical standards power process according to NCTM [3] includes: problem solving skills, reasoning skills, communications skills, connection skills, and representation skills. Mathemathical literacy is an individual's capability to formulate, employ, and interpret mathematics in a number of contexts. It includes mathematical concepts, procedur, facts and tools to describe, explain and predict the phenomena. It assists individuals to recognize the role of OECD

[4]. Lange [5] states that the mathematics literacy is an individual's capacity to identify and understand the role that the human being made, concerned and reflective citizen as well. Mathematical literacy involves seven basic skills that must be had by OECD [6], namely: Communication; Mathematising; Representation; Reasoning and Argument; Devising Strategies for Solving Problems; Using Symbolic, Formal and Technical Language and Operations; Using Mathematics Tools.

Problem solving is a basic ability that should be mastered by students. According to Polya [7], problem solving is an attempt to find a way out of a difficulty in order to gain a goal that is not immediately achieved. While the steps in the problem solving are: read, explore, select a strategy, solve the problem, review and extend (Carson) [8]. The demand for problem-solving abilities is explicitly defined in a competency-based curriculum, as basic competencies that must be developed and integrated into a number of appropriate materials. The importance of problem solving is asserted in NCTM [3] which states that problem solving is an integral part of mathematics learning, so it should be noticed from mathematics learning.

One of the materials containing mathematical problem solving in junior level is simultanenous linear equations. In this material, students learn about problem solving related to everyday situation, so it requires good representation ability in solving problems. One of the indicators of success or failure in the learning process, both formal and informal is the result of learning. There are various factors influence the learning outcomes; one of which is the cognitive style. Uno [9] stated, cognitive style is a typical way of students in learning, both relating to the way of receiving and processing information, as well as the habits associated with the learning environment. Smith [10] suggested that cognitive styles refer to an individual's preferred way of processing information.

In mathematics, the information presented can be in the form of verbal symbols and visual symbols. Acceptance of information in the form of verbal symbols and visual symbols by students may differ depending on the cognitive style. McEwan [11] stated that the cognitive style associated with a person's habits using his sensory apparatus is divided into two groups, namely the *visualizer* and the *verbalizer*. Kozhevnikov [12] also proposed that the cognitive style is divided into two dimensions: the *visualizer* and the *verbalizer*.

Mendelson [13] explains that individuals who have a cognitive style of *visualizer* tend to be more in the picture, more fluid with illustrations and translations, as well as understand and like more visual games, such as puzzles. While individuals who have a cognitive style of *verbalizer* are more likely to say and would prefer to communicate to someone by showing how they do it.

Based on the background, the questions in this study include: how the profile of secondary students' mathematics literacy cognitive style of *visualizer* in solving simultanenous linear equations and how the profile of secondary students' mathematics literacy cognitive style *verbalizer* in solving simultanenous linear equations.

Based on the research question above, the purpose of this research is as follows: to describe the profile of mathematics literacy of secondary students' of cognitive *visualizer* style in solving the problem of simultanenous linear equations and to describe the profile of mathematics literacy of secondary students' of cognitive *verbalizer* in solving problem of simultanenous linear equations.

The expected benefits of this study are as follows; to provide knowledge and insight to readers, especially mathematics teachers about student's mathematics literacy to optimize students' literacy ability in solving problems based on students' cognitive style and as reference for other researchers about secondary students' mathematical literacy in solving problems in terms of cognitive style of *visualizer* and *verbalizer*.

2. Method

2.1. Types of research

This research type is descriptive with qualitative approach. This study aims to describe in depth about the profile of students' mathematical literacy in solving the problem of simultanenous linear equations of cognitive style of *visualizer* and *verbalizer*. Therefore this type of research is qualitative research.

2.2. Research subject

This research was conducted in SMP (Junior High School) IPH Surabaya city of grade VIII with 27 students. The subject selection process began with the provision of cognitive style test and mathematics skills tests. The selection of students 'answers will show students' tendency toward their cognitive style. The cognitive style test is used to determine the subject with the cognitive style of the *visualizer* and the cognitive style of the *verbalizer*. Mathematic skills tests are used to determine students with equal ability. Subjects are also controlled by the same gender; because Kruteski [14] stated that gender have difference output in mathematical. From the test the, the subject obtained was male students. After obtaining the subject, subjects were given problem-solving tests of simultanenous linear equations and interview.

2.3.Data, instruments, and data collection techniques

To obtain the necessary data, data collection techniques were used are test and interview. In this study, the test used was the cognitive style test developed by Mendelson & Thorson [15]. Original instruments were written in English, but firstly it has to be translated into Indonesia language. The instrument consisted of 20 questions with 10 questions about the cognitive style of the *verbalizer* and 10 questions about the cognitive style of the *visualizer*. Furthermore, the subjects were given the mathematics skills tests in the form of mathematics national exam which consists of 10 questions of grade seven and eigth, mathematics skills tests is used in taking subject with equal ability. Furthermore, the provision of problem solving tests is prepared to reveal the profile of mathematics literacy of secondary students'. The problem consists of mathematical problem of simultanenous linear equations.

The interview functions as a guide to explore the subject's mathematical literacy and during the research implementation there was no information passed by.

3. Results and discussion

Results and discussion obtained by researchers can be stated as follows. Each subject is given a problem-solving problem about simultaneous linear related to everyday life. The simultaneous linear problem is as follows:

Wahyu wants to compose some flat waves arranged in hexagon and rectangle shape by attaching them to a carton. He attaches the hexagon and rectangles from top to bottom (lengthwise) so that one side of the adjacent meets. The first flat build was 21 cm height, composed of three regular hexagons and three rectangles. The second flat height is 19 cm, composed of three regular squares and two rectangles. If the third is arranged from a regular hexagon and two rectangles, determine the high of the third flat, if Wahyu wants to make another flat build that is 30 cm in height, with many regulated order rules and many rectangles one at a time, try to interpret which one is more hexagonal or rectangular? Why? Determine the number of regular and rectangular hexagons!, and whether your answer consistent with your interpretation?

3.1. The subject's mathematics literacy profile reviewed from the cognitive styles of visualizer in trouble-shooting simultanenous linear equations

In formulating process, student can identify the mathematical aspects which function to determine the important information. Understand the structure of mathematics, identify the information known and asked in full and pay attention to the build order of the problem. In representing the problem differently, presents all information in the form of an image as shown in figure 1.

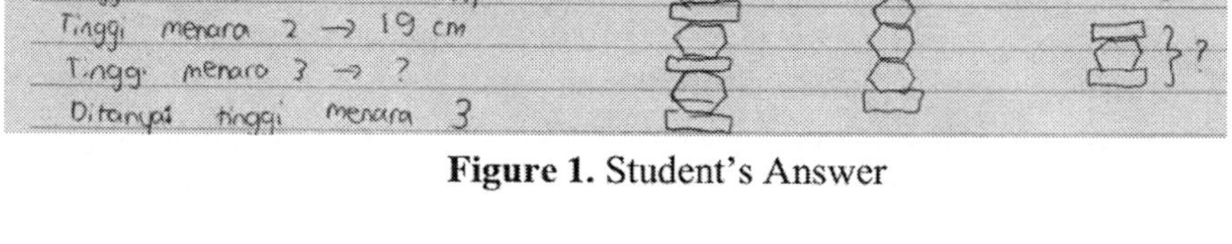

Figure 1. Student's Answer

IOP Conf. Series: Journal of Physics: Conf. Series **947** (2017) 012019 doi:10.1088/1742-6596/947/1/012019

In implementing process student is able to formulate a strategy to find a solution that is by using the elimination method. In applying mathematical rules and structures when searching for solutions, the subject began to define variables by modelling, forming mathematical models and simplifying mathematical models. In using various situations in the process of finding solutions, student simplifies the mathematical equations by eliminating variables with the same coefficients as shown in figure 2.

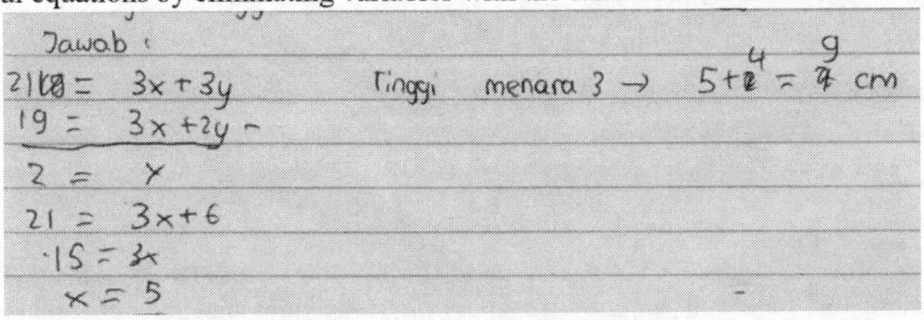
Jawab :
21 = 3x + 3y
19 = 3x + 2y −
2 = y
21 = 3x + 6
15 = 3x
x = 5
Tinggi menara 3 → 5 + 4 = 9 cm

Figure 2. Student's Answer

In interpreting process student is able to re-interpret mathematical results into a real problems. In explaining and evaluating the plausible reasons of a mathematical solution into a real problem, student is unable to give logical reasons, to make limited on trial and connect with the terms given.

3.2. The subject's mathematics literacy profile reviewed from the cognitive styles of verbalizer in trouble-shooting simultanenous linear equations

In formulating process, student is able to identify the mathematical aspects by mentioning the information that is considered important. In understanding the mathematical structure, student mentions information about what is asked and what is not known in detail. In presenting the information in its own language the problem is not sorted. Student pays more attention to information in the form of a sentence from the problem, to get the pattern when starting to read the problem.

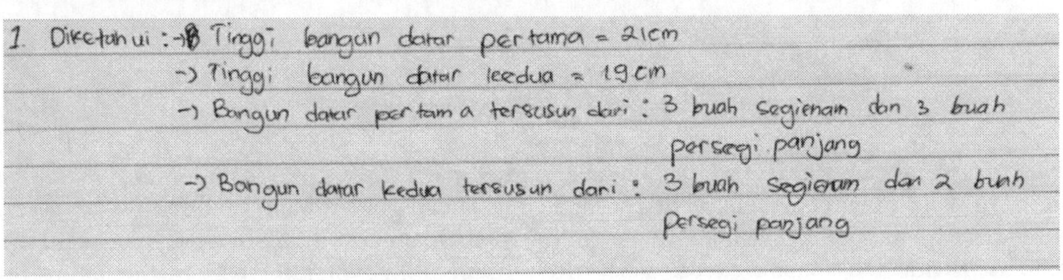
1. Diketahui : → Tinggi bangun datar pertama = 21cm
→ Tinggi bangun datar kedua = 19 cm
→ Bangun datar pertama tersusun dari : 3 buah segienam dan 3 buah persegi panjang
→ Bangun datar kedua tersusun dari : 3 buah segienam dan 2 buah persegi panjang

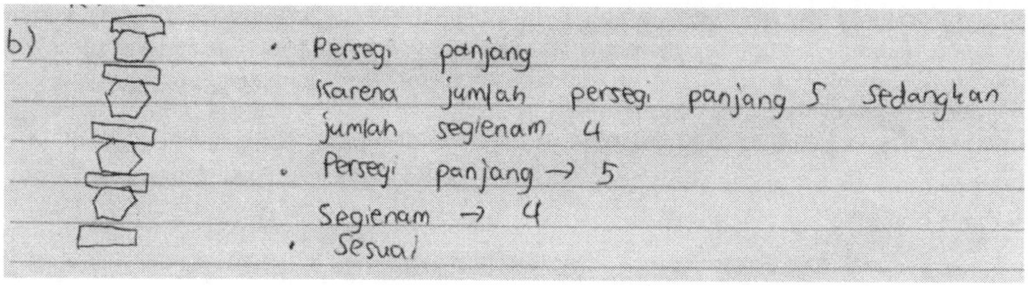

Figure 3. Student's Answer

In applying process, student devises strategy to find solution using the logic mathematical thinking from the pattern that has been acquired. In applying facts, rules and mathematical structures when looking for solutions student was able to connect information to form a pattern. In using various situations in the process of finding solutions, student use algebraic operations.

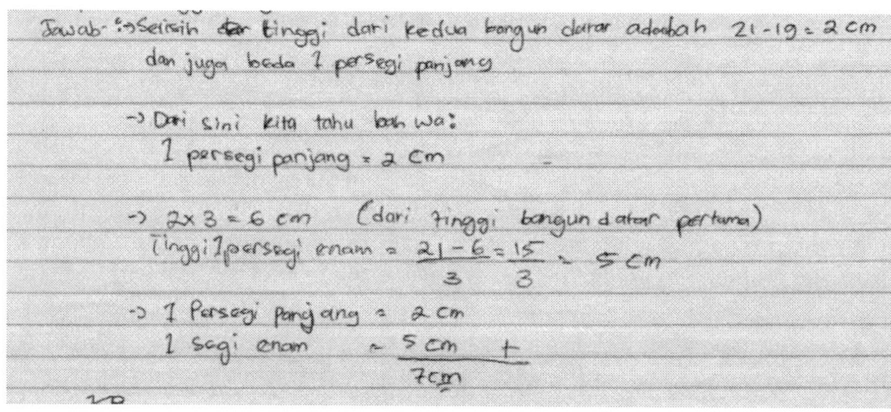

Figure 4. Student's Answer

In interpreting process, student was able to re-interpret mathematics into real problems by looking at the patterns of types of numbers that represent the height of each two-dimensional figure, i.e. even and odd number. In explaining the plausible reasons of the mathematical solution into the real world the subject is able to provide a reasonable excuse as shown in figure 5. In explaining why mathematical results are appropriate or not with the problem, the subject used algebraic operations and mathematical thinking logic in proving.

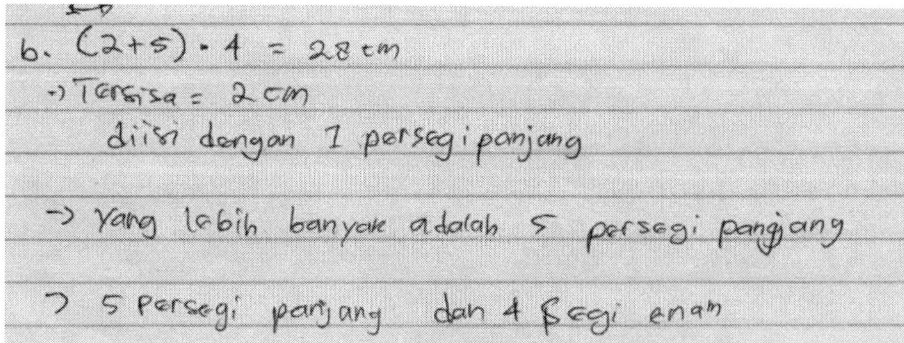

Figure 5. Student's Answer

3.3. Similarities and differences of subjects' literacy profile viewed from cognitive styles of visualizer and verbalizer in trouble-shooting simultanenous linear equations

3.3.1. Similarities of mathematical literacy profiles in solving simultanenous linear equations problems of visualizer and verbalizer subjects

The process of formulating is used to identify the mathematical aspects of both subjects as well as declared the same information about the important information of the question. Both subjects were also able to understand every sentence of the question. In understanding the mathematical structure, both subjects wrote things known and asked in the same way. Both subjects also stated the same about what to be found out to solve the problem. In representing the problem in different ways, both subjects were able to retell the problem without reading the problem even in different words. In the process of appication, finding the second solution, the subjects used the previous information by substituting the value obtained on the information.

3.3.2. Differences of mathematical literacy profiles in solving simultanenous linear equations problems of visualizer and verbalizer subjects

The following is the table of differences in the mathematical literacy profile in solving the simultanenous linear equations problem of *visualizer* and the *verbalizer* subjects.

Table 1. The differentiation of Mathematical Literacy Profiles in Solving Simultanenous Linear Equations Problems which is done by *Visualizer* and Verbalizer Subjects.

Mathematical Literacy	Visualizer Subject	Verbalizer Subject
Formulating	Paying attention to the arrangement of the dimensional figure, presenting information in the form of images, presenting information with the own language, stating known first and then asked, insequence and complete.	Not paying attention to the structure of the building, presenting information in written form, presenting information in own language randomly, declaring asked first and then known but not completely (some information were not mentioned).
Application	Presenting information in the form of mathematical equations by forming coefficients and variables, simplifying the mathematical equations by using elimination.	Presenting information with patterns found from the scratch, using the difference to solve math problems.
Interpretation	Trying to interpret the answer until the answer is obtained in accordance with the conditions given.	Using reason in interpreting the answer according to the conditions given.

4. Conclusions

Profile of mathematical literacy in solving simultanenous linear equations problems subject with the cognitive style of *visualizer* and *verbalizer* have similarities in identifying and understanding the mathematical structure in formalization process. There are differences in representing the problem, in appliying process implies, in designing strategies, in the process of interpreting, and in explaining the logical reasons.

5. Suggestion

Teachers need to pay more attention to the cognitive style of the students in carrying out learning activities by using appropriate approach or strategy related to the learning material.

References

[1] Sri W 2011 *Instrumen penilaian hasil belajar matematika SMP belajar dari PISA dan TIMSS* (Jakarta: Kemendiknas)
[2] Peraturan Menteri Pendidikan Nasional Republik Indonesia Nomor 41 Tahun 2007 Tentang Standar Proses untuk Satuan Pendidikan Dasar dan Menengah. Jakarta
[3] NCTM2000 *Principles and Standards for School Mathematics*(Reston VA: The National Council of Teachers of Mathematics) Inc
[4] PISA 2012 Result: Ready to Learn Students' Engagement aandSelf-Beliefs Vol.III (Paris: OECD Publishing)
[5] Jan D L 2006*Mathematical lliteracy for living from OECD-PISA perspective proc.Conf.*
[6] *PISA 2012 Mathematics Framework*(Paris: PISA OECD Publishing)
[7] George P 1980 *On Solving Mathematical Problems in High School*(New Jersey: Princeton University Press)
[8] Jamin C2007 A problem with problem solving: Teaching Thinking Without Teaching Knowledge*The Mathematics Educator***17**pp 7–14
[9] Hamzah U 2006 *Orientasibarudalampsikologipembelajaran*(Jakarta: BumiAksara)
[10] MarkKS 2010 *Teoripembelajarandanpengajaran*(Jogjakarta: Mirza Media Pustaka)
[11] McEwan R C and Reynold S 2007 *Verbaliser danvisualiser: Cognitive styles are lessthanequal*
[12] Kozhevnikov M, Hegarty M and Robert E M 2002 Revising the Visualiser/verbaliser dimension: evidence for two types of Visualisers *Cognition & Instruction***20**pp47-77
[13] Mendelson L 2004 For Whom Is A Picture Worth A Thousand Words? Effects of The Visualizing Cognitive Style and Attention on Processing of News Photos *Journal of Literacy***24**
[14] Krutetski V A1976*The Psychology of Mathematics Abilities in School Children*(Chicago: The University of Chicago Press)
[15] Andrew L M and Esther T 2004 How Verbaliser andVisualiser Process *The Newspaper Environmet Journal of Communication*

Teacher's Mathematical Communication Profile in Facilitating and Guiding Discussion

R Umami[1], I K Budayasa[1], and St Suwarsono[2]

[1]Mathematics Education Department, Universitas Negeri Surabaya, Indonesia
[2]Mathematics Education Department, Universitas Sanata Dharma, Indonesia

Abstract. Teacher's communication skill plays an important role, one of which is to guide a class discussion for teaching purposes. This study aimed to investigate a teacher's mathematical communication profile in facilitating and guiding a class discussion. This study is qualitative. A junior teacher of high school (i.e., a teacher with 1 to 5 year teaching experience) teaching mathematics at X-Social Class was selected as the subject of this study. The data was collected by observing the teacher's mathematical communication in facilitating and guiding a discussion with the rules of sinus as the teaching material, and it was followed by a deep interview. The result showed that the junior teacher facilitated and guided a class discussion include providing responses (answer) addressing students' difficulties and providing chances for students to propose and explain their ideas and be active in discussion. The junior teacher provides responses at sharp, optimal, and specific manner. In addition, she provides chances for her students to explain their thinking and have a discussion in anticipative, observative, selective, and connective manner. However, the study found that some of high-school teachers develop mathematical communication skills and use them to develop students' mathematical communication skills.

1. Introduction

A teacher-student communication has tight relation with a teaching-learning process. [1] argued that there were five aspects, which involved representing, listening, reading, discussing, and writing. Representing mathematical ideas or thoughts with another concept may help students describe those ideas and help them discover strategies to solve problems, since it needs a deep analysis and actively involves their thinking. In curriculum standard of mathematics, [2], however, representation is no longer applied in mathematical communication. Thus, it is considered excluded. The second aspect is carefully listening to people questions and opinions. This skill should be improved since it may influence students' mastery in comprehending the given material, capturing information needed, and proposing their comments or opinions. Listening to people opinion and other information may help students either see a new connection or clarify their thinking.

Following [1], the third aspect is reading mathematics textbooks as the sources of thinking and information. When students read, they may memorize, understand, compare, analyze, and organize the content of what they read. They must focus on the contents that provide important information, containing references relevant to the concept they are learning on or problems encountered. Another aspect that should exist in communication is discussing, which includes discussing with teacher and classmates in order to reflect their ideas and thoughts. Students who got engaged with a discussion may express and reflect their thoughts related to the material discussed. In a discussion, verbal communication skill is necessary and it may evolve through regular exercises by teachers like giving chances for students to present their ideas, having students work in small groups, and any other activities that encourage students' verbal communication skill. After having a discussion, students may express their mathematical ideas in a writing activity, consciously conducted in order to express and reflect their thoughts. Through writing, students may see the objectives of writing, reflect what they have done and thought, find out the point of problems, and develop their capability to combine mathematical language with their daily language in order to see a connection between mathematics and other concepts in daily life.

Related to the mathematical communication, [2] makes standard of ability that students should achieve. That is, organize and consolidate mathematical thinking through communication; communicate mathematical thinking coherently and clearly to peers, teachers and others; analyze and evaluate the mathematical thinking and strategies of others; and use the language of mathematics to express mathematical ideas precisely. In this research, the mathematical communication profile is a complete picture of teacher communication in building the learning community of mathematics.

To build the learning community of mathematics, teachers' skills are needed in facilitating and guiding class discussions. [3] proposed a teaching strategy to build Mathematical Discourse Community by providing tasks for students to discuss, creating a comfort learning environment for students, and giving chances to propose and explain their ideas and thoughts, and encouraging students to participate actively in a discussion during the class.

Providing tasks to discuss. In this case, the teacher may develop students' mathematical communication skills by selecting mathematical problems and tasks which encourage students' interest to discuss, provide responses (e.g., in the form of questions or ideas/thought making the students actively engaged in a class discussion). [4] argued that "*effective teachers understand that the tasks and examples they select influence how students come to view, develop, use, and make sense of mathematics*". This implied that in designing task to be discussed, a teacher should have assured that it was aimed to improve students' mathematical comprehension in case of the subject matter concerned on. Hence, in order to complete the tasks, students needed to think independently on how to apply particular formulation, algorithm, and procedures related to particular concepts and comprehensions. Besides, in designing tasks, a teacher had to consider on how big chances the students might have to reveal significant ideas of mathematics. Giving mathematical tasks with appropriate level of difficulty might encourage students' development, improve the use of a series of reasoning activities, and enhance students' mathematical thinking.

Increasing a comfort learning environment. It refers to a conducive learning environment for students to propose and share their ideas and it may improve the quality and the quantity of discussion [5]. In a discussion, students would try to explain and provide some reasons that correlate to their ideas and ask their classmates to give responses or clarification on ideas they shared [6]. [7] found that through a small group discussion, a student' intensity to express his/her thought might increase. This might bring big opportunity for students to develop their mathematical communication skills.

Providing chances for students to deliver and explain their ideas and thoughts. [4] asserted that effective teachers would encourage their students to explain and justify and stand their solutions on. With teacher's guidance, students would learn how to use mathematical ideas, language, and methods.

Encouraging students to actively participate in a discussion. An effective and meaningful discussion would require students to carefully listen to, proceed, and understand other students' ideas and thoughts [5]. Students need to be familiar with giving responses over the answer that another participant had shared in order to make the subject matter meaningful to learn [8]. Discussion is an interaction among students or between teacher and students in order to solve a problem or examine a particular concept. Therefore, teachers need to keep the discussion on track based on the learning objectives. [9] presented 5 practices -which may effectively help teachers to use students' responses to improve students' comprehension- in facilitating and leading a class discussion. These five practices stressed on the importance of planning. It noted that through planning, a teacher may anticipate the likelihood of students' participation, prepare any possible responses from students, and make particular decisions related to how to organize students' presentation and make an agenda of mathematics, respectively. The practices involved anticipation, monitoring, selection, arranging, and connecting students' various responses.

Therefore, the researcher was interested in investigating a teacher's mathematical communication in providing chances for students to propose and explain their ideas and be active in a discussion dealt with anticipative, observative, selective, and connective actions. The subject matter selected for this study was trigonometry (the rule of sine). Trigonometry is a very important subject, not only in mathematics but in other fields. That is, to understand the concept of trigonometry requires students' understanding of algebra and geometry. This study aimed to thoroughly describe a teacher's mathematical communication profile in giving chances for students to propose and explain their ideas and be active in discussion. The result would be useful as means to develop other teachers' mathematical communication skills. It was expected to be a reference for teachers to improve their mathematical communication with students and develop students' mathematical communication skills as well.

2. Method

This study is qualitative-descriptive. It aimed to describe a teacher's mathematical communication profile in providing chances for students to propose and explain their ideas and be active in discussion. A high-school junior teacher with 1-5 year teaching experiences. It observed the subject's mathematical communication in a teaching process the subject organized, and then a deep interview was conducted respectively in order to see and further investigate on how the subject's mathematical communication was in providing chances for students to propose and explain their ideas and be active in discussion. The results of both observation and interview were then analyzed and described thoroughly in the form of teacher's mathematical communication profile in providing chances for students to propose and explain their ideas and be active in discussion.

3. Result

In this chapter, the researcher would describe data informing the subject's mathematical communication. It was collected through observation and in-depth interview. It was presented in 2 sections, including the subject' mathematical communication in providing responses (answer) on students' difficulties dealt with accuracy, be optimal, and be specific and providing chances for students to propose and explain their ideas and be active in a discussion dealt with anticipative, observative, selective, and connective actions.

The result of observation and interview showed that the teacher teachers started the discussion by giving tasks to the students. She gave the tasks after explaining on how to derive the rule of sine along with its application. The tasks was in the form of problem containing 5 essay problems which related to the rule of sine (Figure 1). Problem 1 required the students to determine the length of one of the sides of a triangle, Problem 2 and 4 asked them to determine the magnitude of one of the angles of a triangle, Problem 3 asked them to determine the comparison of length for the sides of a triangle, and Problem 5 asked them to determine the cosine of one of angles of a triangle. Each group must do the task based on their group number, thus, it made them complete a different problem for each other. It was aimed to make the students have better understanding on the stages of the rule of sine. The teacher claimed that the task should be completed in group and those who had completed their tasks might come forward and present their works. The teacher provided the tasks for group discussion and the representative of each group had to present their group works respectively. This was aimed to make them familiar with presentation and class sharing, and to improve their comprehension on the given material (i.e., the rule of sine). The group consisted of 5 pupils who were heterogeneous on their mathematical skills. They had been classified into groups since the class started.

After each group got the task sheet and began to discuss the problem, the subject led the discussion with several phases, including (1) walking around each group, (2) asking some groups that seemed ready to complete and present their works, (3) discussing the group works, (4) repeating the 2nd and 3rd phases until all groups had successfully completed and presented their works.

IOP Conf. Series: Journal of Physics: Conf. Series **947** (2017) 012020 doi:10.1088/1742-6596/947/1/012020

The subject walked around each group while monitoring their works. It was aimed to give the students chances for asking anything they did not get yet dealing with the given task. It was possible for groups to ask while the subject walked around the students. Encountering such situation, the subject gave a direct approach. It did happen in this study. A member of a group raised her hand indicating that she had something to ask. Knowing this situation, the subject provided an accurate responses by immediately approaching the student. Dealing with this, the subject immediately approached the student with a belief that, in each group, there must be one smart student who well-understood the given material, some students with moderate level of intelligence and some others who were completely blind with the rule of sine. In addition, the subject also gave a responses toward a student who proposed a question by approaching her in front of the class.

Noted that the subject's responses dealt with the students' questions, which related to the procedures of problem-solving. In this case, the students were encouraged to discover the answer of their question by themselves. Besides, she gave it (explanation) in specific manner –both orally and written- with a consideration that if the students were still confused with spoken explanation, a written explanation would be applied for better understanding.

After several minutes of group discussion, the subject would be selective by giving chances for groups that had completed their tasks to present their works. It was aimed to motivate other groups to present their works as well, in addition to a consideration of time.

Next, the subject asked all the students to pay attention on her and correct the works of their group mates. She showed the answer keys through power point and tried to match the group works with the answer keys. This was aimed to see whether or not the group works were correct before an appreciation and support were given toward groups and students. She continued the process by correcting the procedures of presented works. When it was found that the students' works were less, the subject would give chances for them to correct their works. It was aimed to make them see in which part they got fault on their works. For instance (in another context), the subject provided a correction for a student's work by changing $BC.\frac{\sqrt{2}}{2} = 12.\frac{\sqrt{3}}{2}$ into$BC.\frac{1}{2}\sqrt{2} = 12.\frac{1}{2}\sqrt{3}$. In addition, to introduce the student with the related symbols, she noted $\sin\alpha$ as $\sin A$, $\sin\beta$ as $\sin B$, and $\sin\gamma$ as $\sin C$. After having a discussion, she reminded the students to be careful and accurate in completing the tasks. After each of the groups came forward presenting their works and be corrected, the subject asked the students to make a conclusion on the rule of sine along with its usage in order to make sure that the expected learning objectives had been successfully reached.

Task 1. Determine the length of BC on the following triangle!

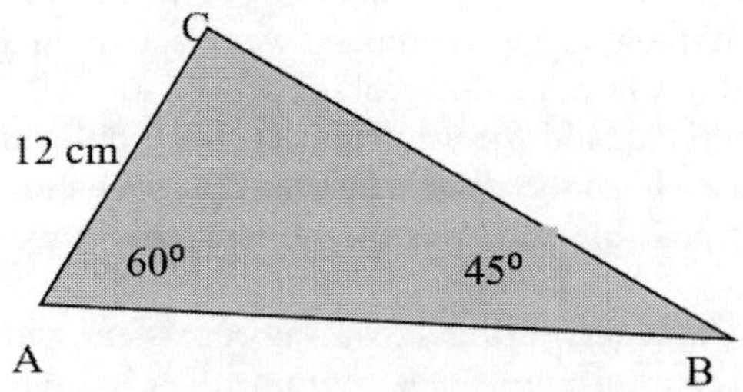

Task 2. Determine the length of BC on the following triangle!

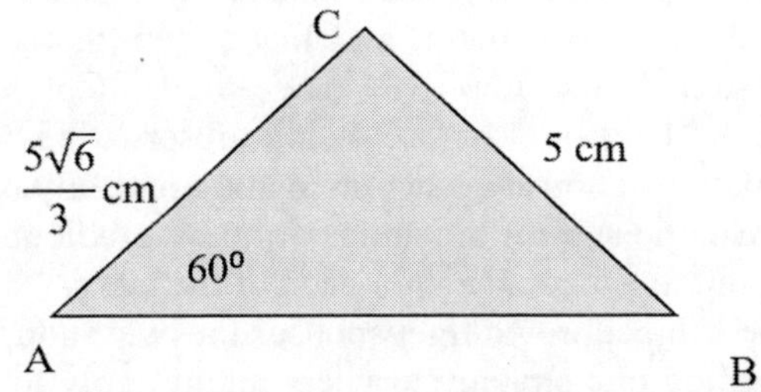

Task 3. Pay attention on the following triangle. Determine the length comparison between AB and BC!

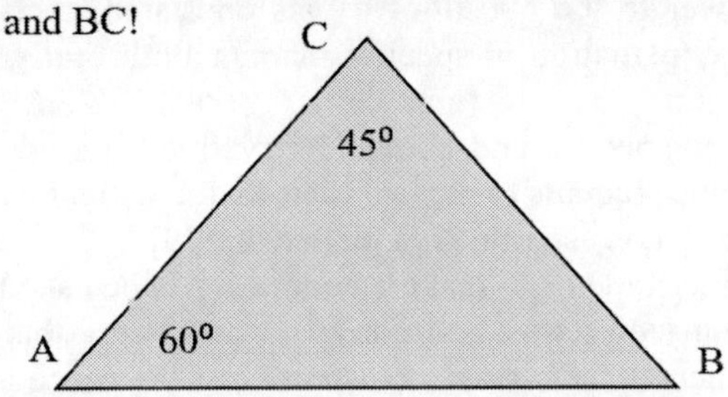

Task 5. Pay attention on the following triangle. Determine the value of cosine C!

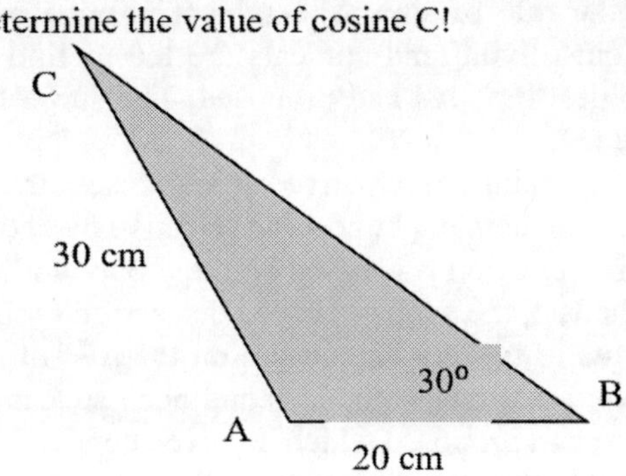

Task 4.
Triangle PQR has three sides (p, q, and r) if the length of $p = 16$ cm, $r = 8\sqrt{2}$ cm, and $\angle R=30°$. Determine the magnitude of $\angle P$!

Figure 1. The tasks was in the form of problem containing 5 essay problems which related to the rule of sine

4. Discussion

This section, the subject's mathematical communication skill in providing responses (answer) on students' difficulties dealt with accuracy, be optimal, and be specific. Accuracy dealt with the teacher's soundness in providing responses. Be optimal dealt with the informed information as a kind of responses on several difficulties the students might encountered. It involved information on students' errors along with the justifications and by considering the scale of priority which referred to particular points on the learning objectives. Be responses dealt with the kinds of ways applied to provide the responses. It could be in the form of oral or written. he subject' mathematical communication in providing chances for students to propose and explain their ideas and be active in a discussion dealt with anticipative, observative, selective, and connective actions. Be anticipative required the teacher to be responsive with various strategies the students might propose for problem-solving. Be observative required the teacher to observe to what extend the students completed the given tasks and to what extend their involvement in a group discussion. Be selective made the teacher enable to select one of the students to present his/her work, and it was based on the assessment and contribution he/she gave to reach the learning objectives. Be connective required the teacher to help students link their works with their classmates.

IOP Conf. Series: Journal of Physics: Conf. Series **947** (2017) 012020 doi:10.1088/1742-6596/947/1/012020

The subject providing chances for students to propose and explain their ideas and be active in discussion, was consistent with [9] that teachers should be observative on students' works along with their involvement during discussion. Teachers might do an observation by walking around while the students did their tasks (whether in group or individual). Such action might encourage students to reveal their ideas, clarify them, and make sure that each of their group members were active in group discussion. It also implicitly made the students consider any points of problems they needed to see closely. Therefore, before being observative, teachers should be anticipative. [9] defined this anticipative action as an action to anticipate any responses from the students toward the given tasks. In this case, the subject did anticipation by predicting any possible solutions the students might make by preparing the answer key for each of the tasks.

The subject providing responses (answer) addressing students' difficulties, was consistent with [10] that argued one strategy teachers might apply to provide a responses was by immediately giving the responses for questions or difficulties the students proposed. The subject immediately approached the student with a belief that, in each group, there must be one smart student who well-understood the given material, some students with moderate level of intelligence and some others who were completely blind with the rule of sine. The subject provide an optimal response which related to the procedures of problem-solving and students are led to find the answer to the question. It was consistent with the strategies that [10] had proposed. The subject giving explanation in specific manner both orally and written.

After being anticipative on responses or students' responses and being observative on students' works, teachers might become selective by choosing some students to present their mathematical works and the selection was based on the group works. This was consistent with a theory that [9].

The last, the subject endeed discussion by inviting the students to make a conclusion. She asked the students to make a conclusion on the rule of sine along with its usage in order to make sure that the expected learning objectives had been successfully reached. This was consistent with the procedures that [9] had proposed which involved connecting various responses or responses from the students as an attempt to help them capture out the correlation between the solutions their solutions and other students' ideas, as well as the idea of mathematics.

5. Conclusion

Based on the analysis of this study, it concluded that the subject's mathematical communication in facilitating and guiding a class discussion was as follows. The Subject's Communication In Providing Responses (Answer) Addressing Students' Difficulties. She gave responses in accurate manner by immediately providing answers for the students' questions. She gave responses in optimal manner by providing explanations that corresponded to the students' questions. She also led the students to find the answers of their questions by themselves. She gave responses in specific manner –both orally and written- with a consideration that if the students were still confused with spoken explanation, a written explanation would be applied for better understanding. The Subject's Communication in Providing Chances for Students to Present and Explain Their Ideas and to be Active in a Discussion. She anticipatively gave chances for students to express their ideas and thoughts. She had estimated the possible solutions they would propose in the form of answer keys for each task. She observatively gave chances for students to express their ideas in order to make them feel comfort to ask about any notion they did not understand yet. She selectively gave chances for students to express their ideas by pleasing some groups who seemed ready to present their works in front of the class. It was aimed to motivate other groups to present their works as well, besides considering the time provided. She connectively gave chances for students to express their thoughts by leading them to make a conclusion on the given material. It was aimed to convince them that the expected learning objectives had successfully been reached out.

IOP Conf. Series: Journal of Physics: Conf. Series **947** (2017) 012020 doi:10.1088/1742-6596/947/1/012020

Acknowledgments
This research has been supported by Direktorat Riset dan Pengabdian Masyarakat, Direktorat Jenderal Penguatan Riset dan Pengembangan Kementerian Riset, Teknologi, dan Pendidikan Tinggi, Corresponding to the research contract No:086/SP2H/P/K7/KM/2016

References

[1] Baroody A J 1993 *Problem Solving, Reasoning, and Communicating (K-8)* (New York: Mc Millan Publishing Company)
[2] National Council of Teachers of Mathematics 2000 *Using NCTM's Content and Process Standards*http://www.math.vt.edu/people/lloyd/curriculum/teaching_activity2.pdf. [Accessed on 4 Oktober 2013]
[3] Clark KK *et al* 2005 Strategies for Building Mathematical Communication in the Middle School Classroom: Modeled in Professional Development, Implemented in the Classroom *Current Issues in Middle Level Education CIMLE* **11** (2) pp 1-12
[4] Anthony G and Margareth W 2009 Characteristics of Effective Teaching of Mathematics: A View from the West *New Zealand Journal of Mathematics Education December* **2** (2) pp 147-164
[5] Brown A L and Campione J C 1995 *Guided Discovery in a Community of Learners* (Cambridge: In K. McGilly Eds Classroom Lessons: Integrating Cognitive Theory and Classroom Practice)
[6] Capacity Building Series (CBS) 2010 *Communication in the Mathematics Classroom* (Ontario: The Literacy and Numeracy Secretariat)
[7] Brenner M E 1998 Development of Mathematical Communication in Problem Solving Groups By Language Minority Students *Bilingual Research Journal,* **22**:2, 3, & 4
[8] Pugalee D A 2001 Using Communication to Develop Student's Literacy *Journal Research of Mathematics Education* **6** (5) pp 296-299
[9] Smith M K and Stein M K *2011 Five Practices for Orchestrating Productive Mathematics Discussions* (USA: The National Council of Teachers of Mathematics, Inc)
[10] Brookhart M S 2008 *How To Give To Your Students Effective Feedback* (USA:Association for Supervision and Curriculum Development (ASCD))

IOP Conf. Series: Journal of Physics: Conf. Series **947** (2017) 012021 doi:10.1088/1742-6596/947/1/012021

Improving Science Process Skills for Primary School Students Through 5E Instructional Model-Based Learning

N L Choirunnisa[1,*], P Prabowo[1], S Suryanti[1]

[1]Primary Education Department, Surabaya State University, Surabaya, Indonesia

*Corresponding author e-mail: nadiachoirunnisa@mhs.unesa.ac.id

Abstract. The main objective of this study is to describe the effectiveness of 5E instructional model-based learning to improve primary school students' science process skills. The science process skills is important for students as it is the foundation for enhancing the mastery of concepts and thinking skills needed in the 21st century. The design of this study was experimental involving one group pre-test and post-test design. The result of this study shows that (1) the implementation of learning in both of classes, IVA and IVB, show that the percentage of learning implementation increased which indicates a better quality of learning and (2) the percentage of students' science process skills test results on the aspects of observing, formulating hypotheses, determining variable, interpreting data and communicating increased as well.

1. Introduction

In this 21st century, the graduates are expected to have double skills of current competence. The four important skills trained by the teacher to the students are 4C (Critical thinking and problem-solving, Communication, Collaboration, Creativity and Innovation) [1]. In science subject, the four skills are integrated to the science process skills. The science process skills can equip students' thinking skills [2]. The basic science process skills consist of observing (calculating, measuring, classifying, finding relationship of space/time), hypothesizing, planning the experiment, controlling variables, interpreting data, drawing conclusions (inference), predicting, applying, and communicating [3].

When viewed from the skills that are included into science process skills, it is clear that the science process skills are important in science learning as the basis for improving the other thinking skills which are much more complex such as 4C that has been explained before. It is also supported by Aydinli (2011) which states that teachers should pay great attention to students' science process skills by providing hands-on activities' task oriented to improve their understanding of a science concept [4]. The science subjects in primary school is less meaningful if it is taught through lecturing method in which teacher actuvely gives information to the students. The curriculum has ordered that the learning process must be actively build concepts that learning is done by the students themselves and the process can be reinforced by a scientific approach. Natural science is a kind of knowledge that has special characteristics to learn the factual of natural phenomena, both events and causal relationships. Natural science is a subject in primary school that encourages students to have knowledge, ideas and concepts that are organized. Thise knowledge, ideas, and concepts are related to the natural surroundings, obtained from experience through a series of scientific processes such as investigation, preparation and

IOP Conf. Series: Journal of Physics: Conf. Series **947** (2017) 012021 doi:10.1088/1742-6596/947/1/012021

presentation of ideas. The scope of natural science materials in the primary school generally covers two aspects, namely scientific work and concept mastery.

The learning process done by students who are actively seeking and finding their own natural science's concepts can certainly not be separated from the science process skills that they have. Students need the science process skills in their activities to find the right science concepts. Science process skills become the driving wheel of discovery and development of facts and concepts as well as the growth and the development of attitudes and values [3]. Students who are active in learning science through the use of process skills, discussions, and experiments, gain more meaningful learning and dispel perceptions of rote learning [5].

The expectations, which are conveyed by various experts, are not in line with the facts occurred in the field. Students do not understand correctly the essence of formulating hypotheses and correlations between variables [6]. Students are still wondering when they should formulate hypotheses and they do not know the reason of doing that. Moreover, they are unable to understand the correlation between the dependent and independent variables [6]. Science process skills are not only about formulating hypotheses and determining variables but also involving other skills such as observing, interpreting data, and communicating that are trained in this study. Science process skills can not be considered as an insignificant thing in science learning. Students need to be familiarized with the science process skills since primary school in order to obtain the correct concepts and to make other high-level thinking skills continues to increase. The higher the students' science process skills, the better the students' conception status (Knowing the Concept) [7].

The low science process skills score of elementary students is an urgent problem to be solved. If it continues, then students cannot improve the other skills that are needed in their life later on. Students need to be facilitated in the learning process that allows them to explore what they want to know freely and actively. The 5E (Engagement, Exploration, Explanation, Elaboration, Evaluation) learning model provides a suitable learning phase for improving the science process skills [8]. The reason of choosing the model is because the 5E instructional model contains structured activities starting from the preparation or agreement in which the initial concept of the students is known. The starting point is then followed by an exploratory phase in which the students perform activities to look for evidence through experiment and at the same time their science process skills are trained in this phase. In the explanatory phase, students explain, in their own language, the results they gain through presentations and discussions. In the elaboration phase, students apply the concept to new issues. In the evaluation phase, students' final abilities are revealed through tests [8]. The constructivist learning of 5E instructional models have a better impact than traditional teaching models [9]. The 5E instructional model-based learning with various methods such as conceptual change, POE (Predict-Observed-Explain), cartoon concept, and animation can improve the concept mastery of students [10]. The integration of science activities with 5E instructional model may cause cognitive conflict that leads the students to try to solve the science problem by using their thinking skills [11].

2. Method

This type of research is a pre-experimental study. The subjects of this research were two classes of 4th graders, class IVA and IVB at SD Negeri 1 Menganti, Gresik. The research design used was One Group Pre-test and Post-test Design:

$$O_1{}^{a} \qquad X^{b} \qquad O_2{}^{c}$$

(a) O_1 = Pre-test, (b) X= 5E instructional model-based learning, (c) O_2 = Post-test

The percentage of the learning implementation was obtained by dividing the number of aspects that were implemented in the learning with the total of aspects multiplied by 100%. The percentage of students' science process skills was obtained by dividing the total score with maximum score multiplied by 100%. The pre-test and post-test scores of science process skills were employed to find the difference by using paired sample t-test. The test was done to be able to determine whether there are significant differences in the science process skills' test results, before and after learning process.

3. Results

3.1 The Implementation of Learning

The implementation of learning is observed by two observers. The lessons were conducted in three meetings. In addition, the lessons obtained a very good percentage. The learning outcomes can be seen in Table 1 and Table 2.

Table 1. Percentage of the Implementation of Using 5E Instructional Model-based Learning in Class IVA

Meeting	Result	
	Implemented (%)	Not Implemented (%)
1	87.50	12.50
2	91.67	8.33
3	95.83	4.17

Based on Table 1 it can be seen that the percentage of learning implementation increases which indicates a better quality of learning.

Table 2. Percentage of the Implementation of Using 5E Instructional Model-based Learning in Class IVB

Meeting	Result	
	Implemented (%)	Not Implemented (%)
1	86,96	13,04
2	95,65	7,35
3	100	0

Based on Table 2 it can be seen that the percentage of learning implementation increases which indicates the quality of learning is getting better as well as what occurs in the class IVA.

3.2 Students' Science Process Skills

The tests of science process skills are conducted twice, before treatment (pre-test) and after treatment (post-test). The difference of those test results can be seen in Figure 1 and Figure 2.

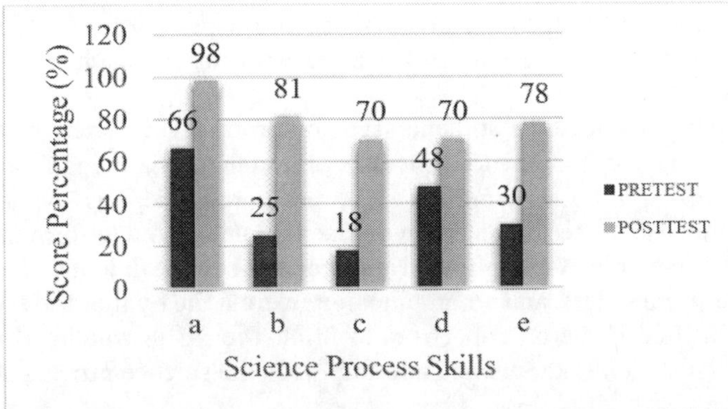

Figure 1. Test Result of Science Process Skills in Class IVA. (a) Observing. (b) Formulating hypotheses. (c) Determining variables. (d) Interpreting data. (e) Communicating.

Based on Figure 1, it can be seen that the percentage of science process skills' score increases from pre-test to post-test in class IVA. In addition, the pre-test average score is 35.25 and the post-test one is 78.10. The post-test average score is better than the pretest one. Based on the result of the t-test, the t-value is -44.361 and it is revealed that the t-table value is 2.093. The value of t-arithmetic is in rejection area H_0. Hence, it can be concluded that there are significant difference of students' science process skill between before and after the treatment involved in learning.

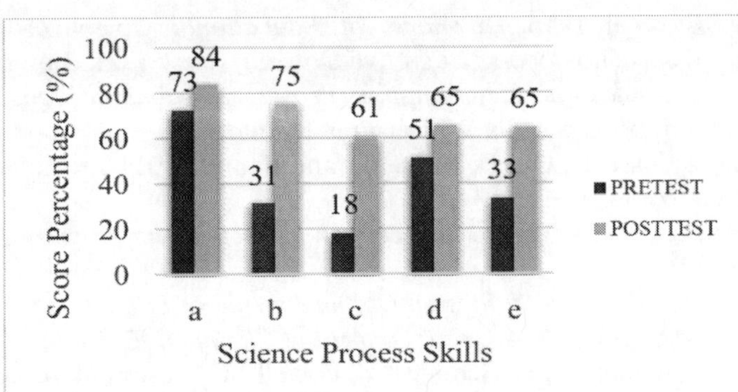

Figure 2. Test Result of Science Process Skills in Class IVB. (a) Observing. (b) Formulating hypotheses. (c) Determining variables. (d) Interpreting data. (e) Communicating.

Based on Figure 2, it can be seen that the percentage of science process skills' score increases from pre-test to post-test in class IVB. The pre-test average score is 38.75 and the post-test one is 80.15. The post-test average score is better than the pretest one. Based on the result of t-test, the t-value is -49.225. In addition, it is also revealed that the t-table value is 2.093. The value of t-arithmetic is in the rejection area H_0. Hence, it can be concluded that there are significant difference of students' science process skill between before and after the treatment.

4. Discussion

Through the results of this study, it can be seen that the percentage of learning implementation increases which indicates better quality of learning both in the class IVA and IVB. It is because the teacher has been given a briefing and training on how to implement the learning based on 5E instructional model. There is no obstacle that occurs, therefore, the teachers can carry out the learning well. The students also look enthusiastic in following the learning process because it provides a fun and enjoyable learning. Students' interests in learning also increases which leads to improvement of students' motivation in learning science [12].

The increase of the percentage of students' science process skills' test results occurred because students were already trained to use science process skills such as observing scientifically, formulating hypotheses, determining variables, interpreting data, and communicating experimental results in the elaboration phase. Moreover, the teacher also gives an example of how to formulate the hypothesis and determine the variables first which then followed and imitated by the students. The science process skills such as observing, interpreting data, and communicating were done by the students so that it only needs to be observed and clarified if there is an error. In brief, the 5E instructional model-based learning successfully enriches the students' knowledge and improves the science process skills [13].

5. Conclusion

Based on the results of the study, it can be concluded that 5E instructional model-based learning is effective to improve students' science process skills seen from the improvement of students' test results.

Acknowledgments
Dr. Hendratno, M.Hum; Prof. Dr. Muslimin Ibrahim; M.Pd, Dr. Wahono Widodo, M.Si; Prof. Dr. Prabowo, M.Pd; and Dr.Suryanti, M.Pd. who have guided and assisted this research.

References

[1] Griffin P 2015 *Assessment and Teaching of 21st Century Skills: Methods and Aprroach* (London: Springer)
[2] Bloom Benjamin S et al 1956 *Taxonomy of Educational Objective: The Classification of Educational Golas Handbook I Cognitive Domain.* (New York: Longmans, Green and Co)
[3] Semiawan C 1992 *Pendekatan Keterampilan Proses: Bagaimana Mengaktifkan Siswa dalam Belajar* (Jakarta: PT Gramedia Widiasarana Indonesia)
[4] Aydinli E Dokme, Unlua Z K, Ozturk N, Demir and Benli E 2011 *Procedia Social & Behavioral Sciences* **15** 3469-3475
[5] Harlen W and Elstgeest 1994 *Unesco Sourcebook for Science in Primary School* (Paris: Unesco Publishing)
[6] Darus F dan Saat R M 2014 *The Malaysian Online Journal of Educational Science* **2** pp 20-26
[7] Choirunnisa, N L and Suyono 2014 *Unesa Journal of Chemical Education* **3** 153-160
[8] Bybee R W, Taylor J A, Gardner A, Scotter P V, Powell J C Westbrook A, and Landes N 2006 *The BSCS 5E Instructional Model: Origins and Effectiveness* (Colorado: Colorado Spring)
[9] Alshehri M A 2016 *IOSR Journal of Research and Method in Education* **6** 43-48
[10] Cepni S and Sahin C 2012 *Eurasian Jounal of Physics and Chemistry Education* **4** 97-127
[11] Lin J L, Cheng M F, Chang Y C, Ying-Chi, Hsiao-Wen Chang J Y, Lin D M 2014 *Eurasia Journal of Mathematics* **10** 415-426
[12] Nur M 1998 *Teori-Teori Perkembangan* (Surabaya: Unesa University Press)
[13] Karsli F and Ayas A 2014 *Procedia-Social and Behavioral Sciences* **143** 663-668

IOP Conf. Series: Journal of Physics: Conf. Series **947** (2017) 012022 doi:10.1088/1742-6596/947/1/012022

Refractive Thinking Profile In Solving Mathematical Problem Reviewed from Students Math Capability

M Maslukha[1,*], A Lukito[1] and R Ekawati[1]

[1] Departement of Mathematic, Universitas Negeri Surabaya
*Corresponding Author: maslukhamaslukha@mhs.unesa.ac.id

Abstract. Refraction is a mental activity experienced by a person to make a decision through reflective thinking and critical thinking. Differences in mathematical capability have an influence on the difference of student's refractive thinking processes in solving math problems. This descriptive research aims to generate a picture of refractive thinking of students in solving mathematical problems in terms of students' math skill. Subjects in this study consisted of three students, namely students with high, medium, and low math skills based on mathematics capability test. Data collection methods used are test-based methods and interviews. After collected data is analyzed through three stages that are, condensing and displaying data, data display, and drawing and verifying conclusion. Results showed refractive thinking profiles of three subjects is different. This difference occurs at the planning and execution stage of the problem. This difference is influenced by mathematical capability and experience of each subject.

1. Introduction

Refraction is an event of change or distortion of direction of the wave as it passes through the boundary plane of medium having different density[1]. Using light metaphor the refraction process is illustrated from reflective thinking toward critical thinking then refractive thinking[2]. Illustrations of refractive thinking processes such as figure 1.

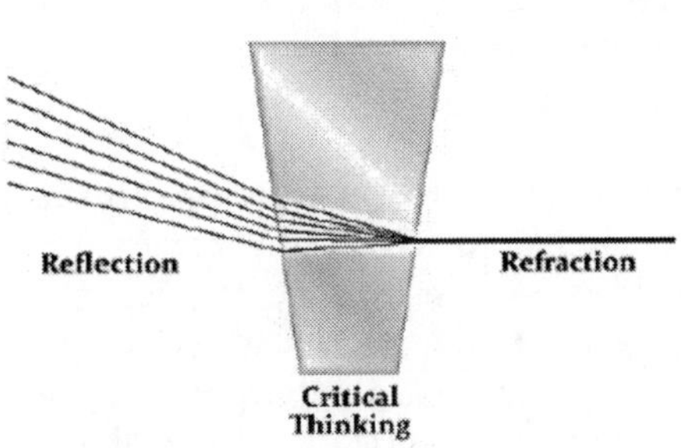

Figure 1. Refraction illustration
Figure was adopted from reference[3]

Based on illustration of figure 1, occurrence of refraction through reflection and critical thinking. Reflection is the first step in the knowledge development cycle. Reflection is the process of constructing experience in dealing with problems to acquire knowledge to solve it[3]. Reflection can be understood as a general term for the intellectual activities in which individuals are engaged to explore their experiences in order to gain a new understanding[4]. Concerning reflective thinking, reflective thinking is a situation that occurs when a person is perplexed and investigates to find a solution to the problem. Perplexity is the uncertainty and difficulty in solving the problem[5].

After reflective thinking, the next process towards a more active mental is called critical thinking. Critical thinking demonstrates capability to evaluate relevant information obtained when performing systematic, cautious, and efficient reflections in solving the problem[3]. The most important thing in critical thinking is to consider and evaluate some of the information obtained so that it is possible to make an decision[6]. While critical thinking is thinking that involves the ability to test, connect and evaluate[7]. In this level of thinking also includes collecting, organizing, remembering and analyzing information. Reflective thinking is part of a critical thinking process that specifically refers to the process of analyzing and making judgments about what happens[5][8][9][10], Reflection is the key of critical thinking, where reflection indicates that a person starts to think at a high level in three taxonomies that involves examining both process knowledge and results and producing new conclusions.

The third step in the knowledge development cycle is refractive thinking[3]. Refractive thinking is a learning transformation in which a person clarifies by providing new facts and evidence. The essence of refraction is the ability to see and identify experiences through different lenses[11]. The term refraction as an instrumental learning that is the process of learning that is done by changing, controlling and experimenting on the environment encountered and studying the resulting reactions[12]. Meanwhile, new knowledge generated from reflection and critical thinking[13]. From these ideas can be concluded that refractive thinking is the thinking process that generates decisions through reflective thinking and critical thinking.

Research related to refractive thinking has been done by Prayitno[3]. This research describes the refractive thinking process of students in solving math problems about decision-making. However in this research, subjects taken only one student and did not include differences in mathematical capability. In fact, differences in mathematical capability have an influence on the refractive thinking process of students in solving mathematical problems.

The purpose of this research was to get a description of refractive thinking profile of high, medium, and low mathematical capability in solving mathematical problems. The troubleshooting steps used in this study consisted of understanding the problem, making completion plan, executing settlement plan, and re-examining. The results of this research are expected to give information about the profile of refractive thinking of students in solving mathematical problems, so that information can be used as consideration for teachers in designing learning activities in accordance with refractive thinking profile with different levels of students' mathematical capability.

2. Method

This research is a qualitative descriptive research. Subjects in this study consisted of three students from the 8th grade of Islamic Junior High School. Three students are high, medium, and low mathematical capability based on mathematical capability tests. Mathematical ability test is taken from the national exam that is relevant to the 8th grade of Junior High School. To avoid having students answer truth by accident, mathematical capability test is made in essay form consisting of 10 questions with 90 minutes time. Description of the refractive thinking profile of students derived from problem-solving tests and semi-structured interviews. The problem-solving tests assigned to the subject are a matter of non-routine mathematical questions related to mathematical material. Interviews are used to clarify written data and explore information that may not be in the written data about students' refractive thinking profiles in solving mathematical problems. Triangulation technique used in this research is triangulation time. The distance between the first and second problem-solving tests is done within a period of about one week. And from the consistency of data collected then the data is analyzed through three stages namely, data condensation, data display, and drawing and verifying conclusion.

3. Results And Discussion

The problem was given to a student is:

Note the following parallelogram ADFE!

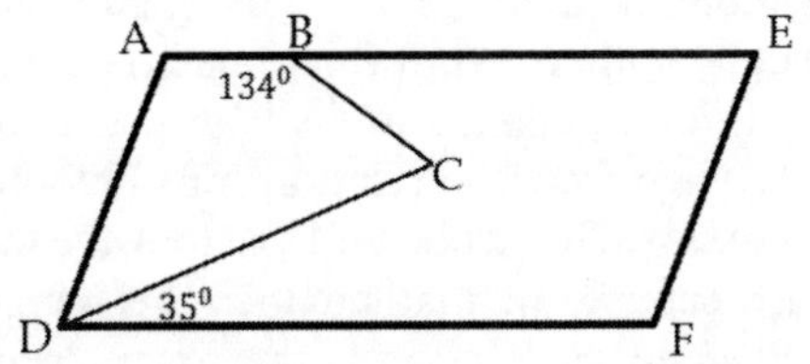

Find $m\angle BCD$.

Figure 2. Problem was given to Students

Based on data analysis results obtained refractive thinking process students with high mathematical capability in solving mathematical problems begins from the stage of understanding the problem. At this stage, students begin to reflect by gathering information and linking prior knowledge in interpreting the problems it receives. Once students are able to interpret the problem furthermore students connect the information they have with information on the matter, which indicates that the student is critical thinking. Furthermore from the information relationships obtained, from the refraction of students have a picture of the idea of settlement that will be used to answer the problem.

Pagano states that refraction occurs because of a reflection signaled by light passing through a medium that triggers the occurrence of critical thinking, so that light coming out of the medium is not the same as reflection[3]. This shows that the components of refractive thinking are reflection and critical thinking. Therefore, the refractive thinking process that students do with high math skills in understanding the problem is in accordance with the process proposed by Pagano.

In making the completion plan, students with high math skills propose more than one settlement strategy. It shows students doing refraction. In accordance with the opinion [3] which states the core of refraction is the ability to see and identify issues and experiences through different view.

Furthermore, the student uses the experience and knowledge possessed beforehand that is the parallelogram and rectangular character to solve the problem given in accordance with the strategy that he previously set. In addition to these methods, students use the concept of angular relationships such as opposite angles, angular positions and number of angles in triangles. This means that at the stage of carrying out the settlement plan, the student does the refraction process because he is able to solve the problem more than one way.

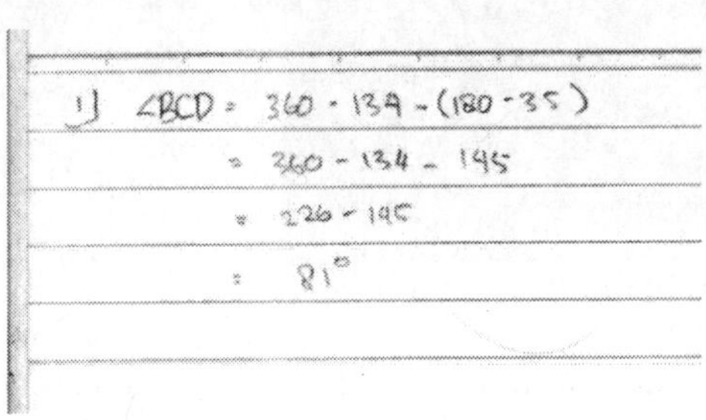

Figure 3. Results of the first completion by student with high mathematical capability

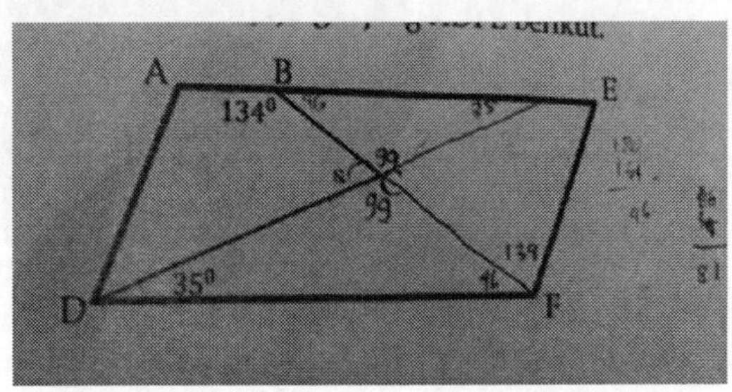

Figure 4. Results of the second completion by student with high mathematical capability

After the student has found the answer, the student checks by comparing the first strategy obtained with another strategy which has the same final result. So he felt very confident that the answer he got was correct. This shows at the re-examining stage, student does refraction.

Refractive thinking profile of medium mathematical capability subjects At the stage of understanding the problem, student begins to reflect by gathering information and connecting it with the knowledge of interpreting the problem. Through critical thinking, students are able to connect the information they have with information on the matter. Then from the relationships obtained, the results of the refraction is that student is able to have an idea of the idea that he will use to solve the problem. Stages of refractive thinking through which students are capable of mathematics are already in line with the phases of refractive thinking put forward by Pagano. Reflection is the first step in the knowledge development cycle [3]. Reflection is a process by which we perceive experience, frame it and gain meaning from it. Critical thinking is the second step in the knowledge development cycle. Critical thinking shows the ability to evaluate pertinent information and opinions gathered in the reflection phase systematically, directionally and efficiently to develop problem-solving skills. Furthermore, refraction is the third step in the knowledge development cycle.

After understanding the problem on the question, students begin to strategize based on the experience and knowledge they have acquired. Students had thought to use quadrilateral to solve the problem. However, he could not continue the idea until the end of the settlement. He had already eliminated the idea and started thinking of other strategies that can be used to solve the problem. Based on his previous experience and knowledge of the nature of triangles and angular relationships, students chose a trial and error to make help in order to find the angle they are asking for. Students tried again and again until finally he got the best solution. The refractive thinking process shown by medium mathematical capability students is in submitting a plan in accordance with [3] stating that refraction occurs when in the face of a problem can not immediately get a solution, allowing one to reflect. Reflection in this situation means connecting the problem it faces with the possible solution of the settlement obtained from the knowledge it possesses. Next, through critical thinking, one will identify, analyze, synthesize, and evaluate some alternative solutions obtained at the reflection stage. On refractive thinking someone offers an proper solution in solving the problem.

In the next stage students use the nature of triangles and angular relationships to execute the strategies that he set before in accordance with the knowledge he has received. Furthermore, students assess every step of the settlement done with a logical consideration. However, at this stage the student is only able to carry out the settlement plan in no more than one way.

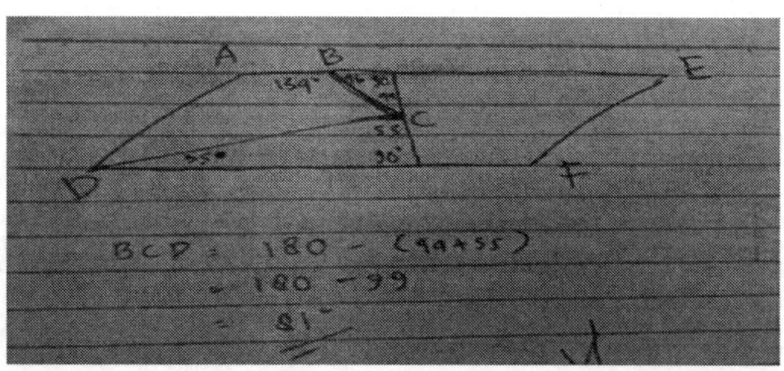

Figure 5. Results of Student with Medium Math Ability

After the student found the answer, the next step was checking the answer. While doing the examination, he found no calculation errors then he fixed it. But at this stage, the student did not doing refraction process because he does not check by comparing the method or first idea obtained by other method or ideas.

Refractive thinking profile of subject has low mathematical capability. At the stage of understanding the problem, students with low mathematical capability need to read the problem until repeatedly to understand the problem on the matter. Then, the student collected the information and connected it to the knowledge or experience it has gained to interpret the information on the problem. Unfortunately, students with low math skills do not have an idea of completion to solve the problem. So, at this stage students have not done the refraction process. He just did the reflection course.

Furthermore, in making the settlement plan, based on his experience, students had thought using the nature of the triangle to solve the problem. Unfortunately, students are unable to continue with this idea until the end of the settlement. So he eliminated the plan. Next he tried to make another strategic plan that is using trapezoidal properties to solve the problem. Because he only proposed one idea of completion, so that the refractive ability of the students did not appear. Students only do the reflection process.

After the student had successfully settled the plan of completion, the student also executed the plan. During the completion of the completion plan, student did not assess every step taken with logical consideration. So that students could not execute a strategy of completion well. So the last answer obtained by students is wrong. This means that students do not refract.

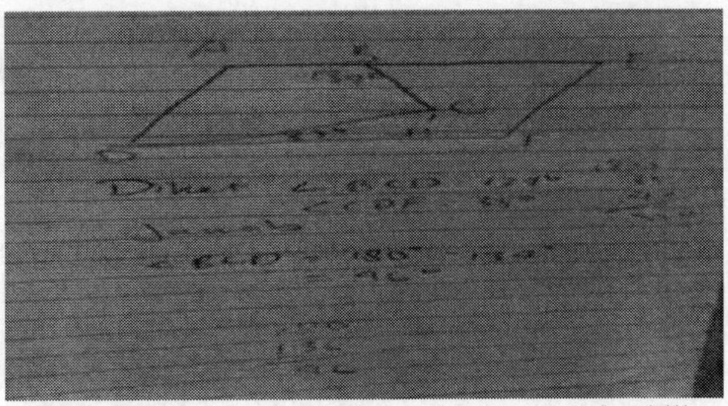

Figure 6. Results of Student with Low Math Ability

In the last stage, student did repetitive examination of the answers. But he was not careful so he did not realize a mistake he made. This shows that student is doing not critical thinking. So at this stage, student with low mathematical capability does not arrive at the refraction stage.

At each stage of problem-solving showed that student with low mathematical capability in solving mathematical problems did not arrive at the refractive phase of thinking. Because there is a critical stage of thinking that has not been passed. Because, process to do refractive thinking is reflection and critical thinking.

In general, the refractive thinking process of the three subjects is different. This difference occurs at the stage of planning and executing problem solving, this difference is influenced by the different levels of subject mathematics ability. Some research found that students' mathematical ability had an effect on the solution[14]. It also explains the need of mathematical skills when doing mathematical manipulations including understanding of concepts and procedural knowledge[15].

4. Conclusion

The results of this research show the different of refractive thinking process of students with high, medium, and low mathematical capability. At the stage of planning and executing problem solving, students with high mathematical capability are able to propose the idea of completion in more than one way and he is able to carry out all the plans he previously set well. While a student with medium mathematical capability is able to propose two ideas of completion. However, only a single idea of completion is carried out well. For student with low mathematical capability, he is only able to pursue one idea of completion and can not carry out the idea properly until the end of the settlement. From this research result, teachers are expected to deliver problem that potentially have many solutions to students with medium and low mathematical capability.

References

[1] Prayitno A, Sutawidjaja A, Subanji, and Muksar M 2015 Karakterisasi Berpikir Refraktif Mahasiswa Menyelesaikan Matematika tentang Data *Prosiding Seminar Nasional Matematika dan Pendidikan Matematika*

[2] Downey G 2005 *How to Guide and facilitate Self Reflective Practice in Re-Entry Programs* (Miami: Presented at CIEE Conference)

[3] Pagano M and Roselle L 2009 Beyond: Refraction and International Experiental Education *Frontiers: The Interdisciplinary Journal of Study Abroad* **18** pp 221

[4] Boud D Keogh R and Walker 1985 *Reflection: Turning Experience into Learning* (London: Kogan Page)

[5] Dewey J 1933 *How We Think A Restatement of the Relation of Reflective Thinking of the Education Process* (Boston: D C health and Company)

[6] Ennis R H 2011 *Critical Thinking* (NJ: Prentice-Hall)

[7] Krulick S and Rudnick J A 1995 *The New Sourcebook for Teaching and Problem solving in Elementary School* (Needam Heights: Allyn and Bacon)

[8] Taylor L 1992 Mathematics Attitude Development from A Vygotskian Perspective *Mathematics Education Research Journal* **4** pp 8-23

[9] Park J Y and Son J B 2011 Expression and Connection: The Integration of the Reflective Learning Process and the Public Writing Process into Social Network Sites *MERLOT Journal of Online Learning and Teaching* **7** pp 170-178

[10] Colley B M, Billics A R, and Lerch C M 2012 Reflection: A Key Component of Thinking Criticaly *The Canadian Journal for the Scholarship of Teaching and Learning* **3** pp 1-19

[11] Mezirow J 1990 *Fostering Critical Reflection in Adulthood: A Guide to Transformative and Emancipatory Learning* (San Francisco: Jossey-Bass)

[12] Sherlock J and Nathan 2004 *Producing Actionable Knowledge: Apllying Mezirow is Theory to the Managerial Learning Context, Academy of Management Best Conference Paper*

[13] Medeni T D 2012 Reflection and Refraction for Knowledge Management systems *International Journal of Ebusiness and Egovernment studies* **4**

[14] Ermawati 2015 *Proses Berpikir Reflektif Siswa SMA dalam Memecahkan Masalah Matematika Ditinjau dari Kemampuan Matematika Siswa* (UNESA: Tesis yang tidak dipublikasikan)

[15] Syaban M 2010 Menumbuh-kembangkan Daya Matematis Siswa *EDUCARE: Jurnal Pendidikan dan Budaya*

IOP Conf. Series: Journal of Physics: Conf. Series **947** (2017) 012023 doi:10.1088/1742-6596/947/1/012023

The Creativity of Reflective and Impulsive Selected Students in Solving Geometric Problems

R N Shoimah[1], A Lukito[2], T Y E Siswono[2]

[1] Primary Education Studies Program, Universitas Negeri Surabaya, Indonesia
[2] Mathematics Departments, Universitas Negeri Surabaya, Indonesia
Corresponding Author. E-mail: retnoshoimah@mhs.unesa.ac.id

Abstract. This research purposed to describe the elementary students' creativity with reflective and impulsive cognitive style in solving geometric problems. This research used qualitative research methods. The data was collected by written tests and task-based interviews. The subjects consisted of two 5th grade students that were measured by MFFT (Matching Familiar Figures Test). The data were analyzed based on the three main components of creativity; that is fluency, flexibility, and novelty. This results showed that subject with reflective cognitive style in solving geometric problems met all components of creativity (fluency; subject generated more than three different right-ideas in solving problems, flexibility; subject generated more than two different ways to get problem solved, and novelty; subject generated new ideas and new ways that original and has never been used before). While subject with impulsive cognitive style in solving geometric problems met two components of creativity (fluency; subject generated more than three different right-ideas in solving problems, flexibility; subject generated two different ways to get problem solved). Thus, it could be concluded that reflective students are more creative in solving geometric problems. The results of this research can also be used as a guideline in the future assessment of creativity based on cognitive style.

1. Introduction

Creativity is one of the most important competencies that must be developed through education. Cropley argued that creativity is the key of success in economy-global era and absolutely necessary in the 21st century [5]. Akgul and Kahveci also stated that creativity is an invaluable skill for the new century [1]. Creativity is the ability to produce any composition, product, or idea that is essentially new and never been used before [7]. In mathematics subjects, creativity is needed in problem solving, because the processes of problem solving are required high thinking levels. Krulik & Rudnick stated that the highest level of thinking is critical thinking and creative thinking [9]. Problem solving and creative thinking are closely related. Problem solving has creative aspects because creative thinking produces novel outcomes, and problem solving involves producing a new response to a new situation, which is a novel outcome [8]. Moreover, Silver suggested three components to assess the creativity of children and adults: fluency, flexibility and novelty [11]. Fluency in problem solving is the ability to find variety of correct solution. Flexibility in problem solving is the ability to use various methods to solve problems, while the novelty in problem solving is the ability to find new and different solutions or methods.

Everyone have different levels of creativity. Cognitive style is one of the factors that influence creativity. Sternberg stated that creativity is related to three psychological attributes; intelligence, cognitive style, and personality/motivation [12]. In addition, Al-Silami stated that creativity as an important element that supports the involvement of cognitive abilities, including problem-solving abilities [2]. The type of cognitive style that many students have is reflective and impulsive cognitive style. The results of Warli's research showed that the proportion of reflective-impulsive children is as many as 73.7% [9]. Rozencwajg & Corroyer's research show that the frequency of reflective-impulsive children is as many as 76% in one class [13].

This research describes the creativity of elementary students with reflective and impulsive cognitive style in solving geometric problems. The data was collected by written tests and task-based interviews. The geometric problem was solved in an open ended matter. Open-ended problem is very suitable for measuring student's creativity because it has many correct answers. Getzels and Jackson suggested one way to measure creativity ability that is open ended problems [6]. The data collected would be analyzed and described based on three components of creativity: fluency, flexibility, and novelty. Fluency is used to check the diversity of problem solving ideas. Flexibility is used to check the diversity of problem solving methods. Novelty is used to check originality of problem solving ideas and methods. Students are considered creative if they met one of the three components.

2. Method

This research is descriptive research with qualitative approach. Before conducting the research, research's instruments such as the geometry problem solving task sheets and interview guides were made and validated. The subjects consisted of two 5th grade students. Process to select research subjects started from the determination of research class, then applied MFFT (Matching Familiar Figures Test) to group students into four cognitive styles based on accuracy and speed (reflective, impulsive, fast accurate and slow inaccurate). Reflective and impulsive subjects are selected from the most reflective and most impulsive groups, the subject is also determined on the consideration that both of them have similar mathematical background and accomplishments, they are same in gender, and have good communication skills in order to succeed in the interviews to get a complete data of their creativity in geometric problem solving.

The data was collected by written tests and task-based interviews. Every subject was given two problem solving tasks and two task-based interviews. The type of written task was an open-ended problem solving with geometry material, those problem have more than one solution. The first problem-solving task asked the subject to make no less than three different garden design with two kind of flowers within an area of 24m^2, while the second problem-solving task asked the subject to make no less than three different pond designs with three kind of fish within an area of 18m^2.

Triangulation and member-check were used to check the credibility of the data. The data of students' creativity in geometric problem solving were assessed based on three components of creativity: fluency, flexibility, and novelty. Fluency in problem solving referred to the number of different right-ideas to get problem solved. Flexibility in problem solving referred to the different ways to get problem solved, while the novelty in problem solving referred to generate ideas or new ways that original and has never been used before.

3. Results

The results of this study describe creativity of reflective and impulsive student in solving geometric problems.

3.1 Reflective Student

Reflective student found five solutions at the first geometric problem solving task and six solutions at the second geometric problem solving task. Fluency component come to be fulfilled if student find no less than three solutions with different ideas. At the first geometric problem solving task, reflective student drew five garden designs with different size and shape. The first design (fs1) was combined by two rectangles with the sizes of 8m^2 and 16m^2, the second design (fs2) was combined by triangle and parallelogram with the sizes of 16m^2 and 8m^2, the third and fourth design (fs3) was combined by triangle and rectangle with the sizes of 16m^2 and 8m^2, the fifth design (fs4) was combined by kite and rectangle with the sizes of 16m^2 and 8m^2. The results of the first geometric problem solving task by reflective student showed on figure 1. At the second geometric problem solving task, reflective student drew six fish pond designs with different size and shape. The first design (fs5) was combined by three rectangles with the sizes of $3m^2$, $6m^2$, and $9m^2$, the second design (fs6) was combined by two triangles and trapezoid with the sizes of $3m^2$, $6m^2$, and $9m^2$, the third and fourth design (fs7) was combined by triangle, parallelogram, and square with the sizes of $3m^2$, $6m^2$, and 9m2, the fifth design (fs8) was combined by three triangles with the sizes of $3m^2$, $6m^2$, and $9m^2$, and the sixth design (fs9) was combined by two triangles and rectangles with the sizes of $3m^2$, $6m^2$, and $9m^2$. The results of second geometric problem solving task by reflective student showed on figure 2. The fluency component of reflective subject is shown by making four designs with different ideas. Reflective subject makes several different designs and uses different ideas on each design. Although the subject made different ideas, all of designs are interconnected. New ideas are always based on previous ideas, starting from a simple idea to more complex idea. So, the relationship between ideas used is very visible.

Flexibility component come to be fulfilled if student find no less than three solutions with different ways. At the first geometric problem solving task, reflective student drew three garden designs with different ways to draw and different ways to determine sizes and shapes of design. At the second geometric problem solving task, reflective student drew four fish ponds design with different ways to draw and different ways to determine sizes and shapes of the design. Flexibility component of reflective subject is shown by using more than two different design methods. The first method gives contribution to the second method.

Novelty component come to be fulfilled if student create new solution that never been known before. Creativity in solving geometry problems in component novelty can be identified from the aspect of ideas and methods. Reflective subject only met novelty on the aspect of idea. The design idea was created by reflective subject that is novel and original. At the first geometric problem solving task, reflective student create four original and new designs (fs2, fs3, and fs4). At the second geometric problem solving task, reflective student create two modification designs (fs7, fs9) and two new designs (fs6, fs8).

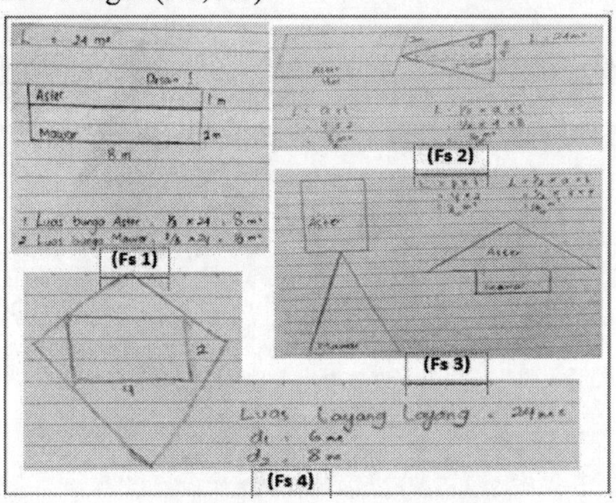

Figure 1. The Result of Cognitive Student Problem Solving Task I.

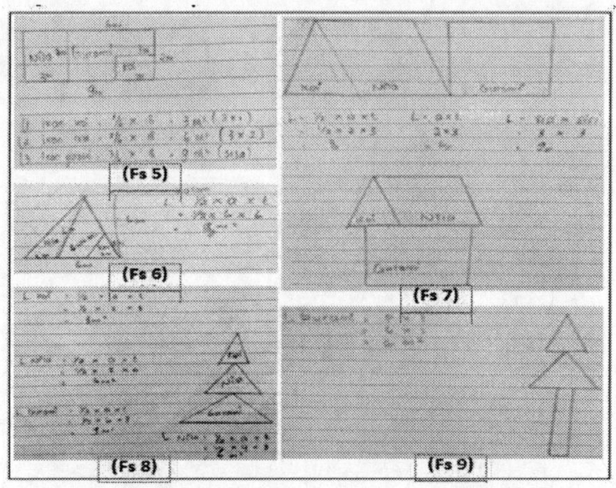

Figure 2. The Result of Cognitive Student Problem Solving Task II.

The summary of creativity of reflective subject in geometric problem solving can be seen in the Table 1.

Table 1. A Summary of Reflective Subject Creativity.

Component of Creativity	Geometric Problem Solving	Description
Fluency	√	Reflective subject made no less than three solutions with different ideas.
Flexibility	√	Reflective subject made no less than three solutions with different methods.
Novelty	√	Reflective subject made different and original design.

Table 1 showed that reflective subject met all component of creativity, she has fluency criteria, flexibility criteria, and novelty criteria in solving geometric problem.

3.2 Impulsive Student

Impulsive student found four solutions at the first geometric problem solving task and four solutions at the second geometric problem solving task. Fluency component come to be fulfilled if student find no less than three solutions with different ideas. At the first geometric problem solving task, impulsive student drew five garden designs with different size and shape. The first and second design (fs1, fs2) was created by rectangle with the size of $24m^2$, the third design (fs3) was created by rhombus with the size of $24m^2$, and the fifth design (fs4) was created by parallelogram with the sizes of $24m^2$. The results of first geometric problem solving task by impulsive student showed in figure 3. At the second geometric problem solving task, impulsive student drew four fish pond designs with different size and shape. The first and second design (fs5, fs6) were created by rectangle with the size of $18m^2$, the third design (fs3) was created by rhombus with the size of $18m^2$, and the fifth design (fs4) was created by rhombus with the size of $18m^2$. The results of second geometric problem solving task by impulsive student showed in figure 4. Fluency component of impulsive subject is shown by making two designs with different ideas. Impulsive subject makes several different designs and uses different ideas on each design. Second ideas based on first ideas, the difference is only in the shape and size of the design.

Flexibility component come to be fulfilled if student find no less than three solutions with different ways. At the first geometric problem solving task, impulsive student drew three garden designs with different ways to draw and different ways to determine sizes and shapes of design. At the second geometric problem solving task, impulsive student drew three fish pond designs with different ways to draw and different ways to determine sizes and shapes of the design. Flexibility component of impulsive subject is shown by using more than two different design methods. The first methods contribute to the second method.

Novelty component come to be fulfilled if student create new solution that never been known before. Creativity in solving geometry problems in component novelty can be identified from the aspect of ideas and methods. Impulsive subject created usual design. The design that was made have been known before.

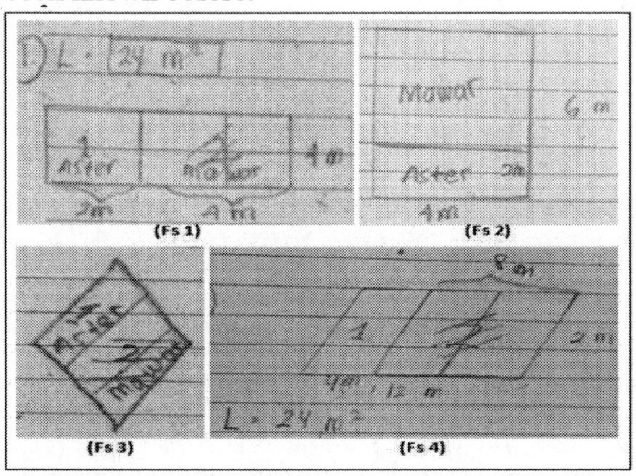

Figure 3. The Result of Cognitive Student Problem Solving Task I.

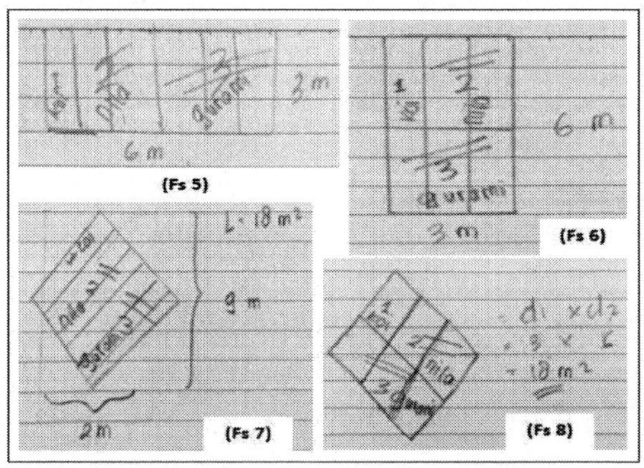

Figure 4. The Result of Cognitive Student Problem Solving Task II.

The summary of creativity of impulsive subject in geometric problem solving can be seen in the Table 2.

Table 2. A Summary of Impulsive Subject Creativity

Component of Creativity	Geometric Problem Solving	Description
Fluency	√	Reflective subject made no less than three solutions with different ideas.
Flexibility	√	Reflective subject made no less than three solutions with

Component of Creativity	Geometric Problem Solving	Description
		different methods.
Novelty	-	Reflective subject made different and original design.

Table 2 shows that impulsive subject met two component of creativity; she has the fluency criteria and flexibility criteria in solving geometric problem.

4. Discussion

The results of this research is in accordance to Al-Silami research which stated that reflective students are more creative than impulsive students, reflective students have a higher matter on the components of fluency, flexibility, originality, and elaboration [2]. The results of Warli research also showed that the students' creativity profile of reflective cognitive style in solving geometry problem tends to be high, while the students' creativity profile of impulsive cognitive style in solving geometry problem tends to be very low.

The results show that reflective students use their cognitive function better than impulsive students, in addition, reflective students are able to provide many different alternative answers and ways, even the solution provided met the criteria of novelty (original idea and has never been used before). Answers with diverse ideas and ways also indicate that reflective students are able to understand the meaning of the problem well. This is in accordance with the statement of Ancillotti which stated that based on cognitive processes, reflective individuals use analytic processing mode, whereas impulsive individuals use holistic processing modes [3]. Borkowski, Peck, Reid, and Kurtz also revealed that reflective individuals showed superior meta-memory compared with impulsive individuals. Reflective individuals student shows cognitive maturity and preference for analytic processing, unlike impulsive individuals student who exhibits no cognitive maturity and indicates preferences for holistic processing [4].

Differences in reflective and impulsive student creativity also indicate that cognitive style is one of the factors that influence creativity. Accordingly, creativity can be investigated from several perspectives: 1) personality differences, 2) differences in cognitive style or ability, and 3) social psychology [13]. Moreover, Sternberg explained that creativity is related to three psychological attributes; intelligence, cognitive style, and personality/motivation [12].

5. Conclusion

Based on the explanation above, it can be concluded that reflective student met all component of creativity in solving geometric problems. Reflective student met fluency component because she found no less than three solutions with different ideas, she also met flexibility component because she found no less than three solutions with different ways, and she met novelty component because she created different solution that is new for her. In other way, impulsive students met only two components of creativity in solving geometric problems. Impulsive student met fluency component because she found no less than three solutions with different ideas, she also met flexibility component because she found no less than three solutions with different ways, and she did not meet novelty because her finding solution never been known by her.

References

[1] Akgul S and Kahveci N G 2016 A Study on the Development of a Mathematics Creativity Scale *Eurasian Journal of Educational Research* **62** pp 57-76 (Doi: 10.14689/ejer.2016.62.5)

[2] Al-Silami T A 2010 *A Comparison of Creative Thinking and Reflective-Impulsive Style in Grade 10 Male Students from Rural and Urban Saudi Arabia* (Retrieved from http://vuir.vu.edu.au/16016/1/Al_Silami_PhD. pdf)

[3] Ancillotti J P 1985 Dimension Reflexive Impulsive De La Personnalité Et Fonctionnement Cognitif (Reflective Impulsivity Style and Cognitive Functioning) *Cahiers de Psychologie Cognitive* **5** pp 667-686

IOP Conf. Series: Journal of Physics: Conf. Series **947** (2017) 012023 doi:10.1088/1742-6596/947/1/012023

[4] Borkowski J G, Peck V A, Reid M K and Kurtz B E 1983 Impulsivity and Strategy Transfer: Metamemory as Mediator *Child Development* **54** pp 459-473

[5] Career Center Maine Department of Labor 2004 *Todays Work Competence in Maine* (Retrieved from http://www.maine.gov/labor/lmis/pdf/Essential WorkCompetencies.pdf.)

[6] Getzels J W and Jackson P W 1962 *Creativity and Intelligence: Exploration with Gifted Children* (New York: Wiley)

[7] Hurlock E 2004 *Psikologi Perkembangan* (Jakarta: Gramedia Pustaka)

[8] Isaksen S G 1983 toward a Model for The Facilitation of Creative Problem Solving *Journal of Creative Behaviour* **17** pp 18-30

[9] Krulik S, Rudnick J and Milou E 2003 *Teaching Mathematics in Middle School: A Practical Guide* (England: Lyx)

[10] Rozencwajg P and Corroyer D 2005 Cognitive Processes in the Reflective Impulsive Cognitive Style *The Journal Of Genetic Psychology* **166(4)** pp 451-463

[11] Silver E A 1997 Fostering Creativity Through Instruction Rich in Mathematical Problem Solving and Thinking in Problem Posing *Electronic Edition ISSN Journal 1615-679X* 29(3)

[12] Sternberg R J 2006 the Nature of Creativity *Creativity Research Journal* 2006 18(1) pp 87-98

[13] Warli 2010 *Profil Kreativitas Siswa yang Bergaya Kognitif Reflektif dan Siswa yang Bergaya Kognitif Impulsif dalam Memecahkan Masalah Geometri* (Surabaya: Unesa)

IOP Conf. Series: Journal of Physics: Conf. Series **947** (2017) 012024 doi:10.1088/1742-6596/947/1/012024

The Physiological Profile of Junior Soccer Players at SSBB Surabaya Bhakti

M Nashirudin[1] and N W Kusnanik[1]

[1]Universitas Negeri Surabaya, Indonesia
Corresponding author: niningwidyah@unesa.ac.id

Abstract. Soccer players are required to have good physical fitness in order to achieve optimum accomplishment; physical fitness stands as the foundation of technical and tactical proficiency as well as the mental maturity during the matches. The purpose of this study was to identify the physiological profile of junior soccer players of SSB Surabaya Bhakti age 16-17. The research was conducted at 20 junior soccer players. This research was quantitative with descriptive analysis. Data were collected by testing of physiological (anaerobic power and capacity including explosive leg power, speed, agility; aerobic capacity: cardiovascular endurance). Data was analyzed using percentage. The result showed that the percentage of explosive leg power of junior soccer players were 30% (good category), speed was 85% (average category), right agility was 90% (average category), left agility was 75% (average category). On the other hand, the aerobic power and capacity of the junior soccer players in this study was 50% (average category). The conclusion of this research is that the physiological profile of junior soccer players at SSB Surabaya Bhakti age 16-17 was majority in average category.

1. Introduction

Recently, physiological profile of team game sport has been an interesting research area. In Indonesia it had concentrated on female field Hockey [1], male volley ball players [2], junior female volley ball players [3], junior high school studemts [4]. But, in Indonesia this research area is still rare concentrated on soccer players. Soccer School (SSB), is a school to learn about the game of soccer and it is a sport organization, especially soccer which serves to develop the potential of the athletes as well as being soccer coaching place for early age [5]. Soccer school coaching is an organization that staged an early age, so they must have some standard components that supported and fulfilled by SSB. The main goal of SSB is to provide a solid foundation on how to play soccer correctly, including forming attitudes, personality, and behavior.

In soccer game activities like kicking the ball, controling the ball, running after ball, chasing opponents, deceiving movement and passing opponents require energy. Bompa [6] states that the sports performance basically depends on several components of which is a physical component. Martens [7] proved that the physical condition of the support towards the success of the implementation of process skills in the sport of soccer, are among other powers (strength), velocity (speed), durability (endurance), dynamic power (power), flexibility (flexibility) and agility (agility) but the Australian Institute of Sport, Rebecca Tanner, Christoher Gore-Physiological Tests for Elite Athletes-2nd Edition of Human Kinetics [8] says that general physical condition is very diverse, but the needs of soccer player ages 17 down as follows: explosive power, speed, agility and aerobic endurance.

2. Methods

This research is quantitative research with descriptive approach method. The instrument used in this research is Jump MD to test power, run 20 meters to test speed, arrowhead agility to test agility, and (Multistage Fitness Test) MFT to test VO2Max.

IOP Conf. Series: Journal of Physics: Conf. Series **947** (2017) 012024 doi:10.1088/1742-6596/947/1/012024

3. Results

Based on the calculation result by using manual calculation and by using computer program, data resulted from the measurements of physical condition at each of physical components can be elaborated further in Table 1.

Table 1. Percentage of Category Explosive Muscle Power Limbs

No	Achievement (*Joule*)	Category	Frequency	Procentage
1	>5450	Very Good	1	5%
2	4940 – 5449	Good	6	30%
3	4450 – 4939	Medium	5	25%
4	3950 – 4449	Less	5	25%
5	3450 – 3949	Bad	3	15%

Based on the table, the category of explosive muscle strength of legs of SSB players Surabaya Bhakti Age of 16-17 is as follows: 1 player (5%) has an explosive muscle limb with the category "Very Good", 6 players (30%) Having a limb power with the category of "Good", 5 players (25%) have explosive limb muscles with the category "Medium", 5 players (25%) have muscle limb explosiveness with the category "Less" and 3 players 15%) has a tension explosive with the category of "bad".

Table 2. Percentage of Category Speed Running 20 Meters

No	Achievement (Seconds)	Category	Frequency	Procentage
1	<2,71	Good	1	5%
2	3,43 - 2,72	Medium	17	85%
3	>3,44	Less	2	10%

Based on Table 2, it is known that the percentage of the speed category of players SSB Surabaya Bhakti Age 16-17 is as follows: 1 player (5%) have running speed with the category of "Good", 17 players (85%) have run speed by category "Medium", 2 players (10%) have running speeds with the category "Less"

Table 3. Percentage of Right Arrowhead Agility

No	Achievement (Seconds)	Category	Frequency	Procentage
1	<7,32	Good	0	0%
2	9,22 – 7,33	Medium	18	90%
3	>9,23	Less	2	10%

Table 4. Percentage of Left Arrowhead Agility

No	Achievement (Seconds)	Category	Frequency	Procentage
1	<7,66	Good	0	0%
2	9,11 – 7,67	Medium	15	75%
3	>9,12	Less	5	25%

Based on Tables 3 and 4, it is known that the percentage of agility category of players SSB Surabaya Bhakti Age 16-17 is as follows: Arrowhead Agility right as many 0 players (0%) have agility with category "Good", 18 athletes (90%) have agility with category "Medium", 2 players (10%) have agility with the category "Less". Arrowhead Agility left as 0 players (0%) have agility with the category of

"Good", 15 athletes (75%) have agility with the category "Medium", 5 players (25%) have agility with the category "Less".

Table 5. Percentage Aerobic Capacity Cardiovascular Endurance Category

No	Achievement (Ml/Kg/Mnt)	Category	Frequency	Procentage
1.	L12 S12 (57,3)	*Excellent*	0	0%
2	L11 S6 (51,9)	*Above Average*	4	20%
3	L9 S2 (43,9)	*Average*	9	45%
4	L7 S6 (38,5)	*Below Average*	1	5%
5	< L7 S3 (< 37,4)	*Poor*	6	30%

Based on Table 5, it is known that the percentage of endurance category of players SSB Surabaya Bhakti Age 16-17 is as many 0 players (0) "Excellent" category, 4 players (20%) have endurance with "Above Average" category, 9 players (45%) have the endurance of the category "Average", 1 player (5%) have endurance with the category "Below Average", and 6 players (30%) have endurance with the category "Poor".

4. Discussion

In this part, we will discuss the results of research on the decomposition of the physiological profile of junior soccer players at SSB Surabaya Bhakti age 16-17 and physical condition measured by tests of anaerobic power and capacity including explosive leg power, speed, agility; aerobic capacity: cardiovascular endurance.

The physical condition is the ability that deal with the physical demands of an exercise to perform optimally [7]. Good physical condition can support good soccer game, because the better level of fitness or endurance player, the better the delivery of oxygen throughout the body when players perform the game of soccer activity. Some components of the physical conditions to support the activities in the game of soccer include the explosive power, speed, agility and endurance.

4.1. Anaerobic Power And Capacity Including Explosive Leg Power

Results of leg muscle explosive power tests conducted on players SSB Surabaya Bhakti as measured by test Jump MD, showed that the average yield of leg muscle players SSB Surabaya Bhakti is 4633.33 Joules. Based on the classification norm criteria, it can be said that the leg muscle explosive power players SSB Surabaya Bhakti fall into the "Medium" category. In soccer, explosive strength is needed for a player, because in soccer they need to jump, dribble and especially kick. Therefore, a player must have enormous power in order to achieve the desired performance.

4.2. Speed

The results of speed tests conducted on players SSB Bhakti Surabaya age 16-17 were measured with a test run 20 M showed that the average speed of the player is 3.05. Based on the classification criteria norm can be said that the player's speed SSB Surabaya Bhakti Age 16-17 fall into the "Medium" category.

4.3. Agility

Agility test results conducted on players SSB Surabaya Bhakti age 16-17 were measured with Arrowhead Agility tests showed that the average soccer player agility SSB Bhakti Surabaya Age 16-17 agility is 8.68 to 8.91 for the right and left agility. Based on the classification criteria norm can be said that the agility of players age 16-17 Bhakti Surabaya SSB fall into the “Medium" category for agility to the right and to the left.

In soccer players not only moves one way and not only ran a straight course. If a player has the agility then the player can make a move to change the direction of a sudden in a variety of positions, it makes opponents hard to keep up, and grab the ball when face to face for a player who has their agility high

IOP Conf. Series: Journal of Physics: Conf. Series **947** (2017) 012024 doi:10.1088/1742-6596/947/1/012024

levels can alter the direction unnoticed by players' opponent. Additionally players who have a good level of agility have the dexterity or agility in moving feet, the ability to change direction with or without the ball quickly, easy to master the techniques of high level. The results showed that the average agility SSB Bhakti Surabaya players Age 16-17 have agility with category "Medium" for agility right and left agility, given the importance of agility in soccer so that players be agile and nimble in running a game to achieve maximum results for myself and the team.

4.4. Aerobic Capacity

The results of durability tests conducted on players SSB Bhakti Surabaya age 16-17 were measured with Multistage Fitness Test (MFT) showed that the average durability player who symbolized VO_2Max is 40.56. Based on the classification criteria norm can be said that the resilience of the SSB Surabaya Bhakti players age 16-17 fall into the category of "Medium". Every soccer player should have a good level of VO_2Max, because as described above. VO_2Max is the body's ability to use and distribute oxygen in the body. The morc a person can distribute oxygen in the body then the longer someone can perform the activity, to perform the activity of the body requires muscles to move and muscles need energy to move the body, the energy obtained from burning substances present in the body and oxygen has a role as fuel for the combustion of such substances so as to energy and eventually the body can move. VO_2Max, the better the level of the better players in the players to overcome fatigue in making the game of soccer with his long time.

Besides getting a good supply of oxygen in the body more easily the player to perform movements that vary in soccer games such as kicking, heading, jumping, defending opponents way, chasing balls, chasing opponents, stealing the ball, herding and others within 2 x 45 minutes. If a player has a VO_2Max of adequate capacity, these players will be able to run farther and do more sprint in one game, and the result is the level of involvement of a player in a game would be great to win the game.

5. Conclusion

Based on the physiological profile of junior soccer players at SSB Surabaya Bhakti age 16-17 who have done so have obtained the following conclusions: Explosive power leg muscle from players SSB Surabaya Bhakti in age 16-17 has average 4633.33 Joules and standard deviation 620.95, in "Medium" category. Speed of players SSB Surabaya Bhakti in age 16-17 can be shown as average 3.05 seconds and standard deviation 0.28, in "Medium" category. Agility players SSB Bhakti Surabaya in age 16-17 on the right agility shows as average 8.68 seconds, and 0.53 standard deviation and on the left agility average 8.91 seconds and standard deviation 0.44, in "Medium" category. Cardiovascular Endurance (VO_2Max) of players SSB Bhakti Surabaya in age 16-17 can be shown as average 40.56 ml / kg / min and 3.82 standard deviation, in "Average" category.

6. REFERENCES

[1] Kusnanik N W 2005 Physical and Physiological Profil of the East Java Female Field Hockey Team *Jurnal Ilmu Keolahragaan* **13** 1
[2] Kusnanik NW 2010 Anthropometric and Physiological Performance of the Indonesia Male Vollyball Players *Journal of Sports Science and Technology* **10** 2
[3] Kusnanik N W 2011 The Profil of anthropometricel and Physiological Performance of the Girls Volleyball Players *Internasional Conference on Scientific melting for Welcoming 26 th South East Asian Games Solidarity for Unity Trough Sports*. Jakarta, Nov. 8.
[4] Kusnanik N W and Hartati H 2017 Physical and Physiological Profile of Junior High School Students in Indonesia *Sport Science* **10** 1
[5] Soedjono et al 1999 *Sepakbola Teknik dan Kerjasama* (Yogyakarta: IKIP Yogyakarta)
[6] Bompa T 1994 *Power Training For Sport* (Canada: York University Toronto Ontario, Second Printing)

[7] Martens R 2012 *Successful Coaching-4th Editions* (Human Kinetics Publishers, INC, Champaign, Illinois)
[8] Tanner R K and Christopher J G 2012 *Physiological Test For Elite Athletes* (Second Edition : Australian Institute of Sport)

Flexibility in Mathematics Problem Solving Based on Adversity Quotient

N A Dina[1], S M Amin[1] and Masriyah[1]

[1]Postgraduate Mathematic Education Program, Universitas Negeri Surabaya, Indonesia

[1]Email: nesadina@mhs.unesa.ac.id

Abstract: Flexibility is an ability which is needed in problem solving. One of the ways in problem solving is influenced by *Adversity Quotient* (AQ). AQ is the power of facing difficulties. There are three categories of AQ namely *climber, camper,* and *quitter*. This research is a descriptive research using qualitative approach. The aim of this research is to describe flexibility in mathematics problem solving based on *Adversity Quotient*. The subjects of this research are *climber* student, *camper* student, and *quitter* student. This research was started by giving *Adversity Response Profile* (ARP) questioner continued by giving problem solving task and interviews. The validity of data measurement was using time triangulation. The results of this research shows that *climber* student uses two strategies in solving problem and doesn't have difficulty. The *camper* student uses two strategies in solving problem but has difficulty to finish the second strategies. The *quitter* student uses one strategy in solving problem and has difficulty to finish it.

1. Introduction

One of the important topics in mathematics education is an effort to increase the ability of student to practice meaningfull learning process. The ability to use the flexible strategy is called flexibility. Flexibility is one of the indicators to measure one's creativity [1]. Flexibility is the ability to select appropriate procedures for particular problems and modify such procedures [2]. The individual's flexibility of mind can possibly be indicated in several ways by means of test. Although they have disappointments to establish a common factor of this type, the concept of flexibility and its probable opposite, rigidity, will not be downed [3].

Flexibility have a strong conection with the ability of solving problem. Creative thinking can be improved with flexibility, and with the use of problem solving techniques. The students should be asked challenging question to improve their flexibility in mathematics. It is commonly believed that divergent thinking exercises should be done in the class [4]. By the problem solving, the students are expected to create many ideas. The applied problems must possess many strategies so that it can arise students' flexibility.

The ability of solving problem seems to be an important one that must be developed through mathematic learning. By learning problem solving in mathematics, students will acquire ways of thinking, habits of persistence and curiosity, and confidence in unfamiliar situations that will serve them better outside of the mathematics classroom [5]. Problem solving is a process of accepting problem as a challenge to solve them [6]. Problem solving steps which is applied in this research including four steps. They are understanding the problem, devising a plan, carrying out the plan, and looking back [7].

Students' successful factor in solving problem is influenced by some ways, one of them is students' difficulty level. The difficulty encountered by two students might be different in the same problem. The difficulty is the students' opportunity to increase their ability. Therefore, the students must motivate themselve and try to overcome their difficulties in solving problem. Because of that, adversity quotient (AQ) is needed in learning process. Adversity quotient is an intelligence to cope with difficulties [2], in addition, adversity quotient can be called as the inteligence to change difficulty to be an opportunity [8]. There are three categories of AQ namely climber (high), camper (medium), and quitter (low) [9].

Through probem solving in this research, students are expected to be able to produce many strategies to solve their problems. The used problems must possess many ways to solve the problems so that it can show the criteria of flexibility. But, students must be categorized based on AQ before they are given problem solving tasks.

Some researches discussed about flexibility. One of them was Arslan & Yazgan. They compared flexibility by using many strategies to solve four of non routine mathematics problems which are given to 6th, 7th, and 8th grade students [10]. Many researches connected flexibility with student ability or student class level, but in this reseach flexibility is connected with adversity quotient. The aim of this research is to describe students' flexibility in mathematic problem solving based on Adversity Quotient. This research can be used as an informative solution for mathematic teachers about students flexibility in mathematics problem solving based on adversity quotient to design student learning and as the reference for other researches who will be having the same topic as this paper.

2. Research methods

This research was done in Junior High School students in Surabaya city, Indonesia. The subject of this research consisted of three students representing three categories of AQ namely climber student, camper student, and quitter student. The instrument of this research involves mathematics problem solving task used to describe flexibility. The data collection was done by way of the interview based on the task. Data analysis in this research consists of three steps namely condensing data, presenting data, and conclusion. The written and interview data are analized according to indicators of flexibility on every steps of problem solving.

3. Results and Discussion

3.1 Results

The result of this research showes that the three subjects were solving problems with different strategy. Climber student succeeded to solve the problem by two strategies. Camper student did the same as climber student did, but she failed to finish the second strategy and the answer of the two strategies weren't correct. Quitter student only used one strategy to solve the problem and her answer wasn't correct.

3.1.1 Climber Student

Climber student was able to implement two strategies namely *sketch and scheme* successfully with the correct answers. In sketch strategy, she understood the problem situation correctly. She understood that the parts of agliculture land among children were connected one another. Then, she tried to ilustrated by picture. She ilustrated rectangle as a whole agliculture land. She devided it based on the rules in the problem. She created variable L as a whole agliculture land and succeeded to finish the answer as seen in figure 1.

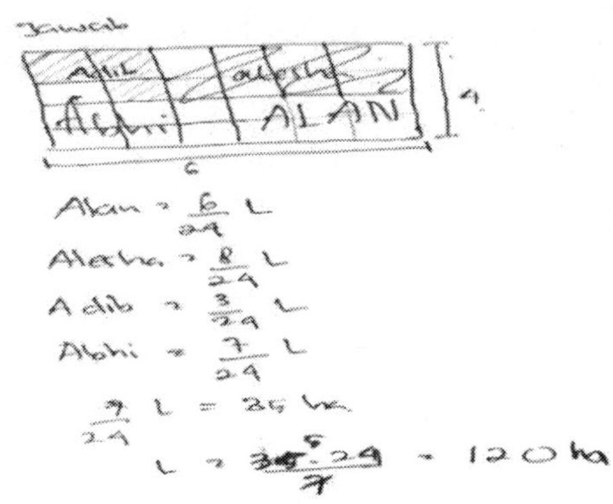

Figure 1. Sketch strategy

In the second strategy, she chose to use scheme to solve the problem. Her understanding about the problem was consistent in that scheme. She created three variables namely x, y, z. X means the whole aglucilture land, y means the rest of first land, while z means the rest of second land. Then, she started to determine the value of the three variables until the final answer, and she got $x = 120$ ha (as seen in figure 2).

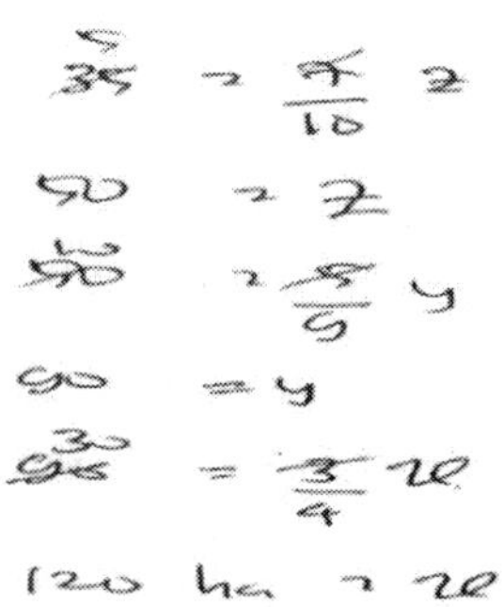

Figure 2. The final answer in scheme strategy.

3.1.2 Camper Student

Camper student was able to make two strategies namely *fraction addition and sketch*. In fraction addition strategy, she can implement it until she finish it but the answer is wrong. In sketch strategy, she was not able to implement it although her ways and understanding are correct. In fraction addition strategy, she didn't succeeded to understand the problem. She thought that the parts of agliculture land among children were not connected one another. Because of her missunderstanding, she answered the wrong answer in the next steps as seen in figure 3 and 4.

.

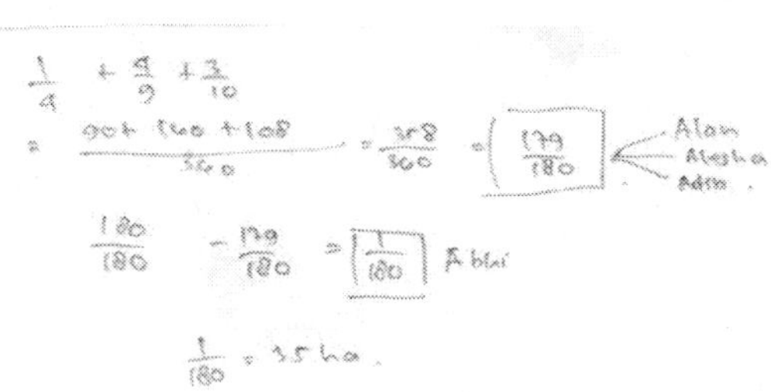

Figure 3. Fraction addition strategy

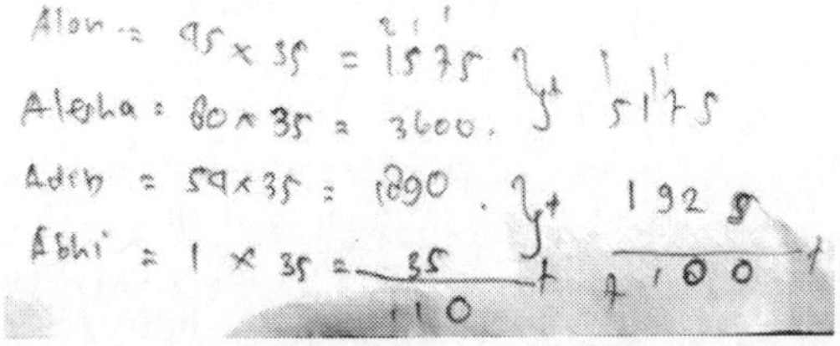

Figure 4. She tried to calculate the area of agriculture land of every child. She got the final answer of 7100 ha.

She tried to use ilustrate the picture in the second strategy. But, her understanding about the problem changed. She succeeded to understand that the parts of agliculture land among children were connected one another. But, she failed to finish the second strategy. She told that she got difficulties to continue because she had no ideas to connect the picture with the problem situation as seen in figure 5.

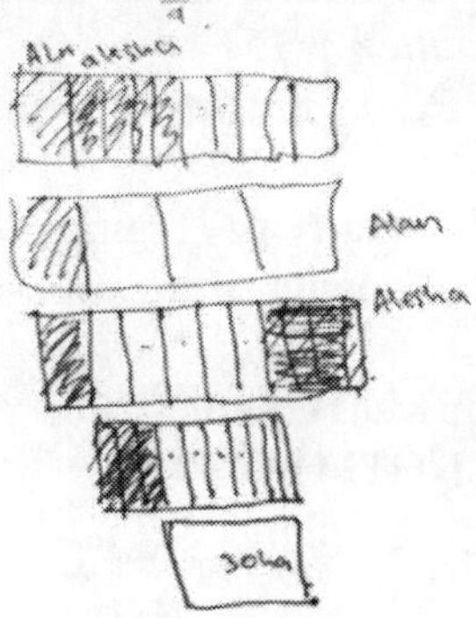

Figure 5. Sketch strategy.

3.1.3 Quitter Student

Quitter student didn't understand the problem. She only used one strategy to solve the problem.She thought that the parts of agliculture land among children were not connected one another. In the next step, she did simple calculation. She added the three part of Alan, Alesha, and Adib and then she determined the result in hectare. After that, she added it with Abhi's and she got 35,994 hectare. She tried to assume the final answer to became 36,4 hectare but she wrong.

3.2 Discussion

Based on the result and the interview from the three students, not all of the subjects encountered difficulties in every step in problem solving. But, the connection between flexibility an AQ become clear in the third steps of problem solving that is carring out the plan.

Student with high level of AQ (climber) was able to show flexibility by using two different strategies to solve the problem. Although at the beginning she tried to change the picture to be scheme, but she succeeded to finish her scheme. The climbers like challenge and wasn't easy to give up when they meet the difficulties [10].

Student with medium level of AQ (camper) was less able to show flexibility because she wasn't able to finish the second strategy. She decided to give up when she find difficulties. The campers like to be in the comfort zone. They feel satisfied when the succeed to achieve something but they refused to get something better [10].

Student with low level of AQ (quitter) wasn't able to show flexibility because she only used one strategy to solve the problem. She didn't want to try another strategy. The quitters didn't like the challenge. They didn't like to try something new in their lives [10].

4. Conclusion

Based on the research, it can be concluded that climber student can show flexibility and is able to cope with difficulties. Camper student is less able to show flexibility because she can't cope with difficulties in the second strategy. Quitter student isn't able to show flexibility because she only use one strategy to solve the problem.

5. References

[1] Siswono Tatag Yuli Eko 2007 *Penjenjangan Kemampuan Berpikir Kreatif dan Identifikasi Tahap Berpikir Kreatif Siswa dalam Memecahkan dan Mengajukan Masalah Matematika* Disertasi PPs UNESA Surabaya p 33

[2] Star J R 2006 Flexibility in the Use of Mathematical Procedure San Diego *American Educational Research Education* p 15

[3] Guilford J P 1967 *Creativity Research Past Present and Future* University of Soyth California p 7

[4] Kandemir Mehmet Ali 2007 Creativity Training in Problem Solving: A Model of Creativity in Mathematics Teacher Education *New Horizons in Education* Vol **55** No 3 Dec 2007 p 2

[5] National Council of Teacher of Mathematics 2000 *Principles and Standards for School Mathematics* (Reston VA: NCTM) p 101

[6] Hudojo Herman 2001 *Pengembangan Kurikulum dan Pembelajaran Matematika* (Malang : JICA) pp 23-25

[7] Polya G 1973 *How To Solve It (2nd Ed)* (Princeton: Princeton University Press) p 25

[8] Sopiatin Popi and Sahrani Sohari 2011 *Psikologi Belajar dalam Perspektif Islam* (Bogor: Ghalia Indonesia) p 27

[9] Stoltz G 2004 Adversity Quotient: *Mengubah Hambatan Menjadi Peluang* (T Hermaya) (Jakarta: PT Gramedia) pp 11-35

[10] Arslan Cigdem and Yazgan Yeliz 2015 Common and Flexible Use of Mathematical Non Routine Problem Solving Strategies *American Journal of Educational Research* 2015 Vol No 12 pp 1519-1523

Process Skill Assessment Instrument: Innovation to measure student's learning result holistically

K N Azizah[1], M Ibrahim[2], W Widodo[3]

[1]Primary Education, Pascasarjana Universitas Negeri Surabaya, Indonesia
[2]Biology Education, Universitas Negeri Surabaya, Indonesia
[3]Science Education, Universitas Negeri Surabaya, Indonesia

E-mail: [1]kharismaazizah@mhs.unesa.ac.id

Abstract. Science process skills (SPS) are very important skills for students. However, the fact that SPS is not being main concern in the primary school learning is undeniable. This research aimed to develop a valid, practical, and effective assessment instrument to measure student's SPS. Assessment instruments comprise of worksheet and test. This development research used one group pre-test post-test design. Data were obtained with validation, observation, and test method to investigate validity, practicality, and the effectivenss of the instruments. Results showed that the validity of assessment instruments is very valid, the reliability is categorized as reliable, student SPS activities have a high percentage, and there is significant improvement on student's SPS score. It can be concluded that assessment instruments of SPS are valid, practical, and effective to be used to measure student's SPS result.

1. Introduction

Science is defined as the nature and all phenomena that was encapsulated in a set of theories or concepts through a series of scientific process [1]. To perform the scientific process, it requires a certain amount of skill in science which was often called science process skills (SPS). SPS included observing, classifying, inferencing, predicting, looking for relationships, communicating, formulating hypotheses, performing experiments, controlling variables, and interpreting data [2]. SPS was a bridge to acquire and organize knowledge about environment and to link ideas that belong to students based on previous experience and new experience [3]. Science learning through SPS was expected to be a vehicle for students to learn about selves and natural surroundings, and further prospect of applying SPS in daily life [5]. Given the importance of SPS for students, SPS should be trained and pioneered through learning process in primary school.

In primary school, there were three domain: knowledge, attitudes, and skills, which should be implemented holistically to improve student's comprehension. However, until the recent development, domain of knowledge still dominated as the benchmark in measuring students' achievement [6]. Learning sciences was not always judged by using test assessment form to measure student's understanding. Teacher should collect information about student comprehension by assessing all three domains to give more accurate information about student's knowledge, skill and attitude. A set of variety assessment methods were needed to measure student's science process skill and conceptual understanding holistically. Then it was essential to develop a set of assessment instrument which not only measured student's material understanding but also student's capability on SPS. The developed assessement instrument was based on Process Skills Approach (PSA). The material chosen in this instrument is the topic electricity as a medium to train SPS. Electricity chapter was chosen because

IOP Conf. Series: Journal of Physics: Conf. Series **947** (2017) 012026 doi:10.1088/1742-6596/947/1/012026

understanding electricity not only requires reading the descriptions of the material, but also doing investigation activities or experiments so that students can understand the material well.

To create a set of assessment instrument, understanding how assessments were made, what type of assessments answered what questions, and how the data from assessments can be used to help teachers, students, parents, and other stakeholders to make decisions about teaching and learning are required [8]. A good assessment instrument can be sort out from these following traits of validity and reliability. First, content validity means that the assessment measures what it was intended to be measured and nothing more. Second, reliability refered to the consistency of the assessment result. Consistency means student result was the same when they take the same test on different occasions and when different but equivalent tests are taken at the same time or at different times. And third, student engagement and motivation means an assessment instrument should persuade student's interest.

There were several relevant research in developing an assessment instrumen for measuring student's SPS. Monica [9] aimed to develop a reliable and convenient test for measuring integrated science process skills competence in high schools. Shahali [10] also did a research to develop a test instrument to measure student's integrated SPS in Malaysia primary school. Temiz [11] developed multiple format instruments to measure both basic and integrated SPS to be applied in four high schools in Turkey. In contrast to the researches mentioned previuously, Rosana [6] developed a non-test assessment instrument of science process skills. This non-test instrument was applied in high school also. Three of four previous researches developed instrument to assess Integrated SPS in high school. However, basic science process skills were important to be trained in primary school as well [12]. And it was only one of four researcher who developed a non-test instruments. A set of assessment insrument was essentially needed to assess student's SPS thoroughly. This research aimed to develop a set of assessment instrument to measure basic science process skills of primary school students. The assessment instruments consisted of test and non-test instrument in order that student's SPS result can be viewed from two sides fairly.

2. Research method

SPS assessment instrument was developed as a product which can measure the level of student's SPS as whole. Assessment instruments in this research was developed using the 4-D model (four D model). There were four stages in the 4-D model, i.e. define, design, develop, and disseminate [13]. In this research, the stages which were done only up to the third stage, because the product was applied to a limited extent. One group pre-test post-test design was used in this research. Before learning process, student's SPS was assessed to know initial level of student's SPS. After treatment, teaching and learning activity using developed material, student's SPS was assessed to know the level of student's SPS after learning process. The research was carried out on Masangankulon Primary School in 2016/2017 academic year. Class used for research is V-D consisting of 25 students, 12 students were male and 13 were female students.

The data collected in this study were validity and reliability of assessment instrument, student's SPS activity, and the student's SPS score. Table 1 conveys variables, data, instruments, and analysis of this study.

Table 1. Data, instruments, and analysis.

Variables	Data	Instrumen	Analysis
Validity	Expert validation results	Validation sheet	Description of validity criteria (Ratumanan& Laurens, 2010).
	Readibility level	Fry Graphic	Description of readibility (Fry, 1977).
Practicality	Percentage of student's SPS activity	Observation sheet of student's SPS activity	Description of student's activity (Arikunto, 2010)

Effectivity	Student's SPS score	Test sheet	Description of gain score (Hake, 1998)

3. Results and discussion

Assessment instrument consisted of worksheet, observation sheet of SPS activity, and test sheet. The instrument was tested to find out its validity, practicality, and effectivity. Validation results included examination, assessment, and advice from the validator. Table 2 conveys the validity and reliability of assessment instruments.

Table 2. Validity and reliability of assessment instruments.

Instrument	Validity	Reliability	Category
Worksheet	4	83%	Very valid. Reliable.
SPS observation sheet	3	96%	Valid. Very reliable.
SPS test sheet	4	86%	Very valid. Very reliable.

Validating instruments needed to be done because it aimed to see if the instruments can be used by students or not [14]. Validation also aimed to find out the deficiencies of the instruments. With a valid instruments, teacher can conduct a good and fair assessment. Results of instruments validation indicated that the instruments is valid, reliable, and can be used with revision. The revision related to technical writing, i.e. fix a few typographical words. In addition, validator adviced to add other experiment tool like plastic ruler or any other material as a comparison of static electricity symptoms.

Readibility level of the instruments also determined its validity. Fry Graph was used to determine worksheets readability level. The main aspect in preparing learning materials was the linguistic aspect/readability [15]. Aspects of language/readability included: using Indonesia language according to Standardized Indonesian Language; the clarity of the language; and the easiness to read. Based on the coordinates (142.11) which were drawn on the Fry Graph, the assesment instruments was readable for fifth grade in primary school. It means that the worksheet can be read and understood by the fifth grade students. Readibility level of the instruments determined learning achievement because if students understood the contents of worksheet, so it means that students can understood the learning material without the help of others.

The practicality of assessment instruments was viewed by observing student activities. Student activities which represented a non-test result were expressed by percentage. Observed activities regarding to SPS included observing, making hypotheses, calculating, experimenting, classifying, inferencing, and communicating. Graphic 1 shows the observation results of student's activities.

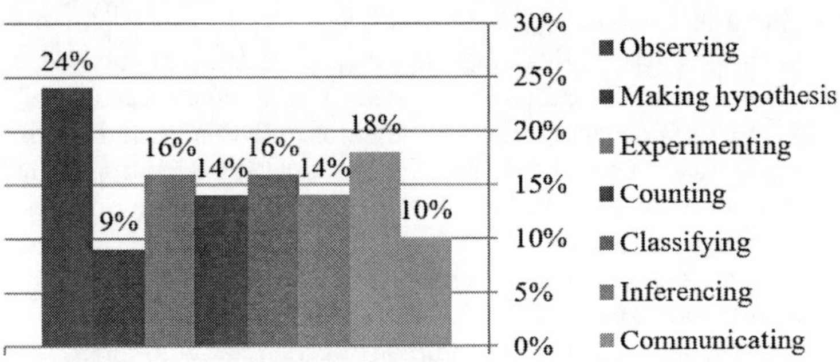

Graphic 1. Student's SPS activities portion

Based on Graphic 1, activity which had the highest occurrence was observing, which achieve 24%. Observing had the highest occurance because it was included into basic process skills. Observing is a response to various objects and events of nature using sensory perception [16]. Observing is often done by students in daily life so the students do not have difficulty in implementing it. The lowest accurance activity was making a hypothesis with an average percentage 9%. Making hypothesis had the lowest occurrence because students were still unfamiliar and have difficulty in making a hypothesis. Student did not fully comprehend the concept of hypothesis. Most of the students had the idea that a hypothesis is formulated after an experiment is conducted [17].

Next, the effectivity of assessment instruments was viewed from student's test result. Students were given a SPS test sheet consist of 7 SPS problems: observing, making hypothesis, experimenting, counting, classifying, inferencing, and communicating. Table 3 conveys student's test results.

Table 3. Score gain of student's SPS test results.

Science Process Skills	Average of Score Gain	Category
Observing	0,78	High
Making hypothesis	0,63	Middle
Experimenting	0,66	Middle
Counting	0,92	High
Classifying	0,96	High
Inferencing	0,65	Middle
Communicating	0,88	High
Average	0,78	High

Based on Table 3, there were four SPS which get high gain score i.e., classifying, counting, communicating, and observing. It is proved that high percentage of SPS occurance during learning process can affect student's test result. When students performed certain process skill activities more often, students will master the skills in the process and improve their test result. It can be said that process skills approach focuses on engaging students actively and creatively in the study [2]. Four skills which had the highest activity percentage also reach a high gain score of test result. Based on the findings, we can measure knowledge domain by using SPS test result and skill domain by using observation sheet of SPS activity. As an impact, assessing student's learning result can be done holistically, authentically, and fairly. This results confirmed that assessing SPS should be supported by a variety assessment methods in order to maximize student's SPS improvemnet [18].

4. Conclusion

Learning science in primary school can not be judged in knowledge domain only, but also skills and attitudes domain. Thus, Student's SPS needed to be assessed holistically by variety methods. The developed SPS assessment instrument consisted of test and non-test method. Non test method can be done by observing student's activity while doing learning process and test method can be done by testing student with SPS problems. Based on results and discussion, it can be concluded: 1) SPS assessment instrument was valid and reliable according to validator judgement and readibility level; 2) SPS observation sheets was practical to measure percentage of each SPS activity during learning process; 3) SPS test was effective to measure student's improvement from pretest to post-test in SPS. The results also showed that SPS which had a high percentage of occurance tended to have a high gain score. therefore, it was proven that skills and knowledge realm were related to one another.

Several recommendations can be given for further research: 1) before applying and measuring SPS, teacher should have fundamental understanding of SPS and how to introduce it to primary school students; 2) this research used electricity material as a media to apply SPS in primary school. Further research can use another material in science or any other subject as long as the topic not only required understanding the concept, but also practicing the skills; 3) SPS can be used along with any learning model to improve student's comprehension.

References

[1] Mariana I and Praginda W 2009 *Hakikat IPA dan Pendidikan IPA* (Bandung: PPPPTK IPA) p 6
[2] Semiawan C, Tangyong A F, Belen S, Matahelemual Y 1986 *Pendekatan Keterampilan Proses* (Jakarta: PT. Gramedia)
[3] Cavendish S, Galton M, Hargreaves L and Harlen W 1990 *Assessing Science in The Primary Classroom: Observing Activities* (London: Paul Chapman Publishing Ltd.)
[4] Khaeruddin and Sujiono E 2005 *Pembelajaran Sains* (Makassar: Badan Penerbit Universitas Negeri Makassar)
[5] Djojosoediro W 2010 *PJJ PGSD Universitas Negeri Surabaya* (Surabaya: Universitas Negeri Makassar) p 20
[6] Rosana D, Supahar S, Dewi D, Esmiyati E, and Sukmasari V 2016 *Proc. Int. Conf. On Research, Implementation, and Education of Mathematics and Science (Yogyakarta)* Vol. **03** p 37-42
[7] McColskey W and O'Sullivan R 2000 *How to Assess Student Performance in Science* (Greensboro: SERVE) p 4
[8] Evans S 2013 *Five Characteristics of Quality Educational Assessments* (Portland: NWEA) p 2
[9] Monica K 2005 A *Development and Validation of A Test of Integrated Science Process Skills for Further Education and Training Learners* (South Africa: University of Pretoria) p 5
[10] Shahali E and Halim L 2010 *Procedia – Social and Behavioral Sciences* Vol. **09** pp 142-146
[11] Temiz B, Tasar M, and Tan M 2006 *Int. Education Journal* Vol. **07** pp 1007-1027
[12] Bati K, Erturk G, and Kaptan F 2010 *Procedia – Social and Behavioral Sciences* Vol. **02** pp 1993-1998
[13] Thiagarajan S, Semmel D, and Semmel M 1974 *Instructional Development for Training Teachers of Exceptional Children* (Indiana: Center for Innovation Teaching Handicapped) pp 5-10
[14] Dahar R 2011 *Teori-Teori Belajar dan Pembelajaran* (Jakarta: Erlangga) p 72
[15] Nur M 2011 *Pemotivasian Siswa untuk Belajar* (Surabaya: Unesa Press)
[16] Dimyati and Mudjiono 2015 *Belajar dan Pembelajaran* (Bandung: Rineka Cipta)
[17] Darus F B and Saat R M 2014 *The Malaysian Online Journal of Educational Sciences* Vol. **02** pp 20-26
[18] Demirbas M and Tanriverdi G 2012 *Sakarya University Journal of Education* (conference.pixel-online.net) pp 1-6

IOP Conf. Series: Journal of Physics: Conf. Series **947** (2017) 012027 doi:10.1088/1742-6596/947/1/012027

Laboratory Activity Worksheet to Train High Order Thinking Skill of Student on Surface Chemistry Lecture

B Yonata[1], H Nasrudin[1]

[1] Chemistry Department, Universitas Negeri Surabaya, Surabaya 60231, Indonesia

Abstract. A worksheet has to be a set with activity which is help students to arrange their own experiments. For this reason, this research is focused on how to train students' higher order thinking skills in laboratory activity by developing laboratory activity worksheet on surface chemistry lecture. To ensure that the laboratory activity worksheet already contains aspects of the higher order thinking skill, it requires theoretical and empirical validation. From the data analysis results, it shows that the developed worksheet worth to use. The worksheet is worthy of theoretical and empirical feasibility. This conclusion is based on the findings: 1) Assessment from the validators about the theoretical feasibility aspects in the category is very feasible with an assessment range of 95.24% to 97.92%. 2) students' higher thinking skill from N Gain values ranges from 0.50 (enough) to 1.00 (high) so it can be concluded that the laboratory activity worksheet on surface chemistry lecture is empirical in terms of worth. The empirical feasibility is supported by the responses of the students in very reasonable categories. It is expected that the laboratory activity worksheet on surface chemistry lecture can train students' high order thinking skills for students who program surface chemistry lecture.

1. Introduction

Humans have a great curiosity and this is what distinguishes it from other beings. Curiosity begins early and is needed throughout life. Curiosity comes from the observation of phenomena or natural phenomena that exist around. If this observation continues to find a solution then this observation becomes an initiation in solving the problem. Habituation to be able to see problems and find solutions need to be given especially from an early age.

The purpose of science education is to assist individuals in using science process skills, in other words to define the problems around them, observe, analyze, hypothesize, experiment, summarize, and apply the information they have [1]. The easiest learning habit is through laboratory activities guided by a science-based laboratory activity worksheet to train students' high-order thinking skills.

High-level thinking describes abstract intellectual processes [2]. High-level thinking is thinking at a higher level than simply memorizing or conveying something to someone just as it is conveyed to others [3]. First, the levels of thinking cannot be unmeshed from the levels of learning; they involve interdependent, multiple components and levels. Second, whether or not thinking can be learned without subject matter content is only a theoretical point. In real life, students will learn content in both community and school experiences, no matter what theorists conclude, and the concepts and vocabulary they learn in the prior year will help them learn both higher order thinking skills and new content in the coming year. Third, higher order thinking involves a variety of thinking processes applied to complex situations and having multiple variables [4].

Physical chemistry in the chemistry department has a course that deals with colloid chemistry and surface namely surface chemistry. There are three materials discussed: viscosity, surface thermodynamics (surface tension and adsorption), and colloids. This subject of students is also required to carry out laboratory activities with four practical titles namely viscosity, surface tension, adsorption, and colloid.

A worksheet is used as a 'prescription' system which consider students only need to carry out practicum procedures that available and not giving students the opportunity to develop their thinking skills. This evidence also has an impact on pre laboratory activities. Students only re-read the practicum procedures and only carry out as listed on the practicum procedure. No student has an initiative to add or replace materials and modify the practicum procedure. Whereas many things can be done for example the title of practicum viscosity, students may add another solute to compare with the viscosity of the universal solvent (water) in determining the relative viscosity. Thinking, for the purposes of this inquiry, is defined accordingly as that operation in which present facts suggest other facts (or truths) in such a way as to induce belief in the latter upon the ground or warrant of the former [5]. Bloom stated that education should focus on mastery subject and the achievement of higher order thinking as a criticism of the use of utilitarian that simply meant learning as a means of transferring words [6]

Based on the description of facts and expectations that have been submitted, it is necessary to trained high-order thinking skills of students through laboratory activity. The laboratory activity worksheet developed should be both theoretical and empirical valid.

2. Methods

The object of this research is the Surface Chemistry Laboratory Activity Worksheet. The development of this laboratory activity worksheet follows the path of the 4-D Model development model [7]. This model consists of 4 development stages of Define, Design, Develop, and Disseminate, but in this research is limited to develop stage. There are 4 topics that will be conducted by students.

First topic is Viscosity. Viscosity is the resistance to flow which is a property of fluids, both liquids and gases. Second topic is surface tension. Surface tension (γ) is measured as the energy required to increase the surface area of a liquid by a unit of area. Then, the next topic is adsorption. Adsoroption occurs when a molecule (adsorbate) that forms a bond to the surface (adsorbent). The last topic is colloid. All materials are capable of scattering light (Tyndall effect) to some extent.

Before conducting experiments, students answered questions that consist of high order thinking skill in pretest. After conducting experiments, students were given posttest that also consist of high order thinking skill. The differences in high-order thinking skills before and after using laboratory activity worksheet were analyzed by calculating the difference in posttest and pretestt (N Gain score), calculated by the formula (1)

$$< g \geq \frac{\%<G>}{\%<G\ max>} = \frac{(\%<Sf>-\%<Si>)}{100\%-\%<Si>} \qquad (1)$$

[8]

Table 1. *N gain* score interpretation

<g>	Criteria
<g> <0.3	Less
0.7><g>≥0.3	Enough
<g>≥0.7	High

[8]

3. Results and Discussion

3.1. Worksheet Development

The learning achievement of Surface Chemistry course is that after graduating from course, the students are able to understand the surface characteristics of capillarity, surface thermodynamics, adsorption, surfactant, detergent, emulsion, base and aerosol, chemisorption and catalyst so as to develop a conceptual framework to formulate the action or Alternative actions in solving chemical problems in life. Conceptual framework can be gained if students already achieve knowledge by their own way. Conceptual framework that they must gain as follows in equations (2), (3), (4) and (5). When a liquid

IOP Conf. Series: Journal of Physics: Conf. Series **947** (2017) 012027 doi:10.1088/1742-6596/947/1/012027

flows through a tube, layers of liquid slide over each other and intermolecular forces cause resistance to flow. Viscosity (η) is the ratio of shear stress (τ) to shear rate (D) as equation (2) [9].

$$\eta = \tau / D \quad (2)$$

For water surface tension can be measured by (3)

$$\gamma_W = \gamma_W^d + \gamma_W^h \quad (3)$$

Associative Adsorption occurs when a gas molecule adsorbs without fragmentation. While dissociative Adsorption occurs when a gas molecule adsorbs and fragmentation occurs.

$$\theta = \frac{N_S}{N} \quad (4)$$

where Ns = number of surface sites occupied by adsorbate and N = total number of substrate adsorption sites. When θ = 1, a mono layer exist.
One characteristic of colloid is Tyndall Effect. The noticeable turbidity associated with colloidal dispersions are a consequence of intense light scattering [9].

$$I_t/I_o = \exp[-\tau l] \quad (5)$$

where I_o is the intensity of the incident light beam, I_t is the intensity of the transmitted light beam, l is the length of the sample and τ is the turbidity.

Task analysis is used to identify the task completion stages in accordance with the subject matter of the Surface Chemistry Course. The task analysis are observing the phenomenon, Formulating the problem, Determining the hypothesis, Defining variables, Defining tools and materials, Developing trial procedures, Analyzing and making Conclusion
Concept Analysis aims to identify key concepts to be discussed, systematically compile and detail relevant concepts. The result of this concept analysis is a concept map of Surface Chemistry subject matter and selected materials that can represent the material being studied.
The learning outcome indicators that will be the basis for the preparation of the laboratory activity worksheet are investigating the relative condensed number of a liquid by using water as a comparison, Measuring surface tension of solution with capillary pipe method, Determining θ according to Freundlich on the acid adsorption process by activated carbon, and Understand the colloidal system and its use in daily life
In addition to the pretest and posttest issues, at this stage of design is also prepared Chemistry Laboratory Activity Worksheet. Preparation of this practice guide is called draft 1. Draft 1 is then reviewed in order to gain Draft 2. Draft 2 of the Surface Chemical Laboratory Activity Worksheet was then validated and applied to students currently enrolling the Surface Chemistry course. Validator assessment results are presented in Table 2

Table 2. Result of Feasibility Assessment of Surface Chemistry Laboratory Activity Worksheet

	Aspect to assess	Scoring range	Percentage range	Criteria
1.	Construct	3 - 4	83.33 – 100.00	Very worth
2.	Content	3 - 4	91.67 – 100.00	Very worth
3.	Language and readability	3 - 4	83.33 – 100.00	Very worth
4.	Conformity of High order thinking skill			
	a. Meet inquiry components	3 – 4	83.33 – 100.00	Very worth
	b. Meet the analytical component	4	100.00	Very worth
	c. Meets component evaluate	4	100.00	Very worth
	d. Meet the component of creating	4	100.00	Very worth

Based on the data in Table 1 it can be concluded that the Surface Chemistry Laboratory Activity Worksheet is very feasible or very worth. In terms of construction, the Surface Chemistry Laboratory Activity Worksheet complies with the students' thinking level. This worksheet can also actively involve students as students are asked to find a solution to the problems presented or the issues they must address. For the Content aspect, the Surface Chemistry Laboratory Activity Worksheet presents phenomena that support the material. Truth content and content updates are also rated between good and excellent.

In terms of language and readability, the Surface Chemistry Laboratory Activity Worksheet uses language appropriate to the student's age, and has used good and correct Indonesian. The terms used are correct, easy to understand, and steady in their use. In general the language and readability aspects get good and excellent judgment from the validators. Approximately 83.33% to 95.83% of students also consider that the language used is easy to understand and uses the term steady, good, and correct.

Aspects of conformity with high-level thinking empowerment get good and excellent judgment from validators. The Surface Chemistry Laboratory Activity Worksheet has been assessed to have fulfilled high-level thinking components including analysis (C4), evaluation (C5), and creating (C6).

In the Surface Chemistry Laboratory Activity Worksheet consists of 4 topics. On the topic of viscosity, students are asked to observe the phenomenon, look for appropriate literature to solve the problem. The next step is to formulate the problem, develop the hypothesis, determine the variables, asks them to determine the variables they will use in the investigation. The determination of these variables should also consider the availability of tools and materials in Physical Chemistry laboratory at Chemistry Department. In other words, that variable can be done in Physics Chemistry laboratory in Chemistry Department. Students work in groups with 3 group members. After students look at the tools and materials available in the laboratory, then they determine the procedures they will do.

On the other three topics, there are some student workflow differences after they have searched for the appropriate literature, they must specify experiment variables before formulating the problem formulation. This is done because the given phenomenon is not as detailed on the topic of viscosity. Finding problems and hypotheses is a very important activity in improving the creativity of science [1]. The sequence of pre-laboratory activities should be done by the students so that they have a stock before carrying out lab work in the laboratory. In this way it is expected to activate their ability to understand problems and make problem solving so that their high-level thinking skills become more honed. This is supported by the revelation that higher-order thinking is thinking at a higher level than simply memorizing or conveying something to someone just as it is conveyed to others [3].

3.2. Higher-Order Thinking Skills

There were 24 students divided into 8 groups consisting of 3 people. In the implementation there were 2 groups that got the topic of viscosity and surface tension, 3 groups got adsorption topics, and 3 groups that got the topic of colloid.

After students got pretest, they received a Surface Chemistry Laboratory Activity Worksheet. Each group must complete each stage listed in the Surface Chemistry laboratory activity worksheet before the practicum takes place. During this time span, each group was allowed to discuss the group's thoughts to the Surface Chemistry lecturer. Once each group was considered ready and the procedures prepared by the students already pay attention to the tools and materials available in the laboratory, then each group could implement the settlement plan in the time set. In Figure 1 shows exemplary part in the worksheet that shows high order thinking skills.

IOP Conf. Series: Journal of Physics: Conf. Series **947** (2017) 012027 doi:10.1088/1742-6596/947/1/012027

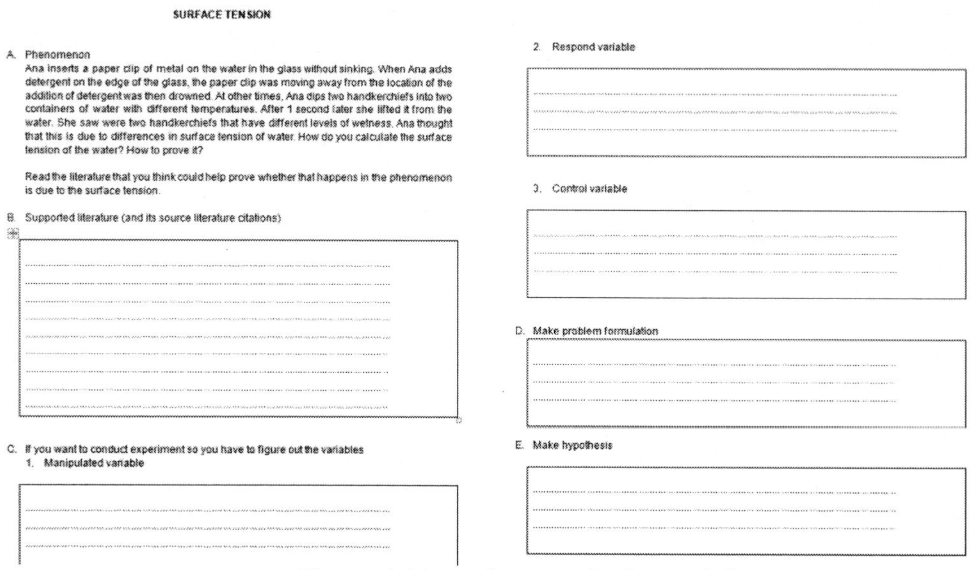

SURFACE TENSION

A. Phenomenon

Ana inserts a paper clip of metal on the water in the glass without sinking. When Ana adds detergent on the edge of the glass, the paper clip was moving away from the location of the addition of detergent was then drowned. At other times, Ana dips two handkerchiefs into two containers of water with different temperatures. After 1 second later she lifted it from the water. She saw were two handkerchiefs that have different levels of wetness. Ana thought that this is due to differences in surface tension of water. How do you calculate the surface tension of the water? How to prove it?

Read the literature that you think could help prove whether that happens in the phenomenon is due to the surface tension.

B. Supported literature (and its source literature citations)

C. If you want to conduct experiment so you have to figure out the variables

1. Manipulated variable

2. Respond variable

3. Control variable

D. Make problem formulation

E. Make hypothesis

Figure 1. Exemplary part in the worksheet

Table 3. Pretest and Posttest Scores

Student No.	Laboratory Topic	Pretestt		Posttest		N Gain	Category
		Score	Mastery	Score	Mastery		
1	Adsorption	29,2	NM	95,8	M	0,94	High
2	Adsorption	25,0	NM	95,8	M	0,94	High
3	Adsorption	20,8	NM	100,0	M	1,00	High
4	Adsorption	25,0	NM	100,0	M	1,00	High
5	Adsorption	25,0	NM	95,8	M	0,94	High
6	Adsorption	25,0	NM	95,8	M	0,94	High
7	Adsorption	29,2	NM	100,0	M	1,00	High
8	Adsorption	12,5	NM	100,0	M	1,00	High
9	Adsorption	29,2	NM	100,0	M	1,00	High
10	Colloid	16,7	NM	95,8	M	0,95	High
11	Colloid	25,0	NM	95,8	M	0,94	High
12	Colloid	25,0	NM	83,3	M	0,78	High
13	Colloid	16,7	NM	83,3	M	0,80	High
14	Colloid	16,7	NM	100,0	M	1,00	High
15	Colloid	25,0	NM	100,0	M	1,00	High
16	Colloid	20,8	NM	100,0	M	1,00	High
17	Colloid	25,0	NM	62,5	M	0,50	Enough
18	Colloid	25,0	NM	87,5	M	0,83	High
19A	Surface Tension	33,3	NM	100,0	M	1,00	High
20A	Surface Tension	20,8	NM	79,2	M	0,74	High
21A	Surface Tension	45,8	NM	100,0	M	1,00	High
22A	Surface Tension	29,2	NM	100,0	M	1,00	High
23A	Surface Tension	41,7	NM	100,0	M	1,00	High
24A	Surface Tension	41,7	NM	100,0	M	1,00	High
19B	Viscosity	70,8	M	100,0	M	1,00	High

Student No.	Laboratory Topic	Pretestt		Posttest		N Gain	Category
		Score	Mastery	Score	Mastery		
20B	Viscosity	54,2	NM	100,0	M	1,00	High
21B	Viscosity	87,5	M	100,0	M	1,00	High
22B	Viscosity	70,8	M	95,8	M	0,86	High
23B	Viscosity	66,7	M	100,0	M	1,00	High
24B	Viscosity	83,3	M	100,0	M	1,00	High

NM= not mastery; M=mastery

Based on Table 3 it can be seen that some students have had a good initial trajectory about high-order thinking skills. From the data it can also be seen that the N Gain values range from 0.50 (enough) to 1.00 (high) so it can be concluded that the Surface Chemistry Laboratory activity worksheet is feasible in terms of empirical. Inquiry studies emphasize the importance of helping students understand the key structures or ideas of a discipline so that students actively engage in learning and learning are through personal discovery [10].

The students' ability in designing an experimental procedure scores between 3 and 4. In the activity of preparing the hypothesis one student got a score of 1, whereas in general the score obtained by students between 3 and 4. Students who scored 1 on the activities of composing the hypothesis did not progress from the pretest score. This is because the student was not active in the learning activities in the classroom. The activities of preparing hypotheses and designing procedures are included in the cognitive domain of C6 (creating) and are included in indicators to measure students' high-order thinking skills. This is in line with the opinion of [11], stating that the indicator in measuring students' high-level thinking skills in the C6 (cognitive) cognitive domain is designing a trial procedure and hypothesizing. This empirical feasibility is also supported by student responses. Table 3 shows the percentage of 'yes' responses from 24 students who were the subject of the study.

Table 4 Student Response to Surface Chemical Laboratory Activity Worksheet

Aspect	**Percentage**
Generating motivation / interest / curiosity.	87.50
Asking you to be creative in solving problems	95.83
The Surface Chemistry Laboratory Activity Worksheet begins with a phenomenon and proceeds to solve the problems contained in the phenomenon	91.67
The phenomenon presented supports the matter.	87.50
You are asked to look for truth content (facts, concepts, principles, laws, theories and scientific processes) through literature and experiments	95.83
Content is up-to-date.	87.50
Beginning with the observation of phenomena in order to understand a concept	100.00
Meet the analytical component	95.83
Meets component evaluating	87.50
Meet the component of creating / creating	100.00
Average	91.01

Based on Table 4 it is known that students respond very well to the Surface Chemistry Laboratory Activity Worksheet. However, 29.17% of students assume that the Surface Chemistry Laboratory Activity Worksheet format (size, etc.) is not appropriate. This is because the column to write literature review of the literature is less broad so that students are still writing on other papers.

4. Conclusions and Suggestion

4.1 Conclusion

The Surface Chemistry Laboratory Activity Worksheet is said to be worthy of its theoretical feasibility and empirical feasibility. This conclusion is based on the findings: Assessment of validator of theoretical feasibility aspects in the category is very feasible with the range of assessment of 95.24% to 97.92%. High order thinking skill of students from N Gain values ranges from 0.50 (enough) to 1.00 (high) so that it can be concluded that the Surface Chemistry Laboratory Activity Worksheet is feasible in terms of empirical. The empirical feasibility is also supported by the responses of the students in very reasonable categories.

4.2 Suggestions

The use of the Surface Chemistry Laboratory Activity Worksheet takes more time when students consult about their flow of thinking in completing the problem-solving procedure. For other researchers expected to anticipate the limitations of time allocation.

References

[1] Aktamis H and Ergin O 2008 The Effect of Scientific Process Skill Education on Students' Scientific Creativity, Science Attitudes and Academic Achievement. *Asia-Pasific Forum on Science Learning and Teaching* 4.
[2] Arends R 2007 *Learning To Teach* New York: McGraw-Hill
[3] Thomas A 2000 *CDL* (CDl: http://cdl.org/resource-library/pdf/feb00PTHOT.pdf)
[4] King F J, Goodson L, and Rohani F 2000 *Higher Order Thinking Skills (*Center for Advancement of Learning and Assessment)
[5] Dewey J 1910 *How We Think. In The Problem of Training Thought* (p. 1). (Boston: D.C. Heath & Co.)
[6] Suyono and Hariyanto 2011 *Belajar dan Pembelajaran, Teori dan Konsep Dasar* (Bandung: Rosdakarya)
[7] Thiagarajan S, Semmel D S, and Semmel M I 1974 *Instructional Development for Training Teachers of Exceptional Children, A Sourcebook.* (Indiana: Eric Publisher)
[8] Hake R R 1998 Interactive Engagement versus Traditional Methods: A Six Thousand Student Survey of Mechanics Test Data for Introductory Physics Courses. *American Journal Physics* **66** 1 pp 64-74.
[9] Shaw D J 2004 Introduction to Colloid and Surface Chemistry 4th Edition (Oxford: Elsevier Science Ltd)
[10] Sugiarto B 2009 *Mengajar Siswa Belajar, Implementasi Guru di Dalam Kelas.* (Surabaya: Unesa University Press)
[11] Krathwohl D R 2002 Revising Bloom's Taxonomy *Theory into Practice* pp. 212-264.

IOP Conf. Series: Journal of Physics: Conf. Series **947** (2017) 012028 doi:10.1088/1742-6596/947/1/012028

Student Teachers' Proof Schemes on Proof Tasks Involving Inequality: Deductive or Inductive?

A H Rosyidi[1], A W Kohar[2]

[1,2] Departement of Mathematics, Universitas Negeri Surabaya, Indonesia

Email: [1]abdulharis@unesa.ac.id

Abstract. Exploring student teachers' proof ability is crucial as it is important for improving the quality of their learning process and help their future students learn how to construct a proof. Hence, this study aims at exploring at the proof schemes of student teachers in the beginning of their studies. Data were collected from 130 proofs resulted by 65 Indonesian student teachers on two proof tasks involving algebraic inequality. To analyse, the proofs were classified into the refined proof schemes level proposed by Lee (2016) ranging from inductive, which only provides irrelevant inferences, to deductive proofs, which consider addressing formal representation. Findings present several examples of each of Lee's level on the student teachers' proofs spanning from irrelevant inferences, novice use of examples or logical reasoning, strategic use examples for reasoning, deductive inferences with major and minor logical coherence, and deductive proof with informal and formal representation. Besides, it was also found that more than half of the students' proofs coded as inductive schemes, which does not meet the requirement for doing the proof for the proof tasks examined in this study. This study suggests teacher educators in teacher colleges to reform the curriculum regarding proof learning which can accommodate the improvement of student teachers' proving ability from inductive to deductive proof as well from informal to formal proof.

1. Introduction

At university level, constructing formal mathematical proof is a vital skill expected to be mastered by undergraduate students at mathematics courses. This is crucial as undergraduate students are demanded to have such skill as the highest level of proving so that they lead to perceiving that proof is an essential component of mathematics which is arguably what differentiate mathematics from other disciplines [1]. However, Clark and Lovric [2] explored numerous challenges that undergraduate students experienced when making the transition to a more formal proof construction at university-level mathematics. This transition requires undergraduate students to change the schemes of reasoning used. For instances, they need to shift from informal to formal language or representation, shift from inductive to deductive inference, understand and apply the theorem, reason from mathematical definitions, justify trueness of propositions using a valid mathematical argument, and to make connections between mathematical objects.

The proof reasoning as discussed above was agreed by experts as an important skill the undergraduate students require to construct a proof [3]. Such reasoning, more particularly, were also needed by undergraduate students who are student teachers. This is due to need of a teacher to have a deep understanding of the nature and the role of proof for the need of instructional practice [1]. In addition, knowledge about mathematical proof is considered as one of the essential components of subject matter knowledge which should be mastered by mathematics teachers [1]. By having this knowledge, as Jones [1] argued, they would have an extremely secure subject knowledge base of mathematical proof if they are to teach it accurately and confidently.

However, many of the proof-related studies on undergraduate students, including for those who were student teachers, indicate their weaknesses regarding knowledge of proof construction. As evidence, they have been observed to experience mathematical proof nearly procedural, regarding the construction and writing of a proof as an algorithm to follow rather than a creative process for solving a problem [4,5]. Such finding was scrutinized to be caused by the finding that they usually used non-formal reasoning when studying at secondary level while at the university level they start to learn formal reasoning in the mathematics courses they took. This influence they way they perceive and construct mathematical proof at the beginning of their studies in undergraduate level. It is clearly important for a student teacher of mathematics to be able to construct, understand, and validate formal mathematical arguments [6], yet research indicates that a lot of students eventually do not succeed in developing skills of mathematical proving, not only in the beginning of their studies, but also by the end of their undergraduate programs. As such, capturing the student teacher's proving schemes in the beginning of their studies is required as one of tools of evaluation for the improvement of courses programs offered in the student teacher curriculum at university level.

This present study concerns on exploring the proof schemes resulted by student teachers who were studying in the first semester of their undergraduate studies in a mathematics education program. The topic examined to the student teachers was inequality since it was offered in Calculus course they took in the first semester.

2. Student Teachers' Proof Schemes on Inequality Task

The classifications of proof schemes as many scholars proposed were based on an underpinning thought that asserted undergraduates' mathematical proving evolved from being inductive-oriented toward deductive-oriented [4]. Miyazaki [7] classified proof schemes from deductive scheme to inductive scheme into four types: Proof A, Proof B, Proof C, and Proof D. Proof A is the type of proof which deductive reasoning is involved and a functional language is used in the course of doing a proof, while Proof B is the type of proof in which deductive reasoning is involved and other languages, drawings, and movable objects are used in the course of doing a proof. While Proof C is the type of proof in which inductive reasoning is involved and other languages, drawings, and movable objects are used, Proof D is the type of proof which inductive reasoning is involved and a functional language is used. Lee [8] further refined such schemes into two types of schemes in terms of deductive-proof construction for true proposition and proof-by-counterexample construction for false propositions. Regarding the former schemes, Lee's [8] schemes present a hierarchical six-level of proof schemes ranging from inductive, which only provides irrelevant inferences, to deductive proofs, which consider addressing formal representation. Table 1 summarizes such schemes.

Table 1. Descriptors for coding students' Deductive-proof Construction (adopted from Lee [8])

Level	Characteristics	Description of students' proofs
0	Irrelevant or minimal engagement in inferences	Do not know how to prove or appeal to conviction in external knowledge sources Unable to relate the antecedent and the consequent with examples
1	Novice use of examples or logical reasoning	Conclude the implication is true (or false incorrectly) using one or more examples or an incorrect example or counterexample Falsify the implication "if P then Q" through erroneous logical reasoning, e.g., "if not P then not Q" Derive a mathematical property unrelated to the implication

IOP Conf. Series: Journal of Physics: Conf. Series **947** (2017) 012028 doi:10.1088/1742-6596/947/1/012028

2	Strategic use of examples for reasoning	Generate examples based on random or case-based sampling, or extreme cases Uses mathematical properties inferred from generated examples to make conclusions
3	Use of deductive inferences with major flaws in logical coherence and validity	Deduce relevant mathematical properties but missing one or two deductive inferences for proving the implication Deduce the implication to be true for some cases of the antecedent but omit other cases
4	Use of deductive inferences with minor flaws in logical coherence and validity	Generate a chain of deductive inferences to justify conclusions but one or two inferences may be interpreted as inductive due to insufficient substantiation or as writing errors Displays disorganization in the chain of deductive inferences
5	Construction of informal deductive proofs	Generate a chain of deductive inferences to justify the implication using informal justifications
6	Construction of deductive proofs using formal representations	Generate a chain of deductive inferences to justify the implication using formal representations

According to Lee [8], level 0 means someone is unable to provide any inferences to the proof he/she creates, level 1 and 2 indicates inductive inferences with different level of attainment, while live 3,4,5,6 shows deductive inferences with the different progress of proof. Table 1 also describes what makes distinguish between one another level as seen in the description of students' proof column.

3. Method

We involved 65 student teachers who were studying mathematics in the first semester of their study at a mathematics education program in our university. The data were collected from their written responses on two proof-related tasks involving algebraic inequality. Thus, we had 130 responses (65 × 2 task responses). Figure 1 shows the proof-related tasks examined to the student teachers. While Lee's [8] research instruments concerned on the complete proposition which required participants to justify the trueness of such proposition and prove the reasoning of proofs they created, the task we developed in this study required participants to (1) choose correct option/s to complete an incomplete proposition and prove their justification, as well as to (2) solve an inequality task involving arbitrary two different real number and justify whether its solution is same with a similar inequality task. For task no 2, some students worked on $(x-a)(x+b) > 0$, while the others worked on $(x-a)(x-b) < 0$

1. If a and b are two different real numbers, then... (i) $(a-b)(b-a) < 0$ (ii) $(a-b)(b-a) = 0$ (iii) $(a-b)(b-a) > 0$ You may choose more than one statements when you think they are correct. Support your option/s by giving complete arguments
2. If a and b are two different real numbers, solve the inequality below $(x-a)(x+b) > 0$ Then, determine whether its solution will be same as the solution of the inequality below. $\frac{(x-a)}{(x+b)} > 0$ Give your reasons.

Figure 1. The proof-related tasks

Data were analyzed by coding each of responses into an appropriate level of proof as proposed by Lee [8] as seen in table 1. To describe, we exemplify the responses of student teachers' proof schemes on those two tasks by presenting examples of each of the levels.

4. Results and Discussion

In general, the levels of student teachers' responses can be described as a progression through four phases: attempts that failed to relate the antecedent and the consequent (level 0), proofs that were based on examples or logical reasoning errors (levels 1–2), proofs that were based on incomplete deductive inferences (levels 3–4), and proofs that were based on coherent deductive inferences (level 5–6). Table 2 shows the frequency distribution of student teachers' deductive-proof construction schemes on the inequality task.

Table 2. Frequency of student teachers' proof schemes on the proof task

Levels	Task 1	%	Task 2	%	Total	%
0	3	4.6%	8	12.3%	11	8.5%
1	30	46.2%	9	13.8%	41	31.5%
2	14	21.5%	12	18.5%	26	20.0%
3	5	7.7%	12	18.5%	17	13.1%
4	6	9.2%	12	18.5%	18	13.8%
5	3	4.6%	7	10.8%	8	6.2%
6	4	6.2%	5	7.7%	9	6.9%
Total	65		65		130	100.0%

The distribution of student teachers' responses as shown in table 2 indicates that most of the responses, which is 41 out of 130, were indicated in level 1 (the use of novice examples), while only 9 responses (6.9%) were indicated in level 6 (formal representation of deductive inference). Regarding deductive and inductive inferences, generally, there were 67 inductive responses (level 1, 2) and 52 deductive responses (level 3,4,5,6). It means the student teachers performed inductive inferences more frequently than deductive inferences.

In the next discussion, we provide each of examples of the students' responses for each level. Figure 2 shows a collective examples of responses for task 1, while figure 3 shows examples for task 2.

a)Level 0

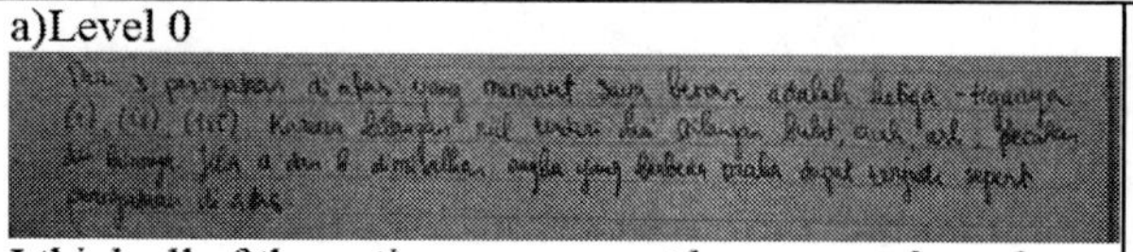

I think all of the options are correct because real numbers consist of integer, whole numbers, natural number, fraction, etc. If ***a*** and ***b*** are different then those three options satisfies

b) Level 1

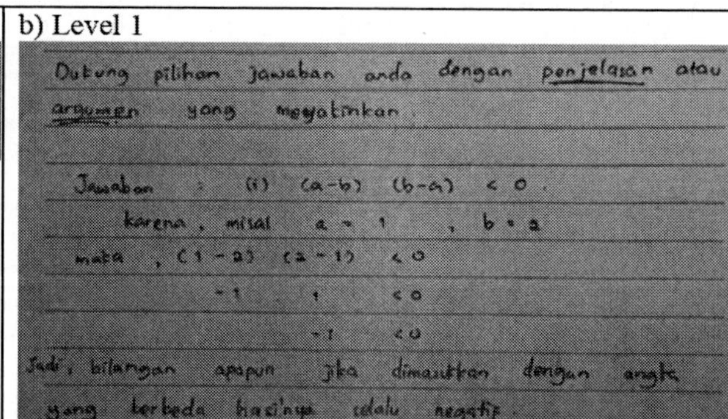

So, if any numbers are substituted into the inequality, it will have a negative number

c) Level 2

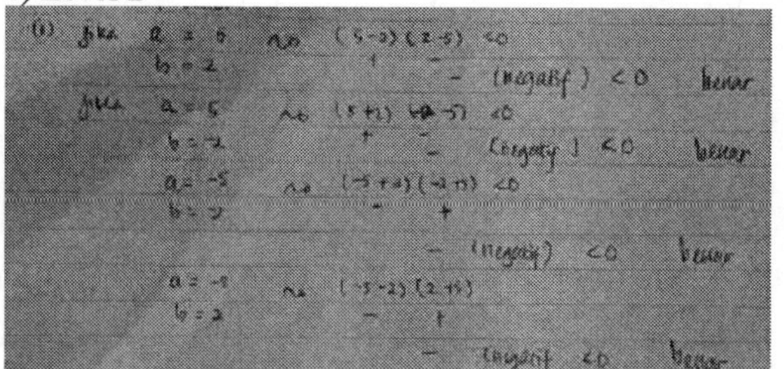

d) Level 3

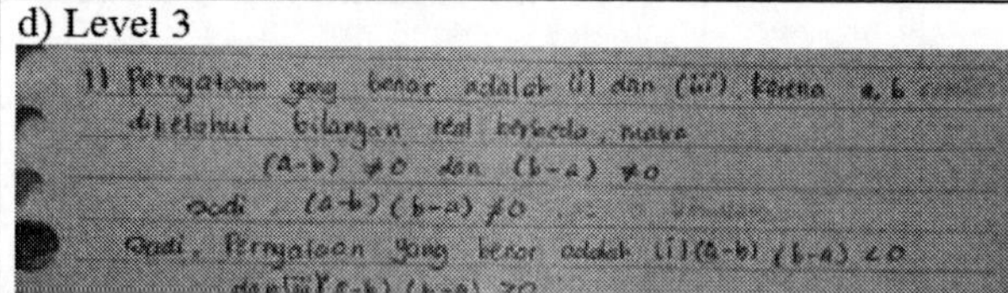

Options (i) and (iii) are both correct because if a, b are different real numbers, then
$(\boldsymbol{a}-\boldsymbol{b}) \neq \mathbf{0}$ **and** $(\boldsymbol{b}-\boldsymbol{a}) \neq \mathbf{0}$.
So, $(\boldsymbol{a}-\boldsymbol{b})(\boldsymbol{b}-\boldsymbol{a}) \neq \mathbf{0}$.

If $a = 5, b = 2$, then $(5-2)(2-5) < 0$ (negative) If $a = 5, b = -2$, then $(5+2)(-2-5) < 0$ (negative) If $a = -5, b = -2$, then $(-5-2)(-2+5) < 0$ (negative) If $a = -5, b = -2$, then $(-5-2)(2+5) < 0$ (negative) All are correct	Accordingly, $(a-b)(b-a) > 0$ and $(a-b)(b-a) < 0$
e) Level 4 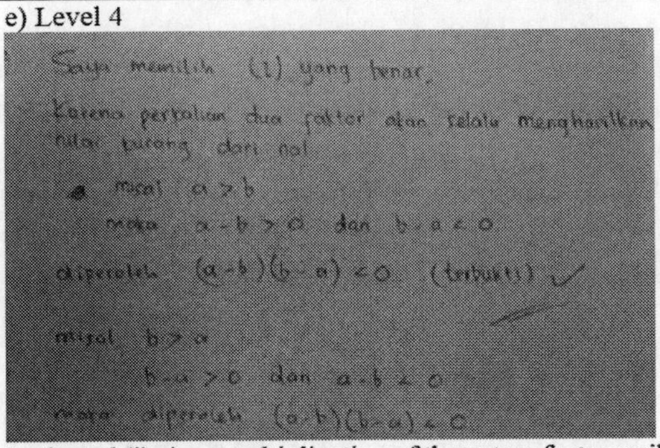 I selected (i) since multiplication of those two factors will always less than 0. Let $a > b$, then $(a-b) > 0$ and $(b-a) < 0$ Obtained that $(a-b)(b-a) < 0$ Let Let $a < b$, then $(b-a) > 0$ and $(a-b) < 0$ Obtained that $(a-b)(b-a) < 0$	f) Level 5 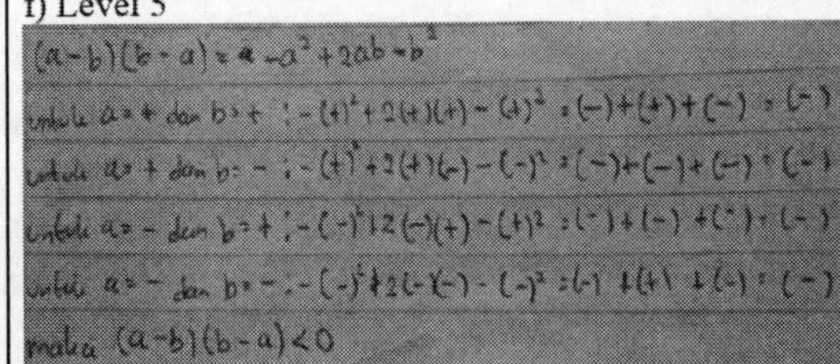 $(a-b)(b-a) = -a^2 + 2ab - b^2$ For $a = +$ and $b = +$, then $= -(+)^2 + 2(+)(+) - (+)^2 = -$ For $a = +$ and $b = -$, then $= -(+)^2 + 2(+)(-) - (-)^2 = -$ For $a = -$ and $b = +$, then $= -(-)^2 + 2(-)(+) - (+)^2 = -$ For $a = -$ and $b = -$, then $= -(-)^2 + 2(-)(-) - (-)^2 = -$ Then, $(a-b)(b-a) < 0$
g) Level 6 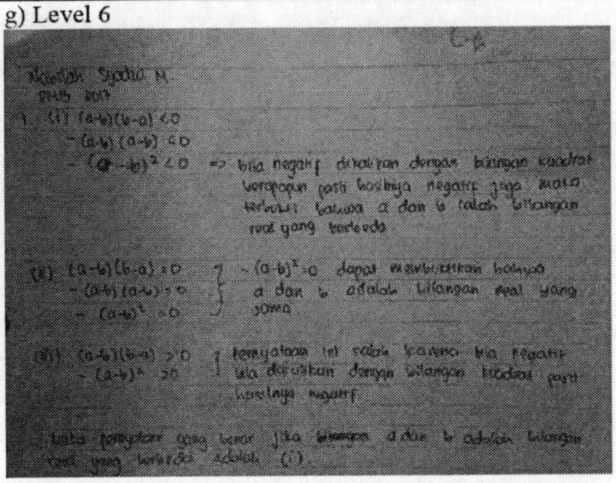	$(a-b)(b-a) < 0$ $-(a-b)(a-b) < 0$ $-(a-b)^2 < 0$ (If negative is multiplied by the square of any number, the product is always negative as well, a and b are two different numbers. This is correct) $-(a-b)^2 = 0$. This happens if a and b are same numbers $-(a-b)^2 > 0$. This is impossible since it always be negative.

Figure 2. Student teachers' responses on task 1

At level 0, some proof schemes referred to justifications given in textbooks, such as arguing that it is based on the student teachers' memory when learning inequality at secondary level. Others, as shown in figure 2a, show irrelevant argument which failed to relate the antecedent and the consequent (the chosen option). At level 1, the proof schemes show novice examples, such as examining the trueness of the proposition after the option/s were selected by substituting random numbers ($a = 1, b = 2$) to the inequlity and conclude that the proposition is correct. At level 2, the proof schemes were observed to have more sophisticated examples since some of the student teachers consider giving case-based sampling to examine their justification. Figure 2c shows that the proof provided all the possibilities of pairs of positive-negative numbers and examined those numbers to the inequality to conclude that the proposition is correct. At level 3, the proofs were labelled as deductive inferences although many incomplete chain of inferences were found. The student teacher on the response shown in figure 2d, for example, consider only one of the cases of possibilities. In other words, instead of considering the the existance of trichotomy property of real number ($a = b, a > b,$ **or** $a < b$) which might have different cases for each of the inequalities offered in the options, the student teacher assumed that there were two possibilties of two different numbers ($a \neq b$), namely $a > b,$ **or** $a < b$, and claimed that those two possibilities were both correct without examining the correctness of each. At level 4, the student teachers constructed almost complete deductive inferences. The characteristic of this level can be observed from the existance of missing step of claim which is not supported by valid arguments. For task 1, we found that the student teacher at figure 2e categorized his/her inferences for the cases of $a > b$ and $b > a$, then by using an algebraic operation, the student teacher claimed that $(a - b)(b - a) < 0$. However, this claim was not supported by any arguments such as arguing why $(a - b)(b - a) < 0$. At level 5, a proof demonstrating logically valid and coherent chain of deductive inferences would incorporate deductive inferences about examining each of cases for the value of a and b which is negative or positive. Figure 2f indicates that for all possibilities of a and b, the results of $(a - b)(b - a)$ is always negative. Thus, $(a - b)(b - a) < 0$. However, it is not indicated that this proof include any formal representation. This is why this proof is categorized at level 5. At level 6, proofs are more deductive and of course are valid. Figure 2g shows that the student teacher used symbolic representations which met relevant theorems of inequality.

<table>
<tr><td>a) Level 0
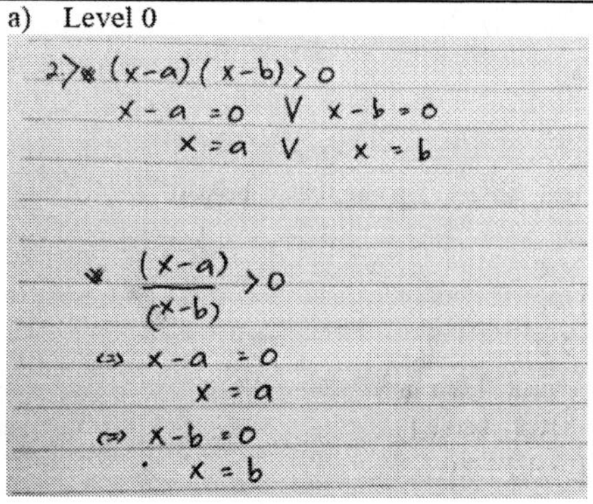</td><td>b) Level 1
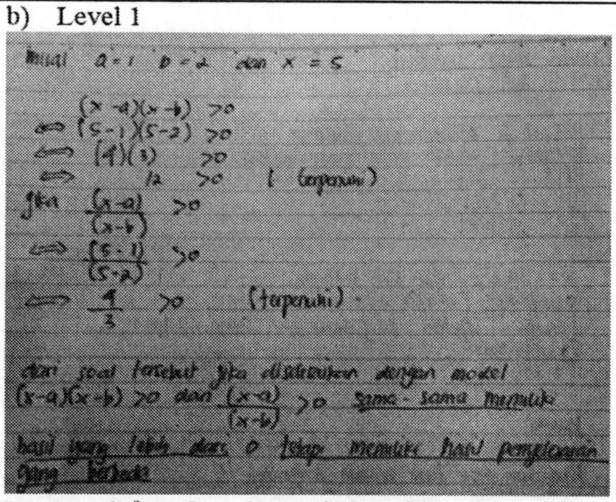
Let $a = 1, b = 2$, and $x = 5$, then $(5 - 1)(5 - 2) > 0$ which means $12 > 0$ (correct). If $\frac{(5-1)}{(5-2)} > 0$ which means $\frac{4}{3} > 0$ (correct). Thus, both $(x - a)(x - b) > 0$ and $\frac{(x-a)}{(x-b)} > 0$ give results more than 0, although have different solution</td></tr>
<tr><td>c) Level 2</td><td>e) Level 4</td></tr>
</table>

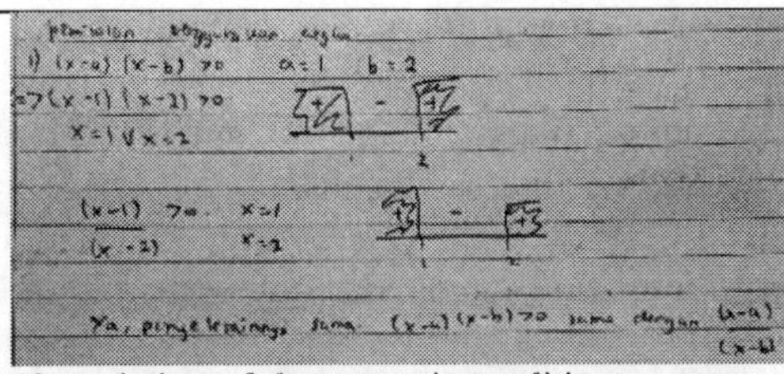 The solution of those two inequalities are same. For every x, with $a = 1$ and $b = 2$ both give solution of $x < 1$ and $x > 1$ *(indicated by the number line)*	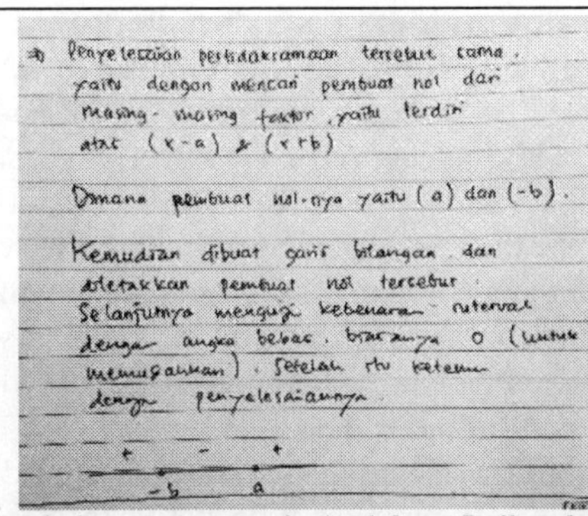 The solutions are same, obtained from finding the zero makers first, namely $(x - a)$ and $(x + b)$ resulting (a) and $(-b)$. Then, make a number line to put those two zero makers. Examine the correctness of interval by arbitrary numbers, usually 0 (to make it easy). It will give the solution
d) Level 3 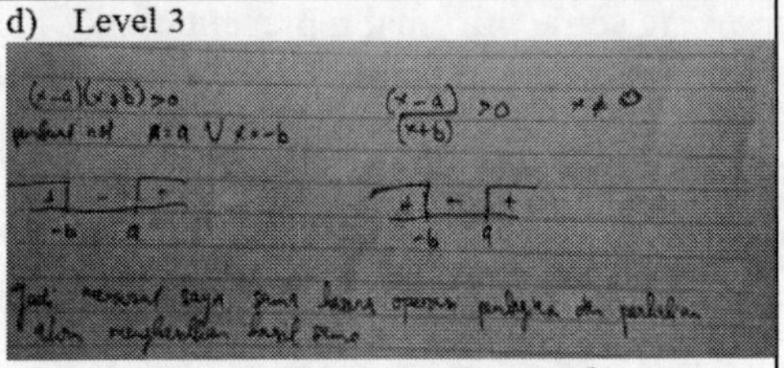 The zero makers are $x = a$ and $x = -b$ *(as shown in the number line created)*. So, both multiplication and division operation yield same results	
f) Level 5 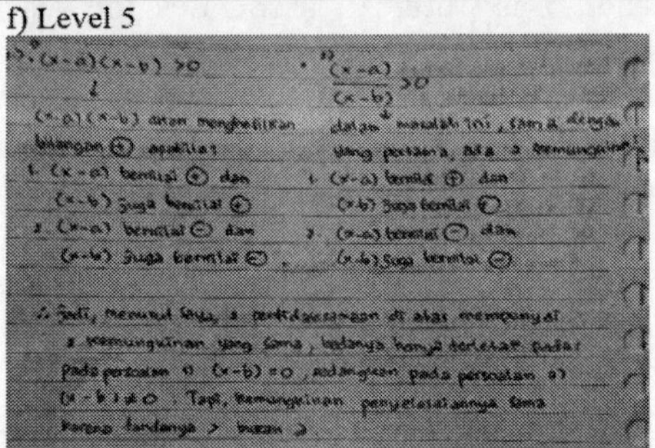 $(x-a)(x-b)$ will be positive, if: $(x-a)$ **is** + **and** $(x-b)$ **is** +; $(x-a)$ **is** − **and** $(x-b)$ **is** − $\frac{(x-a)}{(x-b)}$ will be positive, if: $(x-a)$ **is** + **and** $(x-b)$ **is** +; $(x-a)$ **is** − **and** $(x-b)$ **is** − So, those two inequalities have the same solution. The difference is only about $x - b = 0$ for the first, whereas $x - b \neq 0$. However, it still remaining same because te sign is >, not ≥.	g) Level 6 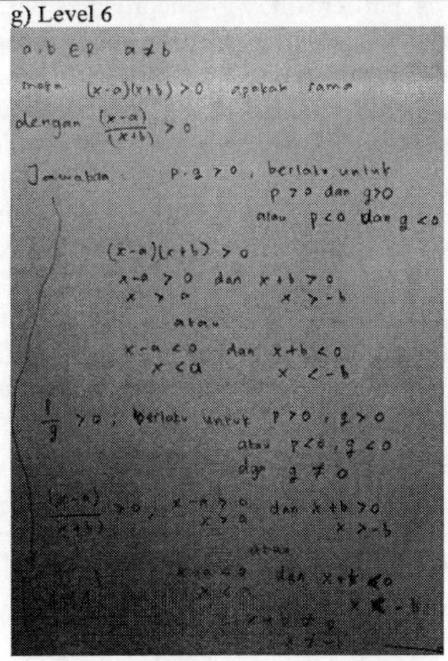

Figure 3. Student teachers' responses on task 2

Figure 3 a shows that the student teachers at level 0 did not use the idea of solving an inequality task correctly since the student teacher seemed mixed up the idea of solving an equation and an inequality. At level 1, we found some student teachers inferred that the implication (the solution of those two inequalities are same) is true based on some examples of numbers for a, b, and x substituted into the inequalities. See figure 3b. A little different from figure 3b, figure 3c indicates that the student teachers at level 2 solved for $\boldsymbol{x}$ by firstly substituting both $\boldsymbol{a}$ and $\boldsymbol{b}$ with certain numbers and verified the other inequality with those numbers to prove that both the two inequalities have the same solutions. Thus, they took a case-based sampling for $\boldsymbol{x}$. At level 3, the student teachers, as seen in figure 3d, no longer deduced by testing some numbers. Instead, they tried to solve each of the two inequalities by keeping $\boldsymbol{x}$, $\boldsymbol{a}$, and $\boldsymbol{b}$ as variables. However, it seemed that they did not consider carefully about the possibilities of values and the position of $\boldsymbol{a}$ and $\boldsymbol{b}$ in the number line they drew. Thus, they made major flaws to validate their inferences. At level 4, the student teachers made logical deductive inferences by arguing that if the open interval representated by the number line shown in figure 3e are examined with arbitrary numbers, then both the two inequalities would have same solutions. However, such claims were not supported by some arguments. Also, the positions of $\boldsymbol{a}$ and $(-\boldsymbol{b})$ in the number line is interchangeable as $\boldsymbol{a}$ can be more than $(-\boldsymbol{b})$ and vice versa. As such, the proof is not completed due to insufficient information related to the proof of the student teachers' claims. At level 5, on the other hand, the student teachers have completed deductive inferences by showing some informal representations. This is shown by Figure 3f pointing out that the proofs were presented in terms of case-based inferences. The student teachers categorized their inferences by considering the probable sign of both factors $(\boldsymbol{x}-\boldsymbol{a})$ and $(\boldsymbol{x}+\boldsymbol{b})$ and examined the value of the product of such two factors whether it is positive or negative. In this case, however, the student teachers used daily language, instead of mathematical representation using symbols, to prove their claims. At level 6, the difference is that the students teachers used mathematical symbols to represent all of their idea. Thus, the student teachers at this level applied deductive inferences as well as formal representations. See figure 3g for the detail work.

The prospective teachers in our study were the student teachers who start learning mathematics in a more formal way since they were still in the beginning in their studies in college when they were involved as participants. In accordance with this fact, Bas¸t¨urk [9] associates the reason for the difficulties experienced by the prospective teachers in performing proof with the fact that they suddenly come across proof in universities. His finding reporting that the difficulties experienced by them are because students in his country do not perform proof at an elementary school and high school levels are similar with what happened in the mathematics curriculum. Therefore, it is suggested that students must encounter proof in their educational lives before beginning their college education. Furthermore, to suggest on how proof is learned in college courses, Güler & Dikici [10] suggest to organized courses about proof by firstly giving samples → theorem → proof or samples → proof → theorem formats with the help of thoroughly selected examples and in a way that allows for understanding theorems, instead of only definition → theorem → proof or theorem → proof → examples. Furthermore, they also suggested that activities oriented towards different methods of proof must be performed in the courses of mathematical proof regarding how to begin the proof in order to increase the proof performances of the prospective teachers. More specifically, because we found many prospective teachers experienced difficulties mainly at the use of definition when proving, it is also suggested that the activities recommended by Güler & Dikici [10] would also be able to give attention on how prospective teachers develop their skill in composing proof by considering the limitation properties of symbols/variables they selected as well as how to use those kinds of functional language properly.

5. Conclusions

We would highlight that more than half of the students' proofs were coded as inductive inferences, which does not meet the requirement for doing proof, which is requiring deductive inferences, for the proof tasks examined in this study. This study suggests teacher educators in teacher colleges to reform the curriculum regarding proof learning which can accommodate the improvement of student teachers' proving ability from inductive to deductive proof as well from informal to formal proof.

6. References

[1] Jones K 1997 Student teachers' conceptions of mathematical proof *Mathematics Education Review* **9** 21-32

[2] Clark M and Lovric M 2009 Understanding secondary–tertiary transition in mathematics *International Journal of Mathematical Education in Science and Technology* **40** 755-776
[3] Blanton M, Stylianou D, David M 2003 The Nature of scaffolding in Undergraduate Transition to Formal Proof *In: Pateman N editors Proceedings of the 27th Conference of the International Group for the Psychology of Mathematics Education Honolulu* **2** 113-120
[4] Harel G and Sowder L 1998 Students' proof schemes: Results from exploratory studies Research in collegiate mathematics education **3** 234-283
[5] Knuth E J 2002 Secondary school mathematics teachers' conceptions of proof *Journal for research in mathematics education* 379-405
[6] Selden A and Selden J 2003 Validations of proofs considered as texts: Can undergraduates tell whether an argument proves a theorem? *Journal for research in mathematics education* 4-36
[7] Miyazaki M 2000 Levels of proof in lower secondary school mathematics *Educational Studies in Mathematics* **41** 47-68
[8] Lee K 2016 Students' proof schemes for mathematical proving and disproving of propositions *The Journal of Mathematical Behavior* **41** 26-44
[9] Basturk S 2010 First-year secondary school mathematics students' conceptions of mathematical proofs and proving *Educational Studies* **36** 283-298
[10] Güler G and Dikici R 2014 Examining prospective mathematics teachers' proof processes for algebraic concepts *International Journal of Mathematical Education in Science and Technology* **45** 475-497

IOP Conf. Series: Journal of Physics: Conf. Series **947** (2017) 012029 doi:10.1088/1742-6596/947/1/012029

The Implementation of Cumulative Learning Theory in Calculating Triangular Prism and Tube Volumes

M Muklis[1], C Abidin[1], M D Pamungkas[1] and Masriyah[1]
[1]Department of Mathematics, Universitas Negeri Surabaya, Surabaya 60231, Indonesia
Corresponding Author. Email: mmuklis16070785008@mhs.unesa.ac.id

Abstract. This study aims at describing the application of cumulative learning theory in calculating the volume of a triangular prism and a tube as well as revealing the students' responses toward the learning. The research method used was descriptive qualitative with elementary school students as the subjects of the research. Data obtained through observation, field notes, questionnaire, tests, and interviews. The results from the application of cumulative learning theory obtained positive students' responses in following the learning and students' learning outcomes was dominantly above the average. This showed that cumulative learning could be used as a reference to be implemented in learning, so as to improve the students' achievement.

1. Introduction

Cumulative learning is one of learning model which is related to knowledge and skill development. In psychology education, cumulative learning is considered be able to consolidate knowledge gained through experience. Then, it is allowed to be reproduced and exploited for subsequent learning situations through cumulative interaction between prior knowledge and new information. Gagné [1] argues that "cumulative learning" is built by the previous learning and relies on a combination of previous and memorable entities as well as their potential. Through their potential, the students are trusted to do cumulative learning. According to Maton [2], cumulative learning means that students are able to follow the new learning method developed based on their previously acquired knowledge or skills. In addition, Wiegand [3] studied whether the performance of a complex science problem was similar to the one used by Piaget in his studies as well as whether Piaget's task could be justified by Gagné's cumulative learning model with children as the participants through providing needed subordinate capabilities in the interval between two times of test. The results indicated that the cumulative effects of the learning on concretely referenced intellectual skills are more conducive to intellectual development than to the adaptation of structures of intellectual growth. Thus, Wiegand [3] described "A cumulative effect of learning within varied stimulus situation deals with the acquisition of skills, strategies, and learning sets that make the individual have to deal with environmental contingencies in more effective way". Furthermore, cumulative learning has effects which offer the potency to investigate the differential impacts of early and later experiences on the formation of lexical and other mental representations [10].

Gagné [1] explains the main theory which proposes that there are eight essential steps of instruction. As the first step or phase, the motivation phase refers to the set of induction process including giving motivation to the students to learn. Motivation is conceptualized as a situational constructive dependence on a one's moment-to-moment experiences and the related interpretation [13]. The second one is the apprehending phase which refers to a phase in which the students pay attention to the stimulus provided by the teacher so that they can catch and interpret the information themselves. It means that the learning is a unique process for each student and it becomes the student's responsibility towards their studies. The next one is the acquisition phase which is a phase in which the students acquire new knowledge by linking the information received from the previous knowledge. In other words, in this phase, the students form associations between the new information and the old information. The process emphasizes the student's ability to construct knowledge rather than to learn through memorizing [15,16]. The next phase is the Storage phase in which, in this phase, the information is stored in the short-term and in the long-term memory. Through repetition of information in short-term memory can be transferred to long-term memory. Next, the recall or retrieval phase, is a phase of remembering or recalling information that has been stored in memory. Sometimes, the information may be lost in memory or there may be a loss of connection with long-term memory. Furthermore, to strengthen the memory, the new and the old information is needed to be arranged in an organized and well-regulated way by groupings into categories. Therefore, the concept will be easier to recall. The next one is generalization phase which is the phase in which the information is transferred into new situations. In order to furtherly improve the memory, students can be assigned to apply certain material with the new information. Generalization itself can be described as the occurrence of functional behavior in response to conditions that are different from the moment when the behavior is trained first [18]. The next one is the appearance phase which refers to the phase in which the students must show a visible appearance after learning something. The feedback phase is the next phase in which the students should be given feedback from what has been shown (reinforcement). In this research, the researcher uses eight phases of cumulative learning in accordance with the theory which has been explained by Gagne including motivation, introduction, acquisition, storage, recall, generalization, performance and feedback. The activities can particularly be conceptualized through productive stimulations in the form of teacher's questioning and meaningful feedback [14]. Firstly, in the motivation phase, the teacher displays the geometrical pictures that are similar to triangular prism and tube. According Serio [20], motivation affects students in their learning process. Secondly, the introduction phase which uses teaching aids for the teacher in introducing the construction of triangular prism and tube. Third is the acquisition phase that is students gain new knowledge on how to calculate the triangular prism and tube volumes after the students do their worksheet. Fourth, the storage phase occurs through working on worksheet in which the students can find the formula of the triangular prism and the tube volume so that students will not forget it easily. Fifth, the recall phase is done when the teacher asks the students and they can answer the questions. Sixth, the generalization phase occurs when the teacher gives the students questions to calculate the volume of the triangular prism and the tube which relates to the real life. Seventh, the performance phase is when the teacher gives the students a chance to present their answers. Eighth, the feedback phase occurs when the teacher gives praise to the students who can show their answers. In order to make the students more active in learning and at the end of the lesson, the teacher gives a test. Cumulative learning means that students are able to use the new learning to develop previously acquired knowledge or skills. It means that students can apply what they have learn during their studies to unfamiliar workplace situations [11]. Although the students need to study in order to learn the subject matter, at school, it is the teacher's job to make the cumulative learning process occurs and goes on productively. Therefore, the teachers need to integrate the subject matter in order to understand it, plan the lessons which relates to it, represent it, demonstrate it, and explain it [17].

Based on Kilpert [4], cumulative learning would be further enabled in the curriculum if the teacher who deals with students' assessments is given more freedom to reward thinking and skills that are beyond the context of the students' general learning". From the statement above, the cumulative learning will be better applied in the learning process when the students are given more freedom to think and to create. However, this study tries to examine students' responses to assessment for evidence of cumulative learning. According to Aebli [5,6,7], the students should play an active role in the learning process by actively build new knowledge based on the existing or prior knowledge structures.

The purpose of this study is to describe the application of cumulative learning theory in calculating triangular prism and tube volume as well as revealing the students' responses toward learning mathematics using cumulative learning theory.

2. Method

This research was a qualitative descriptive research. The data collected was in the form of qualitative data which is a description of the nature of a symptom. The data was presented in accordance with the facts and phenomena that occur during the learning process [8]. The subjects of this research were the students of class VI of MI Subulussalam Blawe Kediri Indonesia. The method used to collect the data in this research was observation which is used to observe the students' activities during the learning process. The test method was administered in order to measure the students' learning mastery. The data of the students' learning result or achievement was obtained by giving test sheet to students. In addition, an interview was administered to confirm student's answer of the test. Moreover, questionnaire was used to know the students' responses toward cumulative learning. According to Arikunto [9] questionnaire is a written statement used to obtain information from respondents in the sense of reports in relation to the person or things that the respondents know.

3. Results

3.1. Observation

The observation was conducted during the learning activities is done by the subject teacher. The observation is done by observing the activities of students during the learning activities took place. The result showed that, in the class, the students were active in the teaching and learning process conducted by the teacher. They followed everything that the teacher instructed and led them to study well.

3.2. Students' responses

The students' responses can be seen from the questionnaire responses that have been distributed and submitted. The researchers distributed questionnaire to 22 respondents, the result of which was shown in Table 1.

Table 1. Students' responses

Category	Percentage
Students are motivated to follow the learning.	81,8%
Students are happy in learning math using media.	86,4%
Students understand and remember easily the concept of mathematics by finding the concept themselves.	77,3%
Students are not bored.	90,9%
Students are happy to be able to discuss with their friends.	95,4%
Students can express opinions in the process learning.	81,8%
Students feel that knowledge that has been learned is useful for the next subject material.	86,4%
Students find it easier to calculate prism and tube volumes after following the cumulative learning.	72,7%
Students are more confident in learning.	63,6%
Students are more interested in doing exercises.	77,3%

From the result of questionnaire of student responses, it is found that the highest percentage is on students feel happy because they can discuss with their friends and does not feel bored when following the lesson. In addition, they are able to follow the learning because it can exchange opinions with friends and not easily get bored. Meanwhile, the lowest percentage is on the students' self-confidence because when in the process of learning, some students feel less confident when the teacher asks them to explain the results of their work.

3.3. Test

The test was conducted at the end of the lesson with twenty-two students. The result showed that the average score was 73.95. Hence, it can be concluded that the students are considered passing the test if the score was ≥ 65. There were 4 of 22 students who did not pass the test.

3.4. Interview

There were five people or subjects who becomes the interviewee. The results are described as follows.
3.4.1. The first subject. The interview result stated that the subject felt very happy with the learning she got through. The subject felt free to express her opinions toward the learning process and the moment of discussing with friends. The subject did not feel difficult in learning and able to do the test by herself without any help.
3.4.2. The second subject. The subject preferred learning by discussion rather than listening to teacher's explanation. The subjects did not find any difficulty in following the learning because students were free to ask directly to the teacher and his friend. In addition, the student was able to do the tasks by himself.
3.4.3. The third subject. The interview result showed that the subject felt happy toward the learning process. The subject felt confident because she could find the step in calculating the volume of prisms and tube herself as well as discuss their ideas. The subject did not find any difficulties in the learning process. She did the test without her friends' help. The subject found the steps when she was joining the teaching leaning process.
3.4.4. The fourth subject. The result showed that the subject felt happy toward the teaching and learning process. The subject wisely used their time very well for discussion in the process of learning by asking to the teacher and his friends. The subject could do the tests by himself. The steps were retrieved from the teaching and learning process.
3.4.5. The fifth subject. The result showed that the subject felt happy toward the teaching and learning process. The subject could freely ask a question if there was a problem although the teaching and learning process was overall fun. The subject did the test given by the teacher by herself. The steps used were as same as what she had revealed in the teaching and learning process.

4. Discussion

From the result of questionnaire, the students' responses in following the cumulative learning were very good or positive. It was because when cumulative learning was implemented, the teacher used eight stages of cumulative learning theory, including when motivating the students and the other phases such as understanding, acquisition, storage, retention, generalization, appearance, feedback [1]. Based on Piaget's Theory of Cognitive Development, children of elementary school age (11-12 years) have not fully been able to think abstractly, therefore, the presence of concrete objects are still needed in the learning [12]. The teacher used props in the form of triangular prism and tube miniatures at the beginning of learning process as a tool to bring students into concrete forms. In addition, based on the results of students' learning achievement, it was found that eighteen students have completed the target while the other four students have not completed it yet. It was because the students made a mistake in performing the operation of calculation. It is revealed from the yield of relatively good long-term retention after a single study session [19].

5. Conclusion

Based on the results of this research, it can be concluded that the implementation of learning by applying cumulative leaning theory in calculating triangular prism and tube volumes on sixth grade students used eight stages or phases of the theory. The result of the questionnaire showed that the students' responses toward the implementation of cumulative learning were very good or positive. Moreover, the result of the test showed that the students' scores increased with the average score 73,95. There were 18 students who passed the test and only 4 students who failed. It showed that the cumulative learning theory can be implemented in teaching and learning process. Through that way, it is expected that the students' learning achievement will improve.

References

[1] Gagné R M 1968 Learning Hierarchies *Educational Psychologist* **6(1)** pp 1–9
[2] Maton K 2009 Cumulative and segmented learning: Exploring the Role of Curriculum Structures in Knowledge-Building *British Journal of Sociology of Education* **30(1)** p 43
[3] Wiegand V K 1970 *A study of subordinate Skills in Science Problem Solving* p 16 (Washington D C: Bureau of Research University of California)
[4] Kilpert L 2011 Lighting Fires from Fires: A Description of The Research Entitled 'Kindling Fires' *The Journal of Independent Teaching and Learning* **6** p 22
[5] Aebli H 1978 *Von Piagets Ent Wick Lungs Psychologie zur Theorie der Kognitiven Sozialisation* (Germany: Kindler)
[6] Aebli H 1983 *Zwölf Grundformen des Lehrens: Twelve Basic Forms of Teaching* (Germany: Klett-Cotta)
[7] Aebli H 1987 *Grundlagen des Lehrens: Basic Forms of Teaching* (Germany: Klett-Cotta)
[8] Riyanto Y 2007 *Metodologi Penelitian Pendidikan Kualitatif dan Kuantitatif* (Surabaya: Unesa University Press)
[9] Arikunto S 2006 *Prosedur Penelitian Suatu Pendekatan Praktek* (Jakarta: Rineka Cipta)
[10] Izura C et al 2011 Age/Order of Acquisition Effects and the Cummulative Learning of Foreign Words: A Word Training Study *Journal of Memory and Language* **64** p 32–58
[11] Kilpert L and Shay S 2013 *Kindling Fires: Examining the Potential for Cumulative Learning in A Journalism Curriculum* Teaching in Higher Education **18(1)** p 40-52
[12] Gintinga H and Suryab E 2017 Development Learning Device Based for Measuring Contextual Critical Thinking Skills Students SD Class VI Mathematical *International Journal of Sciences: Basic and Applied Research* **33** pp 301-310
[13] Kiemer K et al 2015 Effects of A Classroom Discourse Intervention on Teachers' Practice and Students' Motivation to Learn Mathematics and Science *Learning and Instruction* **21** pp 94-103
[14] Jurik V, Groschner and Seidel T 2014 Predicting Students' Cognitive Learning Activity and Intrinsic Learning Motivation: How Powerful are Teacher Statements, Student Profiles, and Gender? *Learning and Individual Differences* **32** pp 132-139
[15] Othman H et al 2010 Applied Mathematics Cans Enhance Employability Skills Through PBL **8** *Social and Behavioral Sciences* pp 332–337
[16] Noor L A et al 2011 Would Problem-Based Learning Affect Students' Generic Competencies? *African Journal of Education and Technology* **1(3)** pp 1–14
[17] Lampert M 2010 Learning Teaching in, from, and for Practice: What Do We Mean? *American Association of Colleges for Teacher Education* **19** p 22

[18] Poncy B C et al 2010 Evaluation of Techniques to Promote Generalization with Basic Math Fact Skills *Oklahoma State's Journal* **420** pp 76-77
[19] Rawson K A and Dunlosky J 2011 Optimizing Schedules of Retrieval Practice for Durable and Efficient Learning: How Much is Enough? *Journal of Experimental Psychology* **37** pp 899-912.
[20] Serio A et al 2013 Impact of an Augmented Reality System on Students' Motivation for A Visual Art Course *Computers & Education* **68** pp 586–596

IOP Conf. Series: Journal of Physics: Conf. Series **947** (2017) 012030 doi:10.1088/1742-6596/947/1/012030

Students' Relational Thinking of Impulsive and Reflective in Solving Mathematical Problem

M A Satriawan[1], M T Budiarto[2], T Y E Siswono[3]

[1]Mathematics Education Program, Postgraduate Program, Universitas Negeri Surabaya, Jl. Ketintang, Surabaya 60231, Indonesia
[2]Departement of Mmathematics Education, Universitas Negeri Surabaya, Jl. Ketintang, Surabaya 60231, Indonesia

E-mail:m.satriawan@mhs.unesa.ac.id

Abstract. This is a descriptive research which qualitatively investigates students' relational thinking of impulsive and reflective cognitive style in solving mathematical problem. The method used in this research are test and interview. The data analyzed by reducing, presenting and concluding the data. The results of research show that the students' reflective cognitive style can possibly help to find out important elements in understanding a problem. Reading more than one is useful to identify what is being questioned and write the information which is known, building relation in every element and connecting information with arithmetic operation, connecting between what is being questioned with known information, making equation model to find out the value by using substitution, and building a connection on re-checking, re-reading, and re-counting. The impulsive students' cognitive style supports important elements in understanding problems, building a connection in every element, connecting information with arithmetic operation, building a relation about a problem comprehensively by connecting between what is being questioned with known information, finding out the unknown value by using arithmetic operation without making any equation model. The result of re-checking problem solving, impulsive student was only reading at glance without re-counting the result of problem solving.

1. Introduction

Thinking activity cannot be separated from daily life. In fact, thinking is an activity which cannot be avoided. Thinking process of each person is different one other. One of thinking processes is relational thinking, where it can be said as the ability to understand analogy between object and event which considered differently and applying an abstract rule in a new situation [1].

Relational thinking involves using fundamental properties of number and operations to transform mathematical expressions rather than simply calculating an answer following a prescribed sequence of procedures [2].This is because of relational thinking considers mathematical proposition which presents relation between two mathematical objects, A and B with second mathematical proposition which presents the relation between concept A and C, and then sum up the relation between B and C. In the process of learning mathematic, problem solving is an important thing for students. Generally, a problem is inexpediency between a fact and expectation. This means a problem is a situation that confronts a person, that requires resolution, and for which the path to the solution is not immediately known [3].

Every individual has his/her own characteristics in solving mathematical problem which makes everyone different one another. Problem solving is an effort to find a way of difficulty, a way around an obstacle, and attaining an aim that is not immediately understood [4]. This difference is caused by several factors; one of which is cognitive style. Cognitive style is understood to be an individual's preferences and habitual approach to organize and represent information [5]. Knowledge about cognitive style gives a lot of information about individual differentiation. Cognitive style functions as an individual's characteristic and consistent approach to organize and process information [6].There are 19 types of cognitive style. In general this type of cognitive style is used to classify learners based on their decision making speed and its differences can be explained by speed, accuracy, and electiveness shown for searching and manipulating the information [7]. Frequency of reflective-impulsive students in class generally counted as much as 76.2%. Besides that, reflective-impulsive style is a cognitive style which is easier to be noticed by teacher in teaching and learning process than any other cognitive styles.

This research focuses on impulsive-reflective students because based on author's observation, the amount of students is bigger than group of fast and accurate students, and group of slow and inaccurate students. Then in grouping this cognitive style is based on students' taken times to solve the problem in a certain level of accuracy from the answer given [8]. Reflective students are defined as students who need more time in solving problem, but their solution or answer given is accurate, it is therefore the given answer tends to be correct. Students who need short time in solving problem but give inaccurate and tend to be having wrong answer are classified as impulsive students. Impulsivity is a cognitive style in which individuals act before they think. Reflectivity refers to a cognitive style in which individuals thinks before they act, usually scanning information carefully and slowly [9].

The objective of this research is to describe relational thinking of junior high school students in mathematics problem solving based on cognitive style of reflective and impulsive. According to cognitive development theory, students of junior high school are in formal operational stage where they start to think abstractly, reason logically, and draw a conclusion from information given [10]. Stephen (2012) stated that relational thinking allows students to develop a strong understanding and foundation of algebra. It is importance to think of relational thinking for the students in learning mathematic because it is related to problem solving. Therefore it is beneficial for teacher to collect information on how the students relate information from the problem given with the concept they have understood. Relational thinking enables students to develop their understanding and strong foundation about algebra [11]. The research did not focus on male or female, but focuses on students' reflective-impulsive cognitive style.

2. Method

This is a descriptive research with qualitative approach. It is a research that uses qualitative data and describes to produce information about students' relational thinking in solving problem. This research is a description towards the result and interview of the research subjects. The researcher learned, analyzed, and explained descriptively about students' relational thinking in solving mathematical problem from the perspective of reflective-impulsive cognitive style from the data of test result in written form and interview.

The procedure of this research was done in several stages starting on arranging additional instrument draft and doing validation of instrument by validation experts of mathematical education field. The experts were given research instrument of contextual exercises algebral materials in a form of validation sheet including Draft I which was going to be validated. The validation sheet is used to collect the data about opinion, suggestion, and comment of the expert about Draft I. The instrument is categorized as valid if the conclusion on validation sheet can be used with a little revision or without revision. In collecting data, the researcher gave a writing test to research subject, then it was continued with interviews. Data analysis in this research is by data reduction, presentation and conclusion. The objective of data analysis is to facilitate the interpretation of data and make the data meaningfull. All techniques were purposed to eleborate the data, configure the data, and manage the data.

This research was conducted in junior high school's students; with research subjects were 2 students of grade VIII. The researcher's consideration to take this class, simply because according to the theory of cognitive development from Piaget, the age of 13-15 is in a stage of formal operational when they start to be able to think abstractly, to give reason logically, and to draw a conclusion from information they that have and able to use their own opinion. The research subjects in this research were two students consisting of one student with reflective style and another one with impulsive cognitive style. Both students had similar ability, were communicative, and had same sex. Communicative students in this research are students who are able to deliver the meaning of what they have written orally and systematically. While, to find students who have similar ability, the researcher had discussed the teacher of the students.

Selecting research subjects was done by giving a test of cognitive style from Kagan (1964). The instrument used was cognitive style test *(matching familiar figure test)*. In a class, the researcher gave a cognitive style test and then students were formed into two groups, reflective group and impulsive group. After both groups were fulfilled, the researcher chose one of the students from each group who could communicate fluently and was willing to be interviewed.

3. Results and discussion

This research purposes to describe a process of relational thinking of students' reflective and impulsive cognitive style in solving mathematical problem. The trends of reflective students are more thoroughly than the impulsive students. For a complete description, it can be seen on table 1.

Table 1. A Summary of students' relational thinking of reflective cognitive style and impulsive cognitive style in solving mathematical problem

Stage of Polya problem solving	Relational Thinking Description	
	Reflective Student	Impulsive Student
Understanding the problem	➢ Relating information which has been got with accumulation, subtraction, division, and multiplication operation. ➢ Relating between known information in a question with something being questioned to answer what is being questioned.	➢ Relating information which has been got with accumulation, subtraction, division, and multiplication operation. ➢ Relating between known information in a question with something being questioned to answer what is being questioned.
Devising a plan	Relating every step which is going to be used in solving problem, such as: using multiplication operation to decide the value which has not been known yet and using substitution to decide what is not known.	Relating every step which is going to be used in solving problem, such as: using multiplication operation to decide the value which has not been known yet and using substitution to decide what is not known.
Carrying out the plan	➢ Making equation and using substitution to decide value which has not been known. ➢ Using multiplication and addition to decide what is not known.	➢ Using multiplication operation and division operation to decide value which has not been known.

Stage of Polya problem solving	Relational Thinking Description	
	Reflective Student	Impulsive Student
		➢ Using multiplication, addition, Subtraction operation to decide what is not known.
Looking back	Reading, calculation, and re-checking every step which has been done in solving problem.	Reading at glance, giving correction, without re-calculating result of problem solving.

3.1 Relational thinking of reflective student

According to the result of the research, it is found that students' relational thinking of reflective can solve mathematical problem when understanding the problem and finding out important elements in solving the problem, that is by identifying what is being asked and finding out the information in a question, and also mentioning what is being asked and write down completely what is known. Reflective student builds relation in every element by relating information which has been known with accumulation, multiplication, division, and subtraction operation. Based on the results of the interview reflective student also built a relation about a problem entirely, that is what has been known has a relation with clues in doing arithmetic operation. It means that information which is achieved by reflective student is linked with accumulation, subtraction, division, and multiplication operation. Furthermore, there is a relation between what is known and what is being questioned; therefore what is being known is used to answer what is being questioned.

Reflective students wrote steps which are going to be used to solve the problem. Students' understanding towards the problem relates with indentified information. However, knowledge has relation with the choice of arithmetic operation which is going to be used in planning problem solving. The operations used are multiplication and addition. Students' understanding about symbols, characteristics or rules to produce a model can be seen when the students make equation model from information which is got in problem solving task. Moreover, student used substitution way to find out the value of each A, B, C, and D, and value of P, Q, and S. Reflective students also build relation between unknown numbers and arithmetic operation, that is by understanding the goal of using addition operation to find out buying price and selling price and using addition and subtraction operation to find out the total price and profit.

In the stage of problem solving plan, reflective students build a relation between numbers and arithmetic operation used and the relation between multiplication and division in executing plan. This can be seen when students solve the problem given, the students use a same way with what has been planned that is by doing multiplication, accumulation, subtraction and substitution operational to solve a question. While in the stage of re-checking, students build relation by reading, re-checking every step and re-calculating the result of operation. Re-calculation has a relation with what is on the question and answer sheet, while the clue has relation with suitability of arithmetic operation used in answering question.

According to the result of problem solving and explanation above, it can be seen that reflective students write down in detail and complete in every step of solving the problem. According to the time needed, reflective students seem to need a longer time to solve the problem. Besides that, during interview, students answer the question clearly and carefully. It is in line with Kagan who stated that reflective student has slow characteristic in answering question even it tends to be a right answer [12]. Furthermore, Kagan stated that people with reflective cognitive style tend to be cautious and use all alternatives. Time needed is longer in responding, but error made tend to be small. Moreover, as stated by Philip et al, reflective students consider various alternative before responding, so the possibility of right answer is high [13].

3.2 Relational thinking of impulsive student

According to the result of the research, it can be seen that impulsive students' relational thinking in solving mathematical problem when understanding problem on indicators find important elements in understanding the problem, that is by mentioning what is being asked and write down various information which have been known. Impulsive students build relation in every element. Information which has got by impulsive student is connected with accumulation, subtraction, division, and multiplication operation. Impulsive students build connection about the entire problem by using information which has been known to answer the question.

Students write down plan steps which are going to be used to solve the problem. Impulsive students build relation between unknown numbers and arithmetic operation by using adding operation to know buying price and selling price and using adding operation to know selling price and using adding and subtraction operation to know total price and profit.

In the stage of problem solving plan, impulsive students combine relation on plan execution, which is used to solve the problem based on the plan which has been made before, such as: solving multiplication, accumulation, subtraction, and division operation to solve the question. Impulsive students only use multiplication and division operation on indicator of making problem model to find out the price of each A, B, C, and D, and each price of P, Q, R, and S. While in stage of re-checking problem solving, impulsive students build relation by reading at glance without re-calculating the result of problem solving.

According to the result of the problem solving and explanation above, it can be seen that impulsive students do not write down in detail every step in solving the problem. There are relationships between refletive-impulsive aspect and relational thingking both students' connection of each information which has been known in exercise of problem solving with aritmetic operation in understanding the problems. There are a lot of numbers appear without any explanation. The time needed in solving the problem is also short and in doing interview, impulsive students tend to give a quick response without firstly analyzing the problem encountered. This is in line with Kagan who stated that impulsive students use alternatives briefly and quickly to solve something [12]. Then, impulsive students also use a short time to give response, but tend to make error because they do not use all the alternatives. Moreover, Philip et al, also stated that impulsive student is a fast-response student in every situation, but the first response given tend to be wrong [13].

4. Conclusion

Relational thinking of reflective and impulsive subjects in solving mathematical problem is different. In the stage of executing mathematical problem solving, reflective subject is really cautious, so when there is any mistake, reflective subject will directly realize and correct it. However, impulsive subjects tend to be quick and less cautious in solving the problem encountered, so when there is any mistake, impulsive subject does not realize it and only know a part of information of question. Therefore, researcher suggests that the teacher give more attention to the students with different cognitive style, in addition, the process of learning should also be considered to receck the result of work to avoid the incorrect answer.

5. Acknowledgements

The researcher would like to thank to the consultants, because of them this journal can be accomplished.

References

[1] Doumas L A A and Hummel J E 2005*Approaches to Modelling Human Mental.* Representations*: What Works, What Doesn't, and Why In Holyoake K J and Morisson. RG (Ed).The Cambridge Handbook of Thinking and Reasoning*(Cambridge: University Press)

[2] Carpenter T P, Frunke M L, Madison, LeviL and Zeringue J K 2005Algebrain elementary school: developing rational thinking *ZDM***37(1)** pp 53-59

IOP Conf. Series: Journal of Physics: Conf. Series **947** (2017) 012030 doi:10.1088/1742-6596/947/1/012030

[3] Posamentier A S, Krulik S 1998*Probelm solving strategies for efficient and elegant solutions: A Resource for the Mathematics Teacher*(Corwin Press, Inc. California USA)

[4] Polya G 1973 *How to Solve it.* 2nd Ed. (Princeton University Press, ISBN 0-691-08097-6)

[5] Michael J, et al 2010*Multiple Perspectives on Problem Solving and Learning in the digital age* (New York: Springer Science + Business media)

[6] Tennant M 1988*Psychology and Adult Learning*(London: Routledge)

[7] Lagana A, et al 2004 Computational science and its applicatio-iccsa 2004 *International Conference AssisiJerman: Springer-Veriag Berlin Heidelberg*

[8] Rozencwajg P and Corroyer D 2005cognitive processes in the reflective-impulsive cognitive *Style. The Journal of Genetic Psychology***166**(4) pp451-463 (Doi.org/10.3200/GNTP.166.4.451-466)

[9] Santrock J W 1998*Adolescence*(Boston: McGraw-Hill)

[10] Genovese J2003 Piaget, pedagogy, and evolutionary psychology *Journal Evolutionary Psychology***1** pp 127-137

[11] Stephens C A 2006 Equivalence and rational thinking preservice elementary teachers awareness of opportunities and misconceptions *Journal of Mathematics Teacher Education***9** pp 249-278

[12] Kagan J1966Reflection-impulsivity: the generality and dynamics of conceptual tempo *Journal of Abnormal Psychology* **71 (1)**pp 17-24

[13] Philipet al 1997 The effects of verbal and material rewards and punisher on the performance of impulsive and reflective children *Child Study Journal***7 2** pp71

IOP Conf. Series: Journal of Physics: Conf. Series **947** (2017) 012031 doi:10.1088/1742-6596/947/1/012031

Creativity of Field-dependent and Field-independent Students in Posing Mathematical Problems

N Azlina[1,3], S M Amin[2], A Lukito[2]

[1]Primary Education Studies Program, Universitas Negeri Surabaya
[2]Mathematics Department, Universitas Negeri Surabaya

[3]Email: nurazlina@mhs.unesa.ac.id

Abstract. This study aims at describing the creativity of elementary school students with different cognitive styles in mathematical problem-posing. The posed problems were assessed based on three components of creativity, namely fluency, flexibility, and novelty. The free-type problem posing was used in this study. This study is a descriptive research with qualitative approach. Data collections were conducted through written task and task-based interviews. The subjects were two elementary students. One of them is Field Dependent (FD) and the other is Field Independent (FI) which were measured by GEFT (Group Embedded Figures Test). Further, the data were analyzed based on creativity components. The results show thatFD student's posed problems have fulfilled the two components of creativity namely fluency, in which the subject posed at least 3 mathematical problems, and flexibility, in whichthe subject posed problems with at least 3 different categories/ideas. Meanwhile,FI student's posed problems have fulfilled all three components of creativity, namely fluency, in which thesubject posed at least 3 mathematical problems, flexibility, in which thesubject posed problems with at least 3 different categories/ideas, and novelty, in which the subject posed problems that are purely the result of her own ideas and different from problems they have known.

1. Introduction

Mann and Singer stated that specifically, mathematics creativity refers to the growth of mathematics as a whole. Therefore, the failure in including creativity in mathematic class will eliminate students' opportunities to enhance their understanding of mathematics [1, 2]. Students' creativity must be facilitated in order to have maximum growth, but previous research conducted by Jellen and Urban revealed that Indonesian students' creativity were in the low level compared to the other countries. Indonesian students' lack of creativity cannot be separated from the educational influences [3]. In addition, the research that was conducted by Mann, Hirsh,and Savic revealed that the learning process and assessment of traditional mathematic class which emphasized on speed and accuracy in counting with limited problem-solving did not give adequate opportunities for the students to develop their creativity in mathematic class [4, 5, 6]. One way to support students' creativity in mathematic class is trough problem-posing activity. Lavy and Shriki, Georgiev and Nedyalkova, Singer, Unal and Arikan state that problem-posing activity can be viewed as a wayto explore students' mathematical thinking in order to build up their mathematical understanding and to foster their creativity [7, 8, 9].

According to Woodman and Schoenfeldt, every student has different creativity from one another [10]. One of the reasons about the differences in their creativity is because they have different cognitive styles. Someone's cognitive style can be differed from many characteristics. Witkin introduced Field-

Dependent (FD) and Field-Independent (FI) as types of cognitive style. Idrisexplained that, generally, the characteristic differences between Field-Dependent individual and Field-Independent individual is on how the surroundings affect their perspective about something [11]. A research about creativity conducted by Rahman revealed that Field-Independent students were better in terms of posing mathematical problems than Field-Dependent student [12]. Another research about creative thinking process conducted by Mudrika revealed that Field-Independent students were better in explaining, verifying and checking their problem-posing procedures than Field-Dependent student [13]. Those researches mentioned earlier showed the superiority of Field-Independent students compared to Field-Dependent ones.

This study aims at describing the creativity of Field-Dependent and Field-Independent students in mathematical problem-posing. The creativity of the posed problems was reviewed from three components of creativity, namely fluency, flexibility and novelty [14]. Fluency in problem posing refers to the number of problems posed. Flexibility refers to the number of different categories of generated problems. Novelty refers to how rare the response was in a set of all responses known by the subject.

2. Research Methods

This study uses descriptive design with qualitative approach. Before conducting the research, several research instruments such as mathematical problem-posing task sheet and interview guidelines were established and validated. The determination of the subjects' cognitive style was done through GEFT(Group Embedded Figures Test). The GEFT results decided two subjects with different cognitive style, one subject is Field-Dependent and the other subject is Field-Independent. Aside from GEFT results, the subjects were also determined through the consideration that both of them have similar mathematic achievements and have good communication skills in order to carry out the interviews in order to get a complete data of their creativity in mathematical problem-posing.

The data were conducted through written task about mathematical problem-posing and task-based interviews. This study used semi-structured type and pre-solution type problem-posing since the students were asked to pose mathematical problems freely based on the given information. The given information in the mathematical problem-posing task sheet were three different plane figures. The given information in the first mathematical problem-posing task consists of the figures of kite, rhombus and rectangle. The given information in the second mathematical problem posing task consists of the figures of right triangle, parallelogram and square.

Time triangulation and member-check were used to check the credibility of the data. The data of students' creativity in mathematical problem posing were analyzed based on three creativity components (fluency, flexibility and novelty). Fluency component is fulfilled if the student posed at least three mathematical problems [14]. Flexibility component is fulfilled if the student posed mathematics problems with at least three different ideas[14]. Novelty component is fulfilled if the student posed at least one mathematical problem that is purely the result of her own ideas and its different from the problems she has known[14].

3. Results

The results of this study are the description of the creativity of Field-Dependent and Field-Independent students in mathematical problem-posing which is described in detail as follows.

3.1. Field-Dependent Student

Field-Dependent student posed five mathematical problems in the first mathematical problem-posing task and also five mathematical problems in the second task. Fluency component is fulfilled if the student posed at least three mathematical problems. Hence, it can be stated that Field-Dependent student's posed problems have fulfilled the fluency component. In the first mathematical problem-posing task, Field-Dependent student posed five problems with four different ideas, such as the area of kite, the volume of cuboid, the area of rectangular, and the similarities. In the second mathematics problem-posing task, Field-Dependent student posed five problems with five different ideas, such as the angle of right triangle, the length of rectangular, the volume of cubic, the circumference of square, and the characteristics of square. Flexibility component is fulfilled if the student posed mathematical problems with at least three different ideas. Hence, it can be stated that the Field-Dependent student's posed problems have fulfilled the flexibility component.

However, among all of her mathematical posed problems, none of them is purelythe result of her own ideas not it is different from the problems she knows. Trough task-based interviews, she stated that she posed problems that are similar to the ones she had solved in the past. She just changed the number but the big ideas were all the same without any modifications that differentiate them from the problems that she knows in the past. Novelty component is fulfilled if student posed at least one mathematical problem that is purely the result of her own ideas and different from the problems she knows. Hence it can be stated that Field-Dependent student's posed problems have not fulfilled the novelty component. The posed problems by Field-Dependent student can be seen in Figure 1.

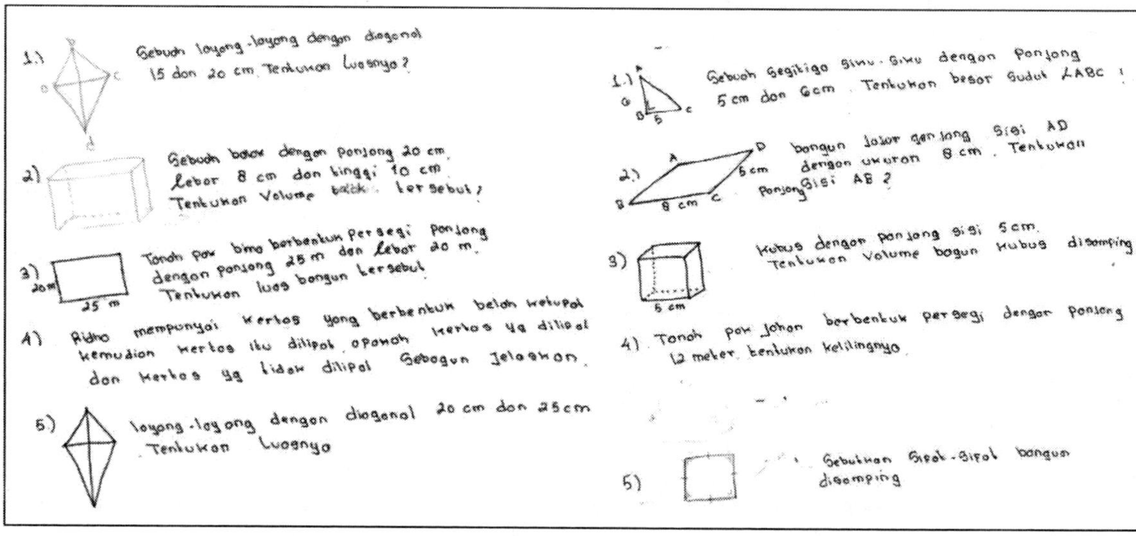

Figure 1. Field-Dependent Student's Posed Problems.

The summary of Field-Dependent student's creativity in posingmathematical problem can be seen in the following table.

Table1. The Creativity of Field-Dependent Student in Math Problem-Posing.

CreativityComponents		**Description**
Fluency	√	Posed at least three mathematics problems.
Flexibility	√	Posed mathematics problems with at least three different ideas.
Novelty	-	Did not posed at least one mathematical problem that is purely the result of her own ideas and different from the problems known to her.

* √ meansfulfilled

3.2. Field-Independent Student

Field-Independent student posed four mathematical problems in the first mathematical problem-posing task and five mathematical problems at the second task. Fluency component is fulfilled if the student posed at least three mathematical problems, hence it can be stated that Field-Independent student's posed problems have fulfilled the fluency component. In the first mathematical problem-posing task, Field-Independent student posed four problems with four different ideas, such as the area of kite, the similarity, the area of rectangular, and the volume of cube. In the second mathematical problem-posing task, the Field-Independent student posed five problems with five different ideas, such as the number of roses (the circumference of parallelogram), the volume of cube, the area of triangle, the size of the plane thatis

IOP Conf. Series: Journal of Physics: Conf. Series **947** (2017) 012031 doi:10.1088/1742-6596/947/1/012031

made from a square sheet that was cut the same size, and the area and circumference of a square. Flexibility component is fulfilled if student posed mathematical problems with at least three different ideas. Hence, in can be stated that Field-Independent student's posed problems have fulfilled the flexibility component.

Among all of her mathematical posed problems, there is one problem in every task that she posed which is purely from the result of her own ideas and different from the problems she has known. In the first mathematical problem-posing task, there is one problem that she posed by modifying the ideas of the problems she knows in the past. She added the ideas of kite's area and the similarity. The problem posed is an open-ended problem that has many solutions. She also posed a problem that is purely from the result of her own ideas and different from the problems she knows in the second problem-posing task. She posed a problem, the idea of which came from many papers that she saw in front of her. The problem is about the size of the plane that is made from cutting or folding a sheet of square paper several times with the same size in every cuts or folds. That problem is also an open-ended problem that has many solutions. Novelty component is fulfilled if student posed at least one mathematical problem that is purely the result of her own ideas and different from the problems she knows. Hence, it can be stated that Field-Independent student's posed problems have fulfilled the novelty component. The posed problems by Field-Independent student can be seen in Figure 2.

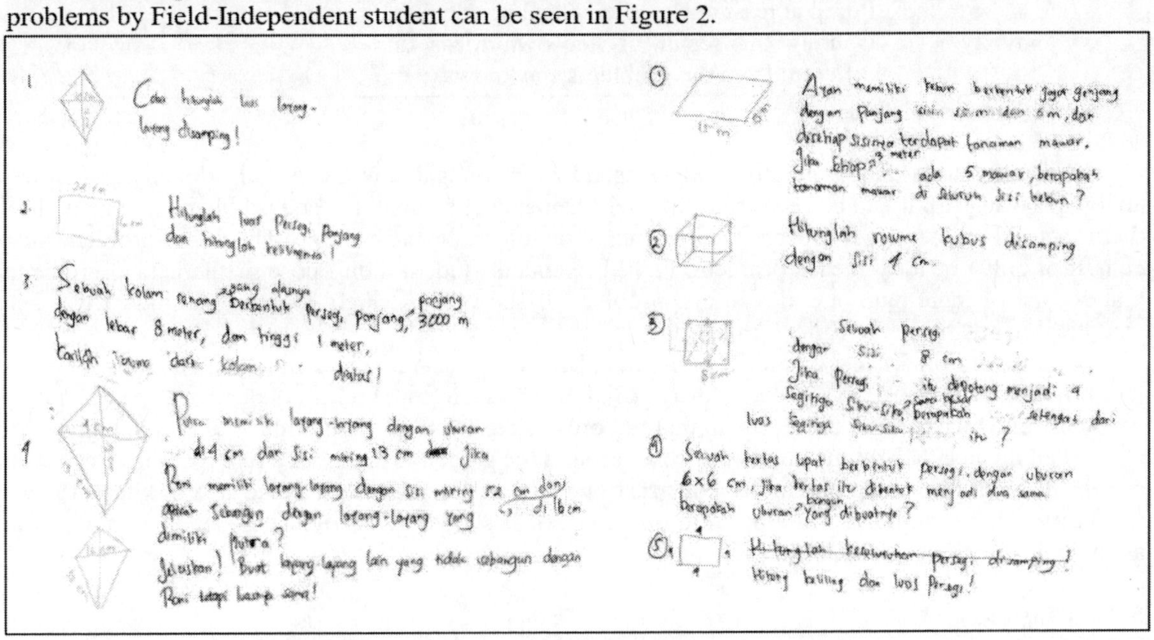

Figure 2. Field-Independent Student's Posed Problems.

The summary of Field-Independent student's creativity in mathematical problem-posing can be seen in the following table.

Table2. The Creativity of Field-Independent Student in Math Problem Posing.

Creativity Components		**Description**
Fluency	√	Posed at least three mathematics problems.
Flexibility	√	Posed mathematical problems with at least three different ideas.
Novelty	√	Posed at least one mathematical problem that is purely the result of her own ideas and different from the problems she knows.

* √ meanssatisfied

4. Discussion

From the previous descriptions, it is revealed that there is a correlation between students' cognitive style and their creativity in posing mathematical problem. Therefore, this study tried draw outline of the creativity of Field-Dependent and Field-Independent Students in mathematical problem-posing. The summary of Field-Dependent and Field-Independent students' creativity in mathematical problem posing were reviewed trough three creativity's componentswhich can be seen in the following table.

Table3. CreativityRecapitulation of FD and FI Students in Mathematical Problem-Posing

Creativity's Components	Indicator	FD Student	FI Student
Fluency	Posed at least three mathematics problems.	√	√
Flexibility	Posed mathematics problems with at least three different ideas.	√	√
Novelty	Posed at least one mathematical problem that ispurely the result of her own ideas and different from the problems she knows.	-	√

* √ meanssatisfied

Brosnan states that, generally, Field-Dependent individual saw the stimuli globally and in less analytical way [15]. It can be seen from the Field-Dependent student's posed problems in which all of them focused on the global aspects such as area, circumference and volume. The posed problems also consist of only one idea in each problem. Field-Dependent student didn't pose mathematical problems that consistsof more than one idea in one problem. It showed that Field-Dependent student was less analytical in posing mathematical problems.

Different from Field-Dependent Individual, Brosnan states that Field-Independent individual saw the stimuli in a structured and analytical way [15]. It can be seen from Field-Independent student's posed problems in which there is one problem that not only asked for the circumference of parallelogram but also asked the detail about the number of roses around the garden if there were five roses in every three meters. There is also one problem that consists of more than one idea, such as the area of kites and their similarity. Those two ideas were put into one context. It showed that Field-Independent student was analytical in posing mathematical problems.

5. Conclusions

From the descriptions explained previously, it can be concluded that Field-Dependent student's posed problems fulfilled only two components of creativity, namely fluency since she posed at least three mathematical problems and flexibility since she posed mathematical problems with at least three different ideas. In the other hand, Field-Independent student's posed problems have fulfilled all the three components of creativity, namely fluency, since she posed at least three mathematics problems, flexibility, since she posed mathematical problems with at least three different ideas, and novelty, since she posed at least one mathematical problem that ispurely the result of her own ideas and different from the problems she knows.

References

[1] Singer F M2012 Exploring Mathematical Thinking And Mathematical Creativity Through Problem Posing*Int. Workshop of Israel Science Foundation*
[2] Mann E L 2005*Mathematical Creativity and School Mathematics: Indicators of Mathematical Creativity in Middle School Students* (Utrecht: Utrecht University). http://www.fisme.science.uu.nl/publicaties/ literatuur/2005_mann_creativity.pdf

[3] Jellen H G & Urban K K 1989 Assessing creative potential world-wide: The first cross-cultural application of the Test for Creative Thinking-Drawing Production (TCT-DP). *Gifted Education International*, **6(2)**78-86.
[4] Mann E L 2006 Creativity: The Essence of Mathematics*Journal for the Education of the Gifted***30(2)** 236–260.
[5] Hirsh R A 2010 Creativity: Cultural Capital in the Mathematics Classroom *Creative Education 2010***1(3)**154-161.
[6] Savic M 2016 Mathematical Problem-Solving Via Wallas' Four Stages Of Creativity: Implications For The Undergraduate Classroom*The Mathematics Enthusiast***13(3)**.
[7] Lavy I & Shriki A 2010 Engaging In Problem Posing Activities In A Dynamic Geometry Setting And The Development Of Prospective Teachers' Mathematical*The Journal of Mathematical Behavior***12(002)**.
[8] Georgiev V& Nedyalkova V 2011 *Group Creativity And Development Of Mathematical Problem Posing And Solving Capabilities* (Pisa: University of Pisa)
[9] Unal H & Arikan E E 2014 Development of The Structured Problem Posing Skills And Using Metaphoric Perceptions*European Journal of Science and Mathematics Education***2(3)** 155-166.
[10] Woodman R W& Schoenfeldt L F 1989 *Handbook of Creativity Third edition*(New York: Spriger Business Media).
[11] Idris N 2006 *Teaching and Learning Mathematics: Making Sense and Developing Cognitive Abilitie)*(Malaysia: Utusan Publication & Distributios SDN BHD).
[12] Rahman A 2010 *Profil Pengajuan Masalah Matematika berdasarkan Gaya Kognitif Siswa* (Surabaya: UniversitasNegeri Surabaya)
[13] Mudrika, N 2015*Proses Berpikir Kreatif Siswa SMP dalam Pengajuan Masalah Matematika ditinjau dari Gaya Kognitif*(Surabaya: Universitas Negeri Surabaya, Surabaya).
[14] Silver E A 1997Fostering creativity through instruction rich in mathematical problem solving and problem posing. *ZDM* **29(3)** 75-80.
[15] Brosnan, M1998 *Technophobia: The Psychological Impact of Information Technology*. (London: Routledge).

The Use of Open-Ended Questions with Giving Feedback (OEQGF) for Effective Mathematic Learning

I Sabilah[1], J T Manoy[2]

[1]Postgraduate Programme in Mathematics Education, Universitas Negeri Surabaya, Ketintang, Surabaya, 60231, Indonesia
[2]Mathematics Department, Universitas Negeri Surabaya, Ketintang, Surabaya, 60231, Indonesia

Email Corresponding Author: [1]imamasabilah16070785050@mhs.unesa.ac.id, [2]janetmanoy@unesa.ac.id

Abstract. Feedback deals with giving information to students related to their task which is done through score of their achievement, reaction, and comments. Considering its hierarchy, task difficulty level consists of low, middle and high levels. The difficulty level of open-ended questions is middle to high. Open-ended question is a good way to train students' knowledge. This research is a descriptive research which aims at describing teacher's learning management, students' activities, students' learning achievement, and students' responses in mathematic learning using OEQGF. The subject was a teacher of mathematics who teaches eighth graders, and students themselves. The research design used is one shot case study. The result shows that: management learning has been very well implemented by the teacher; every students' activity has been carried out by students; the students' learning achievement have reached the criteria of completeness, and the students' responses can be considered as positive. Therefore, it can be concluded that mathematic learning using OEQGF is effective.

1. Introduction

Assessment does not only deal with documents that are done by the students. It can, moreover, affects their learning process as well as their interest in learning [1]. Score is not the most important thing. How the learning activity in the field is implemented is also an important thing to consider. Assessment which results students' learning achievement is used by teachers continuously to monitor the process and the progress of students' learning and to improve the learning to be more effective. A learning is not only about the final score but also about the process of the students in finding out the new knowledge. In other words, learning is a process of gaining knowledge. Somebody learns through a certain process until he finds something which takes them from not knowing to knowing. The open-ended question is a question that is created to provide opportunities to emergence various answers through various strategies. It can be an individual way to develop intellectual potential and experience in finding something new [2]. For learning, the task with moderate to difficult level of difficulty is better than an easy task [3]. It means that a person learns better if the level of difficulty of the tasks also increases gradually. The use of open-ended questions needs to be cultivated in mathematics learning in order to optimize the development of students' potentials [4].

So far, evaluating students' achievement progressively is rarely done. Nevertheless, it is important for teachers to evaluate student learning in order to provide feedback for students as well as consider the effectiveness and efficiency of the learning process carried out [5]. Open-ended and closed questions are frequently discussed by several researchers. Most students preferred to use closed questions because it is easier to answer. However, students will have more opportunities to use their math knowledge and skills in a comprehensive manner in open-ended questions [6]. It means that giving continuous feedback and open-ended questions are important to improve the effectiveness of the learning.

Therefore, the purpose of this study is to describe the effectiveness of teacher's learning management, students' activities, students' learning achievement, and students' responses in mathematics learning using OEQGF. This research is a descriptive research involving 34 junior high school students. The research design used was one-shot case study. In this research, teachers' learning management in the mathematics lesson using OEQGF was good, the students were actively involved in mathematics learning using OEQGF, the students' achievements in mathematics using OEQGF have completed the criteria, and the students' responses towards mathematics learning using OEQGF were positive.

2. Theory

The conceptual framework in this study focused on the effectiveness of the open–ended question and the feedback.

2.1. Open-Ended Questions

The *open-ended questions* are an instrument that is formulated in such a way in order to provide opportunities to emerge several different answers. It encourages a variety of strategies or ways to develop their intellectual potential and experience in discovering something new [2]. *Open-ended* question is a problem that is formulated to encourage multiple correct answers [7].

Learning with the *problem* (the problem), or open learning, deals with presenting problems with various ways of problem-solving *(flexibility)* and the solution can also be varied *(fluency)* [8]. The main purpose of the *open-ended* question provision is not to get an answer, but it emphasizes on how the students reach an answer that is not direct one. Then, the indirect way to reach the answer have be collected to the teacher [9]. Moreover, it is also explained that the "openness" of the problem is considered lost if the teacher only gives an alternative way to answer these problems. Therefore, it can be stated that there are many approaches or method sin getting answers, not only one.

The aspects of transparency dealing with the open-ended questions can be classified into three types, namely: (1) opening the process of completion, if the question has various ways of settlement, (2) opening the ending result, in which it has a lot of correct answers, and (3) opening the development of the sequel, in which students have completed something then they are expected to be able to develop new problems by changing the terms or conditions on a point or item that has been resolved [4].

Based on several opinions, this research dealing with *open-ended question* is a math question having more than one answer which can be gotten through a variety of strategies of completion or having one answer which can be gotten through various strategies of completion or having more than one answer gotten through completion strategy.

2.2. Feedback

Evaluation is a process in education and teaching activities to obtain data of the results of the learning experienced by students and to process or interpret it into a qualitative or quantitative data types according to a certain standard [10]. Furthermore, the assessment does not only document what the students know and what they can do, but it affects the learning and their motivation [1]. The evaluation should be done fairly and wisely in accordance with the progress of learning done by the students.

Feedback is the provision of information obtained from tests or other measuring whether there is improvement on the students' learning achievement [11]. In the process of teaching and learning, feedback is frequently given to boost the students' motivation. Assessment of learning outcomes will make students be able to find out how far they have managed to follow the lessons taught by the teacher [12]. Giving judgment or commenting on the students' achievement in a sustainable manner will encourage them to learn because every child has a tendency or potential to get good results [13]. The feedback given makes students always have challenges and problems that must be faced and solved. Therefore, it encourages them to learn more thoroughly and carefully.

Feedback is considered very influential in the students' learning activities. Responding to the teacher's feedback, the student will soon be able to find out whether the answer they have done is right/wrong, appropriate/inappropriate, or there is something needs to be fixed [14]. Feedback can be both positive reinforcement and negative reinforcement. Through positive reinforcement (excellent, good, right, and so forth), it is expected that the students will do the same activities regularly. Instead, through negative reinforcement (less precise, one needs to be improved, and so on), it is expected that the students will not repeat the same error.

Based on the explanation above, in this study, the feedback is considered as a gift in the form of a score information or comments or responses given on an ongoing basis to avoid any shortcomings or errors that can possibly occur as well as maintain the things that have been already correct.

2.3. Effective Mathematics Learning Using Open-Ended Questions with Gift Feedback

Learning is stated as effective when it fulfils two criteria: (1) Needs analysis of students' learning activities done by observing the correlation between students' capabilities and expectations of the learning process; (2) A description of the test system used which is based on the students' learning needs [15].

There are four aspects to determine the effectiveness of learning [16] which are described as follows:

- The quality of learning, which is the quantity of the presentation of information by teachers and teachers' skills in helping students to learn the material easily.
- Compliance, in relation to the learning rate, which measures the ability of teachers to ensure that students are ready to learn the new material by linking some related knowledges or experiences
- Motivating, which measures the ability of teachers in giving encouragement to the students to understand the material being taught and the tasks given.
- Time, which measures the teacher's ability to allocate time for students to learn the material or work on a job in a certain period of time.

Learning is effective when the students are actively involved in organizing information discovered [17]. Assessment of student learning achievement used to determine the effectiveness[18]. Moreover, as some description of it can be seen, Slavin emphasizes on learning effectiveness in the teacher's management of learning, Eggen and Kauchak emphasize on the effectiveness of learning in student activities, and Sudjana emphasizes on the learning effectiveness on student learning achievement.

In this study, the effectiveness of learning is stated to be able to help students improve their abilities to correspond to the objectives achieved. Learning mathematics using OEQGF is considered effective in terms of four aspects, namely the teacher's learning management, students' activities, students' test results and students' responses.

2.3.1. Teacher's Learning Management. In this study, the teacher's learning management includes the skills of teachers in implementing instructional activities which is appropriate to the lesson plan of mathematic learning using OEQGF. Learning management using OEQGF is stated as good if the average score totally obtained minimum score included in the range $3{,}00 \leq$ Score $< 3{,}50$. Learning management assessment using the score and the assessment categories is presented at Table 1 *[19]*.

Table 1. The Category of Learning Management.

Total Average Score	Categories
$1{,}00 \leq Score < 2{,}00$	Not Good
$2{,}00 \leq Score < 3{,}00$	Less Good

$3,00 \leq Score < 3,50$	Good
$3,50 \leq Score \leq 4,00$	Very Good

2.3.2. Students' Activities

In this study, the students' activities include learning activities in which students listen/ pay attention to the teacher's explanations; ask questions related to the materials / tasks submitted, understand the given problem or the teacher's explanation; express their opinions; work on a given task; discuss with friends; present the results of the task group discussion. In addition, it involves the students to listen, pay attention, ask questions, or give feedback related to assigned presentation results done by other groups; engage in activities that are not relevant to learning, such as sleeping, eating, making noise in class, and so on. In this study, the highest activity rates show a trend of students' activities. If the amount of the average percentage of the first to the second meeting of all the activities of more than or equal to 60%, then the students' activity during learning can considered active.

2.3.3. Learning Achievement

In this study, the learning achievement is assessed only in the realm of knowledge gained through worksheets and tests of student learning achievement. Classical learning completeness is achieved if at least 80% of the students completed it individually. The percentage was adopted from *[20]*.

2.3.4. Students' Responses

The students' responses of study are the responses of students towards the learning process which are obtained through the questionnaire. Questionnaire consisted of ten statements including five favorable statements and five unfavorable statements. The students' responses use OEQGF are considered positive if students' responses in the questionnaire during the course of the first meeting to the last meeting achieved minimum criteria included in the positive (strong) category. The students' response assessment using the score and the assessment categories are presented at Table 2 *[19]*.

Table 2. The Criteria percentage score of student responses.

Percentage of Score	Categories
$0\% \leq Score < 20\%$	Very Weak
$20\% \leq Score < 40\%$	Weaker
$40\% \leq Score < 60\%$	Strong
$60\% \leq Score < 80\%$	Stronger
$80\% \leq Score < 100\%$	Very Strong

3. Methodology

3.1. Participants

The subjects of this research are one teacher teaching eight grade mathematics and 34 eighth graders of Junior High School. All students were invited to fill the learning achievement test and questionnaire. Then, 4 students in 1 group were randomly selected to be the subjects of students' activities.

3.2. Data Collection

3.2.1. Observation

In this research, the observer of students' activities is a research associating fellow mathematics education students of Universitas Negeri Surabaya which has been discussing the procedures to fill the observation sheet. The observers of students' activities were two students namely observer 1 and observer 2 who are denoted B1 and B2. The observers of teachers' learning management are researchers. The observation sheet was adapted from an instrument that has been developed by *[21]*.

3.2.2. Learning Achievement Test

The test used is an open-ended test description that was done individually. The tests given to students at the third meeting. The test wasprepared by researchers.

3.2.3. Questionnaire

The type of questionnaire used is closed questionnaire because it contains statements, the answer of which have been provided answers. Filling in the questionnaire does not affect a student's score. The questionnaire sheet was prepared by researchers to determine the students' response to the learning, their understanding of materials, learning activities, and their interests.

4. Results and Discussion

4.1. Learning Management, Students' Activities, Learning Achievement, and Students' Responses

The learning is done at the first meeting and the second meeting. Based on the observation, the teachers' learning management obtained the results on Table 3.

Table 3. The Results of Learning Management.

	Learning Management Result		Students' Activity Result	
	Total Averages Score	Categories	Percentage (%)	Categories
First Meeting	3,53	Very Good	100	Active
Second Meeting	3,40	Good	90,63	Active

In general, the assessment of teacher's learning management got a score of 3 and 4, while for score 2 was obtained twice in the first meeting and in the second meeting. The score 2 at the first meeting was obtained when the teacher motivated the students and guided them to study the material further. It is because the teachers conveyed the motivation that did not match the learning at the first meeting. At the second meeting, the teachers conveyed the overall motivation. The score 2 at second meeting was obtained when the teacher responded to the discussion in the presentation. Nonetheless, the average of the first meeting and second meeting included both discussions which involved 10 skills assessed by the excellent category, 4 abilities rated good and 1 other categorized as adequate. Therefore, the learning management criteria of the first and second meeting are *good,* then the management of learning mathematics using OEQGF can also be stated as good.

The subject observation for students' activities was done at the second meeting with the same subjects observed at the first meeting. Based on the observation of student's activity, the results on Table 3 was obtained. The students' activity in the first meeting and second meeting are respectively 100% and 90.63%. Overall, the students' activities during the first meeting and the second are good enough. It is as stated by the observer on Observation Sheet for Students' Activities during the study. Based on Table 3, the average number of students' activities reached more than 60%, that is 95, 31%. Thus, the students' activities in mathematic learning activities using OEQGF can be stated as active.

Based on the scores obtained from the worksheets and the test scores of students' learning achievement are quite varied as mathematics learning using OEQGF applied. Based on the worksheet scores and the ntest scores obtained classical learning completeness in which percentage reached 94.12% (32 students scored ≥75). Therefore, we can conclude that classical learning completeness has been reached.

Based on students' response result, the lowest percentage of students' responses is 74.75%, while the highest percentage of student responses is 95.96%. The lowest percentage of students' response that is when the students are asked to respond whether the applied learning is better or not compared to the other study. Meanwhile, the highest percentage of student responses obtained when students were asked to respond to a pleasant atmosphere in the classroom when the mathematic learning using OEQGF was applied. The average of the entire ten statements obtained is equal to 81.62%. From the point, it can be seen that the average percentage of students' responses towards mathematics learning using OEQGF was relatively strong. Thus, the students' response in mathematics learning using OEQGF can be stated as positive.

4.2. The use of OEQGF for Effective Mathematic Learning

Based on the analysis of the data obtained, it can be concluded that the use of OEQGF for effective mathematics learning is reviewed through four aspects. There are (1) teacher's learning management in the learning of mathematics using OEQGF included in both categories with an average score of 3.47; (2) students' activities in mathematic learning using OEQGF is stated as active with the average number of activities undertaken students range from activity 1 to activity 7 reached more than 60%, that 95.31% exactly; (3) The results of student learning in mathematics using OEQGF is considered complete in the classical with the percentage of completeness more than 80%, that is 94.12% exactly; (4) Students' responses in mathematics learning using OEQGF stated as positive with the number of responses for each statement in the questionnaire which obtained 50% in the strong category and 50% in a very strong category. Based on the initial agreement, it can be concluded that the study of mathematic learning using OEQGF can be stated as effective.

5. Conclusion

Based on the analysis and discussion results, the study of mathematic learning using OEQGF can be considered effective. Moreover, it can be concluded that the use of open-ended questions with gift feedback can reached all of effectivity learning aspects.

6. Implication

Based on the result, the learning model can be applied and developed in a teaching and learning activities considering the level of difficulty of the task and the feedback given.

References

[1] J W Santrock 2014 *Psikologi Pendidikan* (Jakarta: Salemba Humanica)

[2] Z Arifin 2010 *Membangun Kompetisi Pedagogis Guru Matematika* (Surabaya: Lentera Cendikia)

[3] I K Budayasa and M Nur 2010 *Teori Pembelajaran Perilaku* (Surabaya: UNIPRESS)

[4] A Mahmudi 2008 *Mengembangkan Soal Terbuka (Open - Ended Problem) dalam Pembelajaran Matematika* [Online]. Available: http://staff.uny.ac.id/sites/default/files/penelitian /Ali%20Mahmudi,%20S.Pd,%20M.Pd,%20Dr./Makalah%2002%20PIPM%202008%20_ Mengembangkan%20Soal%20Terbuka_.pdf. [Accessed 2 October 2015]

[5] A M Wahidmurni and A Ridho 2010 *Evaluasi Pembelajaran: Kompetensi dan Praktek* (Yogyakarta: Nuha Litera)

[6] A Wijaya 2012 *Pendidikan Matematika Realistik: Suatu Alternatif Pendekatan Pembelajaran Matematika* (Yogyakarta: Graha Ilmu)

[7] E Suherman et all 2003 *Soal Terbuka* dalam *Strategi Pembelajaran Matematika Kontemporer* (Bandung: JICA-Universitas Pendidikan Indonesia) p 123

[8] A R Luthfiana 2015 *Kemampuan Berfikir Kreatif Siswa dalam Menyelesaikan Soal Open-Ended ditinjau dari Gaya Belajar VARK* (Surabaya: Universitas Negeri Surabaya)

[9] N Saádah 2013 *Profil Pemecahan Masalah Matematika Open-Ended Siswa SMP Berdasarkan Tingkat Kecerdasan Emosional* (Surabaya: Universitas Negeri Surabaya)

[10] Hamdani 2011 *Strategi Belajar Mengajar* (Bandung: CV Pustaka Setia)

[11] E P Widoyoko 2011 *Evaluasi Program Pembelajaran* (Yogyakarta: Pustaka Belajar)

[12] E P Widoyoko 2014 *Penilaian Hasil Pembelajaran di Sekolah* (Yogyakarta: Pustaka Belajar)

[13] Hamalik 2008 *Proses Belajar Mengajar* (Jakarta: Bumi Aksara)

[14] H B Uno 2007 *Model Pembelajaran Menciptakan Proses Belajar Mengajar yang Kreatif dan Efektif* (Jakarta: Bumi Aksara)

[15] A Jihad and Suyanto 2013 *Bagaimana Menjadi Calon Guru dan Guru Profesional* (Yogyakarta: Multi Pressindo)
[16] E R Slavin 2009 *Educational Psychology: Theory and Practice* (New Jersey: Pearson Education Inc)
[17] P D Eggen and Kauchak 1998 *Strategies for Teacher Teaching Content and Thinking Skills* (New Jersey: Prentice Hall)
[18] N Sudjana 2014 *Penilaian Hasil Proses Belajar Mengajar* (Bandung: PT Remaja Rosdakarya)
[19] Masriyah 2006 *Modul 9 Penyusunan Non Tes* (Surabaya: Universitas Terbuka)
[20] S S Asih 2013 *Efektivitas Strategi Quick On The Draw pada Materi Tabung dan Kerucut di Kelas IX SMP Negeri 31 Surabaya* (Surabaya: Universitas Negeri Surabaya)
[21] J T Manoy 2000 *Efektivitas Pembelajaran Diskusi untuk Matematika Pokok Bahasan Lingkaran* (Surabaya: Universitas Negeri Surabaya)

IOP Conf. Series: Journal of Physics: Conf. Series **947** (2017) 012033 doi:10.1088/1742-6596/947/1/012033

The Effectiveness of Guided Inquiry Learning for Comparison Topics

Asnidar[1], S Khabibah[2], and R Sulaiman[2]

[1] Postgraduate Program in Mathematics Education, Universitas Negeri Surabaya, Ketintang, Surabaya 60231, Indonesia
[2] Mathematics Department, Universitas Negeri Surabaya, Ketintang, Surabaya 60231, Indonesia

Emails: [1]asnidarasnidar@mhs.unesa.ac.id

Abstract. This research aims at producing a good quality learning device using guided inquiry for comparison topics and describing the effectiveness of guided inquiry learning for comparison topics. This research is a developmental research using 4-D model. The result is learning device consisting of lesson plan, student's worksheet, and achievement test. The subjects of the study were class VII students, each of which has 46 students. Based on the result in the experimental class, the learning device using guided inquiry for comparison topics has good quality. The learning device has met the valid, practical, and effective aspects. The result, especially in the implementation class, showed that the learning process with guided inquiry has fulfilled the effectiveness indicators. The ability of the teacher to manage the learning process has fulfilled the criteria good. In addition, the students' activity has fulfilled the criteria of, at least, good. Moreover, the students' responses to the learning device and the learning activities were positive, and the students were able to complete the classical learning. Based on the result of this research, it is expected that the learning device resulted can be used as an alternative learning device for teachers in implementing mathematic learning for comparison topics.

1. Introduction

According to a research conducted by Camenzuli and Buhagiar [1] in 2014, it was indicated that the use of inquiry-based learning in the mathematic classroom can give benefits to students in facing social, emotional and behavioral difficulties through a number of ways. Arends [2] states that inquiry learning is another model of instruction that has been developed for the purpose of teaching students how to think. According to Kuhlthau, Maniotes, and Caspari [3], inquiry is an approach to learning whereby students find and use a variety of sources of information and ideas to increase their understanding of a problem, topic, or issue. The inquiry learning model according to Sanjaya [4] is a series of learning activities that emphasize the thinking process of searching and finding answers to a questionable problem. The thinking process itself is usually done through question and answer between teachers and students. According to Gialamas, Cherif, Keller, and Hansen [5], the guided inquiry method of teaching promotes students' active participation in the learning process. The inquiry learning model is suitable for comparison topics because this topic is one of the topic that deals with the problems in everyday life. It is supported by the opinion of Ratzer and Jaeger [6] stating that inquiry is an investigative process that engages students in developing and answering questions as well as solving real-world problems. Besides, the majority of students indicated that they enjoyed learning mathematics through the inquiry process [7].

There are three types of inquiry studies according to Carin [8], namely guided inquiry, less structured guided inquiry, and free inquiry. In this study, the researchers chose the guided inquiry type. Based on the study conducted by Richard Meyer in 2004, it is known that in the process of inquiry learning, students learn better when they are active in learning activities, although they need guidance in doing those activities [2]. Some students cannot learn well the concept that becomes the target of inquiry. Therefore, an appropriate guidance is required. Appropriate guidance can enable students in low level to follow the ongoing activities and avoid the students in high level in monopolizing activities.

Based on the explanation above, it is clear that the purpose of this research is to produce a well-qualified learning aids using guided inquiry for comparison topics and to describe the effectiveness of guided inquiry learning for comparison topics. This research is a developmental research using 4-D model. The result of learning aids consists of lesson plan, student's worksheet, and achievement test. Based on the result of the research in the experimental class, the learning aids using guided inquiry for comparison topics have good quality. The learning set has met the valid, practical, and effective aspects [9]. The result of the research, in terms of the implementation class, showed that the learning activities with guided inquiry has fulfilled the effectiveness indicators including the ability of the teacher to manage the learning that fulfilled the minimum criterion of good category, the students' activity that fulfilled the minimum criterion of good category, the students' responses to the device and the learning activity that were positive, and the students' completion of classical learning. Based on the result of this research, it is expected that the learning aids, as the final result, can be used as alternative learning aids for teachers in implementing mathematic learning for comparison topics.

2. Methods

The procedure of developing learning aid in this research refers to the 4-D model. The four stages of the 4-D model that includes defining, designing, developing and disseminating [10]. The purpose of the defining stage is to define the requirements required in the lesson. There were five activities in this stage including front-end analysis, student analysis, concept analysis, task analysis, and instructional objectives specification. The designing stage aims at producing the initial product (prototype) learning aids using guided inquiry for comparison topics. The activities at this stage were the preparation of tests, media selection, format selection, and initial design. The developing stage aims at producing learning set with good quality. The activities at this stage were expert validation and developmental testing. The dissemination phase in this study was conducted in the same school. A good learning aids generated from the previous stage will be used in the implementation class. This implementation activity aims at determining the effectiveness of guided inquiry learning for comparison topics.

2.1. Participants

The subjects of this research were the students of class VII of SMP Negeri 1 Lembo, Morowali, Center Celebes in 2016/2017 academic year. Two classes were involved in this study. The selection of classes was randomly assigned from five classes because the placement of students in each class was conducted evenly based on the students' math skill level. There were 22 students in the experimental class and 24 students in the implementation class.

2.2. Data collection

Data were collected from validation sheet, observation sheet, student response questionnaire, and test sheet. The validation sheet and the observation sheet use the assessment criteria with the rules: 4 for very good criteria, 3 for good criteria, 2 for poor criteria, and 1 for bad criteria. The validation sheet is used to obtain the learning device validation data from three validators. The observation sheet consists of the teacher's observation sheet related to how the teacher manage the learning process used to obtain the data of teacher ability in managing the learning and observation sheet which is related the students' activity which used to obtain the data of students' activity. The observations were done by two observers during the learning activity. Questionnaire was used to get the students' response related to the learning

IOP Conf. Series: Journal of Physics: Conf. Series **947** (2017) 012033 doi:10.1088/1742-6596/947/1/012033

devices and activities. The test sheet is an achievement test. This instrument is used to obtain data though pre-test and post-test.

2.3. Data analysis

The analysis is performed through data validation, data observation, analyzing students' response from questionnaire, pretest and posttest result. Learning aids are valid if every aspect of learning aids fulfill valid criterion based on the validator assessment with score 3 or 4. The teacher's ability in managing learning needs to be at least in good level if every observation aspect gets score 3 or 4. The students' activity needs to be, at least, good if every aspect of observation get score 3 or 4. The data from students' responses in students' questionnaire were analyzed by calculating the percentage of the number of students who responded agree on each statement on the questionnaire against the number of students who completed the questionnaire. The students' responses toward learning aids and the activities are considered positive if the percentage is at least 80% for each statement on the questionnaire. The data of the test are in the form of scores. The students' learning completeness is classically obtained if at least 80% of students get the score ≥ 70. The achievement test of learning outcomes meets the criteria of test item validity with [11,12] in minimal if $0.40 < r_{xy} \leq 1.00$, the reliability with [11,13] in minimal if 0.40 <r_11 ≤ 1.00, and the test item is regarded sensitive if the sensitivity index [14] $S \geq 0.30$. The learning aids are considered having good quality if they meet the valid, practical, and effective aspects. Guided inquiry learning for comparative material is valued effective if the teacher's ability to manage the learning meets the criteria of, at least, good level; the student activity meets the criteria of, at least, good level; the students' responses to the learning aids and the learning activities are positive, and the students classical learning are completed.

3. Results and discussion

This section contained the exposure of the results and discussion of the research results obtained.

3.1. Results of the development of good quality learning aids

The learning aids using guided inquiry for comparison topics developed included lesson plan, student's worksheet, and achievement test. A good quality of a set of learning aids was produced through defining, designing, and developing stages. The learning aids have been validated by three validators. The aspects of the assessment covered the aspects of content, format, and language. The experimental learning aids conducted in class VIIA of SMP Negeri 1 Lembo was amounted to 22 students and conducted with the teacher as the partner and two observers. One observer observed the way the teacher managed the learning activities and one other observer observed the students' activities. In the try-out, the researchers collected the data on teachers' ability to manage the learning process, students' activities, students' responses toward the learning aids and the learning activities, as well as the pre-test and post-test results. The data generated from the try-out were analyzed to determine whether the learning aids developed have good quality or not. In this study, the try-out conducted once because the results of the analysis showed that the learning aids had met the valid, practical and effective criteria.

3.1.1. Expert validation results on lesson plan, student worksheet, and learning results test. The validation results indicated that the learning aids developed categorized in valid criteria with score of 3 or 4 for each aspect of the assessment.

3.1.2. Developmental testing results. Based on the developmental testing results, the ability of teachers in managing learning for every aspect achieved core of 3 or 4. It showed that the ability of teachers in managing learning met the criteria of, at least, good level or category. The result of the observation of students' activities during the learning process with guided inquiry, showed that the students' activities met the criteria of, at least, good level. The observation of high, middle, and low-ability students' activities gained score 3 or 4 for each observation aspect. Based on the analysis of questionnaires of students' responses to the learning aids and the learning activities, there were more than 80% of students agree with each statement on the questionnaire. It showed that the student's responses toward the learning aids and the learning activities were positive. The coefficient of test item validity was at $0,59 \leq r_{xy} \leq 0,93$ so that the validity of the test item was categorized into the sufficient minimum criteria. The reliability coefficient of the test was at $r_{11}= 0,74$ with the interpretation of high reliability test. Therefore, the test reliability was included in the criteria of enough. The sensitivity index of the test was at $0,30 \leq S \leq 0,80$ so that the sensitivity of the test item was included in the sensitive criterion with $S \geq 0,30$. Thus, the result of the achievement test was feasible to be used because it has met the criteria of validity of test items which was enough for minimum, the reliability was enough for minimum, and also sensitive. According to the data analysis of post-test result, it was known that 18 students got the score of ≥ 70 so that the obtained percentage of students who completed the study was 81,8%. Based on predetermined criteria, it can be concluded that the students have completed the classical learning.

From the description above, it can be concluded that the quality of learning aids using guided inquiry was good because it has met the valid, practical, and effective criteria with the following details.

Table 1. Achievement of a good learning aids criterion

Criterion	Result of data analysis
Valid	Each aspect of the learning aids met the valid criteria.
	The learning results test met the criteria of validity of test items which was enough for minimum, reliability was enough for minimum, and sensitive.
Practical	The ability of teachers to manage the learning met the criteria, at least, good level.
	The students' activities met the criteria of, at least, good level.
Effective	The students' response to the learning aids and the learning activities were positive.
	Students completed classical learning that at least 80% of students got the score $\geq$ 70.

3.2. The effectiveness of learning

In the disseminating stage, a good quality of learning aids was implemented in the VIIB class of SMP Negeri 1 Lembo, which involved 24 students. This stage aimed at knowing the effectiveness of guided inquiry learning for comparison topics. The data collected included the data of teacher's ability in managing the learning activities, the data of students' activities, the data of students' responses, and the achievement test or post-test results.

3.2.1. The ability of teachers in managing the learning activities. Based on the results of research in the class of implementation, it was known that the ability of teachers in managing the learning criteria got, at least, good with score 3 or 4 for each observed aspect. These results have met the requirements of the effectiveness indicators of learning. In the learning activities, the teacher was able to carry out every aspect of observation well or very well. It indicates that most of the learning activities were consistent with the guided inquiry learning steps contained in the lesson plan.

Table 2. The observation result of the teacher's ability in managing the learning activities

No	Aspect of observation	Meeting	
		1	2
1	The ability to communicate learning objectives and motivate the students	4	4
2	The ability to connect the material to be taught with prerequisite materials	4	4
3	The ability to explain the learning process	4	4
4	The ability to present problems to the students	4	4

No	Aspect of observation	Meeting	
		1	2
5	The ability to direct students to formulate hypotheses as possible solutions to the problems studied	4	4
6	The ability to guide students in the investigation to collect data and test hypotheses	4	3
7	The ability to direct students to make conclusions	4	4
8	The ability to bridge the conclusions of the students during the investigation to become a generalization	4	4
9	The ability to lead students to see what they have done and to analyze their thinking processes during learning	4	4

3.2.2. The Student activity. The observation result of the student activities in implementation class shows that the students' activity has met the criteria, at least, good category with score 3 or 4 for every observed aspect. These results have met the requirements of the effectiveness indicators of learning. In the learning activities, the students were able to carry out each student activity that was observed as good or very good.

Table 3. The observation result of the students' activity

No	Student activities	Meeting	Student Skills		
			High	Middle	Low
1	Listen and pay attention to the teacher explanations	1	4	4	3
		2	3	3	3
2	Read / examine the problems presented in the students' worksheet	1	4	4	4
		2	4	4	3
3	Discuss to formulate hypotheses	1	4	3	3
		2	4	4	3
4	Collect the necessary data and test the hypothesis	1	4	3	3
		2	3	3	3
5	Discuss the result to formulate the conclusions	1	3	3	3
		2	4	4	3
6	Present the results of the group discussion or respond to the results of other group discussions	1	4	4	3
		2	4	3	3
7	Work on the test of individual task	1	4	3	3
		2	4	4	3

3.2.3. Student Responses. The result of data analysis of student responses toward the learning aids and the learning activities showed positive response in which more than 80% of students agree with statement in questionnaire. These results have met the requirements of the effectiveness indicators of learning. The high percentage of the response supported the implementation of learning activities.

3.2.4. Post-test results. Based on the data of the post-test result, it was known that the students has completed the classical learning. There were 21 students got the score $\geq$ 70 so that the obtained percentage of students who completed the study was 87.5%. Those results have met the requirements of the effectiveness indicators of learning.

Based on the description above, it can be seen that the guided inquiry learning met all four effectiveness of learning indicators. Therefore, it can be concluded that the guided inquiry learning for comparison topics was effective.

Table 4. Achievement of the learning effectiveness

Indicators	Result of data analysis
The ability of the teachers to manage the learning activities	At least good
Students' activity	At least good
Students' responses	Positive
Students' classical learning completion	Achieved

The results of the research were in line with the result of Camenzuli and Buhagiar's [1] research showing that inquiry based learning supports student's learning of mathematics and leads to the improvement of students' learning achievement.

4. Conclusion

Based on the result of data analysis and discussion of the results, the development of mathematic learning set can make use of guided inquiry model for comparative material. From the overall results, it can be concluded that: the development of learning aids which includes lesson plan, students' worksheet, and achievement test has met the criteria of good quality learning aids and the guided inquiry learning is effective to teach the comparison topics because it has fulfilled the indicators of the effectiveness of learning.

Acknowledgements
A great of thanks sent to Indonesian Ministry of Education and Culture, Directorate General of Teacher and Education Personnel, Directorate of Basic Education Teacher Development.

References

[1] Camenzuli J & Buhagiar M A 2014 Using Inquiry-Based Learning to Support the Mathematical Learning of Students with SEBD *The International Journal of Emotional Education* **6** (2) Issn. 2073-7629 6
[2] Arends R I 2012 *Learning To Teach Ninth Edition* (New York: The McGraw-Hill Companies)
[3] Kuhlthau C C, Maniotes L K, & Caspari A K 2007 *Guided Inquiry: Learning in The 21st Century.* (London: Libraries Unlimited)
[4] Sanjaya W 2014 *Strategi Pembelajaran Berorientasi Standar Proses Pendidikan* (Jakarta: Kencana Prenadamedia Group)
[5] Gialamas S, Cherif A, Keller S, & Hansen A 2001 Using Guided Inquiry in Teaching Mathematical Subjects *Humanistic Mathematics Network Journal* **25** (11)
[6] Ratzer M B & Jaeger P *Rx for The Common Core: Toolkit for Implementing Inquiry Learning* (California: Libraries Unlimited)
[7] Calder N & Brough C 2013 *Child-Centred Inquiry Learning: How Mathematics Understanding Emerges* http://www.cimt.plymouth.ac.uk/journal/default.htm
[8] Carin A A 1993 *Teaching Modern Science Sixth Edition* (New York: Macmillan Publishing Company)
[9] Nieveen N 1999 *Prototyping to Reach Product Quality: Design Approaches and Tools in Education and Training* (Netherlands: Kluwer Academic Publishers)
[10] Thiagarajan, S., Semmel, D.S., & Semmel, M.I. (1974). *Instructional Development for Traning Teachers of Exceptional Children*. Minneapolis: Indiana University.
[11] Hobri H 2010 Metodologi *Penelitian Pengembangan (Aplikasi pada Penelitian Pendidikan Matematika)* (Jember: Pena Salsabila)
[12] Arikunto S 2013 *Prosedur Penelitian: Suatu Pendekatan Praktik* (Jakarta: Rineka Cipta)
[13] Ratumanan T G & Laurens T 2011 *Penilaian Hasil Belajar pada Tingkat Satuan Pendidikan Edisi 2* (Surabaya: Unesa University Press)
[14] Masriyah 2000 Module Evaluasi Pembelajaran *Disampaikan pada Pelatihan Guru Matematika Proyek Peningkatan Perguruan Agama Islam Tk. Menengah Kanwil Departemen Agama Propinsi Jawa timur*

IOP Conf. Series: Journal of Physics: Conf. Series **947** (2017) 012034 doi:10.1088/1742-6596/947/1/012034

The Development of Inquiry Learning Materials to Complete Content Life System Organization in Junior High School Students

F Mayasari[1], Raharjo[1], Z A I Supardi[1]

[1]Universitas Negeri Surabaya, Surabaya 60231, Indonesia

*Corresponding Author.*Email: raharjo@unesa.ac.id

Abstract: This research aims to develop the material eligibility to complete the inquiry learning of student in the material organization system of junior high school students. Learning materials developed include syllabi, lesson plans, students' textbook, worksheets, and learning achievement test. This research is the developmental research which employ *Dick and Carey* model to develop learning material. The experiment was done in Junior High School 4 *Lamongan* regency using *One Group Pretest-Posttest Design*. The data collection used validation, observation, achievement test, questionnaire administration, and documentation. Data analysis techniques used quantitative and qualitative descriptive.The results showed that the developed learning material was valid and can be used. Learning activity accomplished with good category, where student activities were observed. The aspects of attitudes were observed during the learning process are honest, responsible, and confident. Student learning achievement gained an average of 81, 85 in complete category, with N-Gain 0, 75 for a high category. The activities and student response to learning was very well categorized. Based on the results, this researcher concluded that the device classified as feasible of inquiry-based learning (valid, practical, and effective) system used on the material organization of junior high school students.

1. Introduction

The 2013 curriculum is a curriculum of learners' competence development, which emphasizes on the application of scientific approach in the learning process. Scientific approaches of such learning include: observing, asking, trying, reasoning and communicating for all subjects [1]. The 2013 curriculum demonstrates the existence of simplicity, and integrative thematic efforts that refer to SBC (KTSP). The 2013 curriculum encourages students to be able to observe, ask, reason, communicate what is known in the learning process. This is very much in line with the science subject that emphasizes discovery in searching facts, concepts and principles in natural science. Of course it is long before the 2013 curriculum was applied which emphasizes the scientific approach. In accordance with the 2013 curriculum that emphasizes the scientific approach, researchers as science teachers in junior high schools are also required to apply similar learning which is learning that emphasizes the discovery to find facts and concepts in order to gain better knowledge.

Based on the 2013 curriculum which has just began to be applied in researchers' school, the researchers are interested to conduct research on grade VII in the organization system of life material because researchers consider it as an abstract material that cannot be captured by the five senses without any tools or media. Therefore, researchers want to invite students to see the organization system of life starting from the cell level that can only be seen through a microscope.

The researcher previously initiated pre-research activities to identify the ability of students' knowledge in Junior High School 4 *Lamongan* on life system organization material. The test is given to 28 students of IX E. Students of grade IX who are deemed to have received material about life system organization were questioned about the material. The result is that all students get the mark under minimum-standardize criteria *(KKM)* with highest score of 67, 5 and average score of 36, 63. This means that all students were failed. The results of interviews from randomly selected students explain that the life system organization is a difficult material because they never see the cells directly; they only see the cells from the reading books. That is the reason of why they forget in recalling the material. During this time teachers use more lecture methods and group discussions in the life system organization material and students have never been invited to observe and experience directly on the material.

From the results of pre-research, researchers are encouraged to attempt in completing the learning achievements of Junior High School 4 *Lamongan* on life system organization material. The treatment was done to know the effect of participants on the learning outcomes, attitude and skills. Teachers can observe the development of attitude and learning skill outcomes during the learning activities, while the development of learning outcomes can be identified from the pretest and posttest results.

One of the learning strategies that emphasize student activeness in the learning process is inquiry. The inquiry learning strategy is also in line with the 2013 curriculum that emphasizes learning with a scientific approach. Inquiry is a process for answering questions and solving factual problems from observations [2].

Based on the background, the researchers want to do research of inquiry learning materials development which is applied to complete the learning achievements in junior high on life system organization material.

2. Research Methods

This research is a developmental research because it will develop inquiry learning device on life system organization material by using Dick and Carey development model with one group pretest and posttest design [3].

This research was conducted in Junior High School 4 *Lamongan* in the even semester of the 2016/2017 academic year precisely in March 2017. The subject of the research is inquiry learning material on life system organization material on grade VII as the treatment target.

The variables in this study are the validity of learning materials, the implementation of lesson plan, student activities, learning achievements, and student responses.

The data collection technique which is used in this research is: learning materials validation, observation, test, and questionnaire. While the research instruments which are used in this study include: student effectiveness sheet consisting of attitude assessment journal, assessment sheet of student learning achievements, student skill assessment sheet, and student response sheet.

In this study, the data were analyzed by descriptive quantitative and qualitative analysis which included the learning materials validation, the practicality analysis of the learning materials, and the effectiveness analysis of the learning materials which has been developed.

3. Research and Discussion

Based on feasibility analyses of learning material include validity, practicality and effectiveness of learning materials [4], researchers obtained the following results:

Analysis of validation result of learning material development obtained average of lesson plan validity score are 4.0 for format, 3.9 for content, 4.0 for language, and overall aspect in category of lesson plan is "very valid"[5], with a comparison value of 98.4% so it is categorized as "suitable" [5]. The validation result on learning materials was obtained average score for content is 3.7, language is 3.7, material presentation is 3.6, and all aspects on very valid category [5], with comparison percentage of both valuator on 97% where it is categorized as "suitable"[6]. Validation results on Student Activity Sheet *(LKS)* obtained average score for format 3.4, for language 3.6, for content 3.6 and all aspects on"very valid" category [5] with 96% matched [6]. The result of validation on test obtained the average score, on multiple choice questions obtained content validation of 4.0, the validity of the language of 3.3, while the question description obtained content validity of 4.0 and the validity of the language of 3.5. This indicates that the validity of the test results can be used without any revision.

Practical analyses by applying learning materials that have been developed include analysis of lesson plan implementation and student activity analysis. Based on the observation by two observers, the result of lesson plan is as shown in Figure 1.

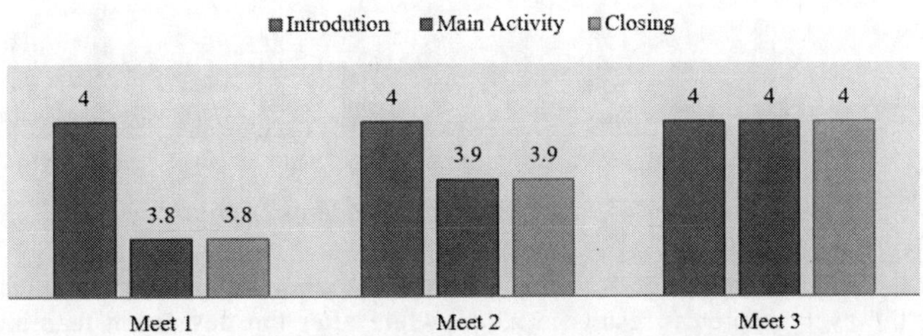

Figure 1. Lesson Plan Implementation Result

Based on the graphics, it can be analyzed that the implementation of lesson plan by using learning materials that have been developed in each syntax shows that the existence of development in each meeting, obtained average value as much as 3.9 with "good" category [6].

On student activeness aspect in three meetings of learning can be described as shown in Figure 2.

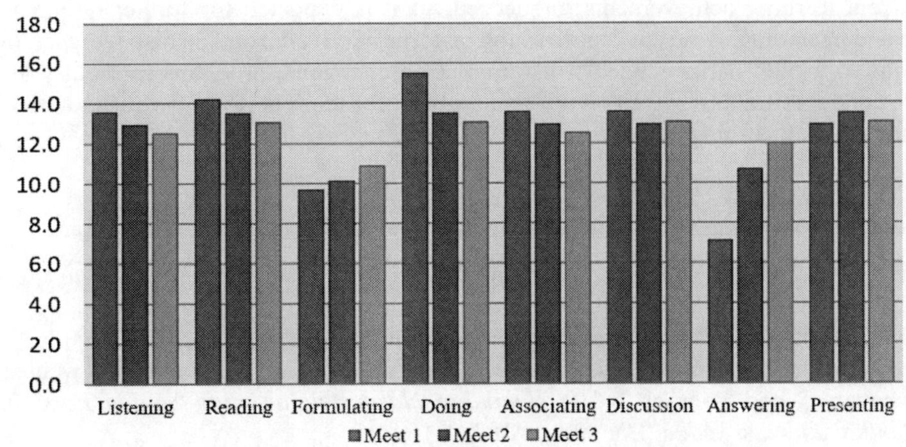

Figure 2. Student' Activity

Based on the graphics, it can be stated that student activities in the highest learning activity is in making observations with a value of 14.0%, and the lowest percentage is in answering teacher questions where it reach 9.9%.

Analysis of the effectiveness in application of learning materials include attitude analysis, analysis of knowledge learning achievements, analysis of skills learning achievements and student responses to learning by applied learning materials that have been developed[7].

Attitude analysis during learning is used to know the development of student attitude during the learning progress and facilitate the growth of student behavior. The development of student attitudes was viewed during teaching-learning process which focused on the value of honest, responsible and confident characters that are recorded in the attitude assessment journal [8].

Analysis of student learning achievements can be seen in Figure 3.

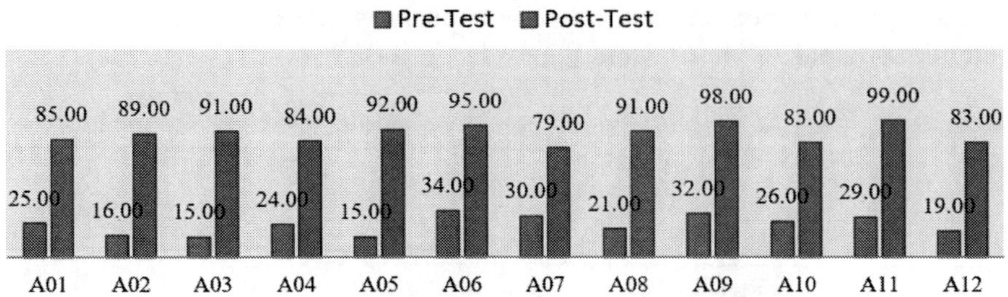

Figure 3. Students' Achievements

Based on the diagram, it can be seen that before the application of developed learning materials was done [9], the average pretest results is 23.83, while after the developed learning materials had just applied, the average posttest results is 89.08 with "passed" category while the minimum student mastery score is 70[10].

4. Conclussion

The conclusion of this research shows an adequate and decent learning material in completing student learning achievements on life system organization material of students in Junior High School 4 *Lamongan*.

5. Suggestion

Inquiry learning materials which are developed by the researchers declared eligible and can possibly lead the student learning achievements to succeed, so it is expected for further research in order to develop the understanding of the concept/learning materials derived from the discovery lesson. A similar study is required to familiarize students in formulating questions /problems in inquiry learning. It is necessary to conduct similar research on the same subject for more perfect learning materials developed in completing student learning achievement

6. References

[1] Permendikbud Nomor 81A Tahun 2013 tentang Implementasi Kurikulum. Jakarta

[2] Kardi S 2013*Model Pengajaran Langsung, Inkuiri, Sains Teknologi Masyarakat* (Surabaya: PPS Unesa)

[3] Sugiyono 2015*Metode Penelitian Pendidikan*(Bandung: Alfabet)

[4] Nieveen 1999*Prototypingto Reach Product Quality*(Enschede: University of TwenteNederlands)

[5] Ratumanan G and Laurent T 2011*Penilaian Hasil Belajar pada Satuan Pendidikan Tingat Pendidikan edisi kedua* (Bandung: Alfabeta)

[6] Borich G 1994*Observation Skills for Effective Teaching* (New York: MacMillan Publishing Company)

[7] Taihuttu J 2016*Pengembangan Perangkat Pembelajaran Model Inkuiri Terbimbing Untuk Meningkatkan Hasil Belajar Fisika Siswa SMA*(Surabaya: Tesis Magister Pendidikan Sains Unesa)

[8] Eggen P and Kauchak D 2012*Strategi dan Model Pembelajaran* (Jakarta: Indeks)

[9] Bayram Z, Oskay O O, Erdem E, Ozgur DS and Sen S 2001Effect of Inquiry based learning method on students' motivation*Procedia_Social and Behavioral Science***106** pp 988-996

[10] Hake 1999*Analyzing change/gain score (online)*. Retrieved September 2016, from (http://www.physics.indiana.edu/sdi/AnalyzingChange-Gain.pdf)

IOP Conf. Series: Journal of Physics: Conf. Series **947** (2017) 012035 doi:10.1088/1742-6596/947/1/012035

Slow Learner Errors Analysis in Solving Fractions Problems in Inclusive Junior High School Class

N Novitasari[1,*], A Lukito[1] and R Ekawati[1]

[1] Mathematics Department, Universitas Negeri Surabaya, Ketintang, Surabaya 60231, Indonesia
**Corresponding Author.* Email: novitasarinovitasari10@mhs.unesa.ac.id

Abstract. A slow learner whose IQ is between 71 and 89 will have difficulties in solving mathematics problems that often lead to errors. The errors could be analyzed to where the errors may occur and its type. This research is qualitative descriptive which aims to describe the locations, types, and causes of slow learner errors in the inclusive junior high school class in solving the fraction problem. The subject of this research is one slow learner of seventh-grade student which was selected through direct observation by the researcher and through discussion with mathematics teacher and special tutor which handles the slow learner students. Data collection methods used in this study are written tasks and semistructured interviews. The collected data was analyzed by Newman's Error Analysis (NEA). Results show that there are four locations of errors, namely comprehension, transformation, process skills, and encoding errors. There are four types of errors, such as concept, principle, algorithm, and counting errors. The results of this error analysis will help teachers to identify the causes of the errors made by the slow learner.

1. Introduction

Education is the right of every citizen, no exception for children who has learning difficulties, including slow learner. A Slow learner has thinking skill below the average than other normal students in his/her age. This is a big challenge for teachers to help slow learners build their understanding of learning in inclusion classes, especially in learning mathematics. Because even average students generally find it difficult and make errors in solving mathematics problems, especially for the slow learner may takes longer time to understand the problem. One of the mathematical topics considered a problem for students is the fractions. Fractions is one of the concepts that is considered difficult to master by some students [5].

A number of researchers have analyzed students' errors in solving fraction problems. One of them is the research of Abdullah et al. [1] indicated that students often make errors in encoding (27.58%), followed by process skills (27.33%), transformation (24.17%) and comprehension (20.92%). However, no field research has analyzed the slow learner error in solving the fraction problem. In fact, the slow learner who is a student with a slow learning ability need more help with various approaches to overcome the difficulty of them and make them easy to understand the topics taught by teachers, especially fractions.

Slow learners are student who has the ability to think below average of their age peers. It does not mean they can not develop the potential that exists within them. It is just slow learner needs more efforts to understand a concept that had been mastered by the majority of class. That is accordance with Ruhela's opinion [10] which reveals that the slow learner has below average cognitive abilities of their age mates and who struggle to cope with the traditional academic demands of the regular classroom. The slow learner intelligence level is between 71 and 89 [3]. However, Chauhan [4] argues that the slow learner has an IQ of between 76 and 89, while Malik [7] describes slow learner IQ being between 70 and 90. From some of the above expert opinions it can be concluded that the slow learner has a level of intelligence between 71 and 89. Their level of intelligence is too high to be considered a child with mental retardation and is too low to be considered a their age mates. In general, a slow learner is a child who has intellectual potential below average but not included as mental retardation. Therefore, a slow learner is a child labeled as borderline mentally retarded, and they are slower to grasp whatever is being taught if it involves abstract or conceptual symbols in the subject matter that should be within easy reach of most children their age [7]. Thus, these slow learning characteristics are the unique characteristics of the slow learner, so they take longer to learn than their peers.

However, Chauhan [4] expresses another characteristic of the slow learner, the first being limited cognitive capacity: the slow learner fails to overcome the learning situation and to give an abstract reason. The second characteristic is the slow learner has a poor memory. The next character is a distraction and lack of concentration: the slow learner can not concentrate too long on the teacher's instructions orally. The last characteristic is an inability to express ideas: slow learner has difficulty finding and combining words, so they are not able to express ideas.

Slow learners are also considered as children with any identifiable combination of dysfunctionalities such as attention deficit disorders (ADD), dyscalculia, dyslexia, dysgraphia, dyspraxia, dysnomia, hyperactivity or other related problems [13]. One of the dysfunctionalities associated with mathematics is dyscalculia. Briefly, dyscalculia is a disturbance in the difficulty of learning mathematics [2]. The slow learner's inability to understand mathematics is shown by his/her weakness in the manipulation of algorithmic computation [6]. This affects the way they solve mathematics problems. In solving mathematics problems, a slow learner has difficulty about following multi-step instructions [9], so they need more specific work instructions. Problems given are also more likely on simple questions. Their weak recall ability makes the slow learner work slowly and sometimes inconsistently. These obstacles often lead to errors in solving mathematics problems, especially on fractions.

The slow learner's errors in solving fraction problem needs to be analyzed further. So that, we can get a clear and detailed explanation about the slow learner's weakness in solving fractions problems, especially regarding the locations, types, and causes of errors. In this research, those errors will be analyzed by Newman's Error Analysis (NEA). Newman identified five errors types in solving written mathematical tasks. They are reading, comprehension, transformation, process skills, and encoding errors [11]. While, regarding the types of errors, Soedjadi [12] associated errors with mathematical object and classified them as fact, concept, principle, and operation errors. However, Mercer and Mercer [8] divide the types of errors into four categories, namely wrong operation, obvious computational error, defective algorithm, and random response. From the two opinions above, the types of errors used in this research are concept, principle, algorithm, counting, and random response errors. The findings of this research can be used as reference for mathematics teachers, especially who teach in inclusion classes to find alternative solutions to improve teaching and learning activities adjusted to the ability and characteristics of a slow learner, especially in solving fractions problems. Thus, the increase in teaching and learning activities is expected to improve learning outcomes or slow learner learning achievement.

The purpose of this research was to analyze the slow learner error in solving the fractions problems that focused on the operation of fractions. This research was conducted to describe the locations, types, and causes of the slow learner error in solving fractions problems.

2. Method

This research is qualitative descriptive. The purpose of this research is to describe the locations, types, and causes of slow learner errors in the inclusive junior high school class in solving the fraction problem. The subject of this research is one slow learner of a seventh-grade student. Due to differences in characteristics and abilities between slow learners and the others. The subject selection was decided by considering the mathematical abilities and communication skills both verbally and in writing. The mathematical ability was viewed based on basic mathematics skill (such as addition, subtraction, multiplication, and division), interest in math, and grade in math. Meanwhile, the slow learner was considered to have good communication skills, if the slow learner was able to communicate with new people, does not avoid eye contact, has good pronunciation (pronunciation can be understood by others), and has handwriting that can be understood or read by others. The subjects was also based on the consideration of mathematics teacher and special tutor. So that we got one slow learner with good mathematics abilities and good communication skills better than the others. Methods of data collection are written tasks and semistructured interviews. The written tasks aim to show and describe the slow learner errors in solving a fraction problem consisting fourteen items. Interviews aimed to clarify written tasks data as well as to determine the causes of the errors made by slow learner. The interviews were conducted after the research subjects conducted a written tasks. The credibility test approach used in this study is the triangulation of time. The data obtained from the test is then analyzed using Newman's Error Analysis (NEA).

3. Results

Based on the results of data analysis including data reduction, data display and conclusion drawing, it was found that the subject made an error as in the location and type of error presented in Table 1.

Table 1. Locations and types of slow learner errors.

Items	Problem	Example of Errors	Locations of Errors	Types of Errors
3	Add it! $2+\frac{1}{9}$	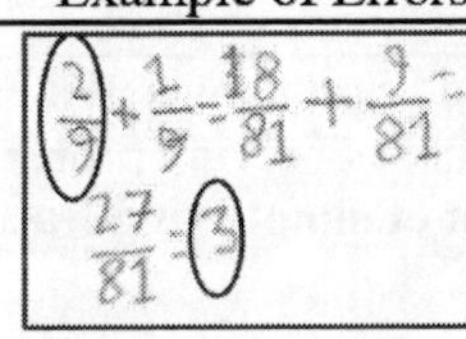 **Figure 1.** Answer of item 3.	*Comprehension error* *Process skills error* *Encoding error*	*Concept error* *Principle error* *Algorithm error* *Counting error*
4	Determine the results! $2\frac{2}{3}-\frac{3}{6}$	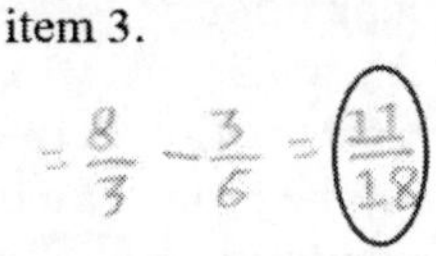 **Figure 2.** Answer of item 4.	*Comprehension error* *Process skills error* *Encoding error*	*Concept error* *Principle error* *Algorithm error* *Counting error*
11	Mr. Dana bought 1 liter of milk. When you want to drink, the milk is spilled, the remaining $\frac{1}{3}$ liter of milk. How many liters of milk spilled?	 **Figure 3.** Answer of item 11.	*Comprehension error* *Process skills error* *Encoding error*	*Concept error* *Principle error* *Algorithm error* *Counting error*
12	Mrs. Rizka bought $\frac{3}{4}$ liter of cooking oil. On the half way the cooking oil was spilled as much as $\frac{2}{3}$ liter. Then, Mrs. Rizka bought another $\frac{5}{12}$ liter. How many liters of cooking oil owned by Mrs. Rizka now?	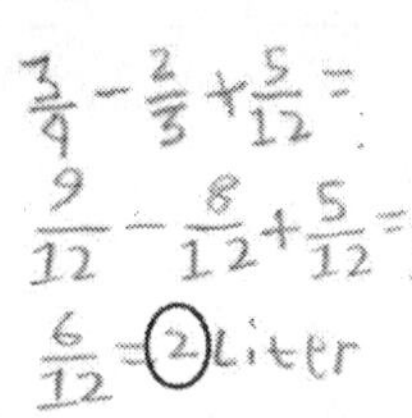 **Figure 4.** Answer of item 12.	*Process skills error* *Encoding error*	*Concept error* *Principle error* *Counting error*

Items	Problem	Example of Errors	Locations of Errors	Types of Errors
14	Mrs. Novi is a tailor. Mrs. Novi received $\frac{2}{3}$ meter of cloth to be handkerchief. For each handkerchief it takes $\frac{1}{6}$ meter. How many handkerchiefs can be made Mrs Novi?	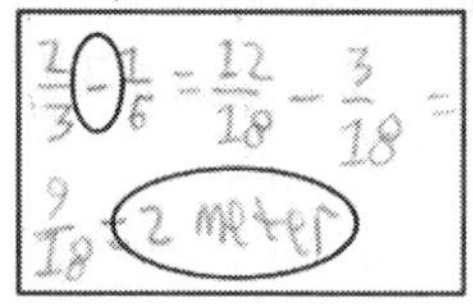 **Figure 5.** Answer of item 14.	*Comprehension error* *Transformation error* *Process skills error* *Encoding error*	*Concept error* *Principle error* *Algorithm error* *Counting error*

Thus, it can be concluded that the results of this research identify four locations of errors and four types of errors made by the slow learner in solving fractions problems. The four locations of errors are the comprehension, transformation, process skill, and encoding errors. While, The four types of errors are the concept, principle, algorithm, and counting errors.

Based on the results of interviews with the subject can be concluded the causes of errors made by the subject when solving the problem of fractions, as presented in Table 2.

Table 2. Causes of slow learner errors in solving fractional problems.

Items	Causes of Errors
3	Did not understand the concept of addition between integers and fractions.[b,p] Incorrect in carrying out the procedure of solving problem.[d,r,s] Did not fully understand the concept of simplifying fractions.[b,d,s] Incorrect in writing the final answer.[e,q]
4	Incomplete or incorrect in carrying out the procedure of solving problem.[d,r,s] Incorrect in doing calculations or computation.[d,s] Incorrect in writing the final answer.[e,q]
11	Did not understanding the concept of subtraction between integers and fractions.[b,p] Incomplete or incorrect in carrying out the procedure of solving problem.[d,r,s] Incorrect in doing calculations or computation.[d,s] Incorrect in writing the final answer.[e,q]
12	Did not fully understand the concept of simplifying fractions.[b,d,s] Incorrect in writing the final answer.[e,q]
14	Incapable to understand and explain the meaning of keywords or information contained in the problem.[b] Incorrect in identifying or determining the operation used.[c,s] Incorrect in carrying out the procedure of solving problem.[d,r,s] Did not fully understand the concept of simplifying fractions.[b,d,s] Incorrect in writing the final answer.[e,q]

[a] Reading Error. [b] Comprehension Error. [c] Transformation Error. [d] Process Skill Error. [e] Encoding Error. [p] Concept Error. [q] Principle Error. [r] Algorithm Error. [s] Counting Error. [t] Random Response Error.

Beside to the causes of error in Table 2, there are another cause that affect the slow learner to make errors in solving the fractions problems that is the poor concentration level.

4. Discussion

Based on the locations and types of errors subject made in this research, it could be seen that subject did not have difficulties in reading. It was shown by how fluent the subject read the problem. The subject could read the keywords or symbols correctly. While in the comprehension part, the subject had difficulties to recognize the keywords in the problem, especially the word problem that require division with fractions. During interview, the subject could answer question regarding what information given in the problem and what was asked to find or to show; but the subject couldn't explain the meaning of the keywords in the problem. As stated by Chauhan [4] that due to limited cognitive capacity, slow learners failed to cope with learning situations and to reason abstractly. Their inability to deal with abstract material resulted in wrong choice of operation to solve the problem. Their poor memory [4], limited comprehension of mathematical language [13], and weak algorithmic manipulation [6] resulted in difficulty to understand abstract mathematical concept. Those caused them to carry out the wrong procedure in solving problem and to make wrong calculation. In the end, they didn't write the answer correctly (encoding).

Furthermore, the subject had difficulty focusing on given tasks [9]. The poor concentration levels also caused errors in solving fractions problems that actually could have solved by them correctly. When the subject poor concentration [4], the fractions problems that should be able to do the subject correctly then the subject more often make errors in solving the problem of the fractions. Due to their poor concentration skills, the subject had difficulty connecting and transferring information given to solve the problems[10].

5. Conclusion

The results showed that on the location of errors the slow learner made comprehension, transformation, process skills, and encoding errors. Based on the type of errors the slow learner made concept, principle, algorithm, and counting errors. Based on the location and type of error can be concluded the cause of slow learner error in solving the problem of fractions. What caused the error made by the slow learner was they didn't understand and couldn't explain the meaning of the keywords or information in given problem. Then, the subject didn't understand the concept of addition and subtraction in integers and fractions. The subject also didn't understand how to simplify the fractions. The next cause was the subject made mistakes by choosing the wrong operation to solve the problem, carrying out the wrong procedure to solve the problem, and making wrong calculation. In the end they didn't write the answer correctly. Another causes was their poor concentration skills. The results of this research can be used as a reference for mathematics teachers, especially who teach in inclusion classes to find a alternative solution for the slow learner in solving fractions problem.

References

[1] Abdullah A M, Zainal N L, and Ali M 2015 *Analysis of Students' Errors in Solving Higher Order Thinking Skills (HOTS) Problems for the Topic of Fraction* vol. **11** pp 133-142

[2] Abdurrahman M 2012 *Anak Berkesulitan Belajar: Teori, Diagnosis, dan Remidiasinya* (Jakarta: Rineka Cipta) p 210

[3] Atik F, Kartini, Supardi, Lusli M M, Damanik T, Rivai and Dipa W 2013 *Panduan Teknis Pelaksanaan Pelatihan Bagi Pelaksana Prosedur Operasional Standar dan Modul Pelatihan: Pendidikan Inklusif Berbasis Sekolah* (Hellen Keller International Indonesia & Direktorat Pembinaan Pendidikan Khusus dan Layanan Khusus Pendidikan Dasar) p 21

[4] Chauhan S 2011 *Slow Learners: Their psychology and Educational Programmes* vol. **1** pp 279-289

[5] Erlinda N and Surya E 2017 *Mathematical Learning Strategy of Fractional form by Using Learning Model of Gagne and Human Figure Line Media* vol. **34** pp 13-22

[6] Herriot S T 1967 *The Slow Learner Project:The Secondary School "Slow Learner" In Mathematics* vol. **5** p 2

[7] Malik S 2009 *Effect of Intervention Training on Mental Abilities of Slow Learners* vol. **1** pp 61-64

[8] Mercer C and D Mercer A R 1981 *Teaching Student with Learning Problems* (Colombus: Charless Merril) p 145

[9] Paul P B 2016 *Coping With Slow Learners* vol. **2** pp 56-58

[10] Ruhela R 2014 *The Pain of the Slow Learner* vol. **4** pp 193-200

[11] Singh P, Rahman A A and Hoon T S 2010 *The Newman Procedure for Analyzing Primary Four Pupils Errors on Written Mathematical Tasks: A Malaysian Perspective* vol. **8** pp 264-271

[12] Soedjadi R 2000 *Hakikat Pendidikan Matematika Di Indonesia* (Jakarta: Direktorat Jendral Pendidikan Tinggi Departemen Pendidikan Nasional) p 13

[13] Yusha'u M A 2012 *Teaching Slow Learners in Mathematics: YUGAL Remediation Model As Alternative Method* vol. **13** pp 108-123

Statistical Literacy: High School Students in Reading, Interpreting and Presenting Data

M Hafiyusholeh[1,2], K Budayasa[2], and T Y E Siswono[2]

[1]Department of Mathematics, Faculty of science and technology, UIN SunanAmpel Surabaya, Indonesia
[2]Department of Mathematics Education, UniversitasNegeri Surabaya, Indonesia

Abstract. One of the foundations for high school students in statistics is to be able to read data; presents data in the form of tables and diagrams and its interpretation. The purpose of this study is to describe high school students' competencies in reading, interpreting and presenting data. Subjects were consisted of male and female students who had high levels of mathematical ability. Collecting data was done in form of task formulation which is analyzed by reducing, presenting and verifying data. Results showed that the students read the data based on explicit explanations on the diagram, such as explaining the points in the diagram as the relation between the x and y axis and determining the simple trend of a graph, including the maximum and minimum point. In interpreting and summarizing the data, both subjects pay attention to general data trends and use them to predict increases or decreases in data. The male estimates the value of the (n+1) of weight data by using the modus of the data, while the females estimate the weigth by using the average. The male tend to do not consider the characteristics of the data, while the female more carefully consider the characteristics of data.

1. Introduction

Statistical literacy has become one of the important life skills Nowadays [1].Being statistically literate is very important for every individual including students as informed citizens who can interpret the statistical messages in various contexts. Thus, statistical literacy, which is becoming a part of the school mathematics curriculum, plays a key role in preparing students to encounter the needs of society when they complete their elementary education [2]. The NCTM for example states, "A knowledge of statistics is necessary if students are to become intelligent consumers who can make critical and informed decisions" [3].

One of the way to improve statistical literacy mastery in students is to include statistical content in the national curriculum. However, based on preliminary research conducted by researchers related to statistical literacy of students, it was found that there are many students in high school and junior high school who have difficulty in reading, interpreting, or presenting data in the form of tables and diagrams. It indicates that the statistical literacy of students is low. A description of the students' with low statistical literacy can also be found in the study[4,5,6,7] which stated "children produced fewer idiosyncratic descriptions of data, possessed intuitive knowledge of center and spread and were constrained in analysis and interpretation by knowledge of data context".

IOP Conf. Series: Journal of Physics: Conf. Series **947** (2017) 012036 doi:10.1088/1742-6596/947/1/012036

Taking into account the previous explanation, the purpose of this reseach is to describe the statistical literacy of high school students in reading, interpreting and presenting data as part of the basic ability that must be mastered by every student.

2. Theoritical Framework

There are several definitions and conceptualizations of statistical literacy found in the literature. Gal's statement clearly identifies a key element: statistical literacy involves the ability to read and interpret the data in tables and graphs published by government statistical associations [8].

Wallman indicated that statistical literacy includes both understanding and critically evaluating statistical messages that are necessary for daily life of individuals and for their decision making, as she highlighted both for the personal and societal needs [9]. Similarly, Gal's conceptualization also included the need for statistical literacy for active citizenship and pointed out that communication with statistics, and interpretation and judging of statistical claims are the key skills of statistically literate individuals [1]. Statistical literacy provides emancipations of citizens through enabling them not only to readdata but also evaluate and communicate with statistical messages [10].

The statistical literacy can be understood by some to show minimum knowledge of basic concepts and statistical procedures [11,12,13]. The statistical literacy as Schield says, focuses on making decisions using statistic as proof. Statistical literacy involves two reading skills; understanding and interpretation [14].

The statistical literacy is an individual ability to understand [9,11,14,15,16]; interpret [15,17]; and create table and graph [15,17,18]; the ability to make decisions using statistics as proof [14].Statistical literacy is the ability of an idividual ability to read (understand); analyze; interpret; and represent a data either in tabular or graphic form [19,20].

3. Method

The subject of the research was determined using purposive sampling technique. The subjects were chosen from the students of SMA Negeri 3 Sidoarjo which consisted of one male student and one female student who had high level of mathematical ability. Collecting data is done through task-based interviews. To obtain the validity of the data, the triangulation was used. The procedure of data analysis in this reseach refers to Miles &Hubberman, with the following steps: (1) data reduction; (2) data exposure; and (3) conclusion and verification [21].

4. Results

The researchers gave two tasks to the subjects to measure and describe the students' statistical literacy. The first task related to the data presentation and the second task related to the ability to read and interpret data.

To describe the students' statistical literacy in terms of reading, interpreting and presenting data, the researcher gave the task of statistical literacy tests as follows:

1. Notice the line diagram below
   a) In what age does the baby weight decrease?
   b) What is the baby weight in 3 weeks age? Explain !
   c) What is the baby weight in 10 month age? Explain !

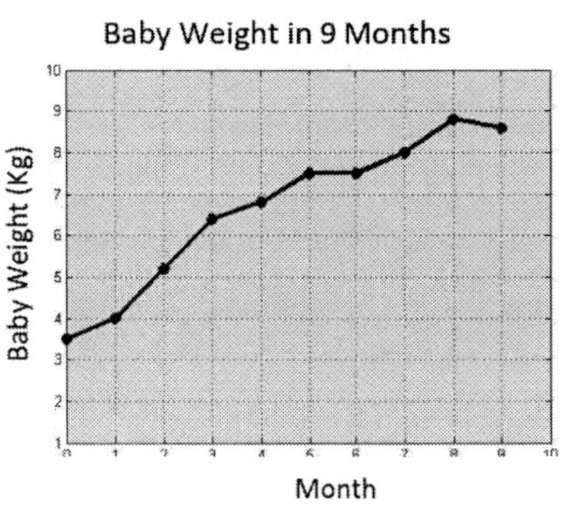

IOP Conf. Series: Journal of Physics: Conf. Series **947** (2017) 012036 doi:10.1088/1742-6596/947/1/012036

2. Consider the following data
   a) The "Literacy" store recorded monthly profits for a year (in million rupiahs) as follows

The .... Month	1	2	3	4	5	6	7	8	9	10	11	12
Profit	2.5	1.8	2.6	4.2	3.5	3.3	4.0	5.0	2.0	4.2	6.2	6.2

   b) The Number of students in a district based on school level in 2016 are:

Education	The number of students
Elementary School	175
Junior high school	600
Senior high school	225

   c) The Mathematics score of 40 students are

68	75	74	72	79	75	79	71	75	78
75	76	74	73	71	74	71	70	74	72
74	77	73	73	70	72	80	70	72	74
73	67	72	72	75	66	69	80	74	74

   Based on given data, present the data in the most appropriate form! Give the reason!

Based on written answers and interviews conducted on the subject of research, and after validation of data, the results wasobtained as follows:

Male and female subjects stated that the provided diagrams were diagrams of infant weight measured from birth to 9 months of age. Subjects observed the relation of points presented in the diagram as the link between age and weight obtained from the relationship between the horizontal axis and the vertical axis. In case of increase and decrease of infant weight as well as the maximum, minimum or constant value obtained by the subject by comparing the trend graph and the value of the other. If the graph is seen decreasing, then it is concluded the weight of the baby decreased. In contrast, if the graph of the diagram increase, it is concluded that the baby's weight has increased. Baby's weight remains if the graph is constant.

The male and female objects at 3 weeks of age, infant weight of 3,875 kg obtained from infant weight at birth 3.5 kg plus average infant weight gain for 3 weeks, 0,375. Baby's weight beyond the data, was determined by adding preliminary data with an average increase or decrease stated by the female subject, while the male subject was hesitant in determining which value. Because according to the male subject, there are many factors that influence the data. Nevertheless, the baby's weight beyond the data was determined by the male subject by using a trend toward increasing or decreasing the baby's weight. The next data of baby's weight value can be determined by formula

$$x_{n+1} = x_n + p$$

where x_{n+1} is data (n+1), x_n show data n-th and p is lowest increase of baby's weight.

Furthermore, in presenting the data, if the subject was given data aboutstore profits every month, the male and female subjects presented it in the form of line and bar chart on the grounds that readers will know the monthly profit store, whether there is an increase or decrease every month. Additional criteria stated by the female subject was to consider whether the characteristic data are discrete or continuous. If the data was continuous, then it was presented in the form of a line diagram and was presented in bar chart if it was discrete data.

Data on the number of students based on the level of school in a district by the male subject was presented in the form of a pie chart in the ground that readers will be easier noticing the number of students in each district. The female subject presented in the form of bar charts in order to compare the number of students for each grade of elementary school, junior high, and high school.

Data on the value of mathematics assignment from 40 students, by male or female subject presented in the form of frequency distribution table. This was seen by both subjects that it was more simple and effective and did not require many space in the presentation.

IOP Conf. Series: Journal of Physics: Conf. Series **947** (2017) 012036 doi:10.1088/1742-6596/947/1/012036

5. Discussion

Based on the explanation of subject data, there are several tendencies of students in reading, interpreting, and presenting data. In the context of reading data, the male and female subjects are able to read the title or topic of the graph, to give meaning to the units of the graph, to find the values or specific units, to read the maximum and minimum value in the graph. It can be inferred that the subject can look for information explicitly expressed in display, recognize graphical conventions, and establish a direct link between the original data and the display as stated in the reading data criterion of [22]. In the context of Aoyama and Stephens, the subjects can be said can read graphs, read simple trends from graph, read important parts of a graph [23].

The interpretation of the data in this context is the determination of a value that can not be explicitly determined directly from the reading data result, which Curcio stated as reading between the data and reading beyond the data [22]. Value can only be determined through a further process through interpolation or extrapolation as well as observations and inferences on general trends of data.

The female subject in estimating the value between the two data or beyond the data begins with considering on the link between the n-th and the (n+1) data, comparing between the two to then decide whether there is an increase, a decrease or a constant. The general trend of the existing chart, serve as a guide in determining future value. In other words, in interpreting the data, the female subject is capable of performing global reading of trends in graphs as well as the criteria of statistical level D stated by Kimura [23]. In determining or interpreting data values, the female subject using the average value of increase or decrease data. If the average data tends to be a positive value, then it is concluded that there is an increase, otherwise if it is negative, then it is concluded there is a decrease. So to estimate value beyond the data, the female subject adds the value of the n-th data with the average increase or decrease to obtain the value of the n-th data (n + 1).

It is somewhat different from the male subject. The male subject in identifying the increase or decrease in data, is tend to look at the data mode. If in the data, there is a lot of increase for each point of data, then the male subject concludes the data tends to increase with the assumption of normal condition. This means that there is no interference that causes the decline in data. In determining the value of the n-th data (n + 1), the male subject defines it from summing the n-th data value with the least increase value of its data rate mode.

Based on these descriptions, it shows that the both subjects in estimating the increase or decrease in data, are guided by one of the data concentration values of mean, median, and mode. These values are used as a reference because the value of data centering gives an overview of how the characteristics of the data set are.

In the context of data presentation, both subjects are equally convinced that the presentation of a data is important because it will affect the understanding and information to be easily accessed by reader. Male and female subjects are able to present data in various forms including line diagrams, bar charts, histograms, pie charts, single data tables and frequency distribution tables. This means that both subjects are able to create tables, graphs and reports as proposed by Ben-Zvi and Garfield. They saidthat one of indicator that people have statistical literate if they are able to organize, compile and display data in tabular form, and able to work with different data representations [15]. Similarly, Schield states that "A core element for producers is the ability to create tables, graphs and reports that are unambiguous and comprehensible" [17].

Each of these forms of presentation is understood by both subjects having particularity in the presentation of the data. If a chart need to be emphasized on the ups and downs of a data quantity, then a line diagram and or bar chart are used. However, if the emphasized information is the quantity of data, then a pie chart is used. If a large number of random data (≥ 30) is obtained, then effective data presentation according to the male subject is the frequency distribution table. A more complex consideration exists on the female subject. Aside from the point of emphasis is on the purpose of the data presented, the subject also emphasizes the characteristics of the data. If the increase or decrease of a data to be presented in the form of continuous data, then the subject of women using the line chart, but if the data in question is discrete data, then the subject used the bar chart.

In the context above, both subjects meet the competency standard of high school graduates in the national curriculum related to the field of statistic ie students are able to read data in the form of tables and bar charts, lines, pie charts, and ogive; able to present data in the form of tables and bar charts, lines, pie charts, and ogive and interpretation.

6. Conclusions

The results showed that the both subjects with high mathematical ability understand information directly from what is written explicitly the students read the data based on explicit explanations on the diagram, such as explaining the points in the diagram as the relation between the x and y axis and determining the simple trend of a graph, including the maximum and minimum point. In interpreting and summarizing the data, both subjects pay attention to general data trends and use them to predict increases or decreases in data. The males estimates the value of the (n+1) of weight data by using the modus of the data, while the females estimate the weigth by using the average. In presenting data, both subjects have ability to present data in various forms. However, the males tend to do not consider the characteristics of the data, while the female more carefully consider the characteristics of data and emphasize the purpose of information that will be presented to the reader.

Suggestions given for future research, it is necessary to study the ability of statistical literacy of students for elementary school level until college. The description of the students' understanding will be used as a foundation in formulating competency standards for each level.

Acknowledgments
Honorable to Rector UIN SunanAmpel who gave researchers the opportunity to continue study

References

[1] Gal I 2004 *Statistical literacy: Meanings, components, responsibilities* In D Ben-Zvi & J Garfield (Eds.), *The challenge of developing statistical literacy, reasoning and thinking* (Dordrecht: Kluwer Academic Publishers) 47–78

[2] Watson J & Callingham R 2004 *Statistical literacy: from idiosyncratic to critical thinking* Curricular Development in Statistics Education: International Association for Statistical Education 2004 Roundtable Voorburg, the Netherlands: International Statistical Institute pp 116-62

[3] National Council of Teachers of Mathematics 1989 *Curriculum and evaluation standards for school mathematics* (Reston, VA: NCTM) p 105

[4] Jones G, Mooney E, Langrall C, and Thornton C 2002 Student's Individual and Collective Statistical Thinking *ICOTS6*. (https://www.stat.auckland.ac.nz/~iase/publications/1.pdf. accessed on September 24th 2013)

[5] Schield M 1998 Statistical literacy and evidential statistics ASA *Proceedings of the Section on Statistical Education* (https://www.statlit.org/pdf/1998SchieldASA.pdf. accessed on June 10th 2015)

[6] Watson J M & Moritz J B 2000 Development of understanding of sampling for statistical literacy *Journal of Mathematical Behavior* **19** 109-36. In Watson J M and Callingham R A 2004 Statistical Literacy: From Idiosyncratic to Critical Thinking *Curricular Development in Statistics Education Sweden* (https://www.stat.auckland.ac.nz/~iase/publications/rt04/4.1.pdf. accessed on May 25th 2017)

[7] Watson, Jane M, Kelly, and Ben A 2008 Sample, Random, and Variation: The Vocabolary of Statistical Literacy *International Journal of Science and Mathematics Education* **6** pp 741-67

[8] Gal I 2003 Functional demands of statistical literacy: Ability to read press releases from statistical agencies *IASE* (https://www.stat.auckland.ac.nz/ ~iase/publications/3/3016.pdf)

[9] Wallman K K 1993 Enhancing statistical literacy: Enriching our society *Journal of the American Statistical Association* **88** pp 1–8

IOP Conf. Series: Journal of Physics: Conf. Series **947** (2017) 012036 doi:10.1088/1742-6596/947/1/012036

[10] Schield M 2011 Statistical literacy: A new mission for data producers *Statistical Journal of the IAOS* **27** pp 173-83
[11] Gal I 2002 Adults' statistical literacy: Meanings, components, responsibilities *International statistical review* **70** pp 1-25
[12] Gal I 2003 Expanding the conceptions of statistical literacy: An analysis of products from statistics agencies *Statistics Education Research Journal* **2** pp 3-21
[13] Schagen I 1998 Statistical literacy is the essential skill for educational managers *Education Journal ISSUE* 98
[14] Schield M 2013 Statistical Literacy: Thinking Critically about Statistics *Journal of Significance* Produced by the Association of Public Data Users. (http://web.augsburg.edu/~schield/MiloPapers/984StatisticalLiteracy6.pdf accessed on October 25th 2013)
[15] Ben-Zvi D and Garfield J 2005 *The challenge of developing statistical literacy, reasoning, and thinking* (Netherland: Kluwer Academic Publisher)
[16] Garfield J and Ben-Zvi D 2008 *Developing Students' Statistical Reasoning: Connecting Research and Teaching Practice* (Germany: Springer)
[17] Schield M 2011 Statistical literacy: A new mission for data producers *Statistical Journal of the IAOS* **27** pp 173-83
[18] Forbes S, et al 2011 Official Statistics and statistical literacy: They need each other *Statistical Journal of the IAOS* **27** pp 113–28
[19] Hafiyusholeh M 2015 *Literasi Statistik Dan Urgensinya Bagi Siswa* Wahana **64** pp 1-5
[20] Hafiyusholeh M, Budayasa I K, Siswono T Y E 2017 LiterasiStatistik: Siswa SMA dalam Membaca, Menafsirkan, dan Menyimpulkan Data *Prosiding SI MaNIs (Seminar Nasional Integrasi Matematikadan Nilai-Nilai Islami)* Vol **1** **Issue 1** pp 79-85
[21] Miles M B & Huberman A M 1994 *Qualitative data analysis: An expanded sourcebook* (2nd.Ed). (London: Sage)
[22] Curcio F R 1987 Comprehension of mathematical relationships expressed in graphs *Journal for Research in Mathematics Education* **18** pp 382-93
[23] Aoyama K and Stephens M 2003 Graph Interpretation Aspects of Statistical Literacy: A Japanese Perspective *Mathematics Education Research Journal* **15** pp 207-25

Problem Solving-based Learning Materials on Fraction for Training Creativity of Elementary School Students

Y N Widhitama[1], A Lukito[2], S Khabibah[2]

[1]Primary Education Department, Universitas Negeri Surabaya, Surabaya, Indonesia
[2]Mathematics Department, Universitas Negeri Surabaya, Surabaya, Indonesia

Email: yukhewidhitama@mhs.unesa.ac.id

Abstract. The aim of this research is to develop problem solving based learning materials on fraction for training creativity of elementary school students. Curriculum 2006 states that mathematics should be studied by all learners starting from elementary level in order for them mastering thinking skills, one of them is creative thinking. To our current knowledge, there is no such a research topic being done. To promote this direction, we initiate by developing learning materials with problem solving approach. The developed materials include Lesson Plan, Student Activity Sheet, Mathematical Creativity Test, and Achievement Test. We implemented a slightly modified 4-D model by Thiagajan et al. (1974) consisting of Define, Design, Development, and Disseminate. Techniques of gathering data include observation, test, and questionnaire. We applied three good qualities for the resulted materials; that is, validity, practicality, and effectiveness. The results show that the four mentioned materials meet the corresponding criteria of good quality product.

1. Introduction

Curriculum 2006 states that mathematics should be taught to all learners in elementary school to equip them with thinking skills, one of them is creative thinking. Creative attribute is important owned by students in accordance with the advance of science, technology, and engineering. It means that every young citizen has to be capable of following the rapid changes of the world, and able to solve the problems they faced in a carefully, properly, and creatively [1]. Mathematics as an intellectual domain stands at or near the top of any hierarchical list of intellectual domains according to the extent to which creativity is evident in disciplinary activity or production [2].

Through learning mathematics, creative thinking can be trained on students, especially elementary school students. Problem solving based learning can be one of the solutions. Problem solving based learning is a student-centered pedagogy in which students learn about a subject by attempting to find a solution to an open-ended problem [3, 4]. In this learning, the quality material availability for the learning supports the achievement of the desired goals. From researchers' observation most teachers use the existing learning materials provided by the government through the Ministry of Education. But those materials do not purposefully support the training of student creative thinking. To support the proposed learning we decided to develop the materials for helping the students and also teachers implement the instruction in an effective way.

In mathematics, solving problems represents the most effective concept to contextualization and re-contextualization of concepts, to operational and basic mathematical knowledge transfer to ensure a sustainable and meaningful learning [5]. This learning approach is relatively new for low grade students.

Previous educational researches relating to creativity had been done in higher grades because lower grade students were generally considered not able to master this higher thinking skills. In addition, fraction topic for primary school students is one of the difficult topic to study. To respond this challenge, we plan to choose the problems that lead to a meaningful learning for the students. One kind of such problems are open-ended problems. With careful consideration, these problems may encourage students to create their own problem-solving strategies [5]. With the helps of mathematics teacher the students develop positive attitudes towards a recognition of the diverse approaches and strategies to the posed problems [7].

The goals of this research is to provide good quality learning materials for training students to be creative in mathematics learning and problem solving. We adopt the criteria for good quality product proposed by Nieveen [8] that includes validity, practicality, and effectiveness.

2. Method

This research is development research following 4-D model by Thiagarajan et al. [9] which consists of Define, Design, Develop, and Disseminate stages. Learning materials being developed include Lesson Plan, Student Activity Sheet, Mathematical Creativity Test, and Achievement Test. The materials were field tested involving real classroom trials with 30 students, and validity testing in another classroom with 30 students. Data collection techniques used are observation, test, and questionnaire. The main gathered data are analyzed qualitatively and the other quantitatively.

3. Results

Before field testing the materials were appraised by some experts for the validity and theoretical practicality.

3.1. Validity

From the data analysis on validity it shows that all the materials meet the minimum requirement of validity. The results are summarized in Table 1.

Table 1. Validation Result of Learning Materials

Tools	Description
Lesson Plan	Valid and can be used with little revision
Student Activity Sheet	Valid and can be used with little revision
Mathematical Creativity Test	Valid and can be used without revision
Achievement Test	Valid and can be used with little revision

Based on a classroom trial, all items of the achievement test satistifies the validity ciriteria as shown in Table 2.

Table 2. Item Validity of Achievement Test

Test Item	r_{table}	r_{count}	Description	Criteria
1	0,361	0,549	Valid	Average
2	0,361	0,643	Valid	High
3	0,361	0,646	Valid	High
4	0,361	0,652	Valid	High

3.2.Practicality
For the practicality, observers reported that both teacher and students were able to do all activities in accordance with the planned activities set in lesson plans and student activity sheets. They agreed on the minimum score of 4 (scale 1-5).

3.3. Effectiveness
Effectiveness of learning materials was determined within three categories; that is, student response, learning achievement, and mathematical creativity.

For eigth aspects of student response, at least 80% of the students are very strongly to agree with the way the learning proceed.

Table 3. Minimum Learning Mastery

No	Aspect	Result	
		Trial Classroom	Implementation Classroom
1	Mastering students (#)	24	27
2	Failing students (#)	6	3
3	Learning mastery (%)	80%	81,97%

From Table 3 we can conclude that both trial and implementation classrooms achieved the classical minimum learning mastery (≥75%). Then result student creativity test show in Table 4.

Table 4. Students' Creativity Scores

Criteria	Trial Classroom	Implementation Classroom
Not creative	0	0
Average creative	13	4
Creative	17	26
Very Creative	0	0

From Table 4, we know that in trial and implementation classrooms, all students met the minimum criteria of creative. So we conclude that the learning materials was indeed able to facilitate students in training student creativity.

4. Discussion
The results of development learning materials have met the criteria of validity, practicality, and effectiveness. All experts have judged that all materials are valid and usable for the fraction learning with little or without revisions. Learning that have been planned carefully gets students positive response as learning run well and achieved the learning objectives. The learning makes sense for the students and teachers presents it orderly [10].

4.1. Validity
Validity has been verified from the content of lesson plan, student activity sheet, achievement test, and mathematical creativity test. The learning materials meet the validity criteria and can be used with little/without revisions.

4.2. Practicality
Practicality has been seen from step-by-step implementation of the lesson plans. Observation on both classrooms, trial and implementation, reported that there is no step that get score under minimal score

is 4 (scale 1-5). Learning will not digress and proceed orderlyly if planned well [11]. In learning, teachers only act as facilitators. Learning was emphasized on how teachers encourage and facilitate students in learning process, not on what should be studied by students [9].

4.3. Effectiveness
Effectiveness of learning is based on student response, learning outcome, and mathematical creativity test result. Students' response in trial and implementation classrooms classify into "Very Strong" category, because the average percentage obtained ≥81% for each aspect. It means that students' response is positive to learn, so students can learn comfortably through problem solving approach and do not feel bored. The students are more actively involved, so more effectively learning occurs [3]. Students from the two classrooms classically achieves the desired learning, as can be seen from the number of students who get learning outcomes higher than minimum learning mastery. Results of students creativity test have met the criteria of Average Creative. Creative students get used to creatively investgting open-ended problems, with many interpretations, ways of problem solving, or answers [2]. Solving open-ended problems allows students to develop creative attitudes and behaviours [6].

5. Conclusion
Based on the results and discussion just described, the developed learning materials have shown to be feasible being used for teaching fraction with problem solving approach to train students creativity.

Acknowledgements
Thanks go to L S Ermawati and Ni M D Susilowati as classroom teachers and also to M L Oktarianto as data retrieval partner in this research.

References
[1] Setyosari P 2012 *Metode Penelitian Pendidikan dan Pengembangan* (Jakarta: Kencana) p 215

[2] Silver E A 1997 Fostering Creativity Through Instruction Rich in Mathematical Problem Solving and Problem Posing *Zentralblatt für Didaktik der Mathematik (ZDM)* **29** Issue 3 pp 75-80

[3] Beetlestone F 2012 *Creative Learning: Strategi Pembelajaran untuk Melesatkan Kreatifitas Siswa* (Bandung: Nusa Media) p 113

[4] Narwanti S 2014 *Creative Learning: Kiat Menjadi Guru Kreatif dan Favorit* (Yogyakarta: Familia Pustaka Keluarga) p 3

[5] Caprioara D2015 Problem Solving-Purpose and Means of Learning Mathematics in School *Procedia-Social and Behavioral Sciences***191**pp 1859-64

[6] Eggen P D and Kauchak D P 2001 *Strategies for Teacher, Teaching Content and Thinking Skill* (Boston: Allyn and Bacon) p 2

[7] Kandemiir M A and Gur H 2007 Creativity Training in Problem Solving: A Model of Creativity in Mathematics Teacher Education *New Horizons in Education* **55** p 3

[8] Akker J v d, Branch R M, Gustafson K, Nieveen N, and Plomp T 1999 *Design Approaches and Tools in Education and Training* (Dordrecht: Springer Science+Business Media B V)

[9] Thiagarajan S, Semmel D S, and Semmel M I 1974 *Instructional Development for Training Teachers of Exceptional Children* (Bllomington: Indiana University)

[10] Slavin R E 2011 *Psikologi Pendidikan: Teori dan Praktik* (Jakarta: Indeks) p 49

[11] Kemp J E, Morrison G R and Ross S M 1991 *Designing Effective Instruction* (New York: Macmillan Collage Publishing Company) p 3

Representation of Students in Solving Simultaneous Linear Equation Problems Based on Multiple Intelligence

Y R Yanti[1,*], S M Amin[2], R Sulaiman[2]

[1]Posgraduate Program in Mathematics Education, Universitas Negeri Surabaya, Ketintang Surabaya 60231, Indonesia
[2]Mathematics Departement, Universitas Negeri Surabaya, Ketintang Surabaya 60231, Indonesia
*Corresponding author: yeniyanti@mhs.unesa.ac.id

Abstract. This study described representation of students who have musical, logical-mathematic and naturalist intelligence in solving a problem. Subjects were selected on the basis of multiple intelligence tests (TPM) consists of 108 statements, with 102 statements adopted from Chislet and Chapman and 6 statements equal to eksistensial intelligences. Data were analyzed based on problem-solving tests (TPM) and interviewing. See the validity of the data then problem-solving tests (TPM) and interviewing is given twice with an analyzed using the representation indikator and the problem solving step. The results showed that: the stage of presenting information known, stage of devising a plan, and stage of carrying out the plan those three subjects were using same form of representation. While he stage of presenting information asked and stage of looking back, subject of logical-mathematic was using different forms of representation with subjects of musical and naturalist intelligence. From this research is expected to provide input to the teacher in determining the learning strategy that will be used by considering the representation of students with the basis of multiple intelligences.

1. Introduction

Representation is a mental picture of a person in solving problems and visualized in verbal, simbolic, or visual form. Representation is a configuration that can represent something else in some manner [1]. Definition of three distinct ways in which people represent the world: (a) through action, (b) through visual images, and (c) through words and language [2]. Representation was categorized in two types namely internal and external representation, internal representation is the process of thinking about mathematical idea that allows one's mind to work on the idea [3]. While external representation is the ability to pour thought in the form of verbal, symbol, visual. The ability to solve mathematical problems depends on one's ability to construct and use mathematical representation in words, graphs, tables, equations, and symbol manipulations [4]. So a complicated problem can be simple if using proper troubleshooting. Multiple intelligences relate to the process of thinking someone to solve the problems faced [5]. Multiple intelligence is the ability to solve problems or create valuable products in one or several cultural and community environments [6].

Based on the observations of researchers during teaching, the process of learning in schools so far students are rarely even never given the opportunity to pour his own representation. Most students simply imitate the teacher's way of solving the problem as a result of the not developing student representation [7]. In addition, students have difficulty in solving the material story of the system of two-linear equation system including difficulties in determining initial conditions, determining equations, solving equations, and translating answer [8]. So in this study is expected that students with mature intelligence is capable of solving problems logically and systematically and has many ways used in problem solving.

This research is descriptive research with qualitative approach. The purpose of this study is to describe representation of students who have musical, logical-mathematic and naturalist intelligence in simultaneous linear equations problem solving. Subjects were selected on the basis of multiple intelligence tests, and selected one subject from each musical, logical-mathematic and naturalist intelligence. The selected subject was given problem solving and continued with the interview. The results show the stage of presenting given information, the stage of devising a plan, and the stage of carrying out the plan and the three subjects using symbolic representation. While the stages of presenting information and the stage of looking back, the subject of logical-mathematic intelligence is using symbolic representation subjects of musical and naturalist intelligence using verbal repression. By this research result, teacher can be expected to determine learning strategy that will be applied by considering representation of student with base of multiple intelligence owned by student.

2. Context and Review of Literature

2.1. Representation
Many experts who define representation said that representation is the ability of students to communicate mathematical ideas/ideas learned in a particular way [8]. Everything students create to externalize and display works is called representation [9]. Three different ways of representing a person are through action, through visual images, and through words [9].

Four basic ideas were used in understanding the concept of representation, namely (a) representation can be viewed as an internal abstraction of Mathematical ideas or cognitive schemes constructed by students through experience, (b) as mental reproduction of the previous mental state, (c) as structurally presented through images, symbols or symbols, (d) as knowledge of something that represents something another [10]. "representations are used by students to think and communicate mathematical ideas" [11]. Representations that are often used in communicating mathematical ideas are diagrams, tables, mathematical statements, written texts or a combination of them all.

Representation is divided into two namely internal and external representations. The internal representation of a person is difficult to observe directly because it is a mental activity of a person in his mind (minds-on). But the internal representation of a person can be inferred or suspected based on external representations in various conditions, for example from the expression through words (spoken), through writing in the form of symbols, drawings, graphs, tables or through a hand-on [12]. So that internal and external representations influence each other in the process of making a simple picture or in expressing mathematical ideas.

Based on the above description, the representation in this study is a mental picture of a person in solving the problem and visualized in the form of verbal, symbol, visual. Verbal forms include words or sentences either written or spoken. Symbol forms include the way of writing in algebraic manipulations and equations. While the visual form includes cartesian and table charts.

2.2. *Mathematical Problem Solving*
Problem solving is a process or individual effort to respond or overcome obstacles or constraints when an answer or answer method has not been apparent [13]. "Solving a problem is finding the unknown means to a distinctly conceived end" (problem solving is an attempt to find something that is not yet known in a suitable way) [14]. Mathematical problem solving is a complex cognitive activity accompanied by a number of processes and strategies [15]. Based on some opinions then understanding problem solving, then problem solving in this research is an effort done by someone (student) to get correct solution of problem faced by using knowledge, skill and comprehension gained before.

The troubleshooting steps used in this research are the troubleshooting steps of Polya. The researcher chooses the polya problem-solving steps because the problem-solving stages proposed are quite simple, and the differences in activities at each stage are clear enough. Here are the steps to solve the problem: (a) Understanding the problem, this stage the students try to understand the problem clearly. By understanding the problem then the student can know what steps will be used to solve the problem at hand. Students should also be able to show what is known and what is being asked. (b) Devising a plan, making the completion plan is not easy, but by collecting information ever received by students. At this stage students need to look at the interrelationships between questions and solutions to be presented, as well as the appropriate ideas or strategies for solving the problems they are facing. (c) carrying out the plan, the students implement the plan already made. The ideas that have been planned will be applied to find solutions from problem solving which are then repressed with tables, words, graphics, or mathematical symbols. (d) looking back, students re-examine the results of the problem solving that has been done. This check is done to decide that the answer is already logical and in accordance with the issues presented.

2.3. *Multiple Intelligence*

Intelligence has more to do with the capacity for solving problems and fashioning products in a contextrich and naturalistic setting [6]. It defines intelligence as the ability to solve problems or create valuable products in one or several cultural and community environments. Intelligence is the ability to learn from experience and to adapt [5]. Based on the above opinion so it is concluded that intelligence is the ability of a person (student) to solve problems or create a valuable product by learning from experience and to adapt. So that intelligence relates to the process of thinking someone to solve the problems encountered.

Gardner defines nine intelligences, namely Linguistic Intelligence, Logical-Mathematic Intelligence, Visual-Spacial Intelligence, Body Kinestetic Intelligence, Musical Intelligence, Interpersonal Intelligence, Intelligence Intelligence, Naturalist Intelligence, and Existential Intelligence. Walter McKenzie classifies nine intelligences in three domains: interactive, analytic, and introspective [16]. Domain used in this research is analytic domain consists of musical intelligence, logic-mathematics, and natural, which is used to analyze data and knowledge. Intelligence in the analytics domain can have social or introspective components and can be used to analyze data and knowledge and incorporate data in existing schemes. This domain is basically a natural hauristic process.

2.4. Relationship between represetation and problem solving based on multiple intelligence

When students construct the correct representation with the problem will be getting the right solution [17]. So in doing problem solving, it takes a person's ability to give consideration to the form of representation involved. It shows that a person's ability to change a representation will affect his or her skills in finding problem-solving solutions. So a complicated problem can be simple if using proper troubleshooting. The representation indicators in solving simultaneous linear equations are shown in Table 1.

Table 1. Representation Indicators in Solving Simultaneous Linear Equations

Problems Solving Steps	Representation	Indicator
Understanding probem	Verbal	Present data or information back to the written text.
	Simbol	Representing data or information on the question of algebraic manipulation.
	Visual	Representing data or information on the matter of a table.
Divising a plan	Verbal	Develop problem-solving strategies using words.
	Simbol	Develop a problem-solving strategy that will be pursued based on information on the problem.
	Visual	Develop a troubleshooting strategy using tables.

Problems Solving Steps	Representation	Indicator
Carrying out the plan	Verbal	Complete the model that has been formulated in the previous step by using written words or text.
	Simbol	Completed the model that has been formulated in the previous step that contains symbols and mathematical formulas according to formal rules.
	Visual	Complete the model that has been formulated in the previous step using the graph.
Looking back	Verbal	Check the results again and summarize the answers obtained in accordance with the question on the problem.
	Simbol	Check back the troubleshooting strategy and recalculate to prove that the answers obtained are correct.
	Visual	Summarizes the answers obtained according to the questions on the problem.

3. Method

The study was conducted on senior high school students in Manyar Gresik Indonesia, with 84 participan and selected 3 students with each subject at different intelligences at musical, logical-mathematic and naturalist intelligence. The subject selection process by means of a multiple intelligence test (TKM) and selected subject with the highest intelligences score from musical, logical-mathematic and naturalist intelligence. Subjects who meet the problem criteria are given the problem of solving simultaneous linear equations problem solving (TPM) and interviewing. The data validation uses triangulation time, so that subjects are given TPM and interviewing twice in differrent time and with equivalent problem. The data obtained in the analysis using representation indicators with the problem solving step. If data from TPM 1 and interviewing valid with data from TPM 2 and interviewing then the data can be used, but if the data is not valid then in the TPM and interviewing again until the data is valid.

4. Result and Discussion

The selected subject is given a problem-solving question twice in a different time, with each question consisting of one question item. After that the subject in the interview to dig up the information is not revealed on the answer sheet. Results of student work with analyzed using indicator in Table 1.

4.1. Repesentation Student of musical intelligence

The results showed the subject of musical intelligence (SMI) solve the problem by using four stages Polya. At the stage of understanding the problem of presenting information that is known by making permisalan then create a mathematical model (equation) and present the information asked by using the written text. At the stage of designing the settlement plan is able to explain the problem-solving plan using words, devise a problem-solving strategy that will be pursued based on the information on the problem. At the stage of carrying out the settlement plan using the substitution method, while at the re-check stage substituting the results of his work to one equation obtained from the problem, without checking again the steps of completion that have been done.

4.2. Student of logical-mathematic intelligence

The results show the subject of logical-mathematical intelligence (SLI) solve the problem by using four stages Polya. SLI presents information that is known by making permisalan then create a mathematical model (equation), presenting the information asked in the form of symbol representation in accordance with the permisalan that has been made. In the stage of designing a settlement plan use your own words and write down each step of the problem. At the stage of implementing the settlement plan using symbol representation, ie in solving the problem using the substitution method. In the re-examination phase, check the completion steps that have been done, then check the results of the answers with the substitution of answers obtained to one equation and SLI write a conclusion in the form of a sentence to clarify what answers were asked on the issue.

4.3. Student of naturalist intelligence

The results showed that the subject of natural intelligence (SNI) was able to solve the problem well. At the stage of understanding the problem of presenting information that is known in the form of symbol representation by making permisalan then create a mathematical model and present the information asked in the form of verbal representation that is by using written text. SNI explains problem-solving plans using words, as well as solving problems using substitution methods by utilizing the equations already obtained from the problem. In the process of re-examining the SNI substitute value of variables that have been obtained previously to one of the equations obtained from the problem, without checking again the steps of completion that have been done later in writing conclusions in the form of a sentence to clarify the answer to what was asked on the problem.

5. Conclusion and Suggestion

The results show that within the stage of presenting given information, devising a plan, and carrying out the plan, the three subjects using symbolic representation. But at the step of presenting the information, SMI and SNI using representation in verbal form that is by using written text, while SLI use representation in symbol form, according to the example tha have been made. In the re-checking stage, SMI and SNI substituted the results of their work to one equation derived from the problem, without re-examining the completion steps that have been done, while the SLI uses representation in the form of symbols, namely by checking back the problem-solving strategies that have been done and Recalculate and by substituting the results of his work into one equation derived from the problem.

For Math teachers subjects are expected to be able to design the right learning strategy so that it can facilitate all plural intelligence especially on SPLDV material learning. So that all students with various kinds of multiple intelligences have good representation. Therefore it is expected that the subject Teachers evaluate the learning process and the way students represent the problem solving.

References

[1] Goldin G A 2002 *Representation in mathematical learning and problem solving In L D English (ED) International Research in Mathematical Education IRME* (New Jersey: Lawrence Erbaum Associates)

[2] Salkind G M & Hjalmarson M 2007 *Mathematical Representations*

[3] Wiryanto 2002 Representasi Siswa Sekolah Dasar Dalam Pemahaman Konsep Pecahan *Prosiding Seminar Nasional Matematika dan Pendidikan Matematika FMIPA UNY* (ISBN: 978-979-16353-8-7)

[4] Brenner M E, Mayer R E, Moseley B, Brar T, Duran R, Smith Reed B, et al 1997 Learning by understanding: The role of multiple representations in learning algebra *American Educational Research Journal* **34** p 663

[5] Gunawan A W 2007 *Genuis Learning Strategy. Petunjuk Praktis untuk Menerapkan Accelerated Learning* (Jakarta: PT. Gramedia Pustaka Utama)

[6] Gardner H 1999 *Intelligence reframed: Multiple intelligences for the 21st century* (New York: Basic Books)

[7] Asyarofi M & Junaedi I 2016 Kemampuan Representasi Matematis Ditinjau Dari Multiple Intelligence Pada Pembelajaran Hybrid Learning Berbasis Konstruktivisme *Unnes Journal of Mathematics Education Research* Vol **5**(1) p 33

[8] Santoso B, Nusantara T, and Subanji 2013 Diagnosis Kesulitan Siswa Dalam Menyelesaikan Soal Cerita Materi Sistem Persamaan Linear Dua Variabel Serta

Upaya Mengatasinya Menggunakan Scaffolding *presented on KNPM V Himpunan Matematika Indonesia* pp 141-502

[9] Hutagaol K 2013 Pembelajaran Kontekstual untuk Meningkatkan Kemampuan Representasi Matematis Siswa Sekolah Menengah Pertama *Infinity Jurnal Ilmiyah Program Studi Matematika STKIP Siliwangi Bandung* **2**(1) p 87

[10] Pape S J & Tchoshanov M A 2001 The Role of Representation(s) in Developing Mathematical Understanding *JSTOR* **40**(2) p 119

[11] Kastberg Signe E 2002 *Understanding Mathematical Concepts: The Case of The Logarithmic Function (Doctoral Disertation)* (University of Georgia)

[12] Wiryanto 2002 Representasi Siswa Sekolah Dasar Dalam Pemahaman Konsep Pecahan *Prosiding Seminar Nasional Matematika dan Pendidikan Matematika FMIPA UNY 943*

[13] Siswono T Y E 2008 *Model Pembelajaran Matematika Berbasis Pengajuan Masalah dan Pemecahan Masalah untuk Meningkatkan Kemampuan Berpikir Kreatif* (Surabaya: UNESA University Press)

[14] Polya G 1973 *How to Solve It: A New Aspect of Mathematical Method. Second Edition* (New Jersey: Princeton University Press)

[15] Montague M 2012 *Math Problem Solving for Middle School Students with Disabilities* (The Access Center: American Institutes For Research 1)

[16] McKenzie Walter 2005 *Multiple Intelligences and Instructional Technology* (Washington: International Society for Technology in Education)

[17] Jones A D 2000 *The Fifth Process Standar: An Argument to Include Representation in Standar*

Students' Relational Understanding in Quadrilateral Problem Solving Based on Adversity Quotient

A N Safitri [1,2], D Juniati [1], Masriyah [1]
[1]Universitas Negeri Surabaya, Indonesia

[2]Email: auliasafitri10@mhs.unesa.ac.id

Abstract. The type of research is qualitative approach which aims to describe how students' relational understanding of solving mathematic problem that was seen from Adversity Quotient aspect. Research subjects were three 7th grade students of Junior High School. They were taken by category of Adversity Quotient (AQ) such quitter, camper, and climber. Data collected based on problem solving and interview. The research result showed that (1) at the stage of understanding the problem, the subjects were able to state and write down what is known and asked, and able to mention the concepts associated with the quadrilateral problem. (2) The three subjects devise a plan by linking concepts relating to quadrilateral problems. (3) The three subjects were able to solve the problem. (4) The three subjects were able to look back the answers. The three subjects were able to understand the problem, devise a plan, carry out the plan and look back. However, the quitter and camper subjects have not been able to give a reason for the steps they have taken.

1. Introduction

Mathematics is one science that is needed in human life. In learning mathematics, students are trained to be critical, logical, systematic, and able to solve problems encountered in everyday life, so that math becomes important to be studied and applied in the learning process from elementary school level to higher school level. One of Mathematics's material is quadrilateral. Quadrilateral is consist of square, rectangle, trapezoid, pararellogram, rhombus and kite. These planes were commonly found in our daily life.

In fact, there are many students have difficulty in studying quadrilateral at school. In addition, students also often make mistakes in determining methods and formulas when applying on quadrilateral. Therefore, the understanding of the concept of a quadrilateral is an important thing that must be mastered by students in learning mathematics.

The understanding of concepts in mathematics learning is very important. Types of understanding are classified into relational understanding and instrumental understandings [1]. Relational understanding is described as knowing both of what and why to do something, whereas instrumental understanding entail rules without reasons. Aspects of understanding that students need in the learning of mathematics is not just know the concept, but a deeper understanding called as a relational understanding. Relational understanding is the ability to deduce specific rules or procedures for more general mathematical relationships [2]. The relational understanding as ability to relate between conceptual knowledge and procedural abilities. Conceptual knowledge contains relationships between a mathematical concepts with other mathematical concepts, whereas procedural knowledge is a related knowledge of symbols to present mathematical ideas and rules in accomplishing mathematical tasks [3]. Relational understanding reached if student is able to link a concept with the other concept and understanding each step that used to solving a problem.

Problem solving is an attempt to find a way out of a difficulty in order to achieve goals that are not immediately achievable [4]. Through this activity, its important aspects of mathematical skills such as the application of rules to non-routine problem, pattern discoveries, generalization, mathematical communication etc. can be better developed. Problem solving as the means by which an individual uses previously acquired knowledge, skills and understanding to satisfy the demands of unfamiliar situation [5].

The stages of problem solving are understand the problem, devise a plan, carry out the plan, and look back [4]. In this study, researchers used the problem-solving steps will be examined how the relational understanding of students in solving problems.

Table 1. Aspects of relational understanding

Polya's Problem Solving Steps	Aspects of Relational Understanding
Understanding the problem	How the subject declare or write down what is known and expressed using it
Devising a plan	How the subject declare whatever concept is given in relation to the given problem and plan the a solution by linking the concept of the existing concept and the reasons
Carrying out the plan	How the subject apply the procedure of mathematical relationships that have been planed for reasons
Looking back	How the subject check the truth of the answer by checking each steps

In general, students will have difficulty in understanding mathematics problems. This is due to each sthudent have different respond ability to solve difficulties of a problem that called Adversity Quotient (AQ). AQ is a measure of knowing one's response to adversity [6]. The categorization of AQ in solving problems are quitter (low AQ), camper (being AQ), and climber (high AQ).

Based on the background of research that has been explained, the purpose of this study is to describe the relational understanding of students with the category AQ quitter, camper and climber in quadrilateral problem solving.

2. Method

This research use qualitative approach that aims to describe how students' relational understanding in quadrilateral problem solving based on Adversity Quotient (AQ). Research subjects were three 7th grade students of Junior High School. Research subject were taken by category of Adversity Quotient (AQ) such quitter, camper, and climber. The instruments of this research involves quadrilateral problem solving tasks used to describe relational understanding. Data collecting was conducted using interviews on the problem-solving task. Data analysis consist of three steps namely condensed data, presented data and conclusion data. The written and interview data are analyzed according to indicators of relational understanding on every problem solving steps.

3. Results and Discussion

3.1. Results

In this section, it was discussed the results of research based on the relational understanding of junior high school students in solving quadrilateral problems reviewed from their Adversity Quotient (AQ).

3.1.1. Relational Understanding Climber Subjects in Solving Quadrilateral Problems.

At the stage of understanding the problem, the subject writes and declares what is known to be important points in the problem, such as the length and width of the rectangle and their size, the area of both is the same. She is able to distinguish both quadrilateral and rectangular shapes by mentioning and drawing them on an answer sheet. She states and writes down what is asked in the given problem. She mentions concepts related to given problems, such as square, square area, rectangular area, and square circumference.

At the next stage, she plans the completion by linking the mentioned concepts, such as square side in variable x, modelling the area of square and rectangle in variable x, making equation of those two areas to obtain the value of x, calculating the square circumference, and giving the reasons for each planned step as seen in figure 1.

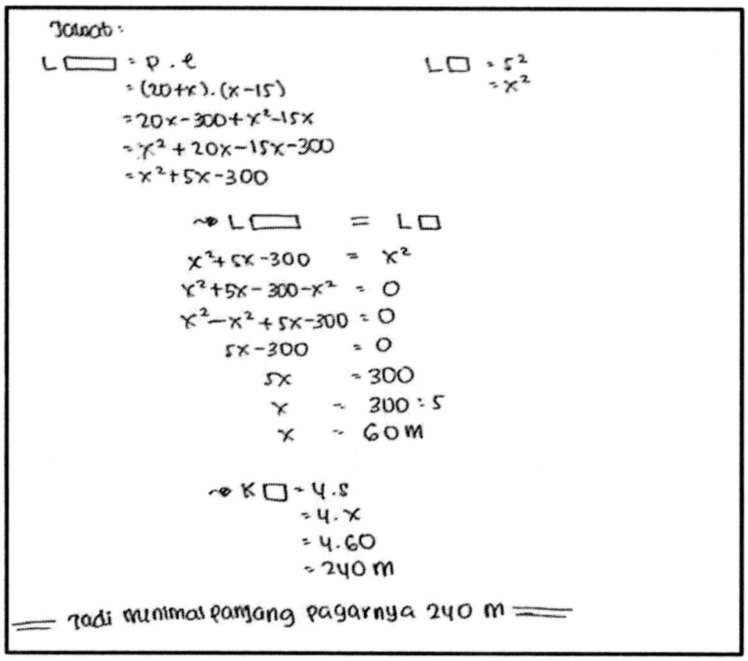

Figure 1. Example of students' complete performance on the task

At the last stage, she examines the answers obtained by recalculating the results of his work, looking back at the answers obtained, looking at the ways that have been done and matching the final answer to the questioned on the problem.

3.1.2. Relational Understanding of Camper Subjects in Solving Quadrilateral Problems.

At the stage of understanding the problem, she writes what is known only in the form of the points only. She states verbally what is known are the length and width of the rectangle along with its size, the area of both is the same, and distinguishes the form of both the building of the square and the rectangle. She mentions the concepts contained in the problem of quadrilateral, square, square area, rectangular area, and circumference of the square.

At the next stage, she plans the solving steps by connecting the mentioned concepts to the problem, such as equalizing the square with the length of the rectangle, finding the side of the square, calculating the circumference of the square. Subjects are not able to give a reason on every step planned.

At the problem solving stage, she searched for the square side by equating the square area with the square area, calculating the square area of the formula, calculating the square area of the formula, using the move of the segment, calculating the circumference of the square. She is incapable of giving any reason for any completion steps, such as not knowing why equating the second area of waking and looking for the sides first, not knowing the purpose of searching around the square. She examines the answers obtained by looking back at the answers and ways that have been done as well as calculating and ensuring the two wake areas are equal.

3.1.3. Relational Understanding of Quitter Subjects in Solving Quadrilateral Problems. At the stage of understanding the problem, she writes and declares what is known to be important points in the problem, such as the length of the rectangle and its size, the width of the rectangle and its size, and the square area equal to the area of the rectangle. In addition, the subject can distinguish the two wakes by drawing a square and square rectangle along with its size. She states and writes down what is asked as to what is being asked about the given problem. She mentions the concepts contained in the problem of the square area, the area of the rectangle, and the circumference of the square.

The next stage of the subject is to plan the settlement by linking the stated concepts of the subject to the given problem, such as determining the length and width of the rectangle, searching for the area of the two fields, looking for a square circumference. She could not give reasons for what is planned, such as why using the step equate the two wake-ups and do not know why look around the square.

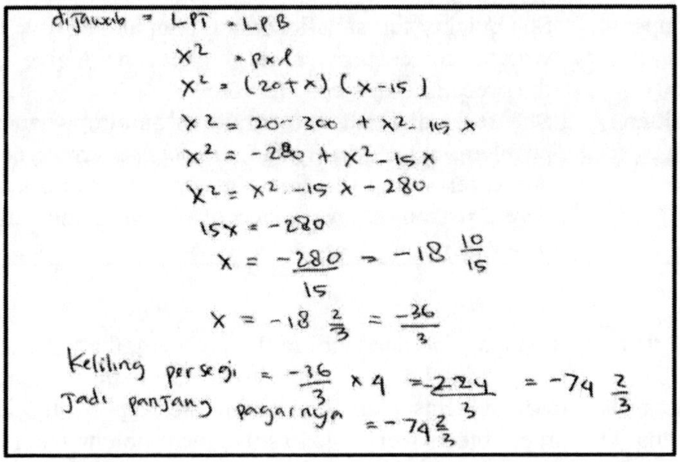

Figure 2. Examples of errors on the task

At the problem solving stage, she determines the square and square lengths, symbolizes the square side by x, calculates the square area equal to the rectangle, finds the square side, calculates the square circumference with the formula. She encounter errors when determining the value of x (the square) and calculate the circumference of a rectangle (**Figure 2**). The subject is incapable of giving reasons for steps taken in solving the problem. The subject checks the answers obtained by looking back at the answers and steps that have been done, aware of errors that have been done but do not want to improve the answer.

3.2. Discussion

3.2.1. Similarities of Junior High School Student Relational Understanding.
Based on the results of problem-solving tasks and interviews on the three subjects obtained some things about the equation of understanding between the three relational. Here are the relational understanding equations of the three subjects. At the stage of understanding the problem, they are both able to declare

IOP Conf. Series: Journal of Physics: Conf. Series **947** (2017) 012039 doi:10.1088/1742-6596/947/1/012039

and write down what is known and asked by using their own language, the three subjects can state the concepts related to the given problem. At the stage of devise a plan, they are both capable of completion by connecting the stated concepts, such as equating the two areas and searching for square circumference using the formula. At the stage carrying out the plan phase, they are equally able to solve the problem by using the square formula and the square formula, looking for the square side, calculating the square circumference with the formula. At the stage of looking a back, they able to re-examine the steps obtained and the answers obtained

3.2.2. Differences in the relational understanding of junior high school students.
In this research, there are some differences of relational understanding in solving quadrilateral problem done by subject with AQ climber, quitter, and camper category. This is as stated by Stolz (2008) that AQ affects a person in response to a problem. Here are the differences of students' relational understanding with the category AQ climber, quitter, and camper.

Table 2. Differences in the relational understanding of Subject based on AQ

AQ Categories	Differences Of Relational Understanding
Climber	She has solved the problem with the correct answer. In addition, the subject has been able to give reasons for every step taken during the planning phase and the stage of solving the problem.
Camper	She quickly felt satisfied when plan and resolve the problem with the correct answer, although it can not give a reason why the procedure or steps are used
Quitter	She feels difficult and afraid of failure when solving the problem, consequently the subject has errors in calculating and determining the final answer. The subject also can not give a reason for every step of planning and settlement that has been done.

4. Conclusion
The results showed that, at the stage of understanding the problem, the three subjects are able to state and write down what is known and asked, and able to mention the concepts associated with the quadrilateral problem. The three subjects plan the solving strategy by linking concepts relating to quadrilateral problems. The three subjects were able to solve the problem using the procedures that have been planned. And the three subjects were able to re-check the answers that have been obtained. Subject climber, camper and quitter able to understand the problem, plan the solving strategy, carry out the plan, and look back. However, the quitter and camper subjects have not been able to give a reason for the steps they have taken.

References

[1] Skemp R R 1976 *Relational Understanding and Instrumental Understanding* (Coventry: University of Warwick)

[2] Skemp Richard 1987 *The Psychology of Learning Mathematics* (New Jersey: Lawrence Erlbaum Associates)

[3] Walle J A, Karp K S and Bay-Williams J M 2010 *Elementary and Middle Schools Mathematics Teaching Developmentally Seventh Edition* (New York: Allyn & Bacon)

[4] Polya G 1973 *How To Solve It* (2nd Ed) (Princeton: Princeton University Press)

[5] Krulik, S and Rudnick J A 1987 *Problem solving: A handbook for teachers* (2nd Ed) (Boston: Allyn and Bacon)

[6] Stoltz P G 2004 *Adversity Quotient: Turning Obtacles Into Opportunities* (Jakarta: PT Grasindo Indonesia)

IOP Conf. Series: Journal of Physics: Conf. Series **947** (2017) 012040 doi:10.1088/1742-6596/947/1/012040

The Effect of Acceleration Sprint and Zig-zag Drill Combination to Increase Students' Speed and Agility

O Bana[1], E Mintarto[1], N W Kusnanik[1*]

[1]Universitas Negeri Surabaya, Indonesia

*Corresponding Author: Niningwidyah@unesa.ac.id

Abstract: The purpose of this research is to analyze the following factors: (1) how far the effect of exercise acceleration sprint on the speed and agility (2) how much influence the zig-zag drill combination to the speed and agility (3) and is there any difference between the effects of exercise acceleration sprint and practice zig-zag drill combination of the speed and agility. This research is quantitative with quasi-experimental approach. The design of this study is matching only design.This study was conducted on 33 male students who take part in extracurricular and divided into 3 groups with 11 students in each group. Group 1 was given training of acceleration sprint, group 2 was given zig-zag training combination drills of conventional and exercises for group 3, for 8 weeks. The data collection was using sprint 30 meter to test the speed and agility t-test to test agility. Data were analyzed using t-test and analysis of variance. The conclusion of the research is (1) there is a significant effect of exercise acceleration sprint for the speed and agility, (2) there is a significant influence combination zig-zag drills, on speed and agility (3) and exercise acceleration sprint have more effect on the speed and agility.

1. Introduction

Sport is a physical activity that is carried out to maintain a healthy and strengthening the muscles of the body. This activity can be done as an activity that entertain, fun or can also be conducted in order to improve performance. One of the desired expectations of a sportsman is to be someone who is well known in public or to achieve maximum performance in sports. Maximum achievement, according to the desired expectations will be achieved through guidance and training directed and carried out effectively and efficiently. According Sukadiyanto & Muluk, (2011), "Exercise is a process of changing to be the better, which is to improve the quality of physical, functional ability of the body equipment, and the quality of child psychological training". One element to achieve in sport is a physical condition [1]. According to Hidayat (2014), the physical condition is a very important element in nearly all branches of sports, therefore doing exercise for physical conditions is necessary for serious concern and systematic so that physical fitness and functional ability will be better [2]. Training methods to improve the speed and agility, drills acceleration sprint and exercises zig-zag drill combination. According Deuster et al (2013), sprint acceleration is a speed development drills, by running fast and gradually increased from jogging, striding, and the sprint is done repeatedly [3]. Meanwhile, according to Mylsidayu & Kurniawan (2015), said that the combination zig-zag drill one method of exercise to improve body agility in changing direction [4]. The researcher' reason to choose the two forms of exercise because the exercise was based on dominating the formation and increased velocity (speed) and agility (agility).

Based on the background above, this study aims to examine and analyze (1) how far influence the workout acceleration sprint on the speed and agility (2) how much the effect of exercises zig-zag combination drills on speed and agility (3) and is there any difference between the effects of exercise acceleration sprint and zig-zag drill combination to speed and agility. This type of research is quantitative research with a quasi-experimental approach (quasi experiment). The study design used is matching only design.The sample in this research is the high school students who participated in extracurricular as many as 33 pupils. The results obtained after 8 weeks researching with training sessions three times a week showed that the experimental group were given exercises acceleration sprint and practice zig-zag drill combination indicated a significant increase. (a) the experimental group I for speed -0.16% and -0.22%, for agility (b) experimental group II of speed -0.12%, and -0.15% for agility, (c) a control group of speed -0.05% and -0.08% for agility. The conclusion of the study are (1) there is a significant effect of exercise acceleration sprint and exercise combination zig-zag drill on the speed and agility, (2) the exercise acceleration sprint have more effect on the speed and agility, (3) the exercise acceleration sprint impacted on the pace and agility.

2. Method

The research is a quantitative study with a quasi-experimental approach. The study design used is matching only design and data analysis using t-test and ANOVA. Data collection used sprint 30 meters measuring the speed and agility whereas t-test is used to measure agility during the pretest and posttest. Furthermore, the data were analyzed using SPSS 20.0 series.

3. Results

The result shows to determine the effect of drills acceleration sprint and exercise zig-zag drill combination on the step test using t-test and ANOVA. The results of processing the data are in the table 1.

Table 1. Test Results of Mean Differences in Paired Samples Free

Test Paired Samples				
		Mean	T	Sig. (2-tailed)
K1	Posttest Speed-Pretest Speed	-0.21909	-6.727	0.000
K2	Posttest Speed-Pretest Speed	-0.12364	-7798	0.000
K3	Posttest Speed-Pretest Speed	-0.05182	-4969	0,001

Table2. Test Results of Mean difference in Paired Samples agility

Test Paired Samples				
		Mean	T	Sig. (2-tailed)
K1	PosttestAgility-Pretest Agility	-0.22818	-10.116	0.000
K2	PosttestAgility-Pretest Agility	-0.15182	-10 201	0,000
K3	posttest Agility-Pretest Agility	-0.08091	-5.709	0.000

The table 2 shows that the level of significance of each variable obtained sig <0.05. It is a significant difference or distinction between pretest and posttest data of each dependent variable (speed and agility), both in the experimental group I, group II and group experiments III. So it can be concluded that there is a significant effect of giving workout, acceleration sprint exercise zig-zag drill combination and in the control group.

3.1 Test Results Mean Inter-group difference (Anova)

The test in the study conducted by Anova (F) and the result is to determine whether the difference of results mean different in each group simultaneously. The test results can be seen in the table 3.

Table 3. Results of ANOVAAnalysis

Source of Variation	Df	F calculated	F Count	Sig.	Sig.	Specification
Inter-group	2	14.866	17.479	0.000	0.000	Significant
In Group	30					
Total	32					

The table 3 shows that significant value from ANOVA test on the speed and agility is sig <0.05. So it can be concluded that there are significant differences in the results of speed and agility in three different groups. Therefore, with the differences between groups, the data analysis can be continued by using test multiple post Hoc-comparations using the Least Significant Diffrence (LSD) in SPSS 20.0 with the aim to determine which independent variables have a significant impact on the increase dependent variable.

Table 4. Test Post Hoc DataSpeed

Variable	(I)	(J)	Sig.	95% Confidence Interval	
				Lower Bound	Upper Bound
Speed	Acceleration Sprint	Zig-zag Combination Drill	0.004	0.0326	0.1583
		Control	0.000	0.1044	0.2301
	Combination Drill Zig-zag	Sprint Acceleration	0,004	-0.1583	-0.0326
		Control	0,027	0,0090	0.1347
	Control	Sprint Acceleration	0.000	-0.2301	-0.1044
		Zig-zag Combination Drill	0.027	-0.1347	-0.0090
Agility	Acceleration Sprint	Zig-zag Combination Drill	0.005	0.0255	0.1272
		Control	0.000	0,0964	0.1982

Zig-zag Combination Drill	Sprint Acceleration	0.005	-0.1272	-0.0255
	Control	0.008	0.0200	0.1218
Control	Sprint Acceleration	0.000	0.1982	-0.0964
	Zig-zag Combination Drill	0.008	-0.1218	- 0.0200

The table 4 shows that there is difference in mean different a significant between group drills sprints acceleration and zig-zag drill combination, or the control group of both variable speed and agility. Based on these results it can be concluded that the exercise acceleration sprint have more influence on the speed and agility compared with the exercise combination of zig-zag drill and control groups.

4. Discussion

4.1 Exercise Experiment of Group I (Acceleration Sprint)

Exercise acceleration sprint is done by building speed, running gradually and slowly increased from running (jogging), length of step (striding) and sprint (sprint), between 50 to 100 meters. Judging from the patterns of movement, workout acceleration sprint is one form of exercises that focuses on improving speed and agility. From the calculation of the mean, it is found that the results is an increased speed and agility after being given the workout. Based on the analysis this exercise will significantly improve the speed and agility in high school male students who take part in extracurricular activities. It is in line with research conducted by Manikanthat that the workout acceleration sprint can increase speed and agility [5]. After testing the significance of the result is significant, it can be said that the administration of workouts acceleration sprint really have positive effect on speed and agility. Kusnanik, (2016) also added that the speed, agility, and acceleration are the component of physical condition which are very important for a sprinter [6]. This is because to be able to sprint quickly athlete must have a component with a good physical condition.

4.2 Experiments Group of Exercise 2 (Combination Drill zig-zag)

Zig-zag drill combination is like a square with a distance of 10 meters on each side. It is done by positioning themselves at the start line and start to run backwards, followed by sprinting intersect then followed with a run backwards and ends with a sprint cross line. The starting point from the patterns of movement that exercises zig-zag drill combination which is one form of exercise that focuses on improving agility in changing direction quickly without losing balance. From the calculation of the mean, it is found that the results have an increase in speed and agility exercises after receiving gifts zig-zag drill combination. Based on the analysis, this exercise will significantly improve the speed and agility in high school male students who take part in extracurricular activities. It is in line with research conducted by Hartati et al, that the exercise combination zig-zag drill can improve agility [7]. After testing the significance, it can be said that the administration of drills combination zig-zag drill truly have positive effect on speed and agility.

4.3 Control Grup

Based on the calculation of the mean group, the results of the exercises shown by the control group had increased the average yield of speed and agility despite the control group allowed to exercise outside of treatment. Exercise under this situation allows a control group to perform other activities that can support the speed and agility, such as: the existence of other activities such as football without no training. In addition to treat specific male student of high school who had exercise can indirectly support the level of biomotoric and skill. The speed and also agility in terms of running and playing ball rise to an increase in proceed from pretest to posttest.

IOP Conf. Series: Journal of Physics: Conf. Series **947** (2017) 012040 doi:10.1088/1742-6596/947/1/012040

5. Conclusion

The achievement of athletes in various sports can be gained through regular training, upgrading the supporting facilities, and the improvement of qualified and experienced trainers. Trainer are required to create an exercise program which is able to support the potential of athletes and the determination of the training method must also be in accordance with the needs of athletes. It can be proven from this study that there is a great significant effect on the acceleration sprint and combination zig-zag drill training on speed and agility.

Acknowledgement

Writing this scientific papers will not be successful without the help and guidance of mentors, so I would like to say many thanks to Mr. Edy Mintarto and Mrs. Nining Widyah Kusnanik who helped and spent some time in the process of writing this paper.

References

[1] Sukadiyanto&Muluk D 2011*Introduction to the Theory and Methodology of PhysicalTraining* (Bandung: PT. LubukAgung)

[2] Hidayat S 2014*Pelatihan Olahraga: Teori dan Metodologi*(Yokyakarta: Graha Ilmu)

[3] Deuster PA et al 2013Effeck of aerobic circuit training and parcours Training on Selected Physiological Variables Among Men college Student*Journal International***11**(1)pp 149-51

[4] Mylsidayu A&Kurniawan F 2015*Coaching Science Basis* (Bandung: Alfabeta)

[5] ManikandanS 2014 Efficacy Of Acceleration Sprint Speed Training On Performance Of Inter Collegiate Athletes*Asian Journal OfPhisical Education And Computer Science In Sports***11**(1) pp4-5

[6] Kusnanik2016 Pelatihan Speed, Agility, and Quickness dan Plyometric pada Sprinter Prima Utama *Jurnal Iptek Olahraga***18**(3)

[7] Hartati WR et al2016 The Effect Of Combination Zig-Zag Practice Drill Forward The Agility Of SSB Yapora Primary U-12 Pekanbaru Player*Student OnlineJournal***3**(2)

Junior high school students' reflective thinking on fraction problem solving: In case of gender differences

M A Rasyid[1,] *, M T Budiarto[2], and A Lukito[3]

[1]Postgraduate in Mathematics Education, Universitas Negeri Surabaya, Ketintang Surabaya 60231, Indonesia
[2,3] Mathematics Department, UniversitasNegeri Surabaya, Indonesia

E-mail: [1]muh.rasyid@mhs.unesa.ac.id

Abstract. This study aims to describe reflective thinking of junior high school students on solving the fractions problem in terms of gender differences. This research is a qualitative approach involving one male student and one female student in seventh grade. The data were collected through the assignment of fractional problem solving and interview, then the data were triangulated and analyzed by three stages, namely data condensation, data display and conclusion. The results showed that the subjects of male and female were reacting, elaborating and contemplating at each stage of solving the fractions problem. But at the stage of devising the plan, the female subject was contemplating, relying more on their beliefs, did not consider their experience, in addition, the female subject didn't use experience of the steps she planned to solve the problem of fractions.

1. Introduction

Reflective thinking is considered as actively, continuously and conscientiously belief or form of knowledge that is taken for granted in light with the reasons that support the inferred conclusion of the truth [1]. Students who think reflectively become aware of what is known and what is needed. It is very important to link the gap in learning situations [2]. The development of reflective thinking has become the focus and importance of mathematics education as it is the core of mathematics education[3]and it is one of the higher-order thinking skills[4]. In addition, reflective thinking is the most important element to learn effectively[5]and can be applied in mathematics learning[6]. There are three components of reflective thinking, they are: reacting, elaborating and contemplating[7].An important role of reflective thinking is to act as a means to activate thinking during a problem-solving situation, as it provides an opportunity to use knowledge and experience related to the problem faced and devised the best strategy to achieve the goal[8]. This shows that reflective thinking of students can be seen or raised through the problem-solving process. In addition, problem solving is an important standard in mathematics learning [9]. Reflective thinking indicators of students in problem solving are based on the Polya problem-solving stages, namely understanding the problem, making a plan, carrying out the plan and looking back[10]in which it integrates the reflective thinking components of Surbeck, Han & Moyer. The problem used in this study is a fractional problem because some experts claim that the fraction is a difficult material[11], since fractions include complex and difficult material for students to learn[12]and none of the fields in school math are as complex and as difficult to teach as fractions[13]. This corresponds is one of the important roles of reflective thinking, namely solving complex problems [14]. Gender difference is a term which is widely used in our everyday lives, including school and education [15]. The difference is likely to have an influence on learning and development in different abilities of male and female, such as research conducted by Zhu [16]and Pargulski & Reynolds [17]show that male surpasses over female in problem solving. In relation to reflective thinking, the research done by Demirel, Derman, & Karagedik [18] and Tuncer & Ozeren [19] showed that they have different findings about the relationship of reflective thinking and gender differences in problem solving. Reflective thinking is a person's mental activity to identify problems and solve problems by using knowledge and experience relating to problems encountered, and performing components of thinking Reflective, namely reacting, elaborating, and contemplating.

Research by Demirel, Derman, & Karagedik [18] showed that in the process of solving mathematical problems using the ability of reflective thinking depends on gender differences. However, the research is a case study using a scale is less in depth about the phenomenon that occurs in the subject of the research; especially the subject is too large. The study aims to examine the relationship between reflective thinking ability of 7th and 8th grade students on their problem solving and attitude toward mathematics, while research conducted by Tuncerv & Ozeren [19] showed that there is no significant difference between reflective thinking of male students and reflective thinking of female students in problem solving. The study used scaling in the data collection and the subjects of the study were prospective teachers.

This study aims to describe reflective thinking of male students and female students in solving fractional problems.

2. Method

This research is a descriptive research with qualitative approach which aims to describe reflective thinking of male students and female students in solving fractional problems.

2.1. Participants

This study involves two people, one male student and one female student in seventh grader of Junior High School average 11-13 years old. Subjects were selected by giving math skills tests to choose high-ability students. After those processes, the researcher look at gender differences by taking into account the social and cultural prospects of research subjects.

2.2. Data collection

Data collection is done by assigning a fractional problem-solving task and then performing task-based interviews.The fractional problem-solving task consists of two questions in the form of a description problem. The problem is designed to express students' reflective thinking. Afterwards, interviews were conducted to see reflective thinking of students in solving the fractional problem.

2.3. Data analysis

After the data had been collected, the triangulation process then was implemented to check the validity of the data and determine whether a data exactly describes the phenomenon in a study. After that, the data is analyzed by three stages, namely; data condensation, data display and conclusion [20].

3. Results and Discussion

3.1. Reflective Thinking of Male Subjects in Fractional Problem Solving

The results showed that the male subjects reacting at each stage of problem solving. In the stage of understanding the problem, the male subject commented on his feelings on the experience of having related information on the issue of fractions. This is evident from the quotation of subject interviews, "This question ... on the fractional matter and the story in everyday life. The problem is easy and difficult and I am interested and challenged to solve it because there is a resemblance to the problem I have ever faced, but this matter is not known the total land area, it is difficult". From the response, so that the male subject can identify and express return information, such as many parts of the land to plant pepper is $\frac{1}{4}$ part, many parts of land to plant cocoa trees is $\frac{2}{5}$ parts, the rest to plant banana trees, the land used to plant banana trees of $140\ m^2$, and searchable area of land. In the stage plan, male subjects commented on his feelings on experience related to strategies / methods/ steps that are planned to solve the problem fractions. Here is the f ollowing the response of the subject, he stated, "Sir, I had never planned such a plan, but there's something like a piece of cake, I subtract a whole bunch of cakes with lots of cakes eaten to find the rest of the cake that has not been eaten..." in the stage of carrying out the plan, the subject commented on his feelings about the experience he has with the strategy/methods/steps that have been done with that have been prepared in solving the fractional problem. Here is the following subject piece of interview, "first, I searched a lot of land for a banana tree and "1" this means an entire field, Sir. Then the subject commented again, "These steps/strategies in my experience are correct and this way I have done before". In the next step, the looking back stage, where the male subject commented on his feelings about the experience he had about examining the solution which is derived from the final answer of the fractional problem. Thus quotation of subject interviews by stating the need to look back at his work, namely "Need Sir, so that no errors in the calculations that I did and first often wrong in the calculation".

Male subject was elaborating at each stage of problem solving. In the stage of understanding the problem, the male subject comparing the reactions to the experiences of information on fractional issues related to the use of concepts and the relationship between information obtained. As the subject mentions the similarity with his experience, "As the fractions are the same ... and equally asked the area of land used". Furthermore, the subject also mentions the difference, "That Sir, if now is not known the total land area and how to look for". In the planning phase, the male subject compares the reactions to the experiences of related strategies/methods/steps planned to solve the fractional problem. As the excerpt of a subject interview about the similarities to his experience, "If you are looking for a lot of land that is used to plant banana trees by subtracting the whole lot of land with lots of parts of the land used for growing pepper and cocoa trees. Then to find the area of land used to grow pepper by multiplying many parts of the land to plant pepper with the total land area". Then the subject mentions the difference with his experience, "It is…Sir, looking for the whole land area. In the past, usually it was known in the matter". In the stage of carrying out the plan, the male subject compares the reaction to the experiences of related strategies/methods/steps that have been done in solving the fractional problem. The subject mentions the similarities with his experience, such as one to find the remaining land by subtracting the overall land with many parts of the land used. The difference with his experience is to find the total land area.Then, in the looking back stage; the male subject compares the reactions to the re-examining experiences and the suitability between the fractional problem-solving results to the given fractional problem and finding the best problem-solving solution. The following is a piece of subject interview about his similarities to his experience, "The similarities are to look at each calculation, the formulas too. And usually if a lot of my time write-off back". The difference is that if this problem I write-off by recalculating the total area of land by finding the area of land to plant cocoa trees, I just sum with the area of land to plant pepper, cocoa trees, and banana trees. I get fixed $400\ m^2$."

Male subject was contemplating at each stage of problem solving. In the stage of understanding the problem, the male subject considering and believing that the information provided is enough to solve the problem of fractions. Like the excerpt of a subject interview, "Yeah sir, I'm sure and I've noticed and thought about it based on my experience". The planning phase, the male subject considers and believes that the planned strategy/method/steps can solve the fractional problem and it is correct based on the mathematical concepts. Like the response of the following subject, "Yes sir, I am sure ....because I used to do something similar like this Sir and it is true". In the stage of carrying out the plan, the male subject considers and believes that the strategy/methods/steps that have been done are correct in solving the fraction problem and evaluating the error and fixing it immediately. The subject's mistake, i.e. excessive "=" writing. Then at the looking back stage, the male student can determine and believe in the conclusions obtained after solving the fractional problem.

3.2. Reflective Thinking of Female Subject in Fractional Problem Solving

The results showed the female subject reacting at each stage of problem solving. In the process of understanding the problem, she commented on her feelings about the experience she had in relation to information on fractional issues. As the following interview quotes, "This question is in the form of a story in everyday life in which there is fractional problems to be solved and I think this problem is a bit difficult but I am challenged to finish it because I enjoy learning fractions". Furthermore, the subject identifies and reveals the information of the given problem. In the planning phase, this subject commented on his feelings about the experience she had about strategy/methods/steps planned to solve the fractional problem. For example the subject revealed that she had never worked out a plan like this, but there were similarities to the problem she had ever done. The remaining part of the land planted with banana trees and look for the area of land used. The stage of carrying out the plan, the subject was commenting on her feelings about the experience she has about the strategy/methods/steps she has done with the ones that have been devised in solving the fractional problem. Next to the looking back stage, she commented on her feelings about the experience of examining the solutions obtained from the final answer to the fractional problem.

Female subject was elaborating at each stage of problem solving. The stage of understanding the problem, she compares the reaction to the experiences of information on fractional issues related to the use of concepts and the relationship between information obtained. The subjects mention the similarities and differences in their experience, as in the interview that, "If it's Sir, like the division of land, equally look for the rest of the land and look for the area to be used for growing rice. In addition, it also uses addition and subtraction operations. The problems I have worked on before, on the matter is already known the total area of land so that can be searched area of land in question such as the area of land to be planted with rice. But on this matter is not known sir, it is too hard Sir". In addition, the subject mentions the concept of addition and subtraction of fractions. In the planning stage, this subject compares the reactions to the experiences of related strategies/methods/steps planned to solve the fractional problem. The subject compares by mentioning the similarities and differences with the experience, as in the interview quote, "Equation, if we look for the remaining part of the land that is subtract many parts of the land with the result of the number of parts of land used or known. It can also with a picture, Sir. If the first problem I ever did Sir, after acquired the remaining lots of land, then multiplied by the total area of land if we want to find the area of land to be used. But this is the problem; it is unknown to the overall size of the land". In the stage of carrying out the plan, she compared the reaction to the experience of doing strategy/methods/steps that have been done in solving the problem fractions. The subject states that the equation of this matter with her experience, that is, both of cases use the addition operation and the reduction of fractions to know the remaining part of the land. Meanwhile, the difference is to look at the stage of land overall. Then checked again, this subject compare the reaction of the experiences to the check and the correspondence between the results of problem solving fractions with fractions given problem and find the best solutions to solve the problems. The Subjects revealed that the expression with their experience of looking back at the formula, the steps and calculations that they did and where the error is usually on multiplication errors or sums. In addition, the subject writes another way using representations with images.

In the next step, the female subject was contemplating at each stage of problem solving. In the stage of understanding the problem, she considered and believed that the information provided was enough to solve the problem of fractions by stating that all information on the matter is important and necessary, and then convinced that the information is sufficient to solve the problem. In the planning stage, the female subject believes that the planned strategy/method/steps can solve the fractional problem and it is correct based on the mathematical concept. In the stage of carrying out the plan, the female subject believes that the strategy/methods/steps that have been done is correct on solving the problem of fractions and evaluates the error and immediately fix it. In addition, the subject reveal a mistake that they made on the writing of the unit area. Then at the looking back stage, the female subject can determine and believe in the conclusions obtained after solving the fractional problem.

Based on the results of the study it can be concluded that the subject of male and female subjects in conducting reacting, elaborating and contemplating using previous experience that has been related to the issue of a given fraction. This is in line with Rodgers argued that reflective thinking is a meaningful process that moves students from one experience to the next with a deeper understanding of their relationship to other experiences and ideas [21]. In addition, based on this experience, the male and female subjects can identify and recount information from the given fractional problem. In line with Sezeropinion, students who think reflectively become aware of what they know and what they need [2]. In addition, the male and female subjects are re-evaluating the results of their work and if any errors are found then immediately fix them. This is a characteristic of someone who thinks reflectively. As Taggart & Wilsonargues, reflective thinking means a person logically deciding something and assessing the consequences of decisions made [22]. In addition, in accordance with the findings of Gurol suggested that, reflective thinking as a process of directional and precise activity in which individuals analyze, evaluate, motivate, get deep meaning and use the right strategy [23].

In general, reflective thinking of male subjects and female subjects have no significant difference, it only differ in the contemplating component for the planning stage. The female subject is limited to belief in what she planned without considering based on her experience. Thus, reflective thinking in problem solving does not depend on gender differences. This is in line with the results of Tuncer & Ozeren research, but unlike the results of Demirel, Derman, & Karagedik research which shows that in the process of solving mathematical problems using reflective thinking skills depends on gender differences.

4. Conclusion

The male and female subject perform all reflective thinking components, namely: reacting, elaborating, and contemplating in solving fractional problems. However, the female subject in drawing up plans, relies more on her beliefs without considering her experience to solve fractional problems.These results indicate the importance of reflective thinking for students, especially helping them in solving math problems, because they can use their experiences related to the problems faced.The finding of this research assumes that little significant difference found in reflective thinking ability of male subject with reflective ability of female subject. Thus, reflective thinking in the fractional problem-solving process does not depend on gender differences.

Acknowledgements

The authors would like to thank to all who involved and assisted in this research, in particular The Directorate General of Teachers and Education Personnel of the Ministry of Education and Culture of the Republic of Indonesia who have provided financial support for this research.

References

[1] Dewey J 1933 *How we think: A restatement of the relation of reflective thinking to the educative process* (Lexington, Massachusetts: D.C. Heath and Company)

[2] Sezer R 2008 Integration of critical thinking skills into elementary school *teacher education courses in mathematics Education* pp 349-362

[3] Meissner H 2006 *Creativity* and mathematics education *Elementary Education Online*5 (1) pp 65-72

[4] King F J, Goodson L, and Rohani F 2013*Higher order thinking skills* (Miami: Cala Press)

[5] Van De Walle J A *2007Elementary and middle school mathematics: Teaching developmentally* 6th ed (Boston: Pearson Education, Inc)

[6] Odafe V U *2008Teaching and learning mathematics: Student reflective adds a new dimension* (Bowling Green State University, Huron, USA)

[7] Surbeck E, Han E P, and J. E. Moyer J E 1991 Assessing reflective responses *Journals Educational Leadership* **48** pp 25-27

[8] Rudd R D 2007 Defining Critical Thinking *Techniques* **82** (7) pp 46-49

[9] NCTM 2000 *Principles and standards for school mathematics* (Reston, VA: National Council of Teacher of Mathematics)

[10] Polya G 1973 *How to solve it* (Princeton, NJ: Princeton University Press)

[11] Fisher J 2009 Fraction, partitioning and the part-whole concept *Teaching and Learning* pp 12-19

[12] Lamon S J 2012 *Teaching fraaction and rations for understanding: Essential content knowledge and instructional strategies for teacher 3rd ed* (USA: Taylor and Francis)

[13] Clarke D M, Roche A and Mitchell A 2008 Practical tips for making fractions come alive and make sense*Mathematics Teaching in The Middle School* **13** (7) pp 372-380

[14] Chelliah K K and *Arumugam* Z 2012 Does reflective practice enhance clinical competency in Medical imaging undergraduates?*Procedia-Social and Behavioral Sciences* **60** pp 73-77

[15] Santrock J W 2011 *Educational psychology 5th ed.*(New York: McGraw-Hill)

[16] Zhu Z 2007 Gender differences in mathematical problem solving patterns: A review of literature *International Education Journal* **8** (2) pp 187-203

[17] Pargulski J R and Reynolds M R 2017 Sex differences in achievement: Distributions matters *Personality and Individual Differences* **104** pp 272-278

[18] Demirel M, Derman I, and Karagedik E 2015 A study on the relationship between reflective thinking skills towards problem solving and attitudes towards mathematics *Procedia-Social and Behavioral Sciences* **197** pp 2086-2096

[19] Tuncer M and Ozeren E 2013 Prespective teachers' evaluation in terms of using reflective thinking skills to solve *problem Procedia-Social and Behavioral Sciences* **51** pp 666-671

[20] Miles M B, Huberman A M, and Saldana J 2014 *Qualitative data analysis: A methods sourcebook 3rd ed.* (*Arizona*: SAGE Publication. Inc)

[21] Rodgers C 2002 *Defining* reflection: Another look at John Dewey and reflective thinking *Teacher College Record* **104** (4) pp 842-866

[22] Taggart G L and Wilson A P 2005 *Promoting reflective thinking in teacher 50 action strategies* (Thousand Oaks, *California*: Corwin Press)

[23] Gurol A 2011 *Determining* the reflective thinking skills of pre-service teachers in learning and teaching process *Energy Education Science and Technology Part B: Social and Educational Studies* **3** (3) pp 387-402

IOP Conf. Series: Journal of Physics: Conf. Series **947** (2017) 012042 doi:10.1088/1742-6596/947/1/012042

The Ability of Students' Mathematical Proof in Determining the Validity of Argument Reviewed from Gender Differences

Feriyanto[1]

[1]*Majapahit Islamic University, KM 07 Jabon Street, Gayaman 61364, Mojokerto, Indonesia*

[1]Email: muhammad.feriyanto25@gmail.com

Abstract. This research aims to describe the ability of students' mathematical proof in determining the validity of argument reviewed from gender differences. The subjects of this research were one male and one female student of the fifth semester of Mathematic Education study program. The subjects were selected based on the highest mathematics ability which was assesed from their previous assignments and tests. In addition, the communication ability of the subjects was also considered in order to facilitate the researcher in conducting interviews. Based on the result of the test with direct and indirect proof, it could be concluded that the subjects were able to: 1) mention all facts/premises and write about what should be shown (conclusion) in direct proof and write additional premise in indirect proof; 2) connect facts/premises to concepts which must be mastered; 3) use equivalent concept to manipulate and organize the proof; 4) use the concept of syllogism and tollens mode to obtain the desired conclusion; 5) construct mathematical evidence systematically, and logically; 6) complement the reason for each step appropriately. The difference was that the male subject wrote the final conclusion, while the female subject did not write the final conclusion on the proof.

1. Introduction

Mathematics has many abstract objects, though not every abstract object is mathematics. Gagne [1] classifies mathematical objects into direct and indirect objects. The direct objects of mathematics includes facts, concepts, principles, and skills. While the mathematics indirect objects consist of the ability of investigate, the ability of problem solving, transfer of learning, self-discipline, and appreciation of mathematical stucture. For this case, the ability to investigate must be completed by providing the evidence. In mathematics, evidence has an important role as a method used to test the knowledge and very different from the way the inductive method works in the empirical activities [2].

According to the Educational Development Center (in [3]), evidence in a mathematical proof is a series of logical arguments that explain the truth of a statement. These arguments can be derived from the premise of the statement itself, definitions and other theorems. Logical means every step in the argument is justified by the preceding steps.

Evidence in mathematics, also called mathematical proof, is one aspect that must be considered in learning mathematics. This is as recommended by NCTM that proof is part of mathematics curriculum at all levels of education. The National Council of Teachers of Mathematics [4] states that students should be able to: know reasoning and proof as fundamental aspects of mathematics; create a conjecture and check the truth of the conjecture; develop and evaluate arguments and mathematical proofs; select and use various types of reasoning and verification methods. Therefore, the ability to do mathematical

IOP Conf. Series: Journal of Physics: Conf. Series **947** (2017) 012042 doi:10.1088/1742-6596/947/1/012042

proving must be possessed by all mathematics students. On the contrary, the fact showed that the students lack of this ability. Some studies ([5], [6], [7], [8]) suggest that many students have difficulty in mathematical proofing. Fadillah and Jamilah (2014) revealed that students have difficulty in proving the very abstract concept of algebra structure. The students are less familiar with the examples related to the concept and not yet accustomed to deductive proof. This is in line with the results of Hidayanto (2010) which states that students have difficulty in performing formal proof from a statement in Calculus I. Taufik' research result also shows that (2014) students have difficulty in proving propositions using mathematical induction. Furthermore, Prihandoko ([8]) reveals that students have difficulty to identify the elements and reasoning patterns in many theorems and definitions. In addition, Prihandoko added that many students are still accustomed to using inductive reasoning patterns and have not been familiar with valid proving types. Students also do not adhere to the basic principles of mathematical logic and set theory, so they can not prove the theorems correctly.

Proof in mathematics has a role of verification and justification, but in education, proof is used for explanation [9]. The verification and explanation in this research is used to determine the validity of the arguments in making a conclusion. Skill in constructing a conclusion requires an understanding of concepts and capabilities connecting the known arguments to learned concepts (definitions, axioms, postulates, theorems and etc.) through systematic and deductive algorithms. Since the mathematical proof is one of the mathematical abilities that must be mastered by students, then it is necessary also to give more attention. One of the important factor is gender differences. Gender is a visible difference between male and female seen from society's value.

Gender differences certainly lead to physiological differences and affect the psychological differences in learning. According to Susento ([10]) gender differences not only result in different abilities in mathematics, but also how to acquire mathematical knowledge. This statement is reinforced by Keitel ([11]) that gender, social and cultural influence mathematics learning. Similarly, the results of the study by Zhu ([12]) suggested that gender differences affect students' problem-solving abilities, in which the problem-solving abilities of male are better than female for middle and upper classes.

There is no research has examined directly the relation of mathematical proof to gender differences. Some researches only relate mathematical abilities in general to gender differences. Therefore, it is important to examine the ability of mathematical proof in determining the validity of the argument viewed from gender differences.

2. Method

This research aimed to describe the ability of mathematical proof in determining the validity of mathematical arguments of both male and female students. The approach of this research was a qualitative approach. The type of this research was descriptive explorative. The subjects of this research were one male and one female student of the fifth semester of mathematics education study program. The reason of choosing the fifth semester students was because the subjects had been taught the introduction to basic mathematics. The way of selecting the research subjects was all of the students given the mathematics ability test and then one male (DZ) and female students (KH) who have the heighest mathematics ability and good communication was selected. Furthermore, DZ and KH were given a mathematical direct proof test and continued with the interview. Then, DZ and KH were given a mathematical indirect proof test and interview. Finally, students' work results were analyzed. This research used triangulation method/ technique to obtain the data. The test and interview were conducted to assure the validity of the data.

3. Results

The following data is the description of mathematical test results done by students DZ and KH

3.1 The description data of mathematical direct proving by KH

Based on the test and interview results of KH in mathematical direct proof, KH has (1) good ability to identified statement that will be proved which showed by KH in mentioning all the facts/ premises also writing what should be shown (conclusion) i.e. $\neg p$; (2) good ability to use the direct proof method that is indicated by using the premises (facts) in making conclusions and relating with the concepts that must be mastered i.e. $p \to q, \neg q \vee r, r \to \neg s$, and s; (3) good ability to organize and manipulate facts known that is shown by the ability of the student to use the concept of equivalent in the proof step i.e. $\neg q \vee r \equiv q \to r$; (4) good ability to make connections between the known facts and elements of conclusions is also good that demonstrated by the ability of students to use the concept of tollens mode (if the premises are $r \to \neg s$ and s, then conclusion is $\neg r$) and syllogism (if the premises $p \to q$ and $q \to r$, then the conclusion is $p \to r$) in obtaining the desired conclusions; (5) good ability to compile the proof that is demonstrated a systematic and logical flow made in constructing mathematical proof but not completed by final conclusions; And (6) good ability to complete the reasons for each step taken.

3.2 The description data of mathematical direct proving by DZ

Based on the test and interview results of DZ in mathematical direct proof, DZ has (1) good ability to to identified statement that will be proved which shown by DZ in mentioning all facts/ premises on the question of proof, and DZ also writes what should be shown (conclusion) ie $\neg p$; (2) good ability to use the direct verification method shown by DZ in using the premises (facts) in providing conclusions and connecting with the concepts that are mastered ie $p \to q, \neg q \vee r, r \to \neg s$, and s; (3) good ability to organize and manipulate the facts that is demonstrated by the ability of the students to use the concept of equivalent in the evidentiary step ie $\neg q \vee r \equiv q \to r$; (4) good ability to make connections between the known facts and elements of the conclusions possessed shown by DZ in using the concept of syllogism (if the premises $p \to q$ and $q \to r$, then the conclusion is $p \to r$) and tollens mode (if the premises are $r \to \neg s$ and s, then conclusion is $\neg r$) in obtaining the desired conclusions; (5) good ability to compile the evidence that is demonstrated a systematic and logical flow made in constructing mathematical evidence but not completed by final conclusions; And (6) good ability to complete the reasons for each step taken.

3.3 The description data of mathematical indirect proving of KH

Based on the test and interview of subject KH in mathematical indirect proof, KH has (1) good ability to read the evidence by mentioning all the facts/ premises on the matter of proof i.e. $\neg p \to r, r \to \neg s$ and $s \vee q$, and KH also write the additional premise that $\neg q$; (2) good ability to use the indirect proof method, she uses the premises and the additional premise in making conclusions and relating to the concepts; (3) good ability to organize and manipulate the known facts, it is demonstrated by the ability of the student to use the concept of equivalent in the proof step; (4) good ability to make connections between the known facts and elements of conclusions, it is demonstrated by the ability of students to use the concept of syllogism (if the premises are $\neg p \to r$ and $r \to \neg s$, then the conclusion is $\neg p \to \neg s$) and tollens mode in obtaining the desired conclusion ie if the premises are $\neg q$ and $\neg s \to q$ then the conclusion is s; (5) good ability to construct evidence is also good, it is demonstrated by the systematic and logical flow made in constructing mathematical proof, but she doesn't write the final conclusion on the proof; And (6) the ability to complete the reasons for each step taken is very good, it is shown in each statement which is accompanied by a logical and precise reason.

3.4 The description data of mathematical indirect proof of student DZ

Based on the test and interview of subject DZ in mathematical indirect proof, DZ has (1) good ability to read evidence, it's proven by mentioning all facts / premises) on the matter of proof i.e. $\neg p \to r, r \to \neg s$ and $s \vee q$, and DZ also writes an additional premise that $\neg q$; (2) good ability to use the direct proof method, using the premises (facts) in showing conclusions and relating with the concepts; (3) good ability to organize and manipulate the known facts, this is demonstrated by the ability of students to use the concept of equivalent in the evidence step; (4) good ability to make connections between the known facts with elements of conclusions, it is demonstrated by the ability of students to use concepts syllogism (if the premises are $\neg p \rightarrow r$ and $r \rightarrow \neg s$, then the conclusion is $\neg p \rightarrow \neg s$) and tollens mode in obtaining the desired conclusion ie if the premises are $\neg q$ and $\neg s \rightarrow q$ then the conclusion is s; (5) good ability to construct evidence, this is demonstrated by a systematic and logical flow made in constructing mathematical proofs, and also accompanied by final conclusions in proof; And (6) good ability to complete the reasons for each step, each statement is accompanied by a logical and precise reason.

3.5 The comparison of the ability to construct mathematical direct proof.

Table 1. The comparison of the Ability to Construct Mathematical Direct Proof of Male and Female Subject

Student KH (Female)	Student DZ (Male)
Mentions all facts / premises (4 premises) on the matter of proof, and KH also write what should be shown (conclusion) that is $\neg p$	Mentions all facts / premises (premises 1, 2, 3, and 4) on the question of proof, and DZ also writes what should be shown (conclusion) ie $\neg p$
Uses the premises (facts) in drawing conclusions and connecting with the concepts that are mastered	Uses the premises (facts) in drawing conclusions and connecting with the concepts that are mastered
Uses the equivalent concept to manipulate and organize the proofs	Uses the equivalent concept to manipulate and organize the proofs
Uses the concept of syllogism and tollens mode in obtaining the desired conclusions	Uses the concept of syllogism and tollens mode in obtaining the desired conclusions
Makes a systematic and logical flow in constructing mathematical proof, but she doesn't write the final conclusions	Makes a systematic and logical flow in constructing mathematical proof, and but he doesn't write the final conclusions
Completes the reasons for each step appropriately	Completes the reasons for each step appropriately

Based on the Table 1, there is no difference in the ability to construct mathematical proof. Both of them are able to make excellent mathematical proofs, either in reading evidence, using proof methods, organizing and manipulating facts, making connections between facts, compiling evidence, or complementing the reasons in each step. It is contrary to the research of Triyadi ([13]) stated female students have mathematical connection, mathematical reasoning, mathematical communication and mathematical problem solving better than female students.

3.6 The comparison of the ability to construct mathematical indirect proof.

Table 2. The comparison of the Ability to Construct Mathematical Indirect Proof of Male and Female Subject

Student KH (Female)	Student DZ (Male)
Mentions all facts / premises (4 premises) on the matter of proof, and write the additional premise that is $\neg q$	Mentions all facts / premises (4 premises) on the matter of proof, and write the additional premise that is $\neg q$
Uses the premises (premises) and the additional premise in showing conclusions and relates with the concepts that are	Uses the premises (premises) and the additional premise in showing conclusions and relates with the concepts that are

mastered	mastered
Uses the equivalent concept to manipulate and organize proofs	Uses the equivalent concept to manipulate and organize proofs
Uses the concept of syllogism and tollens mode in obtaining the desired conclusions	Uses the concept of syllogism and tollens mode in obtaining the desired conclusions
Makes a systematic and logical flow in constructing mathematical proof, but she doesn't write the final conclusions	Makes a systematic and logical flow in constructing mathematical proof, and write the final conclusions
Completes the reasons for each step appropriately	Completes the reasons for each step appropriately

Based on the Table 2, it can be concluded that there is little difference in the ability to construct mathematical proofs, it is on the presence or absence of the final conclusions in mathematical proof. It's contrary to Krusteskii ([14]) that explained the male is better than female in mathematical reasoning, while female is better than male in accuracy, precision and thoroughness think. He also stated that male students have mathematical and mechanical skills better than female students. Likewise the variety sources ([15], [16]) there were gender differences in verbal skills with female outperforming males on many verbal tasks.

4. Conclusions

There are two conclusions that can be drawn based on the results above. The male subject can mention all facts/premises and write what should be shown (conclusion) in direct proof and write additional premise in indirect proof. The subject is able to connect facts/premises to concepts which must be mastered, use equivalent concept to manipulate and organize the proof, use the concept of syllogism and tollens mode to obtain the desired conclusion, construct mathematical evidence systematically, and logically, complements the reason for each step appropriately and he also writes the final conclusion. The female subject also shows the same ability with the male student, except she doesn't write the final conclusion on the proof.

5. References

[1] Gagne R M 1983 Some Issues in the Psychology of Mathematics Instruction *Journal for Research in Mathematics Education* **14 (1)**

[2] Hoyles C 1997 The Curricular Shaping of Students' Approaches to Proof *For the Learning of Mathematics* **17(1)** 7-16

[3] Fahinu 2008 *Meningkatkan Kemampuan Berpikir Kritis dan Kemandirian Belajar Matematik pada Mahasiswa melalui Pembelajaran Generatif* Disertasi (Bandung: UPI)

[4] NCTM 2000 *Principles and Standards for School Mathematics* Reston, (VA: NCTM)

[5] Fadillah, Syarifah and Jamilah 2014 *Pengembangan Bahan Ajar Struktur Aljabar untuk Meningkatkan Kemampuan Pembuktian Matematis Mahasiswa* 106-113

[6] Hidayanto E 2010 *Kesulitan Mahasiswa dalam Melakukan Pembuktian Teorema* (https://www.researchgate.net/publication/274513009)

[7] Taufik A 2016 Diagnosis Kesulitan Mahasiswa di Universitas Kuningan dalam Pembuktian Menggunakan Induksi Matematika Beserta Upaya Mengatasinya Menggunakan Scaffolding *JES-MAT ISSN 2460-8904*. **2(1)** 41-54

[8] Prihandoko A C 2005 *Pendekatan Identifikasi Logik untuk Mengatasi Kesulitan Mahasiswa dalam Memahami Definisi dan Teorema pada Struktur Aljabar Lanjut* Dibiayai oleh Proyek Peningkatan Penelitian Pendidikan Tinggi Direktorat Jenderal Pendidikan Tinggi Departemen Pendidikan Nasional 2005.

[9] Hanna G 1995 Challenge to the Importance of Proof *For the Learning of Mathematics* **15(3)** 42-49
[10] Susento 2006 *Mekanisme Interaksi Antara Pengalaman Kultural-Matematis, Proses Kognitif, dan Topangan dalam Reivensi Terbimbing* Disertasi (Surabaya: Unesa)
[11] Keitel C 1998 *Social Justice and Mathematics Education Gender, Class, Ethnicity and the Politics of Schooling* (Berlin: Freie Universität Berlin)
[12] Zhu Z 2007 Gender Differences in Mathematical Problem Solving Patterns: A Review of Literature *International Education Journal* **8(2)**
[13] Triyadi R 2013 *Kemampuan Matematis Ditinjau dari Perbedaan Gender* Skripsi (Bandung: UPI)
[14] Krutetskii V A 1976 *The Psychology of Mathematics Abilities in School Children* (Chicago: The University of Chicago Press)
[15] Halpern D F 2000 *Sex Differences in Cognitive Abilities (3rd Ed.)* Mahwah (NJ: Lawrence Erlbaum Associates)
[16] Maccoby E E and Jacklin C N 1974 *The Psychology of Sex Differences* (Stanford: Stanford University)

Effect of Exercise Program Speed, Agility, and Quickness (SAQ) in Improving Speed, Agility, and Acceleration

K Azmi[1,2] and N W Kusnanik[1]

[1]Universitas Negeri Surabaya, Indonesia.

[2]Email: niningwidyah@unesa.ac.id

Abstrak. This study aimed to analyze the effect of speed, agility and quickness training program to increase in speed, agility and acceleration. This study was conducted at 26 soccer players and divided into 2 groups with 13 players each group. Group 1 was given SAQ training program, and Group 2 conventional training program for 8 weeks. This study used a quantitative approach with quasi-experimental method. The design of this study used a matching-only design. Data was collected by testing 30-meter sprint (speed), agility t-test (agility), and run 10 meters (acceleration) during the pretest and posttest. Furthermore, the data was analyzed using paired sample t-test and independent t-test. The results showed: that there was a significant effect of speed, agility and quickness training program in improving in speed, agility and acceleration. In summary, it can be concluded that the speed, agility and quickness training program can improve the speed, agility and acceleration of the soccer players.

1. Introduction

An effort, to improve high achievement in every sport that occupied by an athlete, is important. One element or factor which is important to reach an achievement in sport is the physical condition. In some studies also said that to achieve is determined by four factors of the exercise of physical preparation, technical preparation, preparation tactics and mental preparation. The main purpose of physical preparation in training is to improve the functional potential of athletes and develop the bio motoric ability to the highest standards. Speed is one of the basic components necessary bio motor in some sports. Every sports activities both games, competitions, and games always require speed bio motor components. speed is the ability of the complex, because in general, speed is an ability that allows a basketball player to move as quickly as possible at the level of specific resistance [8]. Agility is an important quality in a lot of sports that are played on the field. In badminton, agility demonstrated the ability to move with quick footwork and precise [2]. Linear action such as acceleration and velocity can be influenced by changing the movement mechanism of the arms or legs. Thus, the ability to develop speed in a short time (acceleration) is an important component to support the performance in a wide range of sporting activities [4]. Exercise of speed, agility, and quickness cover the complete spectrum of training intensity, from low intensity to high intensity. Every athlete has a different level. Therefore, the intensity of exercise should coincide with the individual's ability [16].

Exercise involving speed, agility, and quickness is a training method aimed at developing motor skills and body motion control through the development of the neuromuscular system. It aims to improve the athlete's ability to perform multi directional explosive power movements by reprogramming the neuromuscular system, so it can work more efficiently [18]. Exercise of speed, agility, and quickness (SAQ) has become a popular way to train athletes. Speed, agility, and quickness to cover the complete spectrum intensity of exercise, from low intensity to high intensity. SAQ drills can also be used to teach movements, such as heating, or to improve the physical condition of athletes [16]. Exercise of speed, agility, and quickness is a system of progressive exercises and instruction aimed at developing fundamental motor skills to improve the ability of the athlete to be more skilled at faster speeds and with greater precision. This exercise has become a popular way to train athletes in improving the speed, strength, or the ability into maximum potency.

2. Research Methods

This type of research is quantitative study using a quasi-experimental method (quasi-experimental). The research design used is Maching Only Design. The research instrument used in this study are; (1) using a speed test run 30 meters, (2) Test agility using T-Test agility, and (3) Acceleration Tests using the test run 10 meters.

3. Results

Based on the research that has been done, data on speed, agility, and acceleration of each group are obtained. The data is presented as shown in Table 1.

Table 1. Acquisition Data Initial test and Final test Experiment Group I.

Average	Speed		Agility		Acceleration	
	Pretest	Posttest	Pretest	Posttest	Pretest	Posttest
	6.63	7.01	11.15	10.56	4.11	4.36
Standard Deviation	0.48236		0.18566		0.25753	
Increase	5.9%		5.6%		6.0%	

Based on Table 1 in the average pace seen before given exercise and after exercise, the percentage of improvement is happened in increased speed by 5.9%, agility increased 5.6% and acceleration increased 6.0%.

Table 2. Different Test Results Variables Dependent in the Experiment Group

Variable	Pair	t-count	Sig.(2-tailed)	Status
Speed	posttest – pretest	4,577	0,001	Different
Agility	posttest – pretest	-4416	0,001	Different
Acceleration	posttest-pretest	5,199	0,000	Different

Based on Table 2 as a whole there is a difference between before and after the treatment of each of the variables dependent (speed, agility, and acceleration) in the experimental group. This indicates that the level of significance of each variable speed, agility, acceleration is smaller than sig 0:05 or in other words, $P < 0.05$. It can be concluded that there is a difference after being given a workout program speed agility and quickness (SAQ).

In the control group there are also differences, although the difference is relatively small when compared with the experimental group. In details, can be seen in Table 3. Based on Table 3 in the visits average speed before the given exercise and after exercise increased speed by 1.7%, agility increased 1.8%, and increased by 1.6% acceleration

Table 3. Data Acquisition Initial test and Final test control group.

Average	Speed		Agility		Acceleration	
	Pretest	Posttest	Pretest	Posttest	Pretest	Posttest
	6.63	6.74	11.20	11.00	4.10	4.16
Standard Deviation	0.22696		0.06870		0.06349	
Increase	1.7%		1.8%		1.6%	

Table 4. Different Test Results Variables Dependent in Control Group

Variable	Pair	t-count	Sig.(2-tailed)	Status
Speed	Posttest-Pretest	2,752	0,018	Different
Agility	Pretest-Posttest	-3 238	0,007	Different
Acceleration	Posttest-Pretest	2,221	0,046	Different

From the data results Table 4, different test variable dependent in the control group, a whole trend occurred is generated from a significant difference with a variable speed, agility and acceleration under P <0.05.

Based on these two t-test table experimental group and the control group, the t value for speed, agility, and acceleration of the experimental group are respectively 4577, -4416 and 5,199 (see Table 2). While the t value for speed, agility, and acceleration of the experimental group are respectively 2,752, -3238, and 2,221 (see Table 4). When viewed from t each group, it can be concluded that the experimental group were given exercise speed, agility, and quickness significantly improve the speed, agility and acceleration compared with the control group.

Table 5. Results of Analysis Independent Samples Test

	Component Test	T	Sig. (2-tailed)
pretest speed	Equal variances assumed	-.008	.994
	Equal variances not assumed	-.008	.994
pretest agility	Equal variances assumed	.872	.163
	Equal variances not assumed	.872	.163
pretest Acceleration	Equal variances assumed	-.311.	759
	Equal variances not assumed	-.311	.759
Posttest speed	Equal variances assumed	2,117	.045
	Equal variances not assumed	2,117	.046
posttest agility	Equal variances assumed	2,186	.039
	Equal variances not assumed	2,186	.041
posttest Acceleration	Equal variances assumed\	-2236	.035
	Equal variances not assumed	-2236	.035

The data from Table 5 shows a different test results between the experimental group and control group. Based on the table, it can be explained that there are differences the results of the pretest speed, agility and acceleration in both groups. Based on sig. (2 tailed) pre-test bio motor the data of three components such that sig. > 0.05, there is no difference results from all groups. Later in the post-test speed, agility and acceleration both groups there are differences in the results of the third test bio motor components. Furthermore, based on sig. (2 tailed) post-test bio motor of three components such that sig. <0.05, there are significant differences between the experimental group and control group.

4. Discussion

Exercise of speed, agility, and quickness is a complex exercise that enable to improve the physical condition of athletes. Based on the analysis in this study, it suggests that the practice of speed, agility, and quickness can improve the physical condition of the components of speed, agility and acceleration. Exercise *of speed, agility, and quickness* (SAQ) has become a popular way to train athletes to cover a complete spectrum intensity of exercise, from low intensity to high intensity. SAQ drills can also be used to teach movements such as heating or to improve the physical condition of athletes [16]. Exercise *of speed, agility, and quickness* is a system of progressive exercises and instruction aimed at developing fundamental motor skills and improving the ability of the athlete to be more skilled at faster speeds and with greater precision. *Speed, agility, and quickness* has become a popular way to train athletes. The exercise also develops the ability to exert maximum force during movement activity at high speed.

The results of this study consistent with the previous studies conducted by Jovanovic and Sporis [7]. Furthermore, this study finds that, as a whole that the training methods, Speed, Agility, and Quickness (SAQ) significantly improve the performance of football players. In addition, research conducted by Rahul Kumar and Dhapola [14] suggests that exercise SAQ has a significant influence on the variables of physical fitness of speed, agility, reaction time, power, and flexibility cricketers. Therefore the hypothesis proposed in research Rahul Kumar & Dhapola stipulated that the SAQ exercise significant influence on the physical fitness variables. In conclusion Securities SAQ exercise significantly increase reaction time, speed, agility, and flexibility.

Speed is one of the basic capabilities necessary biomotoric in every sport Sukadiyanto and Muluk [17] The speed is very dependent on power because without power, speed cannot be developed. Agility or agility related to gestures that involve footwork and quick changes of position of the body [12]. To be able to provide quality results and quality, one athlete requires agility is good in itself that will affect the performance during exercise so as to improve the performance of athletes. According Kusnanik and Ben [10] speed and agility can be regarded as the most dominant element in a football game. Some way or method of exercise that can be used to improve the speed and agility of athletes among which the speed ladder run and repeated sprints. Overall running technique can be broken down into several phases, the early phase of the block, the acceleration phase, and the velocity step phase. During the acceleration phase, an athlete increase stride length and step frequency. When an athlete reaches the stage of the maximum constant speed, running speed increased through increased stride length and more importantly through the stride frequency. The speed and acceleration are essentials for football athletes because they need to achieve high speed when chasing a ball in a game of football [3]. The same thing was said Johnston, et al. [6] that the acceleration, defined as the rate of change in the speed and with the demands of high intensity movement, is indispensable when important moments in sports field games such as the ability to defeat an opponent in order to get the ball in a football game. Then Kusnanik, [9] also added that the speed, agility, and acceleration as a component of physical condition is very important for a sprinter. This is because, to be able to sprint quickly, athlete must have a good physical condition.

This study also supports the theory of Milanovic [11] that the speed and agility resulted from speed, agility, and quickness (SAQ) exercise as a whole can be considered as a useful tool for speed and agility among young soccer players. Exercise of speed, agility, and quickness training methods can provide a very specific and detailed that will help athletes to support the best performance in sports. In addition, the variety form of drills, speed, agility, and quickness exercises could help to prevent the athlete from bored to participate in training activities.

5. Conclusion

The results of this study suggest that: (1) there is a significant effect of exercise program speed, agility, and quickness (SAQ) to increase speed, (2) there is a significant effect of exercise programs speed, agility, and quickness (SAQ) to increase agility, and (3) there is a significant effect of exercise program speed, agility, and quickness (SAQ) to increase acceleration.

References

[1] Brown L E and Ferrigno V A 2005 *Training for Speed, Agility, and Quickness* (Unites States: Human Kinetics)

[2] Frederick M F A 2014 *Badminton: Specific Movement Agility Testing System* (Malaysia: Movement, Health & Exercise)

[3] Gevat C, Halil T, Fatma A, Alin L, and George S 2012 *The Effects Of 8-Week Speed Training Program On The Acceleration Ability And Maximum Speed Running At 11 Years Athletes* (Romania: Ovidius University Of Constanta)

[4] Haci M S 2014 Relationships Between Acceleration, Agility, And Jumping Ability In Female Volleyball Players *European Journal of Experimental Biology* **4(1)** 303-308.

[5] Hsiu-Ching L and Yee-Pay W 2012 Strength And Agility Training In Adolescents With Down Syndrome: A Randomized Controlled Trial. *Www.Sciencedirect.Com.*

[6] Johnson P and Bujjibabu M 2012 Effect of Plyometric and Speed Agility and Quickness (SAQ) on Speed and Agility of Male Football Palyers *Asian Journal of Phisical Education and Computer Science in Sport* **7 (1)** 26-30.

[7] Erculj F, Bracic M and Jakovljevic S 2011 The Level Of Speed And Agility Of Different Types Of Elite Female Basketball Players *Facta Universitatis: Physical Education And Sport* **9(3)** 283-293

[8] Jovanovic M, Sporis G, Omrcen D and Fiorentini F 2011 Effects Of Speed,Agility,Quickness Training Method On Power Performance In Elite Soccer Players *Journal Of Strength And Conditioning Research* **25(5)** 1285-1295

[9] Kusnanik 2016 Pelatihan Speed, Agility, and Quickness dan Plyometric pada Sprinter Prima Utama Pada Tahun 2013 *Jurnal Iptek Olahraga* **18(3)**

[10] Kusnanik N and Rattray B 2017 Effect Of Ladder Speed Run And Repeated Sprint Ability In Improving Agility And Speed Of Junior Soccer Players *Acta Kinesiological* **11 (1)** 19-22

[11] Milanovic Z, Sporis G, Trajkovic N, James N and Samija K 2013 Effect of a 12 Week SAQ Training Programme on Agility with and without the Ball among Young Soccer Players. *Journal of Sport Science and Medicine* **12** 97-103 http//www.jssm.org.

[12] Mylsudayu A and Kurniawan F 2015 *Ilmu Kepelatihan Dasar* (Bandung: Alfabet)

[13] Nandalal S and Bhupinder S 2016 Twelve Weeks Saq Equipment Training On Selected Motor Fitness Variables Among School Athletes *Periodic Research* **5(1)** 59-62

[14] Prasad Rahul Kumar and Dhapola M S 2014 Effect Of Eight Weeks S.A.Q. Training Programme Onselected Physical Fitness Variables *International Multidisciplinary Research Journal* **3** (7) (ISSN No. 2231-5063)

[15] Richard J J, Mark L W, Matthew J P 2014 Standardisation Of Acceleration Zones In Professional Field Sport Athletes *International Journal Of Sports Science & Coaching* **9(5)**

[16] Sharma S, Dhapola M S 2015 Effect Of Speed, Agility, Quickness (SAQ) Training Programme On Selected Physical Fitness Variables And Performance Ability In Basketball University Players *International Educational E-Journal* **4(3)** 14-22

[17] Sukadiyanto and Muluk, D 2011 *Pengantar Teori dan Metodologi Melatih Fisik* (Bandung: Lubuk Agung)

[18] Vallimurugan V and Vincent J P 2012 Effect of SAQ Training On Selected Physical Fitness Parameters of Men Football Palyers *International Journal of Advanted and Inovation Research* **1(2)** (ISSN: 2278-7844)

Students' Algebraic Reasonsing In Solving Mathematical Problems With Adversity Quotient

F Aryani[1], S M Amin[2] and R Sulaiman[3]

[2] Post graduate student, Universitas Negeri Surabaya
[2,3]Department of Science and Mathematics, Universitas Negeri Surabaya

Email: [1]fennyaryani@mhs.unesa.ac.id

Abstract. Algebraic reasoning is a process in which students generalize mathematical ideas from a set of particular instances and express them in increasingly formal and age-appropriate ways. Using problem solving approach to develop algebraic reasoning of mathematics may enhace the long-term learning trajectory of the majority students. The purpose of this research was to describe the algebraic reasoning of quitter, camper, and climber junior high school students in solving mathematical problems. This research used qualitative descriptive method. Subjects were determined by purposive sampling. The technique of collecting data was done by task-based interviews.The results showed that the algebraic reasoning of three students in the process of pattern seeking by identifying the things that are known and asked in a similar way. But three students found the elements of pattern recognition in different ways or method. So, they are generalize the problem of pattern formation with different ways. The study of algebraic reasoning and problem solving can be a learning paradigm in the improve students' knowledge and skills in algebra work. The goal is to help students' improve academic competence, develop algebraic reasoning in problem solving.

1. Introduction

Reasoning in mathematics is a cognitive process of looking for a reasons and looking for conclusions [1]. Algebra is the way we express generalizations about numbers, quantities, relations and functions [2]. Algebraic thinking or reasoning involves forming generalizations from experiences with number and computation, formalizing these ideas with the use of meaningful symbol system, and exploring the concepts of pattern and function [3]. Problem is a gap or discrepancy between present state and future state or desired goal [4]. A problem is a situation that confronts the learner, that requires resolution, and for which the path to the answer is not immediately known.

AQ is the intelligence to withstand adversity [5]. Stoltz classifies three types of people related to Adversity Quotient (AQ), namely: The Quitter (low AQ), The Camper (moderate AQ), and The Climber (high AQ). Quitters are group of people with less eagerness to accept the challenge of life. Campers are type of people having enthusiasm to face the problems and challenges, but they terminate it due to their incapability. While climbers are people who keep surviving to strive against all sorts of things that will keep crashing, regardless it may be the problem, the challenge, and the obstacle, things that continue to occur every day.

More recently, researchers have begun to explore how reciprocal reasoning when solving equations [6], co-variational reasoning [7], algebraic generalisation strategies [8] and patters as tools for algebraic reasoning [9]. However, since this research is in its junior high school, there are numerous unexplored issues. The research questions are how to describe the algebraic reasoning of quitter, camper, and climber junior high school students in solving mathematical problems.

2. Method

The study was a qualitative research with a descriptive strategy. The participant of this research were the 8 grade students. Students were classified based on the AQ by using questionnaire. From each AQ category, participant were selected by purposive sampling to determine the students' algebraic reasoning. The selection of research participant was performed by considering several criteria, namely: (1) the result of students' writing test on participant selection, (2) the students' communication skills (based on teacher's suggestion in teaching and learning activities), and (3) the students' AQ category. The data in this study was a task-based interviews.

The process of investigating algebraic reasoning consists of three phases: (1) pattern seeking; (2) pattern recognition and (3) generalization [8]. Three Indicator of algebraic reasoning in solving mathematical problem can be shown in table 1 as follows:

Table 1. Indicator of algebraic reasoning in Solving Problems

Indicators	Algebraic reasoning activity observed
pattern seeking	a. Students determine information as something that was known dan asked. b. Students use multiple representations for known things c. Students finding the elements of the pattern.
pattern recognition	a. Students looking at relationships among the sequence has given b. students discover the similarities relationship each constituent element of pattern
Generalization	Students was using strategy of generalization pattern was choosen

3. Result and Discussion

Subjects with the quitter's category was performing of algebraic reasoning such as (1) pattern seeking; (2) pattern recognition and (3) generalization. subject with quitter's category was doing substitution a information that was known and the previous question to ensure that the answers found was correct. It was shown in the figure 1:

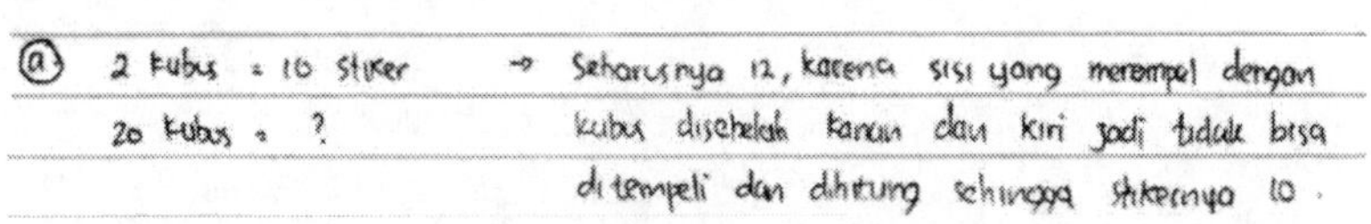

Figure 1. Quitter's answer sheet in explaining information

Figure 2 demonstrates pattern recognition with given situations by experimented to observe relationships and similarities between elements pattern, quitter try to calculate with different numbers so as to find relationships between elements pattern. Quitter explain that in solved problems on text will still use the same pattern. Quitter performed pattern recognition in understanding mathematics problems. Although campers was less persistent in overcoming new things compared to climbers, yet they also possess willingness to try new things including to solve mathematics problems. Their experiences in solving mathematics problems make it easier for them to analyze information of a problem so that they will tend to engage pattern recognition in understanding mathematics problems.

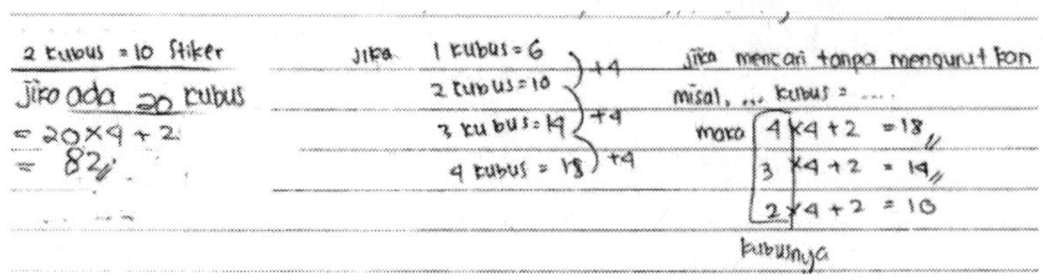

Figure 2. Quitter's answer sheet in explaining pattern recognition

After analyzed the image and finding the pattern formed, the quitter was able to formulate a general formula of the pattern. It was shown in the figure 3:

Figure 3. Quitter's answer sheet in explaining generalization

Subsequently, camper's category was performing of algebraic reasoning such as (1) pattern seeking; (2) pattern recognition and (3) generalization. subject with camper's category was doing substitution a information that was known and the previous question to ensure that the answers found was correct. It was shown in the figure 4:

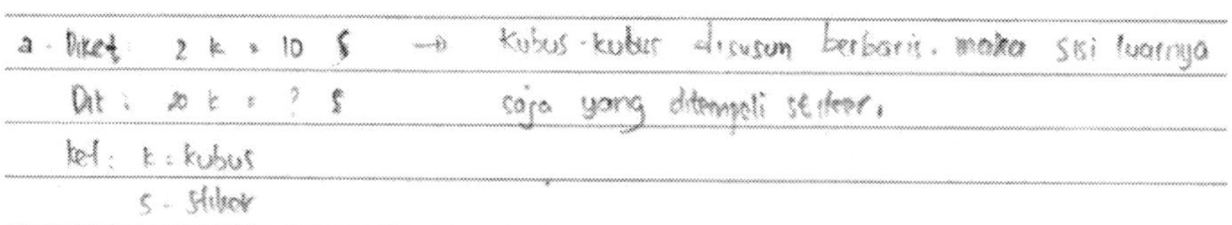

Figure 4. Camper's answer sheet in pattern seeking

It appears that camper could describe the information in the was known and question at once by using camper's own words. Figure 4 demonstrates the representations performed by the student by symbolizing the two inquired numbers with *k* and *s*. Hence, in understanding mathematics problems, camper performed both abstraction. Figure 5 camper finds the elements of pattern by drawing cube rows.

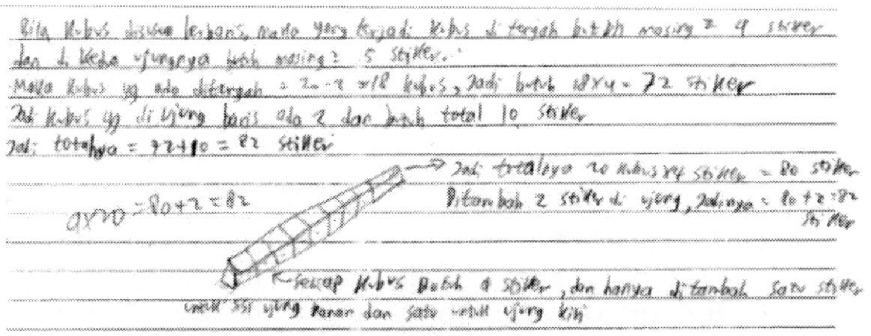

Figure 5. Camper's answer sheet in finds elements of pattern

Figure 5 demonstrates the camper pattern recognition finding relationships between elements of pattern constituents by observing patterns found previously and finds the similarity of relationships each element of pattern. Figure 6 was adjustment strategy to find the generalization formula and using that formula to determine the solving.

IOP Conf. Series: Journal of Physics: Conf. Series **947** (2017) 012044 doi:10.1088/1742-6596/947/1/012044

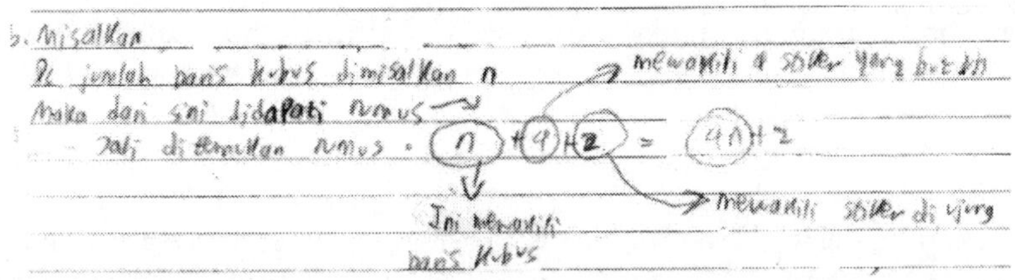

Figure 6. Camper's answer sheet in explaining generalization

Subsequently, climber's algebraic reasoning in solving mathematics problems was obtained through analysis and discussion similar to camper's analysis and discussion. camper's category was performing of algebraic reasoning such as (1) pattern seeking; (2) pattern recognition and (3) generalization. Figure 7 shows an example of climber's answer sheet in explaining the information.

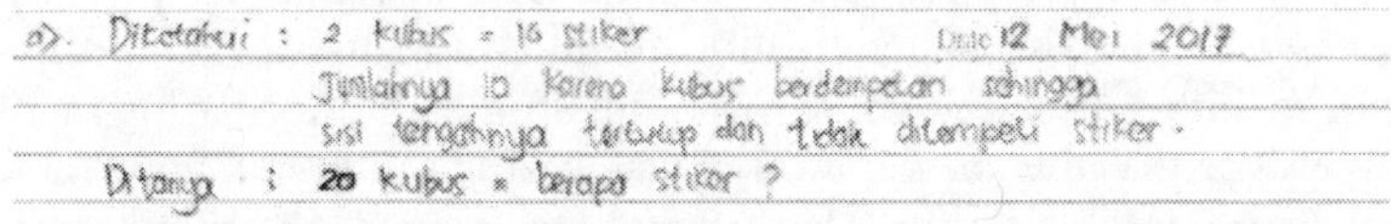

Figure 7. Climber's answer sheet in explaining information

The climber in recognizing the patterns given situation by experimented to observe relationships between elements pattern. Climber finding the number of stickers needed on each number of cube rows by using the formula they find. Climber's realizes that there was a pattern similarity relationship so that in completing tasks use the same way.

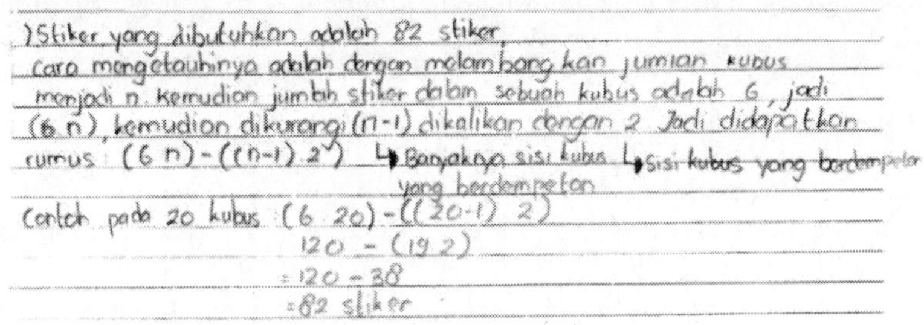

Figure 9. Climber's answer sheet in finds elements of pattern

The climber formulates a general formula of the pattern through analysis and general conclusion from the known. Using a climber image gives a description of the formula generalize. It was shown in the figure 10:

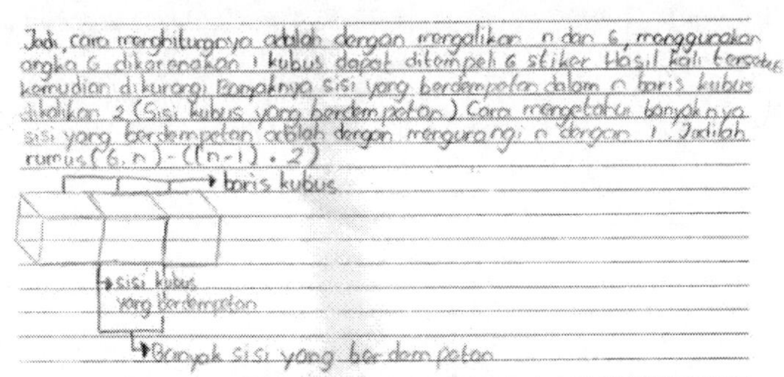

Figure 10. Climber's answer sheet in explaining generalization

The result of algebraic reasoning in this study mostly confirmed the result of previous studies carried out by Istiqomah [10]. It of have studied the algebraic reasoning in senior high school grade in solving mathematics problem.

4. Conclusion

Several conclusion can be derived from the study on students algebraic reasoning based on Adversity Qoutient as follows: 1) in pattern seeking quitter involved was substitution a information that was known and the previous question to ensure that the answers found was correct. In devised a plan of problem solved, climber included the pattern recognition and generalization the solution. 2) in pattern seeking , camper involved was substitution a information that was known and the previous question to ensure that the answers found was correct. In addition, camper also entailed abstraction to understand In planned problem solving, camper finds the elements of pattern by draw cube rows. In implementing the plan of problem solved, camper pattern recognition find relationships between elements of pattern constituents by observed patterns found previously and finds the similarity of relationships each element of pattern was adjustment strategy to find the generalization formula and used that formula to determine the solution. 3) In pattern seeking, climber involved was substitution a information that was known and the previous question to ensure that the answers found was correct, climber finds the elements of pattern by draw cube rows. In implementing the plan of problem solved, the climber formulates a general formula of the pattern through analysis and general conclusion from the known.

5. References

[1] Stacey K 2010 *Mathematics teaching and learning to reach beyond the basics* (Australia: Research Conference University of Melbourne)

[2] Watson and A Key 2007 *Understanding Of Mathematics Learning Paper 6: Algebraic Reasoning* (Nuffield Foundation: University of Oxford) p 3

[3] Van de Walle, J A Karp K, & Bay-Williams J 2011 *Elementary and middle school mathematics: Teaching developmentally* (Boston, MA: Allyn & Bacon) p 262

[4] Ellis H C & Hunt Reed R 2004 *Fundamentals Of Cognitive Psychology* (7th ed.) *Dubuque* (Iowa: Brown & Benchmark Publishers)

[5] Stoltz, Paul G 2005 *Adversity Quontient (Mengubah Hambatan Menjadi Peluang)* (*Jakarta*: Grasindo) p 5

[6] Hackenberg A J & Lee M Y 2013 Relationships between students' fractional knowledge and equation writing *Journal for Research in Mathematics Education* submitted for publication

[7] Ellis A B 2007 The influence of reasoning with emergent quantities on students' generalizations *Cognition and Instruction* **25**(4) pp 439–478

[8] Lannin J & Barker D & Townsend B 2006 Algebraic Generalisation Strategies: Factors Influencing Student Strategy Selection *Mathematics Education Research Journal* **18** (Columbia: University of Missouri) chapter 3 pp 3–28

[9] Herbert K & Brown R H 2000 *Patterns as Tools for Algebraic Reasoning, National council of teachers of mathematics* (Reston VA: Author) pp 123-128

[10] Istiqomah 2016 *Penalaran aljabar siswa SMA dalam menyelesaikan masalah matematika sberdasarkan gaya kognitif.* Thesis (Surabaya: Universitas Negeri Surabaya)

Effects of High Intensity Interval Training on Increasing Explosive Power, Speed, and Agility

F Fajrin[1], N W Kusnanik[2] and Wijono[3]

[1,2,3]Universitas Negeri Surabaya, Indonesia.
Email: [2]niningwidyah@unesa.ac.id

Abstract. High Intensity Interval Training (HIIT) is a type of exercise that combines high-intensity exercise and low intensity exercise in a certain time interval. This type of training is very effective and efficient to improve the physical components. The process of improving athletes achievement related to how the process of improving the physical components, so the selection of a good practice method will be very helpful. This study aims to analyze how is the effects of HIIT on increasing explosive power, speed, and agility. This type of research is quantitative with quasi-experimental methods. The design of this study used the Matching-Only Design, with data analysis using the t-test (paired sample t-test). After being given the treatment for six weeks, the results showed there are significant increasing in explosive power, speed, and agility. HIIT in this study used a form of exercise plyometric as high-intensity exercise and jogging as mild or moderate intensity exercise. Increase was due to the improvement of neuromuscular characteristics that affect the increase in muscle strength and performance. From the data analysis, researchers concluded that, Exercises of High Intensity Interval Training significantly effect on the increase in Power Limbs, speed, and agility.

1. Introduction

Achievements of an athlete is always related to how he/she practices, what is practiced and what its purpose. Selection methods and the right kind of exercise will support the success of the exercise itself. High Intensity Interval Training (HIIT) is one of the methods and the right kind of exercise because it is very effective and efficient way to improve the desired physical components. HIIT workouts can improve cardiac performance and impact on the body's metabolism is also increased sharply. Metabolism herein relates to the body's ability to convert fat into energy. In addition to increased metabolism during practice, at rest also increased the metabolism so that the body remains at rest in a state of energy production [1]. For example, HIIT workout is to combine sprint exercise with jogging. For example, 60 seconds of jogging and then after 60 seconds continue with sprint for 30 seconds and so on.

Some researches have proved the effectiveness of HIIT workouts, which include research conducted by Racil et al. [2] revealed that the addition of exercises plyometric in a HIIT program would be beneficial for weight loss in obese female adolescents. Sperlich et al. [3] revealed that HIIT workout increases VO_2 Max significantly and improve the performance of football players in the games, and research by Pushparajan [4] which is designed to determine the effects of HIIT workout on the lower extremities and the strength of the extremities on the basketball players with the age group 18 to 23 year, revealed that HIIT workouts group gave a significant improvement compared to the control group.

Based on the above brief background, this study aims to analyze how much HIIT influence on the desired physical component, in this case, the Explosive Power, Speed, and Agility. This study used quantitative approach. The method used is a quasi-experimental method to the study design Matching-Only Design. Sample were 30 basketball players aged 18-23 years. Results obtained after doing research for 6 weeks with 3 sessions of exercise a week, showed that the experimental group given a HIIT workout experienced a significant increase. In the dependent variable Explosive Power increased by 15.9%, Speed by 5.9%, and Agility 6.3%. In the control group also experienced increased, but not so great. Explosive Power increased by 3.8%, Speed by 1.3%, and Agility 1.5%. Thus the conclusions of this study are: Exercise of high intensity interval training (HIIT) significantly effects on the increasing Explosive Power, Speed, and Agility.

2. Methods

This study is of quantitative approach with a quasi-experimental method. This study uses - Matching-Only Design, with data analysis using the t-test (paired sample t-test). The process of data collection is done by explosive power tests using Sargent Jump, Speed tests using the Speed Test 30 Meters, and Agility tests using the Illinois Agility Run Test in pre-test and post-test. Furthermore, the data is analyzed using SPSS 20.0 series.

3. Results

The results obtained after the collection and analysis of data showed: (1) There is a significant effect of HIIT to increase Explosive Power, proven results mean explosive power pretest of 97.97 kg-m/sec and the mean posttest of 113.50 kg-m/sec with increase percentage of 15.9% and yield sig <0.05. (2) There is a significant effect of HIIT to increase Speed, proven results mean the posttest of 4.32 seconds is smaller than the mean of the pretest of 4.58 seconds with percentage increase of 5.9% and the sig < 0.05. (3) There is a significant effect HIIT to increase agility, proven results mean the posttest of 15.96 seconds is smaller than the mean of the pretest of 16.97seconds with percentage increase of 6.3% and the sig $<$ 0.05. The following tables display data and results of the t-test (paired sample t-test)

Table 1. Data of Pretest and Posttest Experiment Grop

No	**Initials**	**Explosive Power (kg-m/Sc)**		**Speed (Second)**		**Agility (Second)**	
		Pretest	Posttest	Pretest	Posttest	Pretest	Posttest
1	MI	86.80	109.97	4.35	4.15	16.00	15.12
2	MR	98.75	118.27	4.21	3.70	16.24	15.23
3	RR	90.34	115.12	4.42	4.22	16.21	15.89
4	RRP	85.44	108.88	4.34	4.14	17.31	16.33
5	AS	75.75	107.06	4.91	4.71	16.62	16.23
6	MYS	98.39	113.34	4.57	4.37	16.78	16.32
7	MJ	87.65	108.76	4.51	4.31	17.51	16.45
8	OK	117.82	119.83	4.42	3.68	16.35	15.34
9	INA	91.54	115.74	4.54	4.34	17.52	16.45
10	AA	87.70	108.56	4.91	4.71	17.34	16.32
11	MRK	90.14	110.14	4.70	4.50	17.77	16.23
12	A	111.08	115.77	4.45	4.25	17.55	16.21
13	AM	114.45	115.43	4.75	4.55	16.89	15.12
14	MRM	114.85	115.77	4.79	4.59	17.24	16.11
15	RNF	118.83	119.83	4.81	4.61	17.28	16.11
Mean		97.97	113.50	4.58	4.32	16.97	15.96
% Enhancement		15.9 %		5.9 %		6.3 %	

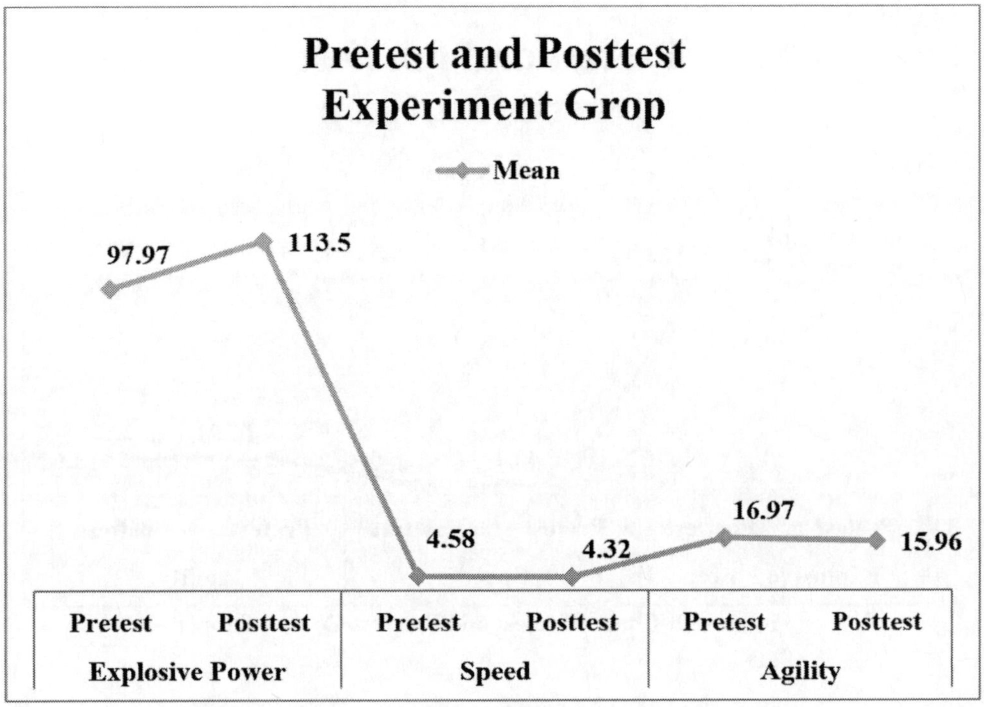

Figure 1. Chat of Pretest and Posttest Experiment Grop

Table 2. Data of Pretest and Posttest Control Grop

No	Inisial	Power Tungkai (kg-m/sec)		Speed (second)		Agility (second)	
		Awal	**Akhir**	**Awal**	**Akhir**	**Awal**	**Akhir**
1	M	105.25	108.25	4.24	4.21	15.63	15.43
2	S	94.15	100.23	4.22	4.22	16.34	16.14
3	RA	94.27	98.78	4.56	4.48	16.25	16.05
4	BP	110.15	110.94	4.19	4.16	16.35	16.15
5	IK	110.81	109.81	4.49	4.47	15.91	15.71
6	MRF	100.01	102.01	4.43	4.38	16.89	16.69
7	MRS	90.96	99.23	4.67	4.61	16.96	16.76
8	SA	110.34	110.89	4.45	4.39	16.57	16.37
9	RRS	105.74	108.74	4.76	4.68	16.30	16.10
10	DY	87.51	99.31	4.87	4.70	17.39	17.19
11	AB	83.33	97.97	4.98	4.95	17.59	17.39
12	FA	95.62	99.11	4.78	4.75	17.51	17.31
13	YRD	108.17	108.17	4.69	4.61	17.59	17.39
14	F	103.78	103.89	4.93	4.85	17.32	17.12
15	ISK	109.95	109.99	4.94	4.86	17.54	17.34
Mean		100.67	104.49	4.61	4.55	16.81	16.56
% Enhancement		3.8%		1.3%		1.2%	

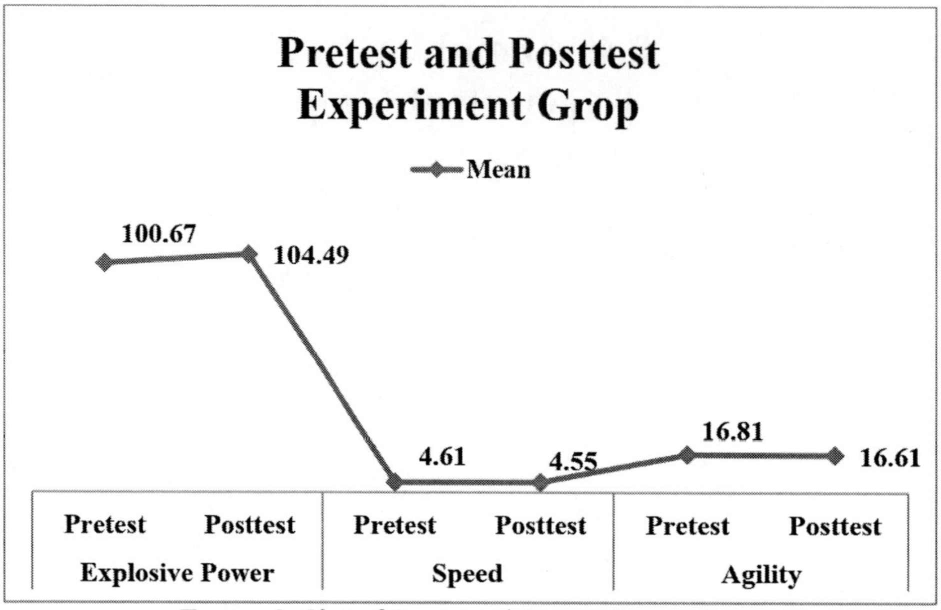

Fugure 2. Chat of Pretest and Posttest Control Grop

Table 3. Paired Samples Test of Experiment Grop

Variable	Pair	Mean	N	t	Df	Sig.(2-tailed)
Explosive Power	posttest - pretest	1.55293	15	5692	14	0,000
Speed	posttest - pretest	-.25667	15	-6.383	14	0.000
Agility	posttest - pretest	-1.01000	15	-9.880	14	0,000

Table 4. Paired Samples Test of Control Grop

Variable	Pair	Mean	N	t	df	Sig. (2-tailed)
Explosive Power	posttest - pretest	3.81867	15	3.213	14	0,006
Speed	posttest - pretest	-.05867	15	-5 609	14	0,000
Agility	posttest - pretest	-.24467	15	-5 478	14	0.000

4. Discussion

This is done on Linux basketball club Dompu, 30 people were divided into two groups and each group consists of 15 people. HIIT workout program used in this study consists of six high-intensity forms of exercise are: Squat Jump, Lateral Jump Over Barrier, Crossover Shuffle, Icky Shuffle, Z-Pattern Run, and Z-Pattern Cut. While mild exercise being used is jogging. Jogging is done at the time of recovery between high-intensity forms of exercise or as an active recovery. Brow (2005) and Chu (2013), in his book explains that the nature of explosive exercises can improve power, speed, and agility[5], [6]. Kusnanik and Ben [7] in his research also revealed some of the ways or methods of exercise that can be used to improve the speed and agility of athletes is with Ladder Speed Run and Repeated Sprint.

The use of other forms of plyometric exercises in a HIIT workout program is very convenient and effective to improve the power, speed, and agility. In addition to adding jogging as an active recovery, working time arrangements and resting time with high intensity (80% -95%) become key to the success of this exercise program. Evidently after the process of data collection and processing, data processing results showed no significant increase in the dependent variable is 3, explosive power, speed, and agility.

The results are consistent with previous studies such as Naimo et al. [8], which called High-Intensity Interval Training Has Positive Effects on Performance In Ice Hockey Players reveal the same thing that the High-intensity Interval Training (HIIT) is able to provide a significant improvement to the output power in athletes. Astorino et al. [9] revealed that HIIT workout can improve endurance and increase the speed of the athlete. Icano et al. [10] revealed that HIIT workout able to increase significantly the ability of speed, agility, power, and performance team sport.

5. Conclusion

Influential biomotor an exercise to the components you want to upgrade cannot be separated from the implementation plan and exercise program is good, true, and using a scientific approach. This study proves that with the preparation of a good exercise program and use the scientific approach is able to provide a significant effect on improving the physical components that are expected. Coupled with the seriousness in the implementation and administration of the exercise menu will be maintaining the success of the exercise program. High Intensity Interval Training (HIIT) workout is prepared using a scientific approach and implemented very seriously. High Intensity Interval Training (HIIT) effects on the increasing Explosive Power, Speed, and Agility.

References

[1] Miramonti Amelia A et al 2015 The effects of four weeks of high intensity interval training and [beta]-hydroxy-[beta]-methylbutyric free acid supplementation on the onset of neuromuscular fatigue *J. of Strength & Conditioning Research*

[2] Racil G, Zouhal H, Elmontassar W, Abderraouf Ben Abderrahmane A B 2015 Plyometric exercise combined with high-intensity interval training improves metabolic abnormalities in young obese females more so than interval training alone. *Applied Physiology, Nutrition, and Metabolism* (pnm-2015-0384.R2)

[3] Sperlich B, Mare M D, Koehler K, Holmberg H C 2011 Effects of 5 weeks of high-intensity interval Training vs volume training in 14-year-old Soccer players *Journal of Strength and Conditioning Research* **25**(5) 1271–1278.

[4] Pusparajan, 2012 Effect Of High Intensity Interval Training (HIIT) On Strength Of Female Basketball Players *International Journal of Innovative Research & Deflopment* (ISSN: 2278 – 0211)

[5] Brown L E and Ferrigno V A 2005 *Training for Speed, Agility, and Quickness* (Unites States: Human Kinetics)

[6] Chu Donald A, Myer Gregory D 2013 *Plyometrics* (United States: Human Kinetics)

[7] Kusnanik N and Rattray Ben 2017 Effect Of Ladder Speed Run And Repeated Sprint Ability In Improving Agility And Speed Of Junior Soccer Players *Acta Kinesiological* **11** (1) 19-22.

[8] Naimo M A, Souza E O 2014 High-intensity Interval Training Has Positive Effects on Performance In Ice Hockey Players *Int J Sports Med* (Doi: 10.1055/s-0034-1382054)

[9] Astorino T A, Alen R P, Roberson D W, Jurancich M 2012 Effect Of High-Intensity Interval Training On Cardiovascular Function, Vo2 Max And Muscular Force *Journal of Strength and Conditioning Research* **26**(1) 138–145

[10] Iacono A D, Eliakim A, Meckel Y 2015 Improving Fitness of Elite Handball Players: Small-Sided Games Vs. High Intensity Interval Training *Journal of Strength and Conditioning Research* **29**(3) 835–843.

IOP Conf. Series: Journal of Physics: Conf. Series **947** (2017) 012046 doi:10.1088/1742-6596/947/1/012046

Guided Inquiry with Cognitive Conflict Strategy: Drilling Indonesian High School Students' Creative Thinking Skills

A F Syadzili[1], Soetjipto[2], Tukiran[3]

[1]Science Education Study Program, Graduate Program, Universitas Negeri Surabaya, Surabaya 60231, Indonesia
[2]Department of Physics, Universitas Negeri Surabaya, Surabaya 60231, Indonesia
[3]Department of Chemistry, Universitas Negeri Surabaya, Surabaya 60231, Indonesia
Corresponding Author. E-mail: [1]asadfurqonsyadzili@mhs.unesa.ac.id; [2]hardjosoetjipto@gmail.com; [3]tukiran@unesa.ac.id

Abstract. This research aims to produce physics learning materials in Indonesian high school using guided inquiry with cognitive conflict strategy to drill students' creative thinking skills in a static fluid learning. This development research used 4D model with one group pre-test and post-test design implemented in the eleventh grade students in the second semester of 2016/2017 academic year. The data were collected by validation sheets, questionnaires, tests and observations, while data analysis techniques is descriptive quantitative analysis. This research obtained several findings, they are : the learning material developed had an average validity score with very valid category. The lesson plan can be implemented very well. The students' responses toward the learning process were very possitive with the students' interest to follow the learning. Creative thinking skills of student before the implementation of product was inadequate, then it is very creative after product was implemented. The impacts of the research suggest that guided inquiry may stimulate the students to think creatifly.

1. Introduction

Teachers are the first factor that determine the success of learning. According to the 2013 curriculum, it is necessary that the competence of educators and education personnel should be in conformity with the curriculum and textbooks [1]. The teachers' preparation become very important because teachers will carry out a mandate of learning that aims to encourage students to be better in doing observations, making questions, reasoning, and communicating what they gain or they know after receiving the learning material [2].

Learning materials have an important role in learning, which can help teaching and learning activities become more effective, and it is expected that students can be drilled to think creatively. Because creative thinking is very important for the students to be drilled in order to find out knowledge by themselves. Learning teachers should prepare the students to have the ability to solve problems, to be critical, and to be creative, with national, regional, and global insight [3].

The use of learning material that emphasizes on the the five points of learning experience aims to prepare the students to cope with the 21st century, learning materials should accommodate diverse skills, such as creative thinking skills. Thinking creative could be identified by four aspects, namely fluency, flexibility, originality and elaboration [4]. In this regard, creative thinking involves the ability to produce original ideas, to feel new relationship and not suspected, or to build a unique and good sequence between seemingly unrelated factors [5]. Creativity is what happens when an individual produces something that is novel as well as appropriate, generative or influential [6].

The most important objective of the creativity test is to identify the creative potential of a child (students). Creativity test is a test to measure divergen thingking skills. Creativity is very important and meaningful in life as its provisions when living in society, nation and state. Creative potential can be measured through several approaches, namely direct measurement; indirect measurement; measuring creativity elements; measuring the characteristics of creative personality; measuring creative potentials in a non-test manner, and assess real creative products [7]. Learning at school should be shifted from being told to be searching actively, and therefore students must be equipped with creative thinking skills.

The results of observation conducted at MA Negeri Lumajang as the Islamic state high school showed that the learning material has not implemented the mandatory of curriculum 2013. The learning process which is reflected in the lesson plan, the students were not taught to do activity observing, asking questions, collecting information, associating, and communicating and the learning is not reflected the activities that involved the students to think creatively, and the learning activities were still dominated by teachers, so that it was observable that the learning activities did not involve the students to do creative thinking. The other facts obtained were that based on the results of the creative thinking tests which were given out and filled out by 12 students, found that the average of students' creative thinking was as much as 40.62% with less creative category. This percentage indicated that the students in the school still had a very low creative thinking skill.

Based on the phenomenon, teachers must posses the principles of learning, the use of instructional media, textbooks, student worksheets, as well as selecting and using learning models [8]. One of the learning models that can stimulate students' active involvement in the learning process is by using guided inquiry learning with cognitive conflict strategy. Guided inquiry can help students to practice in a team, develop competencies in research, knowledge, motivation, writing skills, cooperative learning and social skills [9]. Guided inquiry can also help children in creative thinking. Through creative thinking, students are able to solve problems of everyday life. In line with this issue, guided inquiry learning is constructed with the aim of improving the students' concept understanding and scientific performance [10].

The solutions to reduce the problems and the results of the observations above were to convey anomaly information to create a cognitive conflict in students' mind, of course, by taking into account the students' prior knowledge. The students' prior knowledge is very important to facilitate and link the knowledge that students already have with the knowledge they will receive. Cognitive conflict is a perceptual state in which one notices the discrepancy between one's cognitive structure and the environment (external information), or among the different components (e.g., the conceptions, beliefs, substructures and so on) of one's cognitive structure [11]. Cognitive Conflict generates disequilibrium state which means lack of mental balance. Understanding someone of information, idea, or concept that is plagued by information is considered as contradictory [12]. Cognitive conflict consists of conflicts between conceptions in cognitive structures. The cognitive structure means the structure of knowledge organized in the brain. Conceptions mean a mental representation or a concept understanding in a cognitive structure [13]. The cognitive conflicts presented by teachers are expected to make the students aware of the initial misconceptions. The students would experience a conflict with anomalous phenomena, so that the students would feel challenged to think and question why their initial thinking was wrong. This process will eventually reconstruct the students' concepts to be the correct concepts.

The phases of guided inquiry learning with cognitive conflict strategy adapted as follows: the planning phase with cognitive conflict strategy, retrieving phase (obtaining information with cognitive conflict strategy), processing phase (processing information), creating phase (creating information), sharing Phase (communicating information), and evaluating phase [14]. Those are based on the explanation of guided inquiry, cognitive conflict strategy, and creative thinking skills above. The development of the learning material using guided inquiry with cognitive conflict strategy is expected to be feasible (valid, practical, and effective) to drill the creative thinking skills of Indonesian high school students of physics learning. A good learning material at least have three important framework, those are validity, practicality and effectiveness [15].

2. Method

This research is a development research which aims to produce learning material including: syllabus, lesson plan, students' text books, student worksheet, and tests of creative thinking skills, by using guided inquiry with cognitive conflict strategy on the static fluid learning material. The development of the learning material was refering to the Four D Models which consists of four stages, namely define, design, develop and disseminate [16]. This research only used three stages, namely defining, designing, and developing. The design of research used one-group pretest-posttest design. Pretest aims to determine the knowledge of early students before treatment by guided inquiry learning with cognitive conflict strategy, while posttest aims to determine the knowledge of students after treatment by guided inquiry learning with cognitive conflict strategy.

The subject in this research was learning material of physics in Indonesian high school using guided inquiry with cognitive conflict strategy to drill the students' creative thinking skill which was developed by the researcher, and the subjects of the try-out were the eleventh grade students of MA Negeri Lumajang in the second semester of 2016/2017 academic year as many as 24 students. The variables in this research were: (1) the validity of the learning materials, which include syllabus, lesson plan, students' textbooks, student worksheets, and creative thinking skill test; (2) the practicality of the learning materials, which include the completion of the lesson plan, and students' activities; and (3) the effectiveness of the learning materials refering to the results of creative thinking skill test, and the students' responses toward the learning using guided inquiry with cognitive conflict strategy. The data collection was carried out by using validation, observation, testing, and questionnaires. The research instruments were developed based on the instruments of previous researchers which were adapted to the researcher's needs, and the instruments were validated by experts to obtain input and suggestions before the instruments were used. The data were analyzed using descriptive quantitative analysis by analizing the validity of learning materials, the data of the results of practicality and data of the results of the effectiveness.

3. Results and discussion

3.1. Validation of the learning materials

The results of validation analysis which were assessed by 3 expert validators, showed that the average validation score of the developed learning material was catagorized as very valid. If the validation score is ≥ 2.6, the learning material can be used with little revision [17]. The table 1 shows the analysis of the validation results of the developed learning materials.

Table 1. Validation results of the learning materials.

Device name	Validation score			Average	Category
	1st Validator	2nd Validator	3rd Validator		
Syllabus	4.00	4.00	3.44	3.81	Very valid
Lesson plan	4.00	3.89	3.28	3.72	Very valid
Student worksheet	3.90	3.71	3.29	3.63	Very valid
Student textbook	3.85	3.65	3.25	3.58	Very valid
Creative thinking test	3.69	3.62	3.69	3.67	Very valid
	Average			3.68	Very valid

3.2. The data of the results of practicality

3.2.1. The analysis of the completion of the leesson plan. The completion of the implementation of the learning material was observed by two observers for three meetings. The results showed that all the stages in the lesson plan in the try out was completed, and on the average scores of the completion of the lesson plan 1, lesson plan 2, and lesson plan 3 were 3.61; 3.64; and 3.61 respectively with very good category. The instruments of the completion of lesson plan 1, lesson plan 2, and lesson plan 3 which were observed by two obeservers had average percentage of agreement of 97.02%, 98.04% and 97.02%. The developed instruments can be used if the percentage of agreement is ≥ 75% [18]. This means that all the steps of the learning process which were arranged in the lesson plans could be well implemented by the teacher, and in their assessment, the two observers gave relatively similar scores. The result of the lesson plans implementation analysis is provided in table 2.

Table 2. Lesson plans implementation analysis.

Lesson plan	Observers' score		Average	Category	Percentage of agreement score
	1st Observer	2nd Observer			
1	3.50	3.71	3.61	Very good	97.02%
2	3.57	3.71	3.64	Very good	98.04%
3	3.71	3.50	3.61	Very good	97.02%

The outline of the lesson plan developed mainly consists of three activities. The three stages were organized based on the syntax of guided inquiry with cognitive conflict strategy. Learning activity 1 (introduction) contained mainly on the phase of planning with cognitive conflict strategy. In general, it was about apperception and preparing students to receive learning. Apperception was carried out by providing a phenomenon to generate cognitive conflict in students and asked students to practice their questioning skills by formulating various types of research problems of the given phenomenon. Teachers should provide a confusing situation to trigger the students' curiosity and to motivate investigations [19]. In learning activity II (main activity), there are four steps or phases of guided inquiry with cognitive conflict strategy namely: (1) retrieving phase (gathering information with cognitive conflict strategy), (2) processing phase (processing information), (3) creating phase (creating information), and (4) sharing phase (communicating information). This stage encourages the students to drill their skills in gathering information, expressing ideas they have, and solving problems provided by teachers or problems which come from reading passage or from direct observation by students. In learning activity III (closing) there is one step or phase of guided inquiry with cognitive conflict strategy, namely the evaluating phase. This activity is mainly related to the role of teachers in assisting the students to conclude the results of the learning process, in case of conveying the findings thet they got during the learning process, and the teacher gave evaluation test in the form of creative thinking skill test.

3.2.2. Analysis of students' activity. Based on the results of the observation of the students' activities, the students' activities can be determined from the first meeting until the third meeting. This was proven from the average score of 24 students in the first meeting as much as 42.87; the second meeting 67.31; and the third meeting 68.12; and of the total score of 72. This means that the students' activity have a signifiant increase. It shows that the students' active involvement in the learning process makes them drille to find other alternatives in solving problems found in learning. Intelligence develops because people face new and confusing experiences, and then they try to resolve differences by linking the new knowledge with their prior knowledge and build new meaning [19]. The percentage of agreement average at the first meeting was 96.87%, at the second meeting was 98.75%, and at the third meeting was 98.63%. The intruments that had been developed could only be used when the percentage of agreement was about ≥ 75% [18].

3.3. Data of the results of the effectiveness

3.3.1. Description of students' creative thinking. The instruments used to measure the students' creative thinking skills is creative thinking tests. The test items were open-ended questions, that is a type of questions which has many possible correct answers. The open-ended questions can help improve creativity by producing diverse ideas, and students can solve their own problems in the future [20]. The creative thinking skills measured were fluency, flexibility, originality and elaboration. The creative thinking test was given to the students before the learning material was tried out as the pretest, and after the learning material had been tried out as the posttest.

The pre-test and the post-test of creative thinking skills were then analyzed to obtain the increased value (*N- gain*) of students' creative thinking skills. The results of the analysis of the average increase of the students' creative thinking skills after the three meetings learning process found that the gained score of the creative thinking was 0,79 with high category. If the gain score was > 0.70, then it was categorized as high gain score [21]. The results of the analysis indicated by the gained score showed an increasing case in students' creative thinking skills with high category, with an average of the pretest 29.43 with less creative category, and with an average of the posttest 85.55 with very creative category. This accretion was certainly due to the implementation of guided inquiry learning with cognitive conflict strategy which was capable of drilling the students' creative thinking skills. Guided inquiry also helps students to think creatively. Because of creative thinking, the students are able to solve problems in their everyday life [9].

3.3.2. Students' responses toward the implementation of guided inquiry learning with cognitive conflict strategy. By the end of the learning process, after the three meetings of the learning process, the students were given a questionnaire about the learning activities. It proposed to the students' responses toward the learning model and the learning material that had been developed. The results of analysis of the students' responses toward the guided inquiry learning with cognitive conflict strategy and the learning materials developed. The results of the questionnaire showed that 23 of 24 students responded positively (feel interested, easy, and happy) toward the learning environment and 4% of the students still found some difficulties following the learning activities using guided inquiry with cognitive conflict strategy.

4. Conclusion

Based on the findings of the research and the analysis of the research findings, it can be concluded that the learning material of physics using guided inquiry with cognitive conflict strategy which had been developed is decent (valid, practical, and effective) to drill the students' creative thinking skills. The impact of the research can be strongly said that the strategy can possibly stimulate the students to think creatifly.

5. Acknowledgements

The author would like to give a special gratitude to the advisors. Because of their effort this article can be finished.

6. References

[1] *Peraturan Pemerintah Republik Indonesia Nomor 32 Tahun 2013 tentang Perubahan atas Peraturan Pemerintah Nomor 19 Tahun 2005 tentang Standar Nasional Pendidikan* 2013. Jakarta

[2] *Permendikbud Nomor 81A Tahun 2013 tentang Implementasi Kurikulum.* Jakarta: BSNP

[3] Wardani L K 2003 Berpikir kritis kreatif (sebuah model pendidikan di bidang desain interior) http//digilip.petra.ac.id pp 97-111

[4] Torrance E P and Ball O E 1984 *The Torrance Tests of Creative Thinking Streamlined (revised) Manual, Figural A and B* (Bensenville, IL: Scholastic Testing Service Inc)

[5] Filsaime D K 2008 *Menguak Rahasia Berpikir Kritis dan Kreatif* (Jakarta: Prestasi Pustaka) p 8

[6] Zhanetta G 2011 Unraveling the Mystery Behind Creativity *E-joernal The Journal of Effective Teaching* **3(2)** pp 234-246

[7] Munandar U 2009 *Pengembangan Kreativitas Anak Berbakat* (Jakarta: Rineka Cipta)

[8] Mulyasa E 2013 *Pengembangan dan Implementasi Kurikulum 2013* (Bandung: Remaja Rosdakarya)

IOP Conf. Series: Journal of Physics: Conf. Series **947** (2017) 012046 doi:10.1088/1742-6596/947/1/012046

[9] Kuhlthau C, Maniotes L K and Caspari A K 2007 *Guided Inquiry: Learning in the 21st Century* (USA: Libraries Unlimited. Inc) pp 1-4

[10] Dewi K, Sadia I W and Ristiati N P 2013 Pengembangan Perangkat Pembelajaran IPA Terpadu dengan Setting Inkuiri Terbimbing untuk Meningkatkan Pemahaman Konsep dan Kinerja Ilmiah *E-Journal Program Pascasarjana Universitas Pendidikan Ganesha* **3** pp 20-21

[11] Lee G, Kwon J, Sang-suk P and Jung-whan K 2003 Development of An instrument for Measuring Cognitive Conflict in Secondary-Level Science Classes *Research in Science Teaching* **40(6)** pp 585-603

[12] Sela H and Zaslavsky 2007 Resolving cognitive conflict with peers- there is different between two and four? *Proceeding of the 31st Conference of The International Group for The Psychology of Mathematics Education Seoul-PME* **4** pp 169-176

[13] Kwon J and Lee G 2001 What Do You Know about Students Cognitive Conflict: A Theoritical Model of Cognitive Conflict Process *Proc. of 2001 AETS Annual Meeting Costa Mesa CA* pp 309-325

[14] Branch J and Oberg D 2004 *Focus on Inquiry A Teacher Guide to Implementing Inquiry Based Learning* (Canada: Alberta Education) pp 45-74

[15] Nieveen N 1999 *Prototyping to Reach Product Quality: Design Approaches and Tools in Education and Training* (Netherlands: Kluwer Academic Publishers) pp 125-136

[16] Thiagarajaan S, Semmel D S and Semmel M I 1974 *Instructional Development for Training Teacher of Exceptional Children a Sourcebook* (Bloomington: Center for Innovation on Teaching the Handicaped) p 3

[17] Ratumanan T G and Laurens T 2015 *Penilaian Hasil Belajar pada Tingkat Satuan Pendidikan 3rd Edition* (Yogyakarta: Pensil Komunika) p 181

[18] Borich G D 1994 *Observation Skill for Effective Teaching* (New York: Macmillan Publising Company) p 385

[19] Arends R I 2012 *Learning to Teach 9th Edition* (New York: McGraw-Hill) pp 394-462

[20] Sternberg R J 2012 The Assessment of Creativity: An Investment-Based Approach *Creativity Research Journal* **24(1)** pp 3–12

[21] Hake R R 1998 Interactive-Engagement Versus Traditional Methods: A Six-Thousand-Student Survey of Mechanics Test Data for Introductory Physics Courses *American Journal of Physics* **66(1)** pp 64-74

The Development of Student's Activity Sheets (SAS) Based on Multiple Intelligences and Problem-Solving Skills Using Simple Science Tools

D S Wardani[1,*], T Kirana[1], M Ibrahim[1]

[1]Primary Education Departement, Universitas Negeri Surabaya, Surabaya, Indonesia

*Corresponding author e-mail: duhitawardani@mhs.unesa.ac.id

Abstract. The aim of this research is to produce SAS based on MI and problem-solving skills using simple science tools that are suitable to be used by elementary school students. The feasibility of SAS is evaluated based on its validity, practicality, and effectiveness. The completion Lesson Plan (LP) implementation and student's activities are the indicators of SAS practicality. The effectiveness of SAS is measured by indicators of increased learning outcomes and problem-solving skills. The development of SAS follows the 4-D (define, design, develop, and disseminate) phase. However, this study was done until the third stage (develop). The written SAS was then validated through expert evaluation done by two experts of science, before its is tested to the target students. The try-out of SAS used one group with pre-test and post-test design. The result of this research shows that SAS is valid with "good" category. In addition, SAS is considered practical as seen from the increase of student activity at each meeting and LP implementation. Moreover, it was considered effective due to the significant difference between pre-test and post-test result of the learning outcomes and problem-solving skill test. Therefore, SAS is feasible to be used in learning.

1. Introduction

Currently and in the future, multiple intelligences are needed for everyone to face the increasingly complex challenges of life, including problem-solving skills. Everyone has a unique way to solve the problems by themselves [21]. According to Kurniawan [12], the someone's ability to see a problem, then solve it or make something that can be useful for others is called intelligence. The definition expressed by Kurniawan is in line with Gardner's definition [6] as it is written as follows: "An intelligence as biopsychological potential to process information that is of value in a culture." The essence of this quote is that intelligence is a measurement of how one can solve real problems in life. According to Gardner, a person is regarded to have intelligence when he or she shows a maturity and skill to solve problems and difficulties found in his or her life. To face the increasingly complex challenges of life, it does not only require a single intelligence, but it takes a lot of intelligences [4].

The importance of growing multiple intelligences is also included in the vision of the Ministry of National Education of 2025 [3] that is to result smart and competitive human beings as the outcomes. What is meant by Indonesian people who are intelligent are those having comprehensive intelligences, namely spiritual intelligence, emotional intelligence, social intelligence, kinaesthetic, and intellectual.

To be able to achieve the vision, it is necessary to have an effort in education itself, one of which is through the learning process (Kurniawan, 2016). The learning process should be designed in such a way that each student's intelligence gets an opportunity to be developed or facilitated. The chosen learning method should also be based on multiple intelligences. In other words, the means of delivering the lesson material used by the teacher should facilitate the students in developing the nine kinds of intelligences.

According to Kurniawan [12], one of the ways to do it is that the teacher provides Student's Activity Sheet (SAS) that can be used as a guide for them to conduct various learning activities by accommodating various intelligences in students. Gardner [5] argues that a person is intelligent when he is able to solve problems. Solving problem requires a certain skill. This problem-solving skill by McGregor [16] is very important. According to him, a good thinker should be able to adapt to situations where they find themselves. Inviting students to solve authentic problems with some possible solutions will be able to give them opportunity to develop the key cognitive skills that can support their thinking skills [16].

Researchers have observed the existence of SAS for elementary school students who facilitate nine students' intelligence. The results of this observation suggest that no such SAS has been developed through the research paradigm. That is, it has not been validated by an expert, has not been tested for its practicality, and its effectiveness as required as a learning device [17]. Kurniawan [12] states that multiple intelligence-based learning can be applied to every kind of subject, in both the exact and the social field. Based on that statement, it is necessary to develop SAS based on multiple intelligences and problem-solving skills in science learning.

2. Research Question

Research question of this study is "how is the feasibility of SAS based on multiple intelligences and problem-solving skills using simple science tools?"

3. Method

This study was a research and development (R&D) using 4-D model consisting of defining, designing, developing, and dissemination phases. However, due to the limitations, this study is only achieved at the third phase (develop). The feasibility of SAS was based on its validity, practicality, and effectiveness. The completion Lesson Plan (LP) implementation and student activities were the indicators of SAS practicality. The effectiveness of SAS was measured by the indicators obtained from the increase of learning outcomes and test of problem-solving skills. SAS trials are used to test the practicality and effectiveness implemented to 30 students at Elementary School Muhammadiyah 1 Taman Sidoarjo by using one group pre-test and post-test design. The data was obtained through some ways including validation, observation, test, and questionnaire. The data analysis techniques used were quantitative descriptive and paired sample t-test.

4. Results

4.1. Validity of Student's Activity Sheets (SAS)

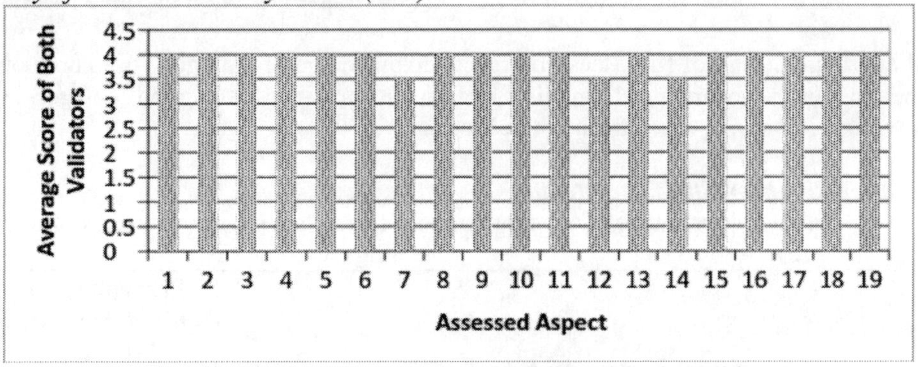

Figure 1. Construct Validity of Both Validators

Based on Figure 1, viewed from the construct validity, the development of SAS obtained expert evaluation with valid categories on 16 aspects. Therefore, construct validity of SAS based on multiple intelligences and problem-solving skills has been scientifically justified.

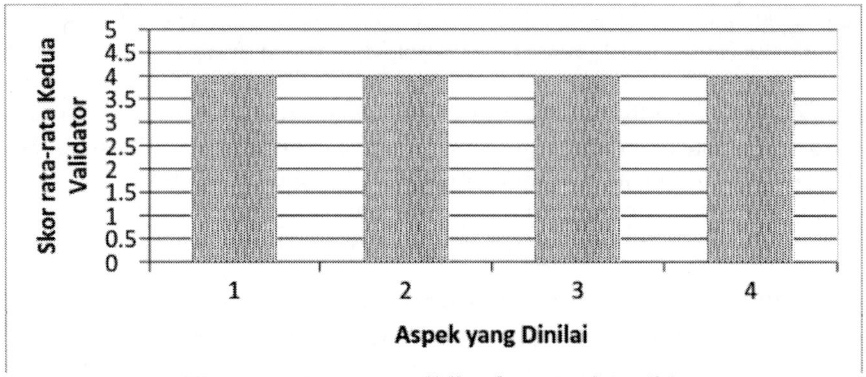

Figure 2. Content Validity from Both Validators

Figure 2 shows that SAS developed by the researchers has obtained valid category on all four aspects viewed from the aspect of content validity. Hence, the content validity of SAS based on multiple intelligences and problem-solving skills has been scientifically justified.

4.2. Practicality of Student Activity Sheets

4.2.1. Observation Results of Lesson Plan Implementation

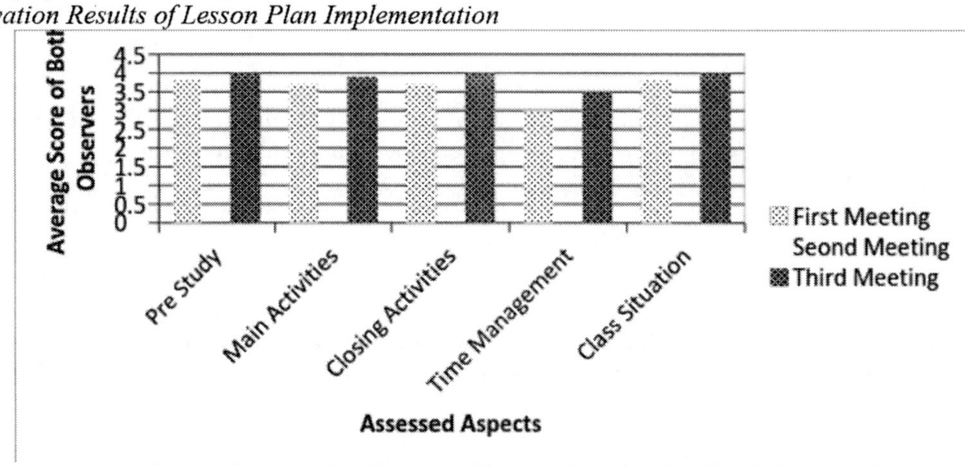

Figure 3. Observation Result of Lesson Plan Implementation in Three Meetings

Based on Figure 3, four of five assessed aspects obtained "good" category from both observers in the first meeting. In the second meeting, all of five aspects obtained also "good" category, so do the last meeting with an increasing score from both observers.

4.2.2. Observation Results of Students Activities

Table 1. Percentage of Student Activity from Data Observation Result in Three Meetings

No.	Observed Aspects	Percentage of Student Activity at Meeting-		
		1 (%)	2 (%)	3 (%)

1	Paying attention to teacher explanations	92	95	97
2	Answering teacher's questions	78	80	85
3	Conducting learning activities that include: a. Verbal-linguistic intelligence b. Logical-mathematics intelligence c. Visual-spatial intelligence d. Musical-rhythmic intelligence e. Body kinaesthetic intelligence f. Intrapersonal intelligence g. Interpersonal intelligence h. Naturalist intelligence i. Spiritual existence intelligence	85	88	93
4	Students identify problems according to the phenomenon	72	78	82
5	Students formulate the problems according to the phenomenon	77	82	87
6	Students formulate hypotheses by reading textbook first	72	73	78
7	Students collect and analyse the data	73	78	83
8	Students make conclusions based on the data	82	85	88
9	Students are in groups and do experiment	100	100	100
10	Students present the results of the discussion	83	87	90
11	Make a conclusion what has been learned today	87	88	92
12	Doing reflection	98	98	100

Based on Table 1, it can be concluded that there was an increase of students' activities at each meeting though the use of Student's Activity Sheets (SAS) based on multiple intelligences and problem-solving skills.

4.3. Effectiveness of Student Activity Sheets

The results of pre-test and post-test of cognitive product of the learning outcomes in the try-out are presented in Table 2.

Table 2. Average Score of Cognitive Product of the Learning Outcomes and Test of Problem-Solving Skills

	Cognitive Product of Learning Outcome		Test of Problem-Solving Skills	
	Pre-test	Post-test	Pre-test	Post-test
Average Score	60.86	87.20	55.67	88.00

Viewed from cognitive product of learning outcomes, the result of t-test of paired sample t-test obtained -t (count) = -22,106 < -t (table) = -2,045. This means that there is a significant difference between post-test and pre-test in the process of learning by using SAS based on multiple intelligences and problem-solving skills.

Viewed from problem solving skills test, result of t-test of paired sample t-test obtained -t (count) = --19.944 < -t (table) = -2,045. It means that there is a significant difference between post-test and pre-test in the process of learning through the use of SAS which is developed based on multiple intelligences and problem-solving skills.

5. Discussion

Student Activity Sheets (SAS) which is developed based on multiple intelligences and problem-solving skills using simple science tools is feasible due to meet three criteria. First, overall, the validity of SAS evaluated by the expert obtained good category and can be used by adding a little improvement. It is considered appropriate because SAS was developed with good and understandable language [15]. In addition, appropriate illustrations and font sizes can improve students' understanding of the passage in SAS [19]. Moreover, SAS aims at providing practice and motivate students to find their own knowledge [20] by using simple tools. It is supported by a research conducted by Siarni et al [22] which states that the utilization of second hand goods as learning media can help in improving students' learning outcomes in science. Second, the criteria of practicality can be seen from the increase of students' activities in each meeting and the LP implementation. The students' activities and implementation of this Lesson Plan (LP) obtained good category because the students gain knowledge or understanding based on the potential respective superior intelligence they have which makes the learning process becomes more fun [7]. This is in accordance with research by Yurt and Polat [24] showing that involving multiple intelligences in learning process creates an exciting learning environment for students' sensory areas. It is supported also by researches of Abdi [1], Kilic [11], Pociask & Settles [18] and Hanafin [9]. Moreover, the problem-solving skills steps are also introduced in this SAS implementation to the students to make them more analytical in making decisions in their lives [10]. In addition, these problem-solving skills can help students improve their analytical power and can help students also to apply the power in various situations [23]. Solving such problems requires deep conceptual knowledge and understanding so that students can apply what they have learned into real life [2]. Third, the criteria of effectiveness can be seen from the significant difference between pre-test and post-test cognitive product of learning outcome and problem-solving skill test. This significant difference indicates that students are already capable of high-level thinking skills. This is in line with Lewy's [14] opinion that the indicators for measuring high-level capabilities include analyzing, evaluating, and creating. In addition, problem-solving skills on the analysis questions also considered as one aspect of High-Order Thinking Skills (HOTS) [8]. This success in achieving better score in post-test was highly related to the role of music instrument during the learning activities. It is in line with Fauzi in Kusbiantoro [13] who argues that music can make the atmosphere more relaxed, so that the brain becomes easier to receive information.

6. Conclusion

Based on the result of the try-out, the findings of research, and the discussion, it can be concluded that SAS is valid with "good" category. Moreover, SAS is considered practical as seen from the increase of students' activities in each meeting and the LP implementation. In addition, SAS is declared as effective due to the significant difference between the pre-test and post-test of cognitive product of learning outcomes and the test of problem-solving skills. Therefore, SAS is feasible to be used in science learning.

Acknowledgements

The author presents her sincere appreciation which goes to her parents. In addition, this paper would not be done without the help, support, and patience of her advisors, Prof. Dr. dr. Tjandra Kirana, M.S., Sp. And. and Prof. Dr. Muslimin Ibrahim, M.Pd.. The researcher thanks them for their supervisions, advices, and guidance from the very early stage of this research as well as the opportunity which gives her extraordinary experiences throughout the past view years until the accomplishment of this paper.

References

[1] Abdi A and Maryam R 2012 The Effect Multiple Intelligences-based Instruction on Students' Creative Thinking Ability at 5th Grade in Primary School *Procedia-Soc and Behav Sc J.* **47** 105-108.

[2] Anderson and David 2001 *Taxonomy for Learning Teaching and Assessing* (New York: Logman)

[3] Depdiknas 2008 *Undang-Undang SISDIKNAS (Sistem Pendidikan Nasional)UU RI no. 20 tahun 2003 (National Education System UU RI no. 20 Year 2003)* (Jakarta: Depdiknas)

[4] Fritz H 2010 Attitudes and Cognitive Organizatio *The Journ of Pscych.* **21** 107-112

[5] Gardner H and Hatch T 1989 Multiple Intelligences Go to School: Educational Implications of the Theory of Multiple Intelligences *Journal of Edcation. Research.* **18**(8) 4

[6] Gardner H 2000 *Intelligence Reframed: Multiple Intelligences for the 21st Century* (New York: Basic Books Inc)

[7] Gozutok 2001 *The Application of Multiple Intelligence Theory to the Schools of Başkent* (University College of Ayşeabla: Ankara Political Publishing)

[8] Gunawan 2003 *Genius Teaching Strategy* (Jakarta: Gramedia Pustaka)

[9] Hanafin 2014 Multiple Intelligences Theory, Action Research, and Teacher Professional Development: The Irish MI Project *Australian J. of Teach Edu.* **39** 126-142

[10] Hudoyo 1988 *Mengajar Belajar Matematika)* (Jakarta: Departemen Pendidikan dan Kebudayaan)

[11] Kilic M S and Hakan S 2014 Primary School 5th Grade Science and Technology Lesson Book's Investigation of Multiple Intelliences Theory *Procedia-Soc. and Behv. Scienc. Journ.* **174** (2015) 2577-2581

[12] Kurniawan A 2016 *Learning with Multiple Intelligences at School*

[13] Kusbiantoro 2010 Desiging Human Error Simulation Game Integrated toward Room Lighting, Temperature and Noise Levels *Thesis Surabaya Institut Teknologi Sepuluh Nopember Available on:* http://digilib.its.ac.id/ITS-Undergraduate-3100010038554/9441 Accessed on Januari 2013

[14] Lewy Zulkardi & Nyimas A 2009 Development of Problem Sheet for Measuring Higher Thinking Skills Subject of Row and Class of Numbers in Class IX Acceleration of SMP Xaverius Maria Palembang *Math Edu J.* **3** (2) 14-28

[15] Majid A 2008 *Perencanaan Pembelajaran* (Bandung PT. Remaja Rosdakarya)

[16] McGregor D 2007 *Developing Thinking: Developing Learning (A Guide to Thinking Skills in Education* (Berkshire: Berkshire Open University Press)

[17] Nieveens & Plomp 2013 *Educational Design: Introduction. From Tjeerd Plomp (eds). Educational &Training System Design: Introduction Design of Education and Training (in Dutch) Utrecht (the Netherlands) Lemma.* (Netherland: Netherland Faculty of Educational Science and Technology University of Twente)

[18] Pociask A & Settles J S 2007 Increasing Student Achievement through Brain Based Studies *Thesis Master in Teaching and Leadership Saint Xavier University Chicago Illinois*

[19] Risky A, Sandra F, Rahmadina F, Karina M 2013 *Pengaruh Pemberian Gambar pada Bacaan terhadap Pemahaman Membaca Siswa Kelas V Sekolah Dasar* (Jakarta: Universitas Indonesia)

[20] Schunk 2012 *Social Cognitive Theory: APA Educational Psychology Handbook Vol 1: Theories, constructs, and critical issues* (New York: Routledge)

[21] Serin B N, Pehlivan A, Serin O, Sahin S 2012 Relationship Between Intelligences Areas and Interpersonal Problem Solving Skills According to Multiple Intelligences Theory *Intern Onl J. of Prim Edu* **1** Issue 2

[22] Siami P M and Rede A 2015 Utilization of Used Goods as Learning Media to Improve Student Learning Outcomes of Grade IV SDN 07 Salule Mamuju Utara *J. Create Onl Tadulako* **3** (2) ISSN 2354-614X

[23] Widjajanti E 2013 Kualitas Lembar Kerja Siswa *Makalah disampaikan dalam Pelatihan Penyusunan LKS Mata Pelajaran Kimia Berdasarkan Kurikulum Tingkat Satuan Pendidikan Bagi Guru SMK/MAK di Ruang Sidang Kimia* FMIPA UNY

[24] Yurt and Polat 2015 The Effectiveness of Multiple Intelligences on Academic Achievement: A Meta-Analysis *Journ. of Soc. Stud. Educ. Research* **6** (1) 84-122

IOP Conf. Series: Journal of Physics: Conf. Series **947** (2017) 012048 doi:10.1088/1742-6596/947/1/012048

Functional Thinking Profile of Junior High School Student in Solving Mathematical Problem Observed by Differences of Sex

A P Siregar[1], D Juniati[2], and R Sulaiman[3]

[1]Postgraduate in Mathematics Education, Universitas Negeri Surabaya, Ketintang, Surabaya, 60231, Indonesia
[2,3]Departement of Mathematics, Universitas Negeri Surabaya, Indonesia

Corresponding author: [1]ardiantosiregar@mhs.unesa.ac.id

Abstract. This study involving 2 grade VIII students was taken place in SMPK Anak Bangsa Surabaya. Subjects were selected using equal mathematics ability criteria. Data was collected using provision of problem-solving tasks and followed by a task-based interview. Obtained data was analysed through the following steps, which are data reduction, data presentation, and conclusions. Meanwhile, to obtain a valid data, in this study, researchers used data triangulation. The results indicated that in the problem number 1 about identifying patterns, the subjects of male and female show a tendency of similarities in stating what is known and asked the question. However, the male students provided a more specific answer in explaining the magnitude of the difference between the first quantity and the increased differences in the other quantities. Related the activities in determining the relationship between two quantities, male subjects and women subject tended to have similarities in the sense of using trial and error on existing mathematical operations. It can be concluded that the functional way of thinking both subjects is relatively identic. Nevertheless, the male subject showed the more specific answer in finding the difference between the two quantities and finding the correspondence relationship between the quantities.

1. Introduction

Mathematics education has an important role in the development of science and technology. In daily life, there are many problems that can be solved by using mathematics. In fact, not all problems faced in daily life are mathematical problems, but to solve those problems a mathematical thinking is required. Krulik & Rudnick [1] said a problem is a situation, quantitative or otherwise, that confronts an individual or group of individuals, that requires resolution, and for which the individual sees no apparent path to the solution. Therefore, the problem in this study is defined as a situation where the student does not know the answer and he/she needs steps to solve it.

Reston [2] suggests that mathematics curricula should emphasis on the development of mathematical concepts rather than procedural knowledge only. In the learning process, children are less encouraged to develop their thinking skills. In fact, the learning process in the classroom tends to make the children memorizing information. One of the important potency that students have is thinking. A potency that needs to be developed precisely and gradually should be more implemented in order to a deal with the students' future life. In learning mathematics, students are expected to reason and think logically, analytically, critically, creatively, and can work together.

There is a point of view that the orientation of math in school highlight more on the "mathematics as the product of thinking" neglecting attention to the thinking process itself. Solso [3] said that thinking is defined as the process of generating new mental representations through information transformation involving complex interactions among mental attributes such as judgment, abstraction, reasoning, imagination, and problem solving. Therefore, it can be concluded that the process of thinking of students in learning mathematics is very important. it is expected that the students are able to understand the mathematics material in depth and can overcome the difficulties that happen gradually.

According to Schawnk [4] in the development of algorithms, student thinking types are divided into two types of thinking processes namely predicative and functional. Functional thinking is indispensable in mathematics because it is one of the main keys in algebraic thinking that contains generalization of functions. In this case, Smith [5] defined functional thinking as "A thinking representation that focuses on the relationship of two or more variations of the number of general-centred relationships that ultimately result in a function". Smith [5] said that there are three methods in developing the functional thinking. They are: (1) recursive patterning involves finding variation within a sequence of values; (2) covariational thinking is based on analyzing how two quantities vary simultaneously and keeping that change as an explicit, dynamic part of a function's description (eg, "as x increases by one, y increases by three"); and (3) a correspondence relationship is based on identifying a correlation between variables (eg, "y is 3 times x plus 2") ". (Blanton & Kaput [6,7]).

One way that can be used to sharpen the functional thinking ability is by solving math problems. Solving mathematical problems requires an understanding of mathematical concepts. The lack of understanding of the relationships among concepts is seen when the students solve math problems. Solving a problem is not only concerned with the final answer, but also the settlement process should also be considered. In solving the problem of mathematics, students are expected through the process step by step in a way that their thinking process can be observed.

Results of a study conducted by Krutetskii [8] stated that in thinking, female students are superior in accuracy, precision, and inequality. On the other hand, male students tend to be less thorough, hasty and tend to finish things in a short way. Furthermore, Schawnk [9] stated that it is very difficult to find a functional thinking subject. However, this finding is not supported in the case of Indonesia and China that the women perform functional thinking well. The study related to the functional thinking profile of students that has been done by McEldoon & Jhonson [10,11] and Tanish [12] encourages researchers to describe the functional profile of students in solving mathematical problems in terms of sex differences. Therefore, this paper would discuss functional thinking profile of junior high school student in solving mathematical problem observed by differences of sex.

2. Methods

This study is descriptive qualitative study that aims to describe the functional thinking activity of junior high school students in solving mathematical problems based on gender differences. This study was conducted at SMP Kristen Anak Bangsa Surabaya. The subject of this study is 1 male student and 1 female student who have equal mathematics ability (score difference no more than 5) selected from 24 grade VIII students consist of 12 male and 12 female students. The main instrument is the researcher himself while the supporting instruments consist of mathematical ability tests, problem solving tasks and interview guidelines.

Data collection techniques in this study were designed in several stages. They are (1) providing problem solving tasks on the subject, (2) asking the subject of think-aloud before, during, and after filing the problem as well as an interview, and (3) recording in audio and visual at the time of the interview. Data analysis is done by reducing data, presenting data, and drawing the conclusion. The functional thinking activities in solving the problems are presented in Table 1.

Table 1. Table of functional thinking activities in solving the problems

No	Mental activity	Aspects observed
1	Identifying patterns	• Identifying the information contained in the problem • Recognizing the difference between two quantities. • Specifying another value for the next data
2	Determining the relationship between two quantities	• Recognizing the one-to-one relationships between the two quantities. • Specifying the value for the data in question
3	Stating common rules	Determining the general rules between two quantities using algebraic notation.

3. Results and Discussion

Subjects in this study were selected based on the results of equivalent mathematical ability resulted from mathematics ability tests (with difference no more than 5) and different gender. The mathematics ability test is adapted from UN problem. Subsequently the subject was given the task of problem solving and followed by interviews. Furthermore, the functional thinking activity of the subject is expressed by referring to each student's functional thinking activity in solving the problem.

3.1 Functional thinking profile of male subject

The following are the findings of functional profile of male subject in solving mathematical problems.

3.1.1. Problem number 1

In identifying patterns, student began by looking at what was known in the matter, as seen from the student's statement of what was known. After student declared what was given in the student's problem, student stated that the problem would be related to the pattern. Furthermore, student stated that every first quantity increased then the other quantity also increased.

In the activity of determining the relationship between the two quantities, the student began by trial and error. That was to try what relationship can be found between quantities one with the other. In the activity of stating the general rule, student wrote the rules that were originally and verbally become algebraic. Furthermore, to test the truth of the rules written algebraically, the students tried the truth by replacing the value of n with some examples and finding the correct results with what was known in the matter.

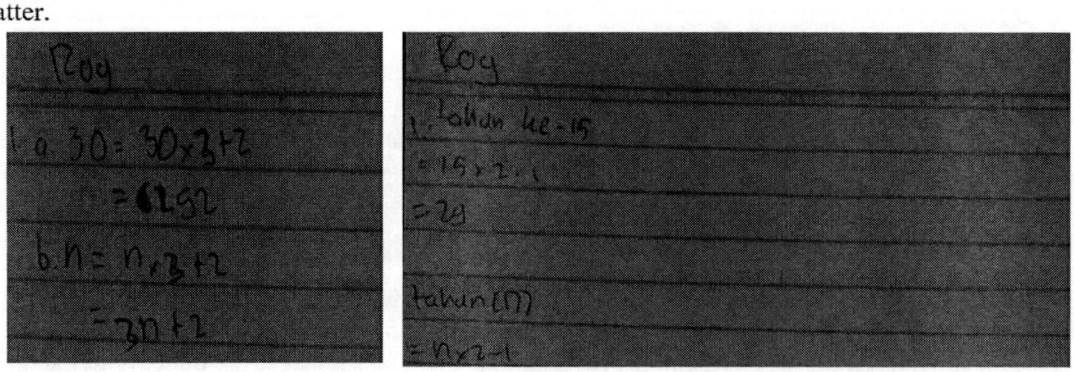

Figure 1. Male performance on problem 1

3.1.2. Problem number 2

In identifying patterns, student began by looking at what was known in the matter, as seen from the student's statement of what was known. After student declared what was known in the student's problem, he declared that the problem would be related to the pattern and the same as the problem of number 1. Next, the student stated that each set made was added then the required material also increased.

In the activity of determining the relationship between the two quantities, the student declared that question 2 was easier to understand than question 1 so the student also stated that in this case everything was very clear. Furthermore, in recognizing the relationship between two quantities, students directly used the multiplication rule by multiplying each quantity with the desired number of sequences.

In the activity of stating the general rules, student wrote the rules that were originally and verbally became algebraic. This could be seen from the results of student work in the problem 2 of problem solving tasks 1 that student wrote to make n series, many flower required was n multiplied 2 plus n multiplied 3 plus n multiplied 1 and on the problem 2 of problem solving tasks 2 student wrote to make n gifts, many fruit required was n multiplied 3 oranges plus n times 5 apples plus n times 4 pears.

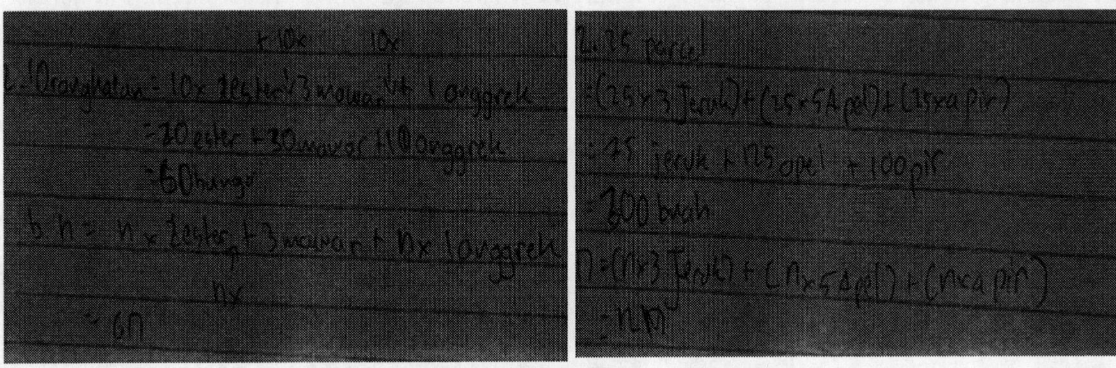

Figure 2. Male performance on problem 2

3.2 Profile of female functional thinking

The following are the findings of functional profile of female subject in solving mathematical problems.

3.2.1. Problem number 1

In identifying patterns, student began by looking at what was known to the problem. After student realized that if the first quantity increased then the other quantity also increased. This was in accordance with the statement of students on problem 1 of problem solving tasks 1 which stated if the pool box of catfish increased then total of the required feed was also increased and on problem 1 of problem solving tasks 2 declared the longer production of bark produced more and more.

In activity of determining the relationship between two quantities, the problem began with a trial and error. That was to try what relationship can be found between quantities one with the other. As in problem 1 of problem solving tasks 2, student initially tried to use multiplication and then proceeded with addition, at first it looked right, but in the next experiment it was not true. Furthermore, the student tried in another way by multiplying and then followed by subtraction and in the end the student found the one-to-one relationship between the two quantities by declaring in verbal language.

In activity of stating the general rule, student wrote the rules that were originally and verbally became algebraic. Furthermore, to test the truth of the rules written algebraically, student tried the truth by replacing the value of n with some examples and finding the correct results with what was known in the problem.

Figure 3. Female performance on problem 1

3.2.2. Problem number 2

In identifying patterns, student began by looking at what was known to the problem. Furthermore, student stated that each set made was added then the required material also increased. In activity of determining the relationship between two quantities, student directly used the rules of multiplication by multiplying each quantity by the desired number of sequences.

In activity of stating the general rule, student wrote the rules that were originally and verbally became algebraic. This was seen from the results of student work as in the problem 2 of problem solving tasks 1, student wrote to make n series, many of the required interest was n multiplied by 6 and on problem 2 of problem solving tasks 2 student wrote to make n the gift, many of the required fruit was n multiplied 12.

Figure 4. Female performance on problem 2

4. Conclusions

Based on results of data analysis above, it can be concluded that the functional thinking activity of male and female subject has been doing in every activity. However, the manner of male and female subjects in the first activity tends to be different. In the activity of identifying pattern, the male subject explained the increasing quantity more specifically. Furthermore, in the activity of determining the one-to-one relationship, the male subject on problem 2 does not directly add the required material, but through multiplication processes of many ranges that will be made with each material for one circuit. Based on results of this study, it can be suggested to educators to provide training and guidance to female students in order to develop their functional thinking.

References

[1] Krulik S & Rudnick J A 1989 *a handbook for senior high school teachers* (USA: Needham Heights)

[2] Reston V A 2000 *Principles and standards for school mathematics* (NCTM: National Council of Teachers of Mathematics)

[3] Solso R L 1995 *Cognitive Psychology* (Needham Heights: Allyn& Bacon)

[4] Schawank I 1998 *European Research in Mathematics Education I.II* (Osnabruck: Corel Ventura)

[5] Smith E 2008 *Representational Thinking as a Framework for Introducing Functions in the Elementary Curriculum. Algebra in the Early Grades* (New York: Lawrence Erlbaum Associates)
[6] Blanton M L and Kaput J J 2004 *Proc. Conf. of the International Group for the Psychology of Mathematics Education* **2** pp 135-142
[7] Blanton M L & Kaput J J 2011 *J. Res. Math. Educ.* **36** 34-42
[8] Krutetskii V A 1976 *The Psychology of Mathematical Abilities in School Children* J Kilpatrick, J Wirszup I (Eds.) (Chicago: Chicago Press)
[9] Schwank I 1999 *European Research in Mathematics Education I.II* (Osnabruck: Corel Ventura)
[10] McEldoon K L, Braswell C C, and Johnson B R 2010 *J. Psyc. Hum. Dev* (USA: Vanderbilt University)
[11] McEldoon K L and Rittle-Johnson B 2010 *Proc. Thirty Second Annu. Meet. North Am. Chapter Int. Gr. Psychol. Math. Educ. Optim. Student Underst. Math.* p 202
[12] Tanish D 2011 *J. Math. Beha.* **30** 206-223

IOP Conf. Series: Journal of Physics: Conf. Series **947** (2017) 012049 doi:10.1088/1742-6596/947/1/012049

The Effectiveness of Guided Inquiry-based Learning Material on Students' Science Literacy Skills

E V Aulia[1], S Poedjiastoeti[1,*], R Agustini[1]

[1] Science Education, Post Graduate Program, Universitas Negeri Surabaya, Jl. Ketintang 30 Surabaya 60231, Indonesia
* Email: sripoedjiastoeti@unesa.ac.id

Abstract. The purpose of this research is to describe the effectiveness of guided inquiry-based learning material to improve students' science literacy skills on solubility and solubility product concepts. This study used Research and Development (R&D) design and was implemented to the 11th graders of Muhammadiyah 4 Senior High School Surabaya in 2016/2017 academic year with one group pre-test and post-test design. The data collection techniques used were validation, observation, test, and questionnaire. The results of this research showed that the students' science literacy skills are different after implementation of guided inquiry-based learning material. The guided inquiry-based learning material is effective to improve students' science literacy skills on solubility and solubility product concepts by getting N-gain score with medium and high category. This improvement caused by the developed learning material such as lesson plan, student worksheet, and science literacy skill tests were categorized as valid and very valid. In addition, each of the learning phases in lesson plan has been well implemented. Therefore, it can be concluded that the guided inquiry-based learning material are effective to improve students' science literacy skills on solubility and solubility product concepts in senior high school.

1. Introduction

Science literacy skills is the ability to engage with science-related issues, scientific ideas, and the capability of becoming a reflective thinking person [1]. The PISA 2015 report reveals that a person who is literate in science must have the competence to explain phenomena, evaluate and design investigations, and interpret data and evidence scientifically. There are four aspects of science literacy skills which are interrelated each other according to PISA 2015, including context, knowledge, competence, and attitudes [1]. Science literacy skills can be trained in chemistry learning because it provides a direct learning experience to the students through the use and development of process skills and scientific attitudes to solve problems in everyday life [2]. Based on the characteristics of chemistry lesson, the suitable method used to develop students' thinking ability and improve students' science literacy skills is guided inquiry learning model. Guided inquiry learning is a scientific process of active exploration that uses critical, logical, and creative thinking skills to answer questions by teacher guidance [3].

Various results of researches showed that the implementation of guided inquiry model which integrated with guided inquiry-based learning material in chemistry lesson is effective and efficient to get learning outcomes with the best quality. Activity and student achievement on solubility and solubility product can be increased if the applied guided inquiry model is equipped with student worksheet [4]. Generally, the students are better to learn through the use of guided inquiry-based learning with laboratory experimental method. Through the guided inquiry learning, significant improvements occur in all aspects of students' attitudes and academic achievements. In addition, there was a significant decrease on students' anxiety in chemical laboratories [5]. In inquiry-based laboratory activities, the students have fewer misconceptions and understand concepts much more meaningfully. The average score of students' performance were also increasing significantly. The students believed that chemical knowledge helps to interpret the phenomenon in everyday life and has an important role in modern life [6]. In addition, implementation of science literacy skills based on inquiry learning in laboratory activities can improve the science literacy skills of high school students in terms of content, context, process and attitudes aspects. Therefore, it can be concluded that guided inquiry learning model can make students more confident and make the learning process more meaningful [7].

The average score of students' conceptual understanding using guided inquiry model is greater than traditional teaching. The guided inquiry learning model is effective to build conceptual relationships and rectifying misconceptions [8]. The students' achievement increases at nominal, functional, and conceptual levels of chemical literacy through the inquiry learning. The highest increase is achieved by conceptual aspect of chemical literacy, that is understanding the relationship between concepts and main ideas [9]. However, the researches did not discuss about learning material which compatible with guided inquiry learning to improve students' science literacy skills and learning outcomes of science literacy on the content, context, processes and attitudes aspects are not discussed in detail. In addition, the science literacy skills on attitude aspect are not discussed and only focus on cognitive aspect. Therefore, this research would develop guided inquiry-based learning material and implemented it in order to improve students' science literacy skills in terms of context, competence, attitude, and knowledge aspects on solubility and solubility product (K_{sp}) concepts.

The results of preliminary study done to the 11th graders of Muhammadiyah 4 Senior High School Surabaya showed that students still lacked of science literacy skills. Based on test of science literacy skills in context, competence and knowledge aspects, the students have not been able to answer the problem well and correctly. There was 92% of the students who could not propose an exact problem formulation in accordance with phenomenon, 84% of the students who could not formulate hypothesis, and 92% of students who could not determine the variables correctly. A lot of students could not answer the questions based on observation table correctly and determine conclusions appropriately. It is because the learning methods used by the teacher during this time always use lecturing method or conventional learning model. It can be seen from the result of student questionnaire that 80% of them agree with the statement above.

Based on questionnaire and interview with chemistry teacher showed that learning material (worksheet) contained questions without experimental activities. A lot of students thought that the summary of the material in the worksheet is not interesting and the language is difficult to understand. The worksheet is not made by the teacher but from the publisher, so it is general and not in accordance with characteristics of students in that school. Based on the description above, the researcher develops a chemistry learning material that are able to fulfill the criteria of validity, practicality, and effectiveness. Generally, the purpose of this research is to describe the effectiveness of guided inquiry-based learning material to improve students' science literacy skills on solubility and solubility product (K_{sp}) concepts.

2. Method

The research subjects of the implementation of guided inquiry-based learning material on solubility and solubility product concepts are 20 students who are 11th graders of Muhammadiyah 4 Senior High School Surabaya in 2016/2017 academic year. The development of learning material uses Research and Development (R&D) design [10]. The developed learning material in this research consists of lesson plan, student worksheet, and science literacy skills tests. The stages in research and development method include preliminary study, model development, and testing [10]. However, this research is limited to the stage of model development. The learning material that is implemented in this research have passed the validity test stage with valid and very valid category.

This research was conducted in 5 weeks starting from April to May 2017 by employing a pre-experimental research, that is one group pre-test and post-test design [11]. Experimental design was used to investigate the effectiveness of guided inquiry-based learning material. The data to seek for the effectiveness was obtained through students' science literacy skills test scores in terms of context, competence, and knowledge aspects (assessment sheet 1) and attitude aspects (assessment sheet 2). Then, the data that were obtained was analyzed descriptively. The scores of students' science literacy skills are categorized based on the following criteria: very less, less, enough, good, and very good [12][13]. The differences in students' science literacy skills scores before and after the treatment were analyzed using N-gain. The N-gain calculation data were then converted using the criteria that presented in Table 1.

$$Normalized\ gain\ (g) = \frac{Post - test\ score - Pre - test\ score}{Maximum\ score - Pre - test\ score}$$

Table 1. Criteria N-gain score was adopted from reference [14]

Score N-gain	Criteria
0.7 < N-gain	High
0.3 ≤ N-gain ≤ 0.7	Medium
N-gain < 0.3	Low

3. Results and Discussion

3.1. Scores of students' science literacy skills in terms of context, competence, and knowledge aspects

Students' science literacy skills were measured using a test instrument consisting of 20 multiple-choice questions and 5 essay questions (assessment sheet 1). Table 2 shows the scores of students' science literacy skills and the increase of N-gain score.

Table 2. Results of N-gain analysis of students' science literacy skills in context, competence, and knowledge aspects

Student	Pre-test		Post-test		N-gain	Category
	Score	Category	Score	Category		
A	34	Less	95	Very good	0.92	High
B	39	Less	70	Good	0.51	Medium
C	30	Less	75	Good	0.64	Medium
D	47	Enough	82	Very good	0.66	Medium
E	33	Less	71	Good	0.57	Medium
F	34	Less	73	Good	0.59	Medium
G	31	Less	87	Very good	0.81	High
H	37	Less	93	Very good	0.89	High
I	35	Less	70	Good	0.54	Medium
J	38	Less	72	Good	0.55	Medium
K	32	Less	82	Very good	0.74	High
L	30	Less	70	Good	0.57	Medium
M	35	Less	72	Good	0.57	Medium
N	29	Less	65	Good	0.51	Medium
O	31	Less	72	Good	0.59	Medium
P	28	Less	76	Good	0.67	Medium
Q	28	Less	70	Good	0.58	Medium
R	36	Less	82	Very good	0.72	High
S	41	Enough	87	Very good	0.78	High
T	35	Less	90	Very good	0.85	High

Data in Table 2 showed that students' science literacy skills scores of context, competence, and knowledge aspects at pre-test ranged from 28 to 47, in which 18 students got scores with less category and 2 students got scores with enough category. After implementing guided inquiry learning using learning material that was developed and conducted post-test, the students' science literacy skills scores ranged from 65 to 95, in which 12 students got scores with good category and 8 students got scores with very good category. The students' science literacy skills are significantly increasing which is stated by N-gain score. There are 7 students (35%) got N-gain score in high category, and 13 students (65%) got N-gain score in medium category. The high increase in N-gain score showed that implementation of developed learning material through guided inquiry learning can improve students' science literacy skills. The increase of N-gain score occurs because the teacher who has been performing all phases of guided inquiry learning implemented in well to excellent category. In addition, the learning steps in the lesson plan were developed in accordance with the syntax of inquiry learning model which includes six phases focusing on the activities, explaining inquiry process, presenting inquiry problem, formulating hypotheses, collecting data, formulating conclusion, and reflecting problems and developing thinking process [15].

Inquiry-based learning is effectively guided to improve students' science literacy skills due to the students' learning stages according to Piaget's cognitive development including the formal operational stage – a stage in which students can already think abstractly or symbolically and solve problems through doing experiment [16]. Visually, the difference between the pre-test and post-test scores of students' science literacy skills is presented in Figure 1.

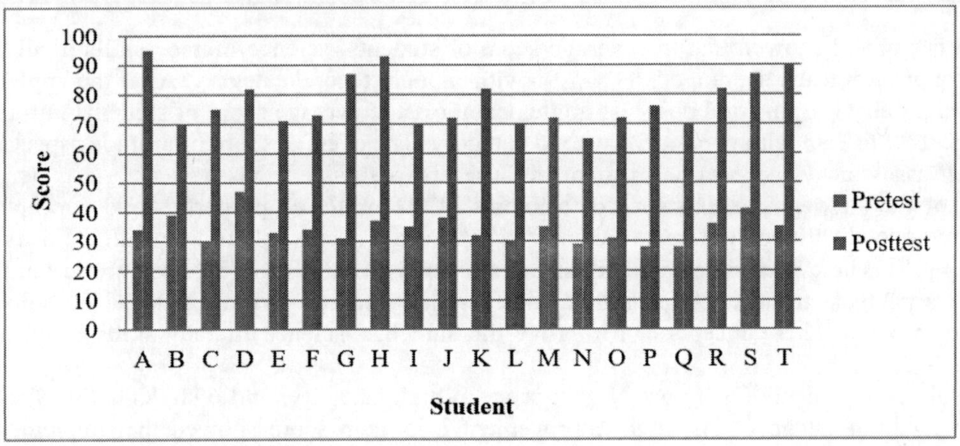

Figure 1. Diagram of pre-test and post-test scores of students' science literacy skills in terms of context, competence, and knowledge aspects

3.2. Scores of students' science literacy skills in attitude aspect

Students' science literacy skills were measured using a questionaire instrument consisting of 30 positive and negative statements. Those statements include 4 indicators of attitudes: students' attitudes toward science, the importance of science in life, students' responses to science as a career, and students' responses to science learning in schools (assessment sheet 2). The assessment sheet 2 consists of two types of assessment: self assessment and peer assessment. The scores of students' science literacy skills and the increase of N-gain score are presented in Table 3.

Table 3. Results of N-gain analysis of students' science literacy skills in attitude aspect

Student	Pre-test		Post-test		N-gain	Category
	Score	Category	Score	Category		
A	2.29	Enough	3.67	Very good	0.81	High
B	2.05	Enough	3.46	Good	0.72	High
C	2.37	Enough	3.60	Very good	0.75	High
D	2.63	Good	3.60	Very good	0.71	High

IOP Conf. Series: Journal of Physics: Conf. Series **947** (2017) 012049 doi:10.1088/1742-6596/947/1/012049

Student	Pre-test		Post-test		N-gain	Category
	Score	Category	Score	Category		
E	2.44	Enough	3.61	Very good	0.75	High
F	2.60	Good	3.60	Very good	0.71	High
G	2.94	Good	3.95	Very good	0.95	High
H	2.36	Enough	3.62	Very good	0.77	High
I	2.96	Good	3.67	Very good	0.68	Medium
J	2.34	Enough	3.56	Very good	0.73	High
K	2.56	Good	3.57	Very good	0.70	High
L	2.73	Good	3.50	Good	0.61	Medium
M	2.49	Enough	3.73	Very good	0.82	High
N	2.54	Good	3.64	Very good	0.75	High
O	2.39	Enough	3.26	Good	0.54	Medium
P	2.78	Good	3.70	Very good	0.75	High
Q	2.67	Good	3.66	Very good	0.74	High
R	2.56	Good	3.45	Good	0.62	Medium
S	2.57	Good	3.32	Good	0.52	Medium
T	2.03	Enough	2.97	Good	0.48	Medium

Data in Table 3 showed that the average score of students' science literacy skills in all aspects of attitude at pre-test reach the range 2.03 to 2.96 with enough to good category. After the implementation of guided inquiry learning and post-test implementation, the average score of students' attitudes reach the range 2.97 to 3.95 with good to very good category. The average score of attitude aspect is derived from self-assessment and peer-assessment because self-assessment has a weakness in terms of the validity of the answer. The existence of elements of the preferred properties and concepts that are considered good by the society cause the answer given is not necessarily in line with the facts practiced everyday [17]. These data indicated that there is an increase in the score between pre-test and post-test in every aspect of attitude. It means that guided inquiry-based learning material on solubility and solubility product (K_{sp}) concepts can improve the students' science literacy skills in all aspects of attitudes.

A total of 14 students (70%) got N-gain score in high category, and 6 students (30%) got N-gain score in medium category. The high improvement in N-gain score showed that implementation of developed learning material through guided inquiry learning can improve students' science literacy skills. The improvement of N-gain score is because teacher has been performing all phases of guided inquiry learning in well to excellent category. In addition, the improvement of the average score on all the indicators of attitude is because the students have been trained to use the scientific attitude through guided inquiry-based learning activities conducted during the three meetings, such as discussing with a group of friends in identifying and formulating problem based on the given phenomenon, preparing hypothesis, determining experimental variables, experimenting, writing experimental data, analyzing experimental results, and formulating conclusions. In guided inquiry, there is a scientific process that employs critical, logical, and creative thinking skills to answer questions posed with scientific attitudes [3].

This results are supported by Ural who states that there is a significant increase in all aspects of students' academic attitudes and achievements through the use of guided inquiry-based laboratory experimental method. In guided inquiry, the students are more confident and the experiment becomes more meaningful [5]. It is also supported by Setowati's research which states that activities and students' achievement on K_{sp} material that include aspects of knowledge, attitude, and skills improve. The assessments of spiritual attitudes, cooperation, honestly, discipline, and responsibility increase and those assessments are classified into very good category [4].

4. Conclusion

Based on the results and discussion of this research, it can be concluded that the developed guided inquiry-based learning material is effective to be used in teaching and learning in order to improve students' science literacy skills on solubility and solubility product (K_{sp}) concepts. This is proven by the increase of N-gain score with medium and high category both on science literacy skills in aspects of context, competence and knowledge, as well as in aspects of attitude. Thus, there is a positive effect between the development of guided inquiry-based learning material and the improvement of various aspects of science literacy skills for senior high school students. Moreover, it is suggested that further researchers can develop similar learning material for other subjects or materials to train students' science literacy skills, experimental skills, as well as to apply chemistry concepts in everyday life. In addition, teachers should prepare various sources of literature, master learning steps, and manage the class well so that guided inquiry-based learning will be more effective in improving students' science literacy skills.

Acknowledgments

The author would like to thank Dr. Utiya Azizah M.Pd., Dr. Harun Nasrudin, M.Pd., and Drs. Agus Subroto, M.Pd. as the validators of the developed learning material. The author also thanks the headmaster and chemistry teacher of Muhammadiyah 4 Senior High School Surabaya who have allowed and assisted this research.

References

[1] OECD 2016 *PISA 2015 Assessment and Analytical Framework: Science, Reading, Mathematic and Financial Literacy* (Paris: OECD Publishing)

[2] Mulyasa 2013 *Menjadi Guru Profesional: Menciptakan Pembelajaran Kreatif dan Menyenangkan* (Bandung: PT Remaja Rosdakarya)

[3] Llewellyn D 2005 *Teaching High School Science Through Inquiry* (Amerika: Corwin Press)

[4] Setiowati H, Nugroho A, and Agustina W 2015 Penerapan Model Pembelajaran Inkuiri Terbimbing (Guided Inquiry) Dilengkapi LKS untuk Meningkatkan Aktivitas dan Prestasi Belajar Siswa pada Materi Pokok Kelarutan dan Hasil Kali Kelarutan Kelas XI MIA SMA Negeri 1 Banyudono Tahun Pelajaran 2014/2015 *Jurnal Pendidikan Kimia (JPK)* **4** pp 54-60

[5] Ural E 2016 The Effect of Guided-Inquiry Laboratory Experiments on Science Education Students' Chemistry Laboratory Attitudes, Anxiety and Achievement *Journal of Education and Training Studies* **4** pp 217-227

[6] Sesen B A and Tarhan L 2013 Inquiry-Based Laboratory Activities in Electrochemistry: High School Students' Achievements and Attitudes *Research in Science Education* **43** pp 413–435

[7] Rakhmawan A, Setiabudi A, and Mudzakir A 2015 Perancangan Pembelajaran Literasi Sains Berbasis Inkuiri pada Kegiatan Laboratorium *Jurnal Penelitian dan Pembelajaran IPA (JPPI)* **1** pp 143-152

[8] Gültepe N and Kiliç Z 2013 Scientific Argumentation and Conceptual Understanding of High School Students on Solubility Equilibrium and Acids and Bases *Journal of Turkish Science Education* **10** pp 5-21

[9] Schwartz Y, Ben-Zvi R, and Hofstein A 2006 The Use of Scientific Literacy Taxonomy for Assessing The Development of Chemical Literacy Among High-School Students *Chemistry Education Research and Practice* **7** pp 203-225

[10] Sukmadinata N S 2015 *Metode Penelitian Pendidikan* (Bandung: Rosdakarya)

[11] Arikunto S 2010 *Prosedur Penelitian* (Jakarta: Rineka Cipta)

[12] Riduwan 2012 *Skala Pengukuran Variabel-Variabel Penelitian* (Bandung: Alfabeta)

[13] Kemendikbud 2014 *Peraturan Menteri Pendidikan dan Kebudayaan Nomor 104 Tahun 2014 tentang Penilaian Hasil Belajar oleh Pendidik pada Pendidikan Dasar dan Pendidikan Menengah* (Jakarta: Departemen Pendidikan dan Kebudayaan)

IOP Conf. Series: Journal of Physics: Conf. Series **947** (2017) 012049 doi:10.1088/1742-6596/947/1/012049

[14] Hake R R 1998 Interactive Engagement Versus Traditional Methods: A Six Thousand Student Survey of Mechanics Test Data for Introductory Physics Courses *American Journal Physics* **66** pp 64-74
[15] Arends R I 2012 *Learning to Teach* (New York: Mc Graw Hill)
[16] Slavin R E 2011 *Psikologi Pendidikan: Teori dan Praktik Edisi Kesembilan* (Jakarta: PT. Indeks)
[17] Andersen L W 1981 *Assessing Affective Characteristic In The Schools* (Boston: Allyn and Bacon)

The Elementary School Students' Mathematical Problem Solving Based on Reading Abilities

R D Wulandari[1,*], A Lukito[2] and S Khabibah[2]

[1] Primary Education Departement, UniversitasNegeri Surabaya
[2] Mathematic Education Departement, UniversitasNegeri Surabaya

*Corresponding Author E-mail: rismawulandari@mhs.unesa.ac.id

Abstract. The aim of this research is to describe the third grade of elementary school students' mathematical problem in solving skills based on their reading abilities. This research is a descriptive research with qualitative approach. This research was conducted at elementary school Kebraon II Surabaya in second semester of 2016-2017 academic years. The participants of this research consist of third grade students with different reading abilities that are independent level, instructional level and frustration level. The participants of this research were selected with purposive sampling technique. The data of this study were collected using reading the narration texts, the Ekwall and Shanker Informal Reading Inventory, problem solving task and interview guidelines. The collected data were evaluated using a descriptive analysis method. Once the study had been completed, it was concluded that problem solving skills varied according to reading abilities, student with independent level and instructional level can solve the problem and students with frustration level can't solve the problem because they can't interpret the problem well.

1. Introduction

Students need Language skills to understand mathematical symbols. Language is the key of cognitive development because language is a communication tools among people. Problems solving means engaging task in which the solution method is not known in advance. Mathematics as a language which uses symbols is used to facilitate a person in doing the analysis so that it can solve the problems of life associated with the use of mathematics. A person will be able to solve math problems if supported by a language understanding with known situations and using previously familiar symbols. In order to find the solution, student must illustrate on their knowledge and through the process to solve the problem; they will often develop new mathematical understandings. Solving problem is not only a goal of learning mathematics but also a major means of doing it. Students should have frequent opportunities to formulate, grapple with, and solve complex problem that require a significant amount of effort and should then be encourage reflecting on their thinking [8]. Model of the problem solving process identifies four sequential phases: problem translation (i.e., utilizing linguistic skill to comprehend what the problem is saying), problem integration (i.e., mathematics interpretation relationships among the problem parts to form a structural representation), solution planning (i.e., determining which operations to use and the order in which to use them), and the solution execution (i.e., carrying out the planned computations to solve the problem) [5].

IOP Conf. Series: Journal of Physics: Conf. Series **947** (2017) 012050 doi:10.1088/1742-6596/947/1/012050

To elementary students, the majority mistakes in math tests are in working on a story problem that requires problem solving. Students' wrong answers are generally caused by the reading ability, understanding, transformation, or carelessness. Students often perform one or more of the four counting operations $(+, -, \times, \div)$ that needed to answer the question, but they do not know which counting operations are used to solve the problem [1] . Students have difficulties in understanding the keywords appearing in problems, moreover, they cannot interpret them into mathematical sentence, students are also unable to figure out what to assume and what information from the problem is necessary to be solved. Whenever students do not understand the problem, they tend to guess the answer without any thinking process, students are impatient and do not like to read mathematical problems, students do not like to read very long problems and students have difficulties in reading and comprehension, unable to understand what important information is in a problem and organize it. In addition students cannot invert the text into mathematical symbols [10].

Language knowledge is essential for mathematical knowledge. As outlined at the beginning, students need a repertoire of literacy skill to cope with school task. Language knowledge supports students in decoding mathematical terms. They need literacy skills to understand words and the structure of sentences used in terms, e.g. operators, which play an important part in action knowledge for the solution of items or relations between words, important for conceptual knowledge [12]. Learning mathematics is particularly intertwined with good technical reading skills, even in adolescence [6]. When solving problems that include a story, students are required to understand the language of the problem and the concrete information presented in the problem, to properly conceptualize the problem in his/her mind based on the information given within the problem, to design and follow a plan and to make the calculations required by the solution process of the problem[11].

Problem solving requires reading, reading comprehension and the use of mathematics, as well as the use of mathematical operations. Students who have trouble reading the problem can't provide the correct answer [4].Problems solving means engaging in task in which the solution method is not known in advance. In order to find the solution, student must draw on their knowledge and through this process; they will often develop new mathematical understandings. Solving problem is not only a goal of learning mathematics but also a major means of doing it. Students should have frequent opportunities to formulate, grapple with, and solve complex problem that require a significant amount of effort and should then be encourage reflecting on their thinking [8].

To read mathematics textbooks, answer questions on mathematics assessments, and understand educator and student communication, students must develop an understanding of the academic language of mathematics [9].Students who experienced difficulties in reading a problem were unable to provide correct answers. In addition, correct vocalization of the problem may not be enough for finding the correct solution to a problem, because the mathematical language involved in the problem must also be understood. There is also a need for students to develop strategies besides understanding the language of the problem and the mathematical language involved in the problem. In addition to the development strategies, how to administer these strategies to new situations then must be mastered. Strengthening teacher-student and student-student relationships is great importance for the understanding of problems [7].

2. Method

This research is a descriptive research with qualitative approach. This research was conducted at elementary school Kebraon II Surabaya in second semester of 2016-2017 academic years. The subject of this research consists of third grade students with different reading abilities that are independent level, instructional level and frustration level. The subjects of this research were selected with purposive sampling technique. To determine the subject of research, the researcher chooses the subject by knowing their reading abilities. In this case students of grade III-B are given a reading text, and then students are asked to read the test aloud. After that, researcher recorded students' reading. When students are reading, researchers mark the areas where students make mistakes and where they correct their mistakes. After that, the researcher calculates how many student errors, then converts into percentage and put into the table IRI (Informal Reading Inventory) the word recognition (word recognition). Then, the researcher gave the problem (contains 10 comprehension questions) related to the text, the students are asked to solve the problem. Then the researcher corrected the students' answer, calculated and converted into percentage, after that the result is inserted into the table of Informative Reading Inventory (Comprehension). From these two results, students can be categorized into independent levels, counseling levels and frustration levels. From the result of the grouping of reading ability, each group of reading ability was chosen randomly and selected all female, so that the difference of gender did not affect on this research

The data of this study were collected using reading narration texts, the Ekwall and Shanker Informal Reading Inventory, problem solving task and interview guidelines. Ekwall and Shankel inventory is used to determine the level of student understanding. Ekwall&Shanker (1988) suggests that in order to make inventory read comprehension using the level of IRI (Informal Reading Inventory), certain instructions are needed, while elementary level students (starting level pre-school to Class I) using reading consisting of 20 to 40 words. The advanced levels (grade II and grade III) use readings ranging from 100 to 150 words and grade VII to grade VII using readings of 250 to 300 words. To measure reading comprehension, it usually uses questions about main ideas, important information, vocabulary and conclusions. Then the collected data were evaluated using a descriptive analysis method that is data condensing, data displaying and drawing and verifying conclusions.

3. Results

3.1 Findings related to reading abilities

Table 1. Findings related to reading abilities of students in the group learning

Students	Reading abilities	Detailed information about reading abilities
Nayla	Independent Level	She can vocalize the text accurately, read the text loudly and rhythmical, she observed the punctuation accurately, when she read the text there was no evidence of lip movement and finger pointing; she read the text at the correct speed and with the correct intonation. She can accurately vocalize the words involved in the word recognition test.
Az-zahra	Instructional Level	She can vocalize the text accurately, read the text loudly and rhythmical, she observed the punctuation accurately, when she read the text there was no evidence of lip movement and finger pointing; she read the text at the correct speed and with the correct intonation. She can't accurately vocalize the words involved in the word recognition test. She made two mistakes in pronunciation.
Abilanissa	Frustration Level	She vocalized the text inaccurately, read the text slowly, abnormally, not rhythmical and read the text word by word. When she read the text, there was lip movement and finger pointing. She observed the punctuation inaccurately. She read the text with incorrect speed and incorrect intonation. She erroneously vocalized the words involved in the word recognition test.

When students are reading, researcher are marking the areas where students make mistakes and where they correct their mistakes. After that, the researcher calculated how many student errors, then converts into percentage and put into the IRI table (Informal Reading Inventory) the word recognition (word recognition). Then, the researcher gives the problem (contains 10 comprehension questions) related to the text, the students are asked to do the problem. Then the researcher corrected the students' answer, calculated and converted into percentage, after those processes the result is inserted into the table of Informative Reading Inventory (Comprehension).

3.2 Findings related to problem solving skills

Table 2. Finding related to problem solving skill when understanding the problem

Students	Detailed information when they understanding the problem
Nayla	She understood the problem by reading the problem loudly, rhythmically and accurately, she observed the punctuation accurately, and read the text twice in order to understand the problem. She could present the problem's idea verbally and based on her memory. She could identify what is required and asked in the problem shortly, clearly and use her language based on her memory. She could also use the reason why she wrote the information as what is known and asked.
Az-zahra	She understood the problem by reading the problem loudly, rhythmically and accurately, she observe the punctuation accurately, has one wrong pronunciation and read the text twice in order to understand the problem. She could present the problem's idea verbally suited to the script (see the script). She could identify what is required and asked in the problem according to the script and could also use the reason why she wrote the information as what is known and asked.
Abilanissa	She understands the problem by reading the problem slowly, non-rhythmically and inaccurately, she read the text word by word, she observed the punctuation inaccurately, had three wrong pronunciations and read the text twice in order to understand the problem. She could present the problem's idea verbally suited to the script (see the script). She could identify what is required and asked in the problem according to the script and could also use the reason why she wrote the information as what is known and asked.

Table 2 shows that Nayla and Az-zahra when understanding the problem, they read the text loudly, rhythmically and accurately and they observe the punctuation accurately. However, Abilanissa read the text slowly, non-rhythmically and inaccurately, she reads the text word by word and she observed punctuation inaccurately. Nayla could present the problem's idea verbally and could identify what is required and asked in the problem shortly, clearly and used her language based on her memory. Az-zahra and Abilanissa could present the problem's idea verbally and could identify what is required and asked in the problem.

Table 3. Finding related to problem solving skill when devising the plan

Students	Detailed information when they were devising the plan
Nayla	She set out the steps to get the solution, then the she used the reason why choosing the strategy to solve the problem and she could give examples of other problems that can be solved with the same strategy.
Az-zahra	She set out the steps to get the solution, then the she used the reason why choosing the strategy to solve the problem and she could give examples of other problems that can be solved with the same strategy.
Abilanissa	She set out steps to get the solution, but students were wrong in the first step. She made a mistake of transforming the mathematical language into a mathematical operation. She used the reason of why choosing the strategy to solve the problem, but the she could give examples of other problems that can be solved with the same strategy.

Table 3 show that Nayla and Az-zahra when are devising the plan, they set out the steps to get the solution then they use the reasons why choosing the strategy to solve the problem and they can give the examples for other problems that can be solved with the same strategy. Abilanissa set out the steps to get the solution but the first step was wrong; she made a mistake in transforming the mathematical language into a mathematical operation. She used the reason why choosing the strategy to solve the problem, but she could give examples of other problems that can be solved with the same strategy.

Table 4. Finding related to problem solving skill when carrying out the plan

Students	Detailed information when they were carrying out the plan
Nayla	She applied the planned steps, then completed the necessary calculations in the planned step and set the final answer.
Az-zahra	She applied the planned steps, then completed the necessary calculations in the planned step and set the final answer.
Abilanissa	She applied the planned steps, then complete the necessary calculations in the planned step and set the final answer, because in the first step, she had been planned incorrectly, the calculation and the final answer of the problem became wrong

Table 4 shows that Nayla, Az-zahra and Abilanissa when are carrying out the plan, they apply the planned steps then completed the necessary calculations in the planned step and set the final answer. But Abilanissa had a first step incorrectly, so the calculation and the final answer became wrong.

Table 5. Finding related to problem solving skill when looking back

Students	Detailed information when they are looking back
Nayla	She looked back the final answers obtained, looked back the precision of the steps that had been used and looked for other ways that might be used to solve the problem.
Az-zahra	She looked back the final answers obtained, looked back the precision of the steps that had been used and looked for other ways that might be used to solve the problem.
Abilanissa	She looked back the final answers obtained, looked back the precision of the steps that have been used, but she did not look for other ways that might be used to solve the problem.

Table. 5 shows that Nayla and Az-zahra in looking back stage, they look back the final answers obtained, looked back the precision of the steps that have been used and look for other ways that might be used to solved the problem, but Abilanissa did not look for other ways that might be used to solve the problem.

4. Discussion

Basic reading skills are imparted to students in elementary school during the first grade. When students proceed through higher classes, acquired skills are developed further and higher skills are inculcated in students. A similar process is followed in the inculcation of mathematical skills. Here, it should be noted that the main skill to be imparted to students, both in mathematics and in all other academic disciplines, is that thinking; this can be realized by means of the language used by the students [4].

When understanding the problem, the students who had reading ability at independent level and instructional level had presented the problem's idea verbally began with reading the text loudly, rhythmically and accurately, they observed the punctuation accurately, and read the text twice in order to understand the problem. She could accurately vocalize the words involved in the word recognition test. This is consistent with the characteristics of students who had the ability to read at the independent level presented by Johnson & Kress that is student who have reading ability at independent level read the text loudly, rhythmically and observe the punctuation accurately. The students with reading ability at frustration level presented the problem's idea verbally began with reading the text slowly, abnormally, non-rhythmically and read the text word by word. When she read the text, there was lip movement and finger pointing. She observed the punctuation inaccurate. She read the text with incorrect speed and incorrect intonation. She erroneously vocalized the words involved in the word recognition test [2].

Students with reading abilities at independent and instructional level can understand the problem, carry out the plan, devising the plan and looking back. Students whose reading abilities at the independent level is able to vocalize problems correctly and summarize what was given and required based on her memory, sure of their answers and able to try different ways of solving the problem.Students whose reading abilities at the frustration level, difficulties are experienced in terms of reading problem. These students were unable to make use of strategies during their problem solving performances. In addition, it can be argued that students were distracted while reading and solving the problem. They were not aware of the mistakes they which had been made while reading and problem solving.

The first step in solving mathematical problems is to understand the math problem itself. The connection between the ability of understanding and problem solving can be emphasized that, if one has the ability to understand mathematical concepts, then he is able to use them to solve problems. Conversely, if one can solve a problem, then the person must have the ability to understand the mathematical concepts that have been studied previously [3].Students that experienced difficulties reading a problem was unable to provide correct answers. In addition, correct vocalization of the problem may not be enough for finding the correct solution to a problem, because the mathematical language involved in the problem must also be understood. There is also a need for students to develop strategies besides understanding the language of the problem and the mathematical language involved in the problem. Furthermore, to the development of strategies, how to administer these strategies to new situations must be mastered. Strengthening teacher-student and student-student relationships is of great importance for the understanding of problems [7]. Therefore, reading level and the problem solving skills of students should be handled together and instructional activities should be focused on the current teaching of these two skills.

5. Conclusion

It can be concluded that: (1) the students with independent reading ability when understanding the problem: reading the problem out loud, fluently and accurately, presenting the idea verbally based on their memory, identifying what is known and asked in a question briefly, clearly and based on their memory; when devising a plan: establishing steps to obtain a solution and using the reasons for choosing a strategy for solving the problem; When carrying out the plan: implementing the planned steps, completing the necessary calculations in the planned steps and establishing the final answer; When looking back: looking back the final answer obtained, looking back the precision of the steps that have been used and looking for other ways that might be used to solve the problem. (2) the students with instructional reading ability when understanding the problem: reading the problem out loud, fluently and accurately, presenting the idea verbally suited to the scripts test, identifying what is known and asked in a question briefly, clearly and suited to the script test; when devising a plan: establishing steps to obtain a solution and using the reasons for choosing a strategy for solving the problem; When carrying out the plan: implementing the planned steps, completing the necessary calculations in the planned steps and establishing the final answer; When looking back: looking back the final answer obtained, looking back the precision of the steps that have been used and looking for other ways that might be used to solve the problem. (3) Students with frustration reading ability when understanding the problem: reading the problem with soft voice, reading word by word and pointing finger, presenting the problem idea on verbally according to the script test, identifying what is known and asked in a question suited to the script test ; when devising a plan : establishing steps to obtain a solution, but having errors in the first step and using the reasons for choosing a strategy for solving the problem; when carry out the plan: implementing the planned steps, completing the necessary calculations in the planned step and establishing the final answer of the question, but the final answer is wrong; when looking back: looking back the final answer obtained, not checking the accuracy of the steps that have been used and looking for other ways that might be used to solve the problem

Acknowledgements
Prof. Dr. Warsono, M.S. as a rector of Universitas Negeri Surabaya
Prof. Dr. IsmetBasuki, M.Pd. as a postgraduate director
Dr. Hendratno, M.Hum as chairman of primary education study program
A principal and teachers of SDN Kebraon II Surabaya
My Parents and My Husband, thanks for the motivation

References

[1] Clements D H, Wilson D C, and Sarama J 2004 Young Children's Composition of Geometric Figures: A Learning Trajectory *Mathematical Thinking and Learning* **6** pp 163-184

[2] Ekwall E E, and Shanker J L 1988 *Diagnosis & Remediation of Disabled Reader* (America: Allyn and Bacon) pp 138

[3] Fauziah A 2010 Peningkatan Kemampuan Pemahaman dan Pemecahan Masalah Matematik Siswa SMP melalui Strategi React *Forum Kependidikan* **30** pp 11-17

[4] Gökhan O, Hayriye G K, and Ahmet C 2015 Evaluation of Students' Mathematics Problem Solving Skills in Relation to their Reading Levels *International Electronic Journal of Elementary Education* **8** pp 113-132

[5] Krawec JL, Huang J, Montague M, Kressler B, and Melia de Alba A 2013 The Effects of Cognitive Stratgy Instruction on Knowledge of Math Problem Solving Process of Middle School Students with Learning Disabilities *Learning Disabilities* **36** pp 80-92

[6] Kyttala M and Bjorn P M 2014 The Role of Literacy Skills in Adolescents' Mathematic Word Problem Performance: Controlling for Visuo-spatial Ability and Mathematics Anxiety *Learning and Individual Differences* **29** pp 59-66

[7] Mercer N and Sams C 2006 Teaching children how to use language to solve maths *problems Language and Education* **20** pp 507-528

[8] National Council of Teacher of Mathematics 2012 *Principles and Standards for School Mathematics* (Reston V.A: NCTM) p 67

[9] Powell S R, Driver M K, Roberts G, and Fall A M 2017 An Analysis of The Mathematics Vocabulary Knowledge of third- and fifth-grade Students: Connections to General Vocabulary and Mathematics Computation *Learning and Individual Differences* **57** pp 22-32

[10] Prathana P, Suwimon W, and Siridej S 2014 An Analysis of Elementary School Students' Difficulties in Mathematical Problem Solving *Procedia-Social & Behavioral Sciences* **116** pp 3169-3174

[11] Thompson D R, Kersaint G, Richards J C, Hunsader P D, and Rubenstein R N 2008 Mathematical *Literacy. Helping Students Make Meaning in the Middle Grades* (Portsmouth NH: Heinemann) p 135

[12] Walzebug A 2014 Is there a Language-Based Social Disadvantage in Solving Mathematical Items? *Learning, Culture and Social Interaction* **3** pp 159-69

IOP Conf. Series: Journal of Physics: Conf. Series **947** (2017) 012051 doi:10.1088/1742-6596/947/1/012051

Algebraic Thinking in Solving Linier Program at High School Level: Female Student's Field Independent Cognitive Style

N Hardiani[1], I K Budayasa[1], D Juniati[1]

[1]Mathematics Education Department of Universitas Negeri Surabaya

Eemail: nurhardiani@mhs.unesa.ac.id

Abstract. The aim of this study was to describe algebraic thinking of high school female student's field independent cognitive style in solving linier program problem by revealing deeply the female students' responses. Subjects in this study were 7 female students having field independent cognitive style in class 11. The type of this research was descriptive qualitative. The method of data collection used was observation, documentation, and interview. Data analysis technique was by reduction, presentation, and conclusion. The results of this study showed that the female students with field independent cognitive style in solving the linier program problem had the ability to represent algebraic ideas from the narrative question that had been read by manipulating symbols and variables presented in tabular form, creating and building mathematical models in two variables linear inequality system which represented algebraic ideas, and interpreting the solutions as variables obtained from the point of intersection in the solution area to obtain maximum benefit.

1. Introduction

Algebra is one part of mathematics which is an extension of arithmetic [1][2] since algebra is represented by the language of symbols of the unknown variable [3][4]. Algebra can also be used to solve a problem in everyday life, so it makes algebra important and must be learned for all students [5][6].

Algebra standard for the school level corresponds to [5], that by learning algebra from kindergarten to high school level students are expected to be able to understand patterns, relationships and functions; students are able to present and analyze mathematical situations and structures that use algebraic symbols; students are able to use mathematical models to present and understand quantitative relationships; and students are able to analyze changes in various situations. Similarly in Indonesia, the standard of mathematical basic competences that must be achieved by high school students based on the curriculum of 2013 are using the concept of algebra to apply the exponent and logarithm rules that match the characteristics of the problems to be solved; using algebraic concepts to analyze absolute values in equations and inequalities; using algebraic concept to predict the sequences patterns and arithmetic and geometry series or other sequences through observation and reasoning; using the concept of algebra to describe various forms of expression which can be converted to quadratic equations; using algebraic concept to formulate the rules and the natures of algebraic function limits through observation of examples; using algebraic concept in solving problems in everyday life. One of the algebraic materials that must be studied by high school students is linier program material. Linier program is an optimization problem solving program, which maximizes or

IOP Conf. Series: Journal of Physics: Conf. Series **947** (2017) 012051 doi:10.1088/1742-6596/947/1/012051

minimizes a linier function that depends on linier constraints. This constraint can be an inequality or a linier equation. The problem of linier program in school is presented in narrative question form so that from the narrative question the students are expected to solve the problem by representing in the form of a linier function which contains symbols or variables that its value based on algebraic agreement is unknown. In that process, students do algebra thinking activities.

Some experts define algebraic thinking. According to [7] algebraic reasoning involves representing, generalizing, and formalizing patterns and regularity in all aspects of mathematics. In line with [8], algebraic thinking is using mathematical symbols and tools to analyze different situations by; representing that information mathematically in words, diagrams, tables, graphs, and equations; interpreting and applying mathematical findings, such as solving for unknowns, testing conjectures, and identifying functional relationships to the same situations and the new related situation. Meanwhile, he described three activities of algebraic thinking: generational activities, transformational activities, and global meta-level activities. The same thing [10], defines algebraic thinking as the ability of generalization, abstraction, analytic thinking, dynamic thinking, modeling, and organization. But according to [11], in his article entitled "Just What is Algebraic Thinking" that algebraic thinking has two main components, namely the ability to think mathematically and algebraic basic idea of. The ability to think mathematically is the ability to solve problems, the ability of representation, and the ability of quantitative reasoning while the algebraic basic idea is algebra as arithmetic generalization, algebra as the mathematics language, and algebra as a tool for a function and mathematical modeling.

To describe students' algebraic thinking, it needs to be viewed from many dimensions because as human beings, students are unique human beings, in which one student is different from other students in many dimensions. Basically each individual is different one another. The dimensions of individual differences include intelligence, logical thinking ability, creativity, cognitive style, personality, values, attitudes, and interests. Cognitive style is different from intelligence and other ability dimensions. [12] divides the two cognitive styles namely field independent and field dependent. He distinguishes individuals who have field independent and field dependent cognitive style through the ability to answer Group Embeded Figure Test (GEFT) test within the specified time as well as certain criteria. Another dimension of individual differences is gender.

Based on the description, the researcher is interested in describing algebraic thinking of female students' field-independent cognitive style in high school level in solving linear program problem. The researcher describes that algebraic thinking of female students' field-independent cognitive style in solving linear program problem has 5 indicators that involve reading, writing, manipulating numbers and symbols using algebraic agreements; variable interpretation and variable expression; representing ideas for using expressions containing variables such as equations, inequalities, and functions; using the equivalent symbolic representation to manipulate formulas, expressions of equations, inequalities or functions; interpreting the solution. These five indicators are adapted from [11].

2. Method

The type of this research was descriptive qualitative. The data collection method used observation, documentation, and interview. Participants in this research were 216 female students of 11th grade high school students from State Senior High School 2 Mataram and State Senior High School 5 Mataram. This research instruments consisted of main instrument namely the researcher and supporting instrument consisting of 4 kinds of instrument namely GEFT Test, Mathematics ability problem, Linier Program Problem, and Interview Guidance. First 216 female students were given GEFT Test and Mathematics Ability Problem. Of the 216 female students, there were 149 students who had field independent cognitive style. Based on parameters that had been determined by the researcher, the results of GEFT test and the results of the mathematical ability of 149 students, there were 7 students selected to be given linear program problem and interviewed. The linier program Problem was "the flower shop 'PUJI' sells two kinds of hand bouquet that are cheap but interesting. The first series consists of 4 roses, 2 dahlia flowers, and 3 tulips while the second series consists of 2 roses, 3 dahlia flowers, and 1 tulip flower. The seller only has stocks of 48 roses, 36 dahlia flowers, and 33 tulips. If the profit in the first series is Rp.15.000, - and that in the second series is Rp.12.000, - then do the following calculation.

- Determine all the possible inequalities models to express the need of roses, dahlias and tulips that are used to make two kinds of flower series, based on the data.
- From the result of the inequality system (a), show the graph and the shaded area of the solution in a system of plane Cartesian coordinate.
- Through the interpretation of the graph, what is the maximum profit that is gained by the seller? And how many series of the first flowers and the second flower that need to be made so that the seller can get the maximum profit?

3. Results and Discussion

Data collection on the research subject in working linier program problems and interview was conducted on March 13th, 2017. Based on the answers of seven male students of field independent cognitive-style, the algebraic thinking in solving linier program problem was described. Descriptions of their algebraic thinking were described as follows. Based on the first indicator, they could think about the problems they had read and then they could explain the general purpose of the narrative question and the questions given. They also found material related to the issues already read. Next they wrote out the essential elements for expressing algebra on the problems already read into a table. Two different tables are presented by them (see Figure 1 and see Figure 2). This was in accordance with [3], that algebraic is a language. Based on both tables, they could manipulate the existing elements in the problem into table [11].

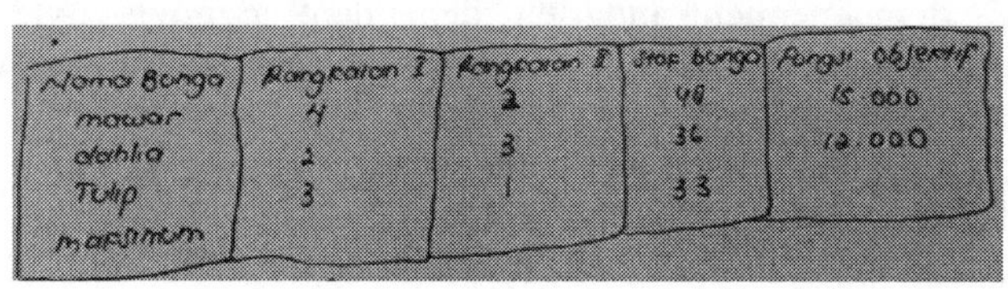

Figure 1. The first table form which is generated based on the first indicator of algebraic thinking

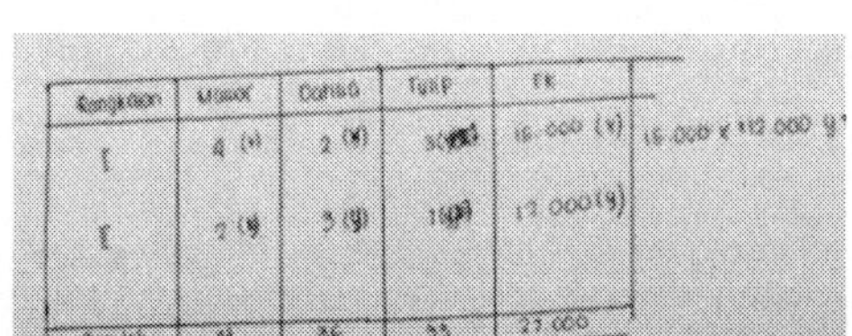

Figure 2. The second table form which is generated based on the first and second indicators of algebraic thinking

In the second indicator, they can interpret the meaning and difference between the constants and the variables in the mathematical model that was already made (see Figure 2) and they could determine the constants for variables x and y and could interpret the function of constraint and purpose function [13]. The following is the interview with a research subject.

Researcher : *Do you understand what you wrote in this table?*
Subject : *Yes, I do.*
Researcher : *Do you also understand these symbols? (while pointing at the table)*
Subject : *Yes, maam.*
Researcher : *Please explain it to me!*
Subject : *Yes, maam (while nodding). Because in the problem there are two flowers series then I suppose that the first series is x and the second series is y.*
Researcher : *Why are they supposed x and y?*
Subject : *Because those are the names of variables.*
Researcher : *May they be changed with other letters?*
Subject : *Yes, they can.*
Researcher : *How about this one (pointing at table 2 in column 2)?*
Subject : *Ohhh ..... it is coefficient.*
Researcher : *What is the meaning of 4x and 2y?*
Subject : *4 shows the number of roses for the first series and 2 shows the number of roses for the second series.*

Female students with field independent cognitive style could link the information available to the narrative question by writing and manipulating symbols such as variables and constants as representational capabilities [1][14][15][16].

Next to the third indicator, they could build a two-variable linier inequality system to create two types of flower series of three types of flowers. They could formulate and create all functional constraints in the form of linier inequalities and purpose functions in the form of equations for flower series (see Figure 3).

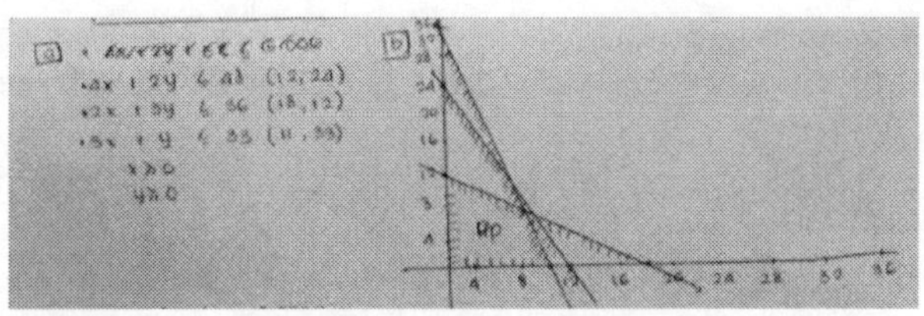

Figure 3. Photo results for the third and the fourth indicator of algebraic thinking

The fourth indicator, they could find all the intersection points of the linier inequalities pair. They could find and complete the completion set of the linier inequality system that had been created by elimination. They substituted on a linier inequality system by using basic operations (addition, subtraction, multiplication, division) between linier inequalities to obtain a cutoff point in the Cartesian plane graph of two pairs of known linier inequalities. They attempted to describe the results of algebraic analysis in the form of a graphic interpretation in the Cartesian coordinate system (see Figure 3). They were able to determine the shaded area of the linier inequality system as the solution area of the linier inequality system (see Figure 3).

The last indicator, they could interpret three cutoff points in the solution area on the graph that had been made then they tested the three points by substituting in the equation of the objective function that had been made to obtain the maximum profit (see Figure 4). The cutoff point that generated the maximum profit was called the maximum point and they could interpret the numbers at the maximum point obtained as the value of the questioned variable as in the following interview.

Researcher : *What is the meaning of maximum profit Rp.207.000, - at the point (9,6)?*

Subject : *So the seller gains maximum profit as much as Rp. 207,000, and ... hmm ... (while looking at the table that had been created). Well ... the meaning of point 9 is the number of the first flower series because it was exemplified before by x and the second flower series was exemplified by y, so the number of both flower series that can be made by the seller are 6.*

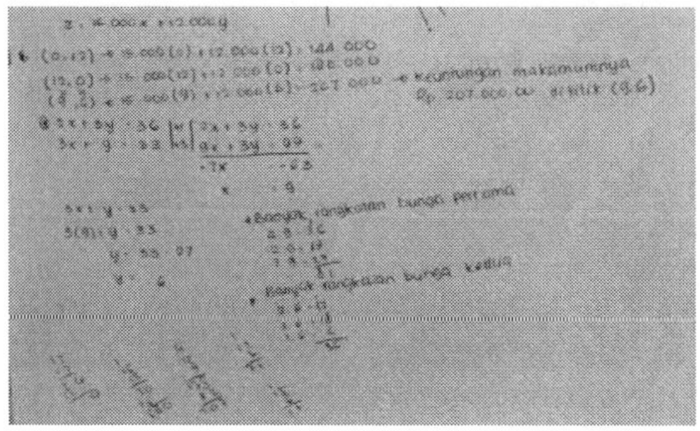

Figure 4. Photo results for the fifth indicators of algebraic thinking

4. Conclusion

The research result states that algebra thinking from 7 high school female students in field independent cognitive style can solve linier program problem. They have an ability to represent algebraic ideas from the narrative questions that have been read by manipulating symbols and variables presented in two different table forms. From these two tables they could create and construct a mathematical model in a two-variable linier inequality system based on algebraic agreement as a constraint function and objective function. They could interpret meaning and difference between constants and variables on constraints function and objective functions that have been made. They were able to complete a two-variable linier inequality system to search for all possible cutoff points to illustrate the results of the algebra analysis of the completion area into a graph of the Cartesian coordinate system. Then they interpret and test the cutoff points of maximum profit after being substituted on the objective function, and they are able to conclude the maximum point as the value of the questioned variable.

References

[1] Kaput J J 2008 *What is Algebra?What is Algebraic Reasonning?* (New York: Lawrence Erlbaum Associates)

[2] Vance J H 1998 Number Operations from An Algebraic Perspective *Teaching Children Mathematics* **4** pp 282-285.

[3] Usiskin Z 1997 Doing Algebra in Grades K-4 *Teaching Children Mathematics* **3** pp 346-356

[4] Blair L 2003 It's Elementary: Introducing Algebraic Thinking Before High School *SEDL Letter* **15**(1) pp 16-20

[5] National Council of Teachers of Mathematics 2000 *Principles and Standards for School Mathematics* (Reston, VA: NCTM)

[6] Chambers D 1994 The Right Algebra for All *Educational Leadership* **51** p 85

[7] Van de Walle J A 2004 *Elementary and Middle School Mathematics: Teaching Developmentally 5th ed* (New York: Longman)

[8] Herbert K & Brown R H 1997 Patterns as Tools for Algebraic *Reasoning Teaching Children Mathematics* **3**

[9] Kieran Carolyn 2004 Algebraic Thinking in the Early Grades: What Is It? *The Mathematics*

IOP Conf. Series: Journal of Physics: Conf. Series **947** (2017) 012051 doi:10.1088/1742-6596/947/1/012051

Educator Vol. **8**(1) pp139-151
[10] Lew Hee Chan 2004 Developing Algebraic Thinking In The Earlier Grades: A Case Study of The South Korean Elementary School Mathematics Curriculum *The Mathematics Educator* **8**
[11] Kriegler S 2008 *Just what is algebraic thinking* UCLA: Department of Mathematis. Available online also at: http://www. math. ucla. edu/~ kriegler/pub/algebrat. html [accessed in Manila, Philippines: December 5, 2015].
[12] Witkin H A, Moore C A, Goodenough DR, and Cox P W 1975 Field-dependent and field-independent cognitive styles and their educational implications *ETS Research Report Series* **2** pp 1-64
[13] Battista M & Van Auken Borrow C 1998 Using Spreadsheets to Promote Algebraic Thinking *Teaching Children Mathematics* **4** pp 470-478.
[14] Wagner S 1983 What are These Things Called Variables? *Mathematics Teacher* **76** pp 474-479
[15] Skemp R R 1987 *The Psychology of Learning Mathematics* (Hillsdale NJ*:* Lawrence Erlbaum Associates)
[16] Schoenfeld A H & Arcavi A 1988 On The Meaning of Variables *Mathematics Teacher* **81** pp 420-427.

.

IOP Conf. Series: Journal of Physics: Conf. Series **947** (2017) 012052 doi:10.1088/1742-6596/947/1/012052

Effectiveness of Cooperative Learning Instructional Tools With Predict-Observe-Explain Strategy on the Topic of Cuboid and Cube Volume

Nurhuda[1], A Lukito[2], Masriyah[3]

[1,2,3]Mathemathics Educations Program, Universitas Negeri Surabaya, Surabaya, Indonesia

E-mail: [1]nurhudanurhuda@mhs.unesa.ac.id

Abstract. This study aims to develop instructional tools and implement it to see the effectiveness. The method used in this research referred to Designing Effective Instruction. Experimental research with two-group pretest-posttest design method was conducted. The instructional tools have been developed is cooperative learning model with predict-observe-explain strategy on the topic of cuboid and cube volume which consist of lesson plans, POE tasks, and Tests. Instructional tools were of good quality by criteria of validity, practicality, and effectiveness. These instructional tools was very effective for teaching the volume of cuboid and cube. Cooperative instructional tool with predict-observe-explain (POE) strategy was good of quality because the teacher was easy to implement the steps of learning, students easy to understand the material and students' learning outcomes completed classically. Learning by using this instructional tool was effective because learning activities were appropriate and students were very active. Students' learning outcomes were completed classically and better than conventional learning. This study produced a good instructional tool and effectively used in learning. Therefore, these instructional tools can be used as an alternative to teach volume of cuboid and cube topics.

1. Introductions

Instructional tool is a collection of learning resources used by teachers and students in conducting learning activities. Slavin stated that in the cooperative learning model, students will sit together in groups of four to understand the material that presented by teacher [1]. In cooperative learning, students were situated to work in small groups consisting of 4-6 people who have different achievement, race, cultural and gender capabilities in a collaborative to achieve the goals learning. According to Dalziel [2], that predict-observe-explain (POE) strategy was developed by White and Gunstone in 1992 which aims to reveal the ability of students in making individual predictions. The POE strategy is based on constructivism learning theory which assumes that with activities of prediction, observation, and explain the observation result, the students are able to build their knowledge. In addition, POE strategies can be used to detect and correct students' misconceptions about a concept [3] [4] [5]. Liew, Chong-Wah;

Treagust, and David state that POE strategy is effectively used to express students' abilities and ideas [6] [7].

Indriana, Arsyad, Mulbar in his research stated that learning with POE strategy can improve students' creative thinking ability and give positive response from students. However, this study does not explain the effectiveness of learning using the POE strategy [8]. Jannah in her research said that POE strategy can increase student activity, teacher activity and student learning outcomes consisting of three aspects attitude, knowledge, and skill. Esra Kele and Pinar Demirel [5] in their research said that the strategy of POE can be used to detect and overcome students' misconceptions about a learning material. However, this study does not explain the improvement of student learning outcomes.

2. Methods

This research aims to develop a good cooperative instructional tools with predict observe explain (POE) strategy and know the effectiveness of learning with instructional tools were developed. This research method consists of development research and experimental research. The developmental research carried out two stages, the instructional tools design and the development stage. In this research, the development model used was based on the design of learning according to Morrison, et al in their book "Designing Effective Instruction". Then, the instructional tools that has been designed, then developed by performing expert validation and revision based on input from the validator. Then, the instructional tools was implemented in the classroom to know the quality of the instructional tools was good. Instructional tools were said to be of good quality if meet the criteria of validity, practicality, and effectiveness [8].

The experimental research was conducted by using two-group pretest posttest design, that was implementing instructional tools in the experimental class which aimed to compare students' learning outcomes in control class to know the results of student learning in the experimental class was better. To know the effectiveness of learning, the developed instructional tools with good quality will be applied to experimental research. While the effectiveness of learning according to Slavin, consists of four indicators that are: quality of instruction, appropriate levels instruction, incentive, time [1]. Thus learning was said to be effective when the implementation of learning by the teacher was well, the implementation of learning was active in student activities, students' responses to the learning were positive, learning completeness was achieved in classical and student learning outcomes were better than the student learning outcomes on conventional learning.

The subjects of this experimental research were 8th grade of Junior High School. This experimental research method was two-group pretest post-design, one experimental class and one control class. In the experimental class, learning used instructional tools that have been developed. While in the control class, learning used instructional tools designed by the teacher of mathematics itself. The data of learning effectiveness analysis were obtained from the experimental class. The better learning outcomes were obtained by comparing the students learning outcomes on 'experimental class with students' learning outcomes on control class.

3. Results and Discussion

3.1. Developmental Research

The developmental research was carried out in two phases: design stage and development stage. The development research process as follows.

3.1.1. Design stage

The design stage was done in accordance with Morrison et al design by doing the design process elements that are: Instructional Problems, Learner Characteristics, Task Analysis, Instructional Objectives, Content Sequencing, Instructional Strategies, Designing the Message, Development of Instruction, and Evaluation Instrument. Based on the data analysis, it was concluded that students had difficulties about the material of cuboid and cube volume as they saw the cuboid and cube in two dimensions. So it needs the right strategy so that students can understand the volume of the cuboid and cube well. Thus, the designed instructional tools was a cooperative instructional tool with predict observe explain (POE) strategy for cuboid and cube volume. The instructional tools consist of Lesson plans, POE Task, and Learning Tests. Lesson plans and POE task consist of two meetings, and the learning Tests used are Pretest dan posttest.

3.1.2. Development stage

Instructional tools that have been designed, then developed for a good quality instructional tool by performing expert validation and testing of instructional tools in the classroom. Based on the validation of experts obtained that instructional tools were valid and can be used in learning. Further, instructional tools applied to the classroom to determine the practicality and effectiveness of instructional tools. Based on the results of the experiment, it was found that instructional tools practical and effective to be implemented in learning. The result of learning test that has been designed was valid, reliable and sensitive. The data of development of instructional tools can be seen in Table 1.

Tabel 1. The data of development of instructional tools

No	Criteria	Information	Conclusion
1	Expert validation	Valid	Valid
2	The Learning Test	Valid, Reliable, and Sensitive	
3	Implementation of learning by teachers	Good	Practice
4	Students activities	Active	
5	Completily learning outcomes	Complete	Efective
6	Students responses	Positive	

Based on Table 1, Expert validation is valid from the validation of experts, and based the pretest and posttest results show that the test designed was valid, reliable and sensitive criteria. Thus, the designed instructional tools in valid category. Instructional tools that are applied in the classroom show practical use both by teacher and student have caused implementation of learning by teachers is good and students are active. Learning in the classroom is also effective, because of student learning outcomes is complete clasically and and students give a positive response. It was found that instructional tools developed were in good quality because they met criteria of valid, practical and effective. This was in accordance with the statement of Nieveen that the instructional tools were said to be of good quality if meet the criteria of validity, practicality, and effectiveness [12].

3.2. Experimental Reasearch

Experimental research was conducted to determine the effectiveness of learning using the developed tools. The effectiveness of learning according to Slavin, consists of four indicators that are: quality of instruction, appropriate levels instruction, incentive, time [3]. The results of all data in the experimental study were collected and analyzed descriptively. The data in question was the implementation of learning by the teacher in managing the class was well, based on the observation sheet with a minimum value of 3 (good) and activities and time in accordance with the expected in the instructional tool, student activity has been active in accordance with the expected and follow the activities that were on POE task. The result of students' respond show that 87.91% of students tend to enjoy the learning process, while the students' learning outcomes classically reaches 85.7%.

Students' learning outcomes in the experimental class and students' learning outcomes in control class were analyzed by covariance analysis (anacova). This analysis was conducted to find out whether the students' learning outcomes in the experimental class were better than the students' learning outcomes in the control class. The result of covariance analysis were: experimental class regression model was $YE = 62{,}343 + 0{,}471X$ and control class regression model was $YK = 47{,}570 + 0{,}686X$. The regression model both experiment and control class shows there is significant influence of students' initial ability with students' learning result. Both regression models satisfied the linear regression model, and both of them were parallel. This showed that there were differences in students' learning outcomes in cooperative learning with POE strategy (experimental class) with students' learning outcomes in conventional learning (control class). Based on regression model for each experiment class and control class it was shown that regression line constant for experiment class was 62,343 bigger than control class that was 47,570, so geometrically regression line for experiment class was above control class. This meaned that students' learning outcomes that follow cooperative learning with POE strategy was better than students' learning outcomes used conventional learning. Thus, students' learning outcomes used cooperative learning with POE strategy on topic of cuboid and cube volume show better improvement.

Thus the result data from the implementation of learning in the experimental class and inferential analysis results can be seen in Table 2.

Tabel 2. Achieved the Criteria for Effectiveness of Learning

No.	Criteria	Information
1.	Implementation of learning by teachers	Good
2.	Students activities	Aktive
3.	Students responses	Positive
4.	Completily learning outcomes	Completed
5.	Inferential Analysis Results	Better

Based on Table 2, in the experimental class shows that It was found that learning using developed instructional tools can be said to be effective. Implementation of learning by teachers

4. Conclusion

Based on the results and discussion, it can be concluded that the developed instructional tools was cooperative instructional tools with POE strategy that are: Lesson plans, POE Task, and Learning Tests. Developed instructional tools were good quality because it met the criteria of valid, practical and effective. Instructional tools were said to be valid because of the validation results by the expert obtained valid and the learning test of met the criteria valid, reliable and sensitive. Instructional tools were said to be practical based on the implementation of learning by teachers and students were in accordance with the expected. Instructional tools were said to be effective based on positive student responses to instructional tools and completeness of students' learning outcomes that were achieved classically. Implementation of learning using developed instructional tools very effectively. The effectiveness of learning based on the ability of teachers in managing learning was well, student activity during the learning process was active, student responses on learning was positive and completeness of students' learning outcomes achieved in classical. Based on the results of inferential statistical analysis of covariance analysis (anacova), it was found that the students' learning outcomes in the experimental class using cooperative learning with predict observe explain (POE) strategy in the topic of cuboid and cube volume was better than the result of the students' learning in the control class. This suggested that cooperative learning with a predict observe explain (POE) strategy was effectively used in teaching the topic of cube and cuboid volume.

5. Acknowledgements

The authors would to thank for the Ministry of Religious in Republic of Indonesia who has provided scholarships to continue studying education to islamic school teachers in order to improve the quality of education. Hopefully the knowledge gained during learning in postgraduate can be useful, especially for islamic school who have delivered and provided this learning task.

6. References

[1] Slavin R E 2006 *Educational Psychology Theory and Practice 8th ed* (Boston: Person Education)

[2] Dalzie J 2010 *Practical eTeaching Strategies for Predict-Observe-Explain, Problem Based Learning and Role Plays* (Australia: LAMS International)

[3] Ipek H, Kala N, Yaman F, and Esra K 2010 Using POE strategy to investigate student's teacher understanding about the effect of substance type on solubility *ScienceDirect Procedia Social and Behavioral Scences* **2** pp648-653.

[4] Keles E and Demirel P 2010 A study towards correcting students' misconceptions related to the color issue in light with POE technique *ScienceDirect Procedia Social and Behavioral Sciences* **2** pp3134-3139

[5] Chong W L and Treagust D F 1998 The effectiveness of Predict-Observe-Explain Tasks in Diagnosing Students' of Science and identifying *American Educational Research* p 24

[6] Kala N, Yaman F, and Ayas A 2013 The Effectiveness of Predict-Observe-Explain Technique in Probing Students' Understanding about Acid-Basa Chemistry: A case for the Concepts of PH, POH, and Strength *International Journal of Science and Science and Mathematics Education* pp555-574.

[7] Nieveen N 1999 Prototyping to Reach Product Quality *Design Approachesand Tools in Educationand Training* (Netherlands: Kluwer Academic Publisher)

[8] Tekin S and Isiksal B 2011 An Investigation into the Performance, Solution Strategies and Difficulties in Middle School Students' Calculation of the Volume of a Rectangular Prism *Journal for Research in Mathematics Education*.

[9] Kibirige I, Osodo J, and Tiala K M 2014 The Effect of Predict-Observe-Explain Strategy on Learners' Misconceptions about Dissolved Salt *Mediterranean Journal of Social Sciences* **5** 4 pp300-310.

[10] Chung Y H, Chin C T, and Jyh C L 2011 Facilitating Preschoolers' Scientific Knowledge Construction via Computer Games Regarding Light and Shadow: The Effect of the Prediction-Observation-Explanation (POE) Strategy *J Sci Educ Technol* pp482-493.

[11] Indriana V, Arsyad N, Mulbar U 2015 Penerapan Pendekatan Pembelajaran POE (Predict-Observe-Explain) untuk Meningkatkan Kemampuan Berpikir Kreatif Siswa Kelas XI IPA-1 SMAN 22 Makasar. *J Inov Pend Mat* **3** 1

IOP Conf. Series: Journal of Physics: Conf. Series **947** (2017) 012053 doi:10.1088/1742-6596/947/1/012053

Error Analysis of Indonesian Junior High School Student in Solving Space and Shape Content PISA Problem Using Newman Procedure

U Sumule [1,*], S M Amin[2], and Y Fuad [2]

[1]Postgraduate in Mathematics Education, Universitas Negeri Surabaya, Ketintang, Surabaya,60231, Indonesia
[2]Mathematics DepartmentUniversitas Negeri Surabaya, Ketintang, Surabaya, 60231, Indonesia

[*]Corresponding author: ulessumule@mhs.unesa.ac.id

Abstract. This study aims to determine the types and causes of errors, as well as efforts being attempted to overcome the mistakes made by junior high school students in completing PISA content space and shape. Two subjects were selected based on the mathematical ability test results with the most error, yet they are able to communicate orally and in writing. Two selected subjects then worked on the PISA ability test question and the subjects were interviewed to find out the type and cause of the error and then given a scaffolding based on the type of mistake made.The results of this study obtained the type of error that students do are comprehension and transformation error. The reasons are students was not able to identify the keywords in the question, write down what is known or given, specify formulas or device a plan. To overcome this error, students were given scaffolding. Scaffolding that given to overcome misunderstandings were reviewing and restructuring. While to overcome the transformation error, scaffolding given were reviewing, restructuring, explaining and developing representational tools. Teachers are advised to use scaffolding to resolve errors so that the students are able to avoid these errors.

1. Introduction

PISA problem items are aimed for measuring the ability of mathematical literacy. Mathematical literacy is the ability to use mathematical content (concepts, facts, procedures and tools) in real life situations [1]. Literacy refers to students' capacity to apply knowledge and skills in key subjects to analyze, reason and communicate effectively as they identify, interpret and solve problems in a variety of situations [2]. There are four categories of mathematics content in PISA i.e. change and relationship, space and shape, quantity and uncertainty, and data [4]. The ability of Indonesian students is in low level in solving PISA problems related to geometric content, particularly in the understanding of space and shape. For example on solving PISA Problem 2000 as presented in figure 1.

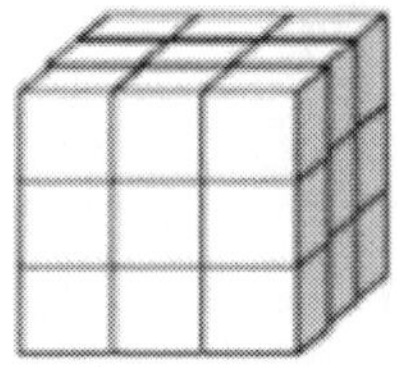

Figure 1. The big cube in the left side is painted and then cut into three parts of three different directions. It becomes small cubes as shown. How many small cubes are produced?

Only 33.4% of students answered correctly and the rest 58.79% failed to solve [3]. This fact proves that the level of students' mathematical ability, especially geometry is still low. This low ability indicates that students have difficulty. The Difficulty is the cause of error [5]. The error is a mistake or negligence [6]. Math errors students need to get attention because if not addressed immediately, the error will impact in a row to the next math problem. To correct the error, students need to find the causes of their error [7]. Tracing the errors requires some analysis. Analysis can mean the procedure of solving the whole component into small parts to see how it effects [8].

Errors can occur at any time, therefore error analysis is a major step to be taken to avoid causing errors [9]. Newman proposed five activities to find the errors of the students in solving the given problem or called Newman's error categories, namely reading errors, comprehension errors, transformation errors, process skills errors and encoding errors [10]. In the problem solving process there are two types of obstacle that can prevent students from finding the right answer, namely: the first obstacle is the language problem and the conceptual understanding associated with the ability to read and understand the meaning of the given problem, and the second obstacle is the problem of mathematical processes consisting of transformation, process skills and accuracy of writing the final answer [11]. The Newman procedure presents a popular approach to teachers and offers a framework that is easy to understand and supported by study to find out the mistakes that students make in solving the problem [9]. The Newman procedure is a method of analyzing errors in the troubleshooting process [11]. Newman hierarchy provides a fundamental framework for sequencing important steps in the process of solving problems [12]. If errors and causes of errors are known then an attempt is needed to resolve the errors of the student in solving the problem. In education, scaffolding is defined as a help provided by a more competent friend or adult [13]. Scaffolding is a support strategy for children's convergent development areas [14]. Three levels of scaffolding which are support in mathematics learning are Level-1 called Environmental Provisions, Level-2 which consist of explaining, reviewing, and restructuring and Level-3 which are developing conceptual thinking [15].

This study found that the kind of mistakes made by students is a mistake of comprehension and transformation error. Comprehension errors occur because students can not understand and interpret key words in the matter and students do not read the questions carefully so that there is unreadable information. Attempts to overcome the errors of comprehension is to provide the scaffolding that is reviewing and restructuring. The transformation error occurs because the student can not determine the formula or plan the solution to solve the problem. Efforts to overcome the misunderstanding is to provide scaffolding in the form of reviewing, restructuring, explaining, developing representation tools [15]. For that teachers are advised to use scaffolding to overcome errors so that later errors do not happen again.

2. Methods

This study used the qualitative method. This study aimed to identify the type of error and the cause of student error in solving the problem PISA content space and shape.

2.1 Subject of study

This study was conducted in Junior High School. Two students (EI and VA) were selected from 23 students of class VIII to be subjected to study. The selection of study subjects was conducted based on the results of the ability tests given to all students. The subjects of the study selected 2 low-ability students in completing the ability test.

2.2 Study instruments

The three instruments used in this study were ability test items, PISA test items, and interview guidelines. The instrument is described as follows.

2.2.1 Test of Ability. The ability test is used to select a study subject. The problem of the ability test is taken from the question of National Examination of geometry material [15]. Five problems were given to the students in the form of a description.

2.2.2 PISA Test Items. PISA test items that was used is taken from PISA 2012 of space and shape content. The test consist of four problems. The first problem is the personal context, the second issue is the general context, the third problem is in the scientific context, and the fourth problem is about the work context [15]. The type of the question are multiple choice and essay. From the student's answer to the problem is expected to be revealed the types of errors and causes of error.

2.2.2 Interviews Guidelines. Interviews were conducted on the subject after the students completed the PISA problem. The interview guidelines used were adapted from the five Newman questions [9]. The interview form in this study is an unstructured interview. Interviews were conducted to find out and dig deeper the types of mistakes and causes of student error in solving the Pisa problem, as well as to provide assistance in the form of scaffolding to solve the missed problem.

2.3 Collection and processing of data

2.3.1 Data collection. Data collection was done by giving the PISA test. After the results of the test obtained an unstructured interview was conducted to identify the types and causes of mistakes made by students in solving the problems of PISA. All interview activities were recorded for transribed.
2.3.2 Provision of Scaffolding. The interview results and student answer sheets are then analyzed to determine the type of error and the cause of the error, as a basis for providing scaffolding. Scaffolding to students is based on the type of student error in completing the PISA test.
2.3.3 Data processing. The recording of the interview is transcribed to make the transcript, as well as to facilitate the categorization of the type of mistake made by the subject in solving the problem. Error analysis is done from the students' result and the subject interview in answering the question.

3. Results and Discussion

Students' errors make in solving the PISA problem of space and shape content are presented in Table 3.1 below.

Table 3.1 The types of error that students make

Subject names	Problem			
	1	2	3	4
EI	T	T	C	C
VA	C	C	C	C

Note: C = Comprehension; T = Transformation;

Based on Table 3.1 it can be seen that the type of mistakes made by students in solving the problem is comprehension and transformation error, the results of study in accordance with study Wijaya, A., van den Heuvel-Panhuizen.M, Doorman. M, & Robitzch.A [20], other than that Ismail Hj Raduana[19].The following discussion of the types of mistakes made by students in this study and efforts to overcome these errors.

3.1Comprehension Error

The comprehension error is done by the VA subject on all questions and EI subjec is done on item 3 and number 4.

3.1.1 Comprehension error done by VA on question number 1. For example, the VA subject's Comprehension Error can be seen in Figure 2

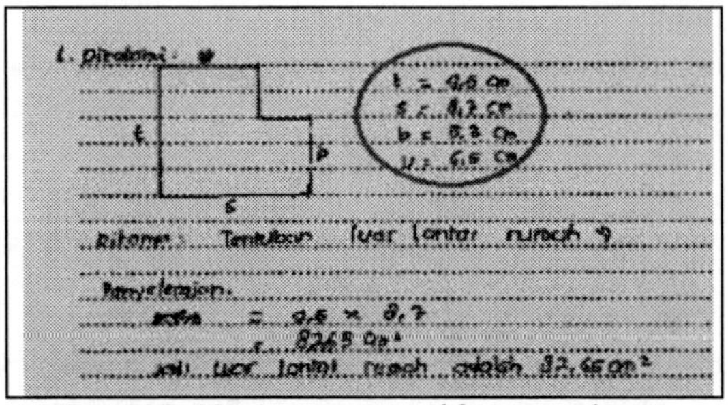

Figure 2. VA error on problem number 1

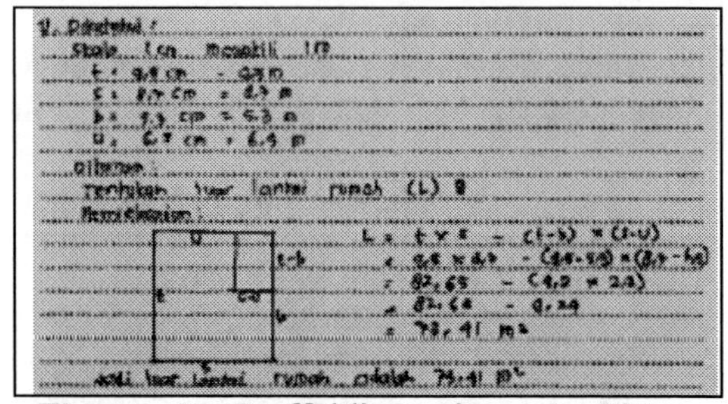

Figure 3. Scaffolding given to VA on question number 1

Based on Figure 2, it can be seen that the VA error is caused by not carefully observing the image so it can not read the scale of 1cm representing 1m contained in the picture. To overcome the misunderstanding of comprehension is given scaffolding in the form of reviewing and restructuring, Reviewing is done by asking subjects to re-read the problem slowly and carefully observe the image to find the scale on the picture, and asking the students to write what is given and what is asked from the problem. Restructuring is done by doing question and answers method for compiling and rebuilding knowledge about a wide rectangle. Developing representational tools that ask students to re-draw the rectangle with the help of adding particular segment in order to understand that on the figure actually there are two different rectangles. Explaining how to determine the floor area of the house is to reduce the area of the two rectangles also be one of the scaffolding given to the student.

3.1.2 Comprehension errors done by on second problem. Based on the results of VA work on second problem it can be seen that the VA does not write the height of the platform board contained in the picture. The error is caused because the VA is not careful observing the image. To solve the VA comprehension error in question number two is given scratching in the form of reviews to ask students to re-read the problem slowly and observe the image carefully in order to find the height of the platform board on the picture. Restructuring is done by doing question and answer to reorder knowledge about the radius of the circle, find the relation of radius, the height of platform board with the location of point M above river surface. Reviewing is done by asking student to re-read what the question asked then make a conclusion from the calculation.

3.1.3Comprehension errors done by VA and EI on the third problem. Based on the results of the work EI and VA on the answer sheet it can be seen that the VA and EI made a mistake of comprehension caused by not writing on the answer sheet that the angle in one round is 360^0. To overcome the errors EI and VA are given scaffolding. The scaffolding is reviewing that is asking students to re-read the problem slowly and carefully then emphasize important information that is circle. Restructuring is to remind the subject of the angle formed in one round, then connect it to the number of the doors, the question and answer to direct the students to determine the angle formed by the two wings of the door by dividing 360^0 by three. Reviewing is asking students to re-read what is asked and make conclusions related to the calculation results.

3.1.4 Error of comprehension done by EI and VA in question number 4. Based on the results of the work of EI and VA on the answer sheet it can be seen that VA and EI made a mistake of comprehension caused by not writing down the size of the sies contained in the question. To overcome the error, then given scanning in the form of reviewing to read slowly and observe images carefully and thoroughly to find important information used to solve the problem. Explaining is to explain the

IOP Conf. Series: Journal of Physics: Conf. Series **947** (2017) 012053 doi:10.1088/1742-6596/947/1/012053

relationship formed by the sides of a right triangle in order to determine the size of one side if there are two known sides. Developing representational tools is to ask students to draw the shape of the garage roof which is the same two rectangles, and the restructuring is the Q & A to rebuild the understanding that the area of the garage roof is the width of the rectangle multiplied by two.

3.2 Transformation error

Based on Table 3.1 it is known that transformation error is done the subject of EI in first and the second questions.

3.2.1 Error transformation done by EI in problem number 1. The result of EI's work on the first problem can be seen in figure 4.

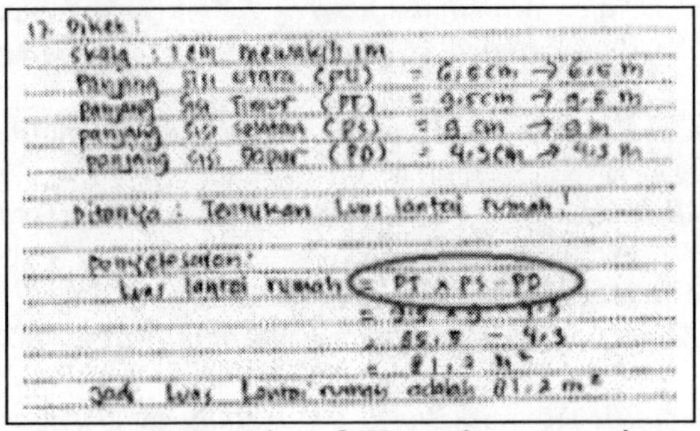

Figure 4. Results of EI work on question number 1

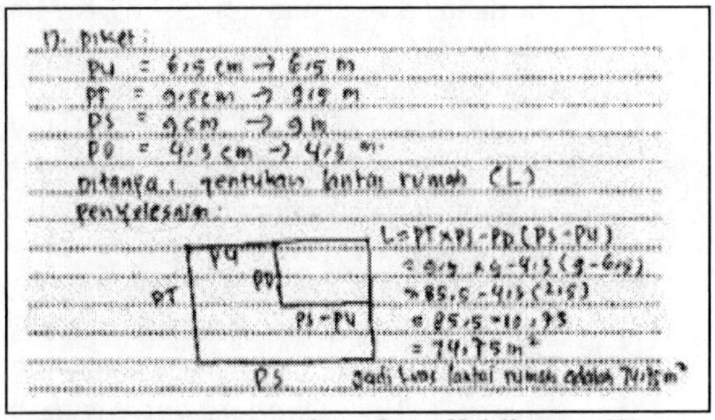

Figure 5. The results of scaffolding to EI on question number 1

Based on the results of the work of EI in Figure 4 it is known that the EI misunderstood that caused the wrong method or formula to determine the floor area of the house. To overcome the transformation error, EI given scaffolding in the form of reviewing that is asking the subject to write what is known and asked again and observed the picture. Restructuring is a question and answers for compiling and rebuilding knowledge about a wide rectangle. Developing representational tools that ask students to draw back the plan with the help line in order to understand that on the plan there are two different rectangles. Explaining how to determine the floor area of the house to reduce the area of the two rectangles is also one of the scaffolding. Restructuring is a question and answer to guide students to do the calculation process. Reviewing is asking students to re-read what is being asked from the question. Restructuring is asking the subject to make a conclusion regarding the results obtained. The result of work after giving the scaffolding can be seen in figure 5.

3.2.2Error transformation done by EI on second problem. Based on the result of the work of the EI subject it can be seen that the EI subject made a transformation error caused by the subject EI incorrectly determining the method for determining the location of point M with the surface of the stream. To overcome the error is given scaffolding in the form of Reviewing is asking the subject to write down what is known and asked from the problem. Restructuring is question and answer to reorder knowledge about the radius of the circle, find the relation of radius, the height of platform board with the location of point M above river surface. Reviewing is done by asking student to re-read what the question asked then find the meaning of the calculation results. Restructuring is done by asking student to make a conclusion from the calculation results.

4. Conclusion

This study found that the kind of errors made by students are comprehension and transformation error. Comprehension errors occur because students can not understand and interpret key words in the problem and students do not read the questions carefully so that there is unreadable information. Attempts to overcome the errors of comprehension is to provide the scaffolding that are reviewing and restructuring. The transformation error occurs because the student can not determine the formula or plan to solve the problem. Efforts to overcome the transformation errors is to provide scaffolding in the form of reviewing, restructuring, explaining, developing representation tools. The implication for teachers is teachers are suggested to use scaffolding to overcome errors so that later errors do not happen again.

5. References

[1] Stacy K & Turner R 2014 *Assessing Mathematical Literacy. ThePISAExperience* (New York: Springer Cham Heidelberg Dordrecht)
[2] OECD 2014 *Education at a Glance 2014: OECD Indicators* (OECD:OECD Publishing)
[3] Wardhani S & Rumiyati 2011 *Instrumen Hasil Belajar Matematika SMP: Belajar Dari PISA dan TIMSS* (Yogyakarta: Pusat Pengemabangan dan Pemberdayaan Pendidikan dan Tenaga Kependidikan (PPPPTK) Matematika)
[4] OECD 2016 *PISA 2015 Assesment and Analytical Framework: Science, Reading, Mathematics, and Financial Literacy* (Paris: OECD Publishing)
[5] Soedjadi R 1996 *Diagnosis Kesulitan Siswa Sekolah Dasar Dalam Belajar Matematika* (Team Basic Science LPTK Dikti)
[6] Rajasah S 2003 *Kamus Lengkap Bahasa Indonesia* (Surabaya: Mitra Cendekia)
[7] Subanji 2015 *Teori Kesalahan Konstruksi Konsep dan Pemecahan Masalah Matematika* (Malang:Universitas Negeri Malang)
[8] Ritcehy T 1996 *Analysis and Synthesis on Scientific Method-Based on a Study by Bernhard Riemann* (Online) https://www.swemorph.com/pdf/anaeng-r.pdf diakses Sabtu, 21 Januari 2017, pkl 19.00
[9] White A L 2005 Active Mathematics in Classrooms: Finding Out Why Children Make Mistakes *Square One* **15** (4) pp 15-19
[10] Newman M A 1977 An Analysis of Sixth-grade Pupils' Error on Written Mathematical Task *Victorian Institute for Educational Study Bulletin* **39** pp 31-43
[11] Prakitipong N & Nakamura S 2006 Analysis of performance of grade five students in Thailand Using Newman Procedure CICE Hiroshima University *Journal of International Cooperation in Education* **9** (1) pp 111-112
[12] Clements M A Ken & Ellerton N F 1996 *The Newman Procedure for Analysing Errors on Written Mathematical Tasks* (The University of Newcastle: Faculty of Education)
[13] Slavin R E 2000 *Educational Psychology Theoryand Practice* (United States: A Person Education Company)
[14] Chang K, Chen I, & Sung Y 2002 The effect of concept mapping to enhance textcomprehension and summarization *The Journal of Experimental Education* **71**(1) pp 5-23
[15] Anghileri J 2015 *Scaffolding Practices that Enhance Mathematics Learning* (Cambridge: University of Cambridge)
[16] PISA released items-mathematics-OECD*www.oecd.org/**pisa**/38709418.**pdf*** Terjemahkan lamanini (diakses Jumat, 20 Januari 2017, pk 20.00).
[17] Kementerian Pendidikan dan Kebudayaan Republik Indonesia 2015 *UjianNasionalTahunPelajaran 2014/2015* (Jakarta: BNSP)
[18] Singh P, Rahman A A, Hoon T S 2010 *The Newman Procedure for Analyzing Primary Four Pupils Errors on Written Mathematical Tasks: A Malaysian Perspective. University Technology MARA, Shah Alam, Malaysia*
[19] Raduana I 2009 Error analysis and the corresponding cognitive activities committed by year five primary students in solving mathematical word problems *Mathematics Department, Institute of Teacher Training Dato' Razali Ismail Campus 21030 Kuala Terengganu Malaysia*
[20] Wijaya A, an den Heuvel-Panhuizen M, Doorman M, & Robitzch A 2014 *Difficulties in solving context-based PISA mathematics task: An analysis of students' error* (Montana: The Mathematics Enthusiast)

Developing Media Module Proposed to Editor in Editorial Division

A Kristanto[1], Mustaji[1], A Mariono[1], Sulistiowati[1], and D W Nuryati[1]

[1] Department of Educational Technology, Universitas Negeri Surabaya, Indonesia

Email: andikristanto@unesa.ac.id

Abstract. In this era of technology in Indonesia, various publishers introduce themselves and participate in advancing the quality of education through the publication of various books as the learning sources. One of the publishers is PT. JP Press. In compiling the learning sources, we found some problems that are left unresolved by the editor. The purpose of this research is to overcome the problems that exist in PT. JP Press by developing media module. This development research uses the ADDIE model. The types of data used in this study are qualitative and quantitative data obtained based on the results of structured interviews with material experts and media experts, as well as the editorial response questionnaire provided for individual try-out and small group try-out. Therefore, it can be concluded that the medium of elementary school supplementary module proposed to the editors of PT. JP Press is valuable to be used in the teaching and learning activities.

1. Introduction

Publishers are the people in the industry that concentrate on producing and reproducing a literature and information or an activity of making information publicly available. Publishers on its systems are distinguished as public (conventional) publishing and self-publish system or indie publishing system if the author is the publisher.

In the development of the technological era in Indonesia, various publishers introduced themselves and participated in advancing the quality of education through the publication of various books as a source of learning. One of the participating publishers is PT. JP Press. JP Publisher became one of the book publishers for elementary to senior high school. The books published by PT. JP Press are the result of the performance of the editors and the author with various educational background. The editor is the party that represents the publisher and is obligated to ensure that the printed text becomes a book. The editor at JP publishers also has task of compiling books according to their respective fields.

The editor at JP has the following tasks: First, the editor is assigned as a team that has to compile the book based on the script. Second, the editor must check the content of the text whether it is in line with the applicable educational curriculum. Third, the editor need to be able to finish writing the book based on the specified time. Fourth, the editor must compile the book to be in production stage based on the quality control.

Based on the results of interviews with the editor-in-chief of PT. JP Press, Mr. Fifkaindi, the researchers found some difficulties in book production activities, especially the production of elementary school supplementary materials. The editorial division has 15 editors in total. Among them, 11 people have the task of preparing primary school supplementary materials. Given that number, nearly

50% of the editors, six editors have difficulty in preparing primary school supplementary materials. The editor has some difficulties such as, determining the structure of the supplementary arrangement, determining the contribution of the item test, and writing the word structure. It makes the elementary school supplementary materials that have been drafted by the editors to be re-edited through the quality control and editing activities takes longer time.

The impact is that the elementary school supplementary materials production is delayed, so the distribution of primary school supplementary materials decreases. Primary school supplementary materials cannot be produced and distributed on time. The impact of production delay is that the editorial's decision in decreasing the quantity of production that later on will cause losses to the company.

Based on the background description, the media needed to overcome the existing problems. Educational Technology is a study and practice to facilitate learning and improve performance through three domains of educational technology areas: creating, managing and using, technology-appropriate processes of resources. To solve learning problems, educational technology comes with one way to facilitate learning and improve performance through creating domains. In the related field, creating refers to research, theory, and practice in the making of learning materials, learning environments, and learning systems in different settings including formal and non-formal ones.

The appropriate media to overcome the above problems is the module. It is because the module is in the form of printed material designed for the learners is that the learners are able to study it independently. There are some advantages of using the module in the learning process. Because the module focuses on the individual capabilities of the editor, there is control over the learning outcomes of the editor through the use of competency standards in each module done by the editor. From that point, the editor is able to know the relationship between learning and learning outcomes obtained.

The module developed by the editor can be used for learning, according to the editor's needs. Based on the interview with the chief editor of PT. JP Press, it is known that there is no module that presents the basic school supplementary module. Therefore, the researchers considered it is needed for editor to develop elementary school supplementary module in the editorial division of PT. JP Press.

2. Context and Review of Literature

2.1. Media

A medium, conceived is any person, material or event, that establish condition which enable the learner to acquire knowledge, skill, and attitude [1]. Media is anything that can be used to channel the message (learning materials), so it can stimulate the attention, interest, thoughts, and feelings of students in learning activities to achieve learning objectives [2].

2.1.1. Media utilization. Generally, the media has a usage to clarify the message which is not conveyed verbally. In other words, it is knowing the message without knowing its meaning. It is also used to overcome the limitations of space, time, energy, and sense power; which cause passion of learning as well as direct interaction between editors with learning resources. Moreover, it allows self-study editors according to their visual, auditory, and kinaesthetic talents and abilities. Lastly, it provides the same stimulus, equate experience, and generate the same perception [3].

2.1.2. Media classification. The types of media are divided based on three main elements, namely: sound, visual, and, motion. Based on the three elements, the media is divided into several groups, namely: audio media, print media, silent visual media, visual media motion, semi-motion audio media, semi-motion media, silent audio-visual media, audio-visual media motion [4].

2.1.3. Media selection. The selection of media should pay attention to several principles, namely clarity of intention and selection of media, in which for the purposes of entertainment, general information, learning, and so forth are involved. In addition, the media relation, which involves knowledge of the

IOP Conf. Series: Journal of Physics: Conf. Series **947** (2017) 012054 doi:10.1088/1742-6596/947/1/012054

nature and characteristics of the media to be selected are also accounted. A number of media can be compared because of some options that are more appropriate to the learning objectives [5].

2.2. Media module

A module is a systematically organized resource with an easy-to-understand language by the editor according to the knowledge level and age of the editor. Thought that way, the editors can learn independently with the help or minimal guidance from the tutor [6]. A module is a complete measuring tool and a unity of a program that can measure goals [7]. The module is a unified overall item consisting of a series of learning activities and has actually resulted an effective learning outcomes in achieving clearly defined specific objectives [8].

2.2.1. Module superiority

A module has the upper hand and advantage when it is used in learning activity. The use of module is focusing on individual editor skill. There is also control on every result achieved by the editor using the competency standard on each module that should be attained by them. The curriculum relevancy is showed by the goals and the way to achieve it, so that the editor can know the relation between learning activity and the results that have been achieved [9].

2.2.2. Module outlining process

Module is a learning media that works the same way as a teacher in a direct learning activity. Thus, module-drafting needs to be standardized based on the learning principles, the way the teacher/tutor teaches, and what the editor learns. The principles of the module outlining are explained based on the learning principle. Learning is the process of behaviour changing which is caused by stimulus coming from the environment [10].

2.2.3. Media module eligibility

Media feasibility of the module carried out by conducting a feasibility test by media experts, material experts, and editors. Material experts and media experts examine media modules before the media is tested or tried out to the editor. Meanwhile, the editor feasibility tests are performed on individual try-out and small group try-out. Individual try-out is conducted to determine the response of the editor as a user before it is tried out to users in small groups [11].

3. Method

The development model used is the ADDIE developmental model. The ADDIE developmental model has several stages: Analysis, Design, Development, Implementation, Evaluation.

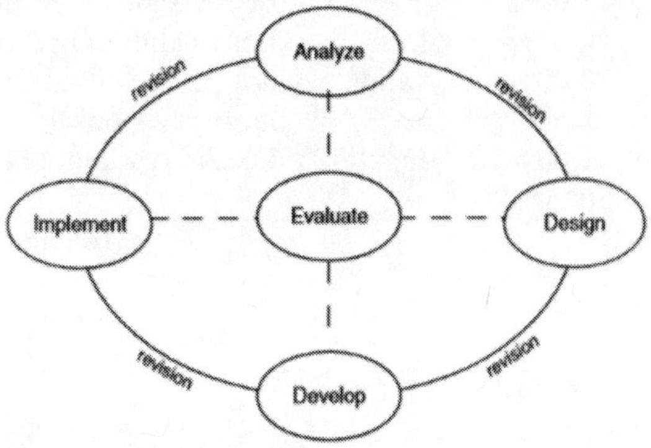

Figure 1. Developmental Phase of ADDIE Model

IOP Conf. Series: Journal of Physics: Conf. Series **947** (2017) 012054 doi:10.1088/1742-6596/947/1/012054

The stages of the development of the ADDIE model:

Table 1. Developmental Stages of the ADDIE Model

Phase Development	Activity
Analyze	Performance analysis and needs analysis.
Design	Design and manufacture of draft modules.
Develop	Turning design specifications into physical form that produces prototype products.
Implement	Implementing the module and carrying out try-out in writing activities for editors.
Evaluate	Carrying out assessment by using formative evaluation and summative evaluation.

The subjects of this study were material experts, media experts, and writer editors used in the try-out consisting of 11 people. The data collection methods used in this research and development are structured interviews, questionnaires, and documentation. The data analysis technique used is module validation analysis and module feasibility test. The module is analyzed by using rating scale with four optional answers: "very appropriate", "appropriate", "less appropriate", and "inappropriate". The level of eligibility of the module is measured using the criteria of the validity of teaching materials. It can be seen in Table 2.

Table 2. Criteria of material validity

No.	Percentage (%)	Criteria	Level of validity
1.	85.01%-100.00%	Very valid	Very valid or usable but needs to be revised.
2.	70.01%-85.00%	Valid	Valid or usable but needs to be revised.
3.	50.01%-70.00%	Less valid	Less valid and recommended not to be used because of large revisions.
4.	01.00%-50.00%	Invalid	Invalid or cannot be used.

(Adapted from [12])

4. Discussion

The developmental stages of the ADDIE model:

4.1. Analyze

Stage analysis is a process of defining what will be learned by the editor, that is to analyze needs and identify problems. Based on the result of interview with the editor-in-chief of PT. JP Press namely Mr. Fifkaindi, the researchers found some difficulties in the book production activities, especially the

IOP Conf. Series: Journal of Physics: Conf. Series **947** (2017) 012054 doi:10.1088/1742-6596/947/1/012054

production of elementary school supplementary materials or modules. The editorial division has totally 15 author editors. Among them, 11 people have the task of preparing primary school supplementary materials. Given that number, nearly 50% of the editors or around six editors have difficulty in preparing primary school supplementary materials. The editor has some difficulties such as determining the structure of the supplementary arrangement, determining the contribution of the item, and writing the word structure. It makes the elementary school supplementary materials that have been drafted by editors to be re-edited by quality control and editing activities takes longer time.

4.2. Design
This stage is done to design the design. The researchers compiled the material to determine the supplementary structure of the structure, determining the contribution of the item, and writing the word structure.

4.3. Development
Development is a process for making the designed design into reality. In this step, it is related to the compilation of the module material up to the printed modules that suit your needs. One important step in this development stage is the validation to the material experts and media experts before being implemented.

4.4. Implementation
Implementation is a tangible step to implement the created media. It means that, at this stage, the module is implemented according to the scenario or initial design. Once the product is ready, it can be piloted through small groups then evaluated and revised. Then, the test can be done in a large group then re-evaluated and revised.

4.5. *Evaluation*
The evaluation stage can occur in each of the four stages. Evaluation occurring in each of the above four stages is called formative evaluation because the evaluation objective aims at the need of revision. The type of evaluation assesses the overall effectiveness of the material or product.

This development produces a product in the form of media module that can help the authorized editor in PT. JP Press for preparing elementary school supplementary materials. After going through several stages of developmental and try-out usage, this media module is considered feasible to be used in learning activities. Here is a discussion based on the data obtained from the try-out and revision results obtained, including:

The data obtained through the material expert gets a percentage of 73.75% from the material expert 1. Based on the criteria of the validity of the teaching materials, the percentage is categorized valid because the media is feasible although it needs some parts to be revised. While the material expert 2 gets a percentage of 90%, based on the criteria of the validity of teaching materials, the percentage is in very valid category, so the media is worth using. The revisions made to refine the module medium include providing an overview of the supplementary elementary school composition and adding some sample questions based on the basic competency translation form.

The data obtained through media experts obtained a percentage of 88, 64% from media experts 1 and 86.53% for media experts 2. Based on the criteria of the validity of teaching materials, the two percentages fall into the very valid category, so the media is feasible to be used and ready to be utilized in the related field for learning activities. The media produced is feasible to apply but there are some things that need to be revised to improve it. The revision is about changing the colour of the cover and module layout in accordance with the identity of JP, adding Unesa logo, adding the author's name, biography, and module summaries, changing the location of the target audience, changing the shape of the cover image, creating the companion, and improving the systematics of module writing. Therefore, based on the results of the combined validity between media experts and material experts, it obtained a percentage of 84, 69% which is included in the valid category. Therefore, the media produced is feasible to be used but some parts still need to be revised.

The data obtained from individual try-out have a combined percentage of eligibility of 80, 21%. Based on the criteria of the validity of instructional materials, the percentage is in valid category, so the media is feasible to be used but still needs to be revised. The data obtained from small groups try-out have a combined percentage of eligibility of 85, 61%. Based on the criteria of the validity of teaching materials, the percentage is in very valid category. Therefore, the media produced is feasible to be applied in the learning activities. The data obtained on a large group try-out have a combined percentage of eligibility of 88, 91%. Based on the criteria of the validity of teaching materials, the percentage is in very valid category. Therefore, the media produced is feasible to be applied in the learning activities.

Based on the data obtained above, the primary school supplementary module media for the editor of the author in the editorial division of PT. JP Press that has been developed can solve the problem exposed in chapter I, that is, developing elementary school supplementary media module for the editor in the editorial division of PT. JP Press which is important for the editors of PT. JP Press.

5. Conclusion

Based on the results of research and media discussion of basic school supplementary module for the authorized editor in the editorial division of PT. JP Press can be concluded that the elementary school supplementary module is developed by following steps of developmental research from ADDIE model which is done by using five steps of development of ADDIE model that includes 1) analysis, 2) design, 3) development, 4) implementation, and 5) evaluation. Therefore, the final product design in the form of elementary school supplement module for the editor of PT. JP Press, which is in accordance with the needs, is proven to be valuable to be used in the teaching and learning activities.

Acknowledgments

Thanks to Prof. Dr. Mustaji, M.Pd who gave the first review for this paper, Dr. Yuni Sri Rahayu who gave the chance to participate in scientific writing training.

References

[1] Gerlach V S and Ely D P 1980 *Teaching and Medias: A Systematic Approach* (New Jersey: Prentice-Hall, INC, Engliwood Cliffs)
[2] Kristanto A 2016 *Media Pembelajaran (*Surabaya: Bintang Surabaya) p 5
[3] Susilana R and Riyana C 2008 *Learning Media* (Bandung: Curriculum and Technology Education, FIP, UPI)
[4] Warsita B 2008 *Learning Technology, Platform and Its Application* (Jakarta: Rineka Cipta)
[5] Mahnun N 2012 Learning Media: A Study of Media Selection Measures and Its Implementation in Learning" *Journal of Islamic Thought* Vol **37** No 1 pp 27-33
[6] Prastowo A 2012 *Creative Guide Creating Innovative Instructional Materials* (Yogyakarta: Diva-Press)
[7] Sukiman 2012 *Development of Learning Media* (Yogyakarta: Pedagogia)
[8] Mbulu J 2001 *Pengajaran Individual* (Malang: Yayasan Elang Emas)
[9] Mulyasa E 2007 *Competency Based Curriculum:* Concepts, Characteristics, and Implementation (Bandung: PT. Remaja Rosdakarya)
[10] Asyar H R 2012 *Creative Developing Learning Media* (Jakarta: Reference Jakarta)
[11] Ratnasari A 2016 Development of Interactive Learning Media Occupational Safety and Health (K3) in Electricity Study Program at Vocational High School *E-Journal Yogyakarta State University* Vol **6** No 1 pp 83-91
[12] Pambudiono A and Suarsini E 2016 Development of Research-Based Biotechnology Textbooks Bioremidiation of Heavy Metal Cadmium For Undergraduate Students of Biology State University of Malang *Journal of Education: Theory, Research and Development* **1** pp 1077-1085

IOP Conf. Series: Journal of Physics: Conf. Series **947** (2017) 012055 doi:10.1088/1742-6596/947/1/012055

The blind student's interpretation of two-dimensional shapes in geometry

Andriyani[1,4], I K Budayasa[2] and D Juniati[3]

[1] Mathematics Education, Universitas Ahmad Dahlan, Yogyakarta, Indonesia
[2,3]Mathematics Education, Universitas Negeri Surabaya, Surabaya, Indonesia

E-mail: [4]andriyaniandriyani@mhs.unesa.ac.id

Abstract. The blind student's interpretation of two-dimensional shapes represents the blind student's mental image of two-dimensional shapes that they can't visualize directly, which is related to illustration of the characteristics and number of edges and angles. The objective of this research is to identify the blind student's interpretation of two-dimensional shapes. This research was an exploratory study with qualitative approach. A subject of this research is a sixth-grade student who experiencing total blind from the fifth grade of elementary school. Researchers interviewed the subject about his interpretation of two-dimensional shapes according to his thinking.The findings of this study show the uniqueness of blind students, who have been totally blind since school age, in knowing and illustrating the characteristics of edges and angles of two-dimensional shapes by utilizing visual experiences that were previously obtained before the blind. The result can inspire teachers to design further learning for development of blind student geometry concepts.

1. Introduction

In elementary school, visualization becomes one of the important ability used to help students in understanding spatial concepts, shapes, sizes, and distances. However, blind students lose visual aid to learn the concept of geometry. Losing of visual experience in blind students causes some difficulties in obtaining the concept of geometry directly.

According to Thinus-Blanc [1], blind students take a long time to construct mental representation of the spatial concept so that geometry is not easy to be taught. In the other result of the study, Vianna [2] also showed that students with visual impairment such as blind students have difficulty of geometry images understanding.

The difficulty of learning geometry is not only experienced by students, but also experienced by teachers who teach this subject. Although using tools in physical models, many teachers have difficulty to instruct geometry to blind students who can't use visual sense (Pritchard) [3].

On the other hand, the difficulty of blind students in geometry also caused by the limitations of geometry books in Braille format as well as geometry learning media that suits the needs of blind students. In fact, the difficulty of gaining access to information also still experienced by the blind as a minority (Abner) [4]. Even though, blind students are granted to access of information just like a normal student, like Augusto and Colford's statement [5, 6].

Several experience of difficulty by the blind students does not mean eliminating opportunities for visually impaired students to study geometry at the school since it is mentioned by Government which rule number of 72 in 1991, that Indonesian mathematic curriculum at *SLB (physical defect school)* is the same as the Indonesian mathematic curriculum with the normal students. With the same curriculum as normal students, teachers need specific strategy to teach mathematic that have abstract concept in the blind learning. For that reason, teachers must know how the interpretation of blind students related to their understanding of the geometry's concepts. The question of how the blind student's interpretation of the two-dimensional shapes as one of geometric concepts is interesting for the researcher. Therefore, this study aims to identify the interpretation of blind students of two-dimensional shapes in geometry, whereas blind students loss visual sight to observe images or geometry models. From the interpretation of blind students, it is expected to know how the understanding of blind students and obstacles are experienced, so that teachers can find strategies and learning media in accordance with the needs of geometry learning for blind students. Furthermore, from the student's interpretation, it is expected to identify how the abstract concept of two-dimensional shapes has been related with real objects that blind students experience it every day.

2. Theoretical Background

2.1 Interpreting on Cognitive Processes

According to Anderson [7], students are classified to understand a concept when they can construct the meaning of the concept that has been learned involving some cognitive processes. One of the most important cognitive processes in understanding is interpreting.

In the cognitive process of interpreting, a student is classified to be able to interpret an idea if he can interpret information or idea by communicating it to a different form of communication through the way of language. Interpreting an idea means explaining what is in his mind about the idea so that the student's interpretation can be an explanation that represents his mental picture. This interpretation depends on the relevant knowledge that the student has. Students who can interpret each part of the idea with their knowledge will be able to relate these sections to a more complete thinking activity. This student's interpretation can be seen from an internal representation that describes something in his mind as suggested by Izsak [8]. In line with that, Michaelidou [9] also proposed that representation has an important role to represent students' mathematical ideas in understanding mathematical concepts of students.

To be able to interpret a concept which is learnt, each student has a different ability that is influenced by his or her cognition, especially conception. The factors that affect individual cognition are: the physical and social environment, physiological structure, willingness and goals, also experiences in the past Krech [10].

2.2 Bind Student's Concept Development

Visually impaired blind physiology is the loss of the senses of vision that impact on differentiation in the development of the concept. According to Ashkenazy [11], visual impairment will make a visually impaired person unable to develop his social behaviour just like any person who uses visual experience.

Concerning to the development of the concept of the blind according to Dick and Kubiak [12], the development of a blind concept depends on the level of blindness and age of blindness so that a blind person who had a visual experience before they blind, can at some point utilize his or her visual experience to understand a particular concept. A blind student will develop his conception (especially for the concept of physical objects) more through tactual experience, while the normal students through visual experience (Lowenfeld) [13].

For visually impaired students, tactual experience is gained by using palpable as a surrogate sensitivity that has a special sensitivity in the acquisition of visual information, such as a shape of the image or exploring the physical objects. The tactile or tactile capabilities involved in information processing by a visual impairment are influenced by tactual perceptions: synthetic perceptual perceptions and perceptible analytic perceptions as Lowenfeld's statement [13].Perception of the blind

IOP Conf. Series: Journal of Physics: Conf. Series **947** (2017) 012055 doi:10.1088/1742-6596/947/1/012055

students is obtained by the functioning of the sense of touch. This palpable sense is the widely use sensory impairment to help him modify and manipulate nearby objects as conveyed by McLaughlin [14].

The importance of tactual perception as a special form of belonging to this blind student is in line with the results of the research conducted by Argyropoulos [15] which showed that in the geometric stage of blind students, there is a complexity of tactual perception as a uniqueness in thinking activities of blind students related to the development and understanding of a concept formed. Furthermore, the suitability of the geometry of thinking stages of students related to their understanding of a concept of form, such as planar shapes, with Van Hiele's theory, in accordance to the research from Gray and Tall [16]. From the results of his research related to students' activities in understanding the concept of planar shapes, students recognize the build-up triangle or quadrilateral by observing the characteristics of the same or different from of the object in the form of the planar shapes. While in the other hand Tilman [17] showed that blind students are less able to integrate all types of facts that they have learned so that any information received is as if stored in a separate frame of reference from other information because they lose sight that facilitates them to connect different experiences - relationships and relationships that help them selectively harness experiences when the formation of perceptions happened, especially regarding the understanding of objects.

The activity of interpreting in this study refers to the process of cognition involved in understanding of the concepts proposed by Anderson [7]. In interpreting, the student explains the mental picture that is in his mind through his own way and language.

3. Method

This research was an exploratory study with a qualitative approach to identify the blind student's interpretation of two-dimensional shapes in geometry. The question of interviews consisted of questions about subject's interpretation of two-dimensional shapes especially square and rectangle. Subjects chosen were employed as a source of qualitative data, and they are fourth-grade student who experiencing a total blind since the fifth grade of elementary school from SLB-A YPAB *Tegalsari* Surabaya city. Data on the blind student's interpretation of the two-dimensional shapes were collected by means of task-assisted interview and the interviews were carried out two times by using equivalent problems on two different times. Student is assigned to give examples of the two-dimensional shapes that they know and to give the reason. When he mentioned the examples, student asked to illustrate the example in their own way, for example by drawing the shapes. The data collected from the interviews were then classified, reduced and validated by using triangulation methods to yield credible data. The credible data were analysed by using qualitative research analysis data method including data display interpreting and conclusion drawing as Miles and Huberman's statement [18].

4. Results

From the results of interviews and observations the researchers conclude that the subject analogized the square's shape with objects that they knew every day, namely the box. According to the subject, square was a flat shape, shape which has four edges with equal length and right angles. In detail, the subject explained that the four edges of the equal length were one on the left, one on the right, one on the top, and one on the bottom. The end of each edge met the other vertex. The edges that the subject recognizes as characteristic of the shape were interpreted as the shape's boundary, whereas the right angles were interpreted as an angle formed by two edges; the perpendicular and horizontal edges.

When the subject was asked about square's example that's he knew, the subject said that the floor tile surface was a square's example because the four edges of the tile are the equal length and angle of the tiles is right angles. Next, the researcher asked the subject to illustrate his interpretation of the two-dimensional shapes as described earlier. He illustrated his interpretation by using a paper and a pen to draw a square. The subject drew a vertical line from top to bottom along the edge of his index finger with a size of about 3.5 cm. Then the subject drew a horizontal line that starting from the end of the

IOP Conf. Series: Journal of Physics: Conf. Series **947** (2017) 012055 doi:10.1088/1742-6596/947/1/012055

previous line from left to right and the length is almost the same. In the same way, the subject drew the third and fourth lines, as shown in Figure 1.

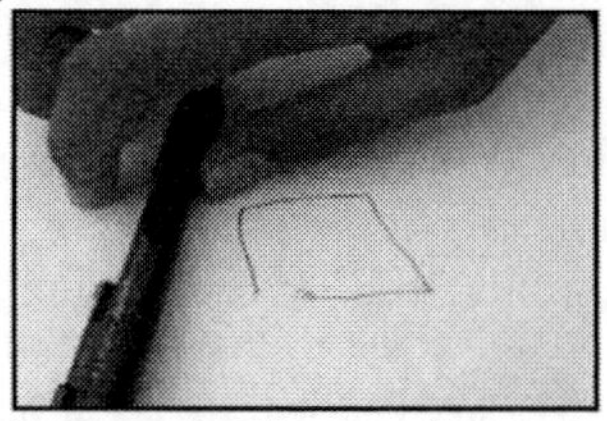

Figure 1. Figure of square

The subject explained his reason for illustrating a square like figure 1 because the shape in the figure has characteristics of the edges with equal length and angles as a square. Researcher asked the subject to show the characteristics of the edges and angles of the subject referred to in Figure 1. The subject fingered his drawing that he made previously and back to draw because there was line that did not intersect at one the vertex, as in figure 2.

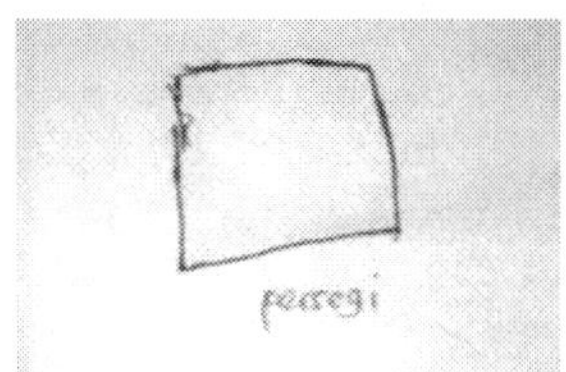

Figure 2. Figure of square after revised

The two-dimensional illustrations which are illustrated by the subject showed that the subject knew a square as a simple closed curve having right angles.

From the interviews and observations, the researcher also found that subject interpreted the rectangular shape as a flat shape, it has two opposite edges of equal length and two opposite edges of equal short, also a right angles. Each end the vertex of length edges met the adjacent end to the vertex of another short edge. The two opposite edges of equal length and equal short were interpreted as edges having sizes of equal length and parallel as shown by the subject by hand in Fig. 3.

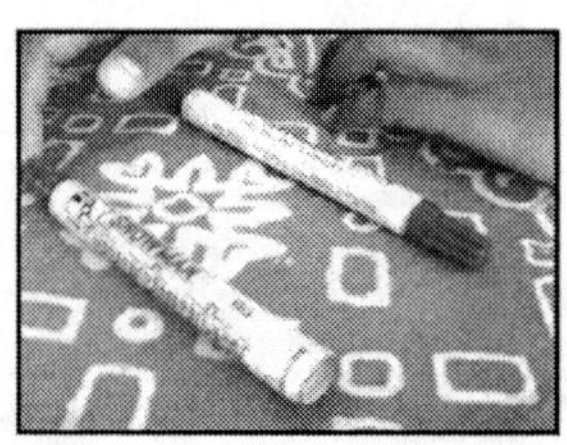

Figure 3. Figure hand-on of opposite edges by board-markers

The rectangular angle is interpreted as an angle formed by two adjacent perpendicular edges to each other as the subject represents by hand on Figure 4.

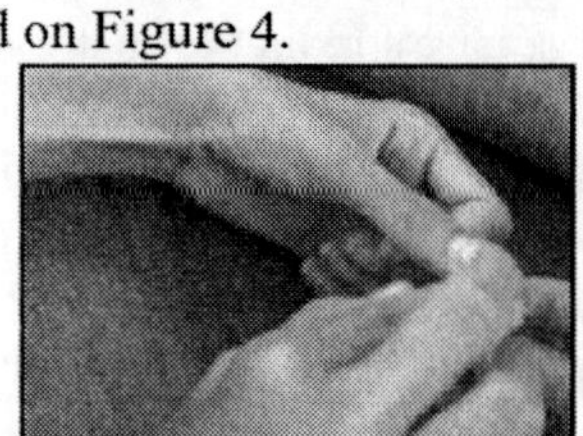

Figure 4. Figure hand-on of right angle by fingers

When the subject was asked to give the example of rectangular which has known, subject said that the door surface was a rectangular example because two edges of the door surface were the same length and the other two edges were equally short. Besides that, angles of the door surface were right angle. Next, the researcher asked subject to illustrate his interpretation of the two-dimensional shapes as described earlier in another way. The subject asked for a paper and a pen to draw rectangle. Subject drew a vertical line from top to bottom along the edge of his index finger with a size of about 4 cm. Then the subject drew a horizontal line that start from the end vertex of previous line from left to right with a size of about 2.5 cm. In the same way, the subject drew the third and fourth lines, as shown in Figure 5.

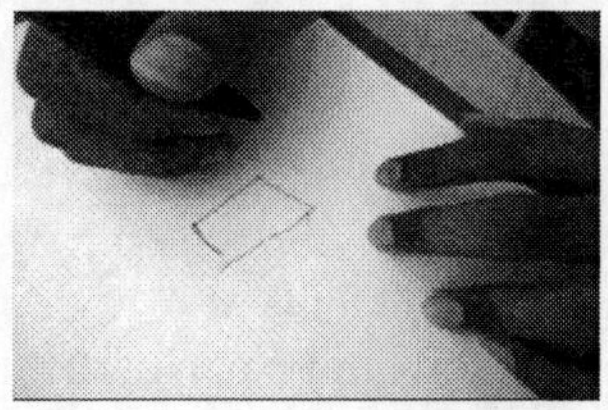

Figure 5. Figure of rectangle

The subject explained the reason to illustrate the rectangle as in figure 5 because the shape of the image has the same characteristics edges and angle as the rectangle. The researcher asked subject to show the characteristics of the edges and angles that referred to in Figure 5. He fingered the drawing and improved the previous drawing because there was a line that wasn't intersect at one vertex or does not form an angle, like showed in figure 6.

Figure 6. Figure of rectangle after revised

The two-dimensional shapes illustration showed that the subject knew about rectangular. It is a simple closed curve that having four angles likes a rectangular.

5. Discussion and Conclusion

From the results of the study it is shown that by using the language and the way the students blind totally blind after birth interpret the two-dimensional shape, especially the square and rectangle, as flat, flat, high, and have certain side and angle characteristics. The activity of interpreting this phenomenon according to the cognitive process interprets the concept of understanding which is expressed by Anderson [2].

In interpreting a two-dimensional form, student associates a real object that has similar characteristics to that two-dimensional form. This shows that students' perceptions are influenced by certain knowledge or experiences that they gained as presented by Krech [10].

Student representation by using props to illustrate the characteristics of two-dimensional shapes also indicates that students have gained certain experience with the characteristics of the represented form. In addition, a detailed explanation of students in describing the characteristics and examples of real objects that are known to form two-dimensional, also shows the role of certain experiences that affect the perception and development of student concepts.

The experience of the student is not only affects his perceptions of a particular object, but also influences the development of blind student's concepts that are highly depend on the level of blindness and age of blindness as presented by Dick and Kubiak [7]. This is an evident from the results of research showing that the visual experience of a blind student with total blindness since he born has an important role in the way he illustrates or describes a two dimensional shapes. The student draws the up-side along the index finger assumed as a ruler. The experience of using a ruler, obtained by blind students before he experienced blindness. This is a unique case that the blind student has to measure the length of the side and creates a straight line that is interpreted as the side. The shape and the length of the subject line explored by using tactual ability through its sense of touch as a substitute for the sense of sight as was presented by McLaughlin [12]. According to Argyropoulos[3], tactual perception that is used by blind students is a uniqueness in blind student's thinking activity related to its development and understanding to a concept of form.

Based on the results of the study, it can be concluded that students with total blindness after birth interpret the two-dimensional form as a flat form, having no height, having characteristics of side length and certain angle type. The representation students use to represent their interpretation of the two-dimensional form shows that the perception and development of the concept is influenced by the experience before blinding, and the representation shows the uniqueness of the blind students in understanding a two dimensional shapes.

References

[1] Thinus-Blanc C and Gaunet F 1997 *Psychological Bulletin***121**20-42
[2] Vianna C S, Barbosa P M, Rocha D F and Silva B 2008 *Teaching Geometry for Blind and* Visually*Impaired Students*
[3] Pritchard C K and Lamb J H 2012 *Journal of Mathematics Teacher***106**pp22-27
[4] Abner G H and Lahm E A 2002 *Journal of Vision Impairment and Blindness* **96**p 2
[5] Augusto C and Schroeder P 1995 *Journal of Visual Impairment & Blindness***89**p 9
[6] Colford M 2000 *Library Journal* **125** p 122
[7] Anderson L W and David R K 2001 *A Taxonomy for Learning, Teaching, and Assessing: A Revision of Bloom's Taxonomy of Educational Objectives* (Boston: Pearson Education Group)
[8] Izsák A 2003*Journal for Research in Mathematics Education* **34** pp191–227
[9] Michaelidou N, Gagatsis A, and Pitta-Pantazi D 2004 *Journal for Research in Mathematics* Education**38**pp173-192
[10] Krech D, Cruchfield R S, and Ballachey E L 1982 *Individual in Society* (Berkeley: McGraw-Hill International Book Company)
[11] Ashkenazy E C A, Ophir-Cohen M, and Tirosh E 2005 *Journal of Visual Impairment and Blindness* **99**pp478-485
[12] Dick T and Kubiak E 1997 *The Mathematics Teacher* **90**pp344-349
[13] Lowenfeld B 1973 *History of the Education of Visually Handicapped Children* (New York: JohnDay Co)
[14] McLaughlin M, Hespanha J, and Sukhatme G 2002 *Touch in Virtual Environments: Haptics and* the *design of interactive systems* (Upper Saddle River, NJ: Prentice Hall)
[15] Argyropoulos, V S 2002 *British Journal of Visual Impairment***20** 1 7-16
[16] Gray EandTall D2007 *Mathematics Education Research Journal***19** 23-40
[17] Tilman M H 1969 *International Journal for the Education of the Blind* **3** 5-7
[18] Miles M B and Huberman A M 1994*Qualitative Data Analysis 2nd an Expanded Source Book* (California: Sage Publication)

IOP Conf. Series: Journal of Physics: Conf. Series **947** (2017) 012056 doi:10.1088/1742-6596/947/1/012056

The Clinical Test of Nano gold Cosmetic for Recovering Skin Damage Due to Chemicals: Special Case

T Taufikurohmah [1], A P Wardana[2], S Tjahjani[3], I G M Sanjaya[4], A Baktir[5], and A Syahrani[6]

[1,2,3,4]Faculty of Mathematics and Science Universitas Negeri Surabaya, Indonesia
[5]Chemistry Department, Airlangga University, Surabaya, Indonesia
[6]Pharmacy Depatment, Airlangga University, Surabaya, Indonesia

E-mail: titiktaufikurohmah@unesa.ac.id

Abstract. Manufacturing of Nano gold cosmetics was done at PT. Gizi Indonesia. Clinical trials to cosmetics data supported that cosmetics are able to treat skin health which has been reported partially. For special cases, the recovery process of facial skin damage should also receive attention including cases of facial skin damage caused by chemicals such as *phenol, HCl, aqua regia* or other harsh chemicals. The problem determined whether the Nano gold is able to recover skin damage due to the harsh chemicals. This clinical trial data on the forms of early skin damage caused by *phenol* was delivered in the forms of facial photos patients. The recovery progress of facial skin condition was obtained every week for two months. The data included the forms of widespread wounds during the recovery process. This statement supported by *anova statistical* analysis of the widespread wound changing every week for 8 times. The conclusion is skin damage due to Phenol impregnation can be recovered with the use of Nano gold cosmetics for 8 weeks. This results support the manufacturing of Nano gold cosmetics for the needs of society. It also suggest that Nano gold material can be used for medicine manufacturing in the future.

1. Introduction

The empirical data of Nano gold as essential material in anti-aging cosmetics has been investigated in past researches. For example, the data of Nano gold activity as free radical scaffolding used diphenyl picryl hidrazil (DPPH) free radical test material [1, 2]. Pre-clinic test of Nano gold using mice as animal test showed that Nano gold increases fibroblast proliferation and collagen biosynthesis [3]. The activity test of Nano gold at organ wound healing showed that Nano gold increases collagen quantity and cell proliferation in liver, kidney and brain [4].

The fact that Nano gold supports enzyme as co-enzyme that catalyses many enzymatic reaction in collagen biosynthesis and cell proliferation [5]. This statement can be used as a basic of wound healing including skin aging, skin destroying caused heating, disease or exposed chemical material [6]. Recently many crime cases that occur related to chemical material usage [7]. That was correlated to easiness of having dangerous chemical materials. Criminals may sprinkel the chemical material in the face skin to attack the target or victim. Human as target of the crime was not ready to protect his/her body or skin [8]. If the handling of this case was not fast, it may cause the chemical mamany layer of tissue. The dangerous material are acid of chloride, sulphate, nitric, aqua-regia and phenol [9].

The accident with dangerous chemical often occurs in laboratory and many other places. This accident leaves scars that cannot recover by conventional medicine, hence causes the problem. It is necessary to receive something or material for resolving the bad effects. This research aims to solve the problem of skin damage caused by the phenol. Nano gold in the cosmetics formulation can be used for this skin recovery to normal condition [9]. Medical reports of volunteer wound healing are the main research data. The data analysis follows the picture analysis with imaging program. The fast picture of volunteer face is used as standard picture that the next picture will be compared with. The next pictures will be collected at periodical time in the range one week. The decrease area of wound indicates that recovery process occurs. The significant decrease area of wounds is concluded by statistical analysis Anova.

2. Method

Materials used in this research are standard materials in cosmetics formulation; that is, Stearic Acid, Lexemul CS-20, Beeswax, Laurex, Olive Oil, Dimethicone 100 cps, Methyl paraben, Propyl paraben, TiO_2, Butyl Hidroxy Toluen (BHT), Emulgent T, Mono Propilen Glycol (MPG), PG Helm, Aquadest, Aloe Vera, anti-UV B ; Octhyl para-methoxy cinnamate (OMC) and Nano gold.

For preparing the Nano gold face cream we follow the sequence of steps: (1) put Stearic Acid, Lexemul CS-20, Beeswax, Laurex, Olive Oil, Dimethicone 100 cps, Methyl paraben, Propyl paraben, TiO_2, and Butyl Hidroxy Toluen (BHT) in a bowl, (2) melt this mixture and boil the aquadest, (3) prepare Emulgent T, Mono Propilen Glycol (MPG), PG Helm and add the boiling aquadest, (4) mix the melted mixture and solution in a big bowl and stir these until forming a cream, and (5) after the cream is warm, add Aloe-vera extract, anti-UVB; Octhyl para-methoxy cinnamate (OMC) and Nano gold.

The volunteer was a man who exposes a phenol at face area. His face divides into three areas, left cheek, right cheek and the neck. Nano gold cosmetics or cream used were day cream and night cream. Day cream was applied in the morning in all face area and the neck. Then in the night applied the same areas with night cream. This treatment would be done every day and the pictures of the face was taken every week for 4 weeks [3]. The fast picture was used as a benchmarking picture. The collected data were pictures of volunteer face being taken every week. The pictures then were analyzed with imaging program. The quantitative data that resulted from imaging program then were computed statistically with SPSS program.

3. Result and Discussion

The data from week-0, week-1, week-2, week-3 and week-4 are shown in Picture 1. The area of wound healing process is marked and its area is computed. The pictures then are analyzed with imaging program. The basic of imaging program method is to compute the needed inclusive area by measuring the number of pixels of the picture. The computed areas change every day because recovery process occurs. Recovery process showed decreasing areas of the wound. Wound healing that occurred in this research showed by comparing quantitative areas every time. The analysis was done every week until four weeks. The decreasing areas indicated that wound healing process occurred. This progress can be shown Figure 1 that at the beginning picture at week-0 column faster than that at week-4 column.

By using the program, the pictures resulted in quantitative data shown in Table 1. From these decreasing areas we compared the areas of week-0 with week-1, then week-1 with week-2 and so on. These decreasing areas were very clear, but we still used statistic program to show the significant level of data. The quantitative data in Table 1 that resulted from imaging program was then computed using statistic with SPSS program.

Area	Week-0	Week-1	Week-2	Week-3	Week-4

IOP Conf. Series: Journal of Physics: Conf. Series **947** (2017) 012056 doi:10.1088/1742-6596/947/1/012056

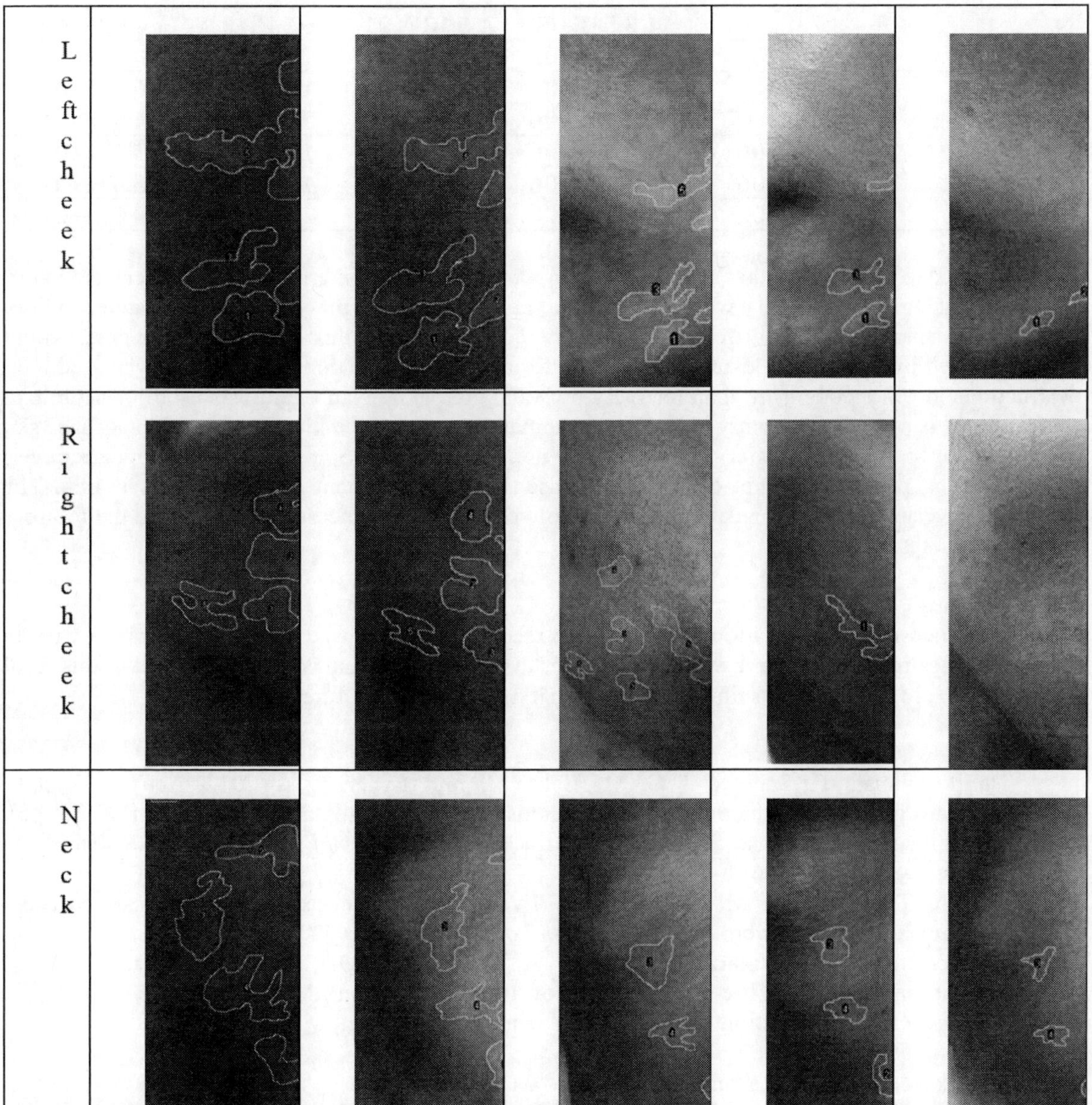

Figure 1. Wound healing process using Nano gold cream

Table 1. Week-based wound healing areas of volunteer face skin

Area mm^2	Week-0	Week-1	Week-2	Week-3	Week-4
Left cheek	9.73E+02	8.88E+02	1.53E+02	1.21E+02	0.00E+00
	9.00E+02	6.71E+02	8.16E+01	0.00E+00	0.00E+00
	4.83E+02	4.54E+02	5.35E+01	0.00E+00	0.00E+00
	5.14E+02	3.99E+02	6.14E+01	0.00E+00	0.00E+00
	4.82E+02	2.99E+02	3.85E+01	0.00E+00	0.00E+00
Right cheek	4.06E+06	5.37E+02	8.14E+01	5.19E+01	1.23E+02
	3.25E+06	3.61E+02	6.13E+01	3.56E+01	1.16E+02
	4.11E+06	7.38E+02	9.23E+01	7.28E+01	0.00E+00

	3.73E+06	3.51E+02	8.12E+01	6.54E+01	0.00E+00
	1.05E+07	9.98E+02	2.10E+02	1.31E+01	0.00E+00
Neck	5.07E+06	2.38E+06	4.90E+05	1.02E+02	6.35E+01
	3.36E+06	1.98E+06	2.40E+05	5.28E+01	4.56E+01
	7.13E+06	4.38E+06	7.69E+05	1.65E+02	7.39E+01
	7.58E+06	2.99E+06	3.87E+05	1.57E+02	5.67E+01
	5.07E+06	2.74E+06	4.45E+05	7.03E+01	3.01E+01

By SPPS statistic it shows that at 95% confidence with 0.05 margin error there is a decreasing areas with different significance every week at neck data. Left cheek and right cheek give the same condition giving decreasing areas with different significance occurred too in this area. The other case, wound healing caused by mercury exposure in cosmetics formulation had be done [10]. Mercury caused more destructions in cell metabolism than recovery by Nano gold [11]. Nano gold was used for neurotoxic recovery that is caused by mercury [12]. This computation supports the fact that the volunteer face skin condition that has recovered after 4 weeks is near normal condition. Spots that ever follow after every wound healing process may not occur in this case. It is an important notice in this research. This statement is very important to ensure that Nano gold can be used in medicinal material in the future as impact of this research.

4. Conclusions

Nano gold in cosmetic formulation can be used to recover wound caused by phenol exposure. Step by step wound healing process for 4 week recovery results in a near normal condition. Impact of this result is that Nano gold is a material suitable for medicinal material in the future.

Refferences

[1] Taufikurohmah T, Sanjaya I G M, Baktir I, and Syahrani A 2012 Activity Test of Nano gold for Reduction of Free Radicals, a Pre-Assessment Utilization Nano gold in Pharmaceutical as Medicines and Cosmetics *Journal of Material Science and Engineering* **2** 12 pp611–617,

[2] Ananthi T and Chitra M 2013 In vitro Evaluation of Antioxidant Activity of Michelia champaca (L .) Flowers *American Journal of Advanced Drug Delivery*

[3] Taufikurohmah T, Winarni D, Baktir A, Sanjaya I G M, and Syahrani A 2013 Histology Study: Pre-Clinic Test of Nano gold in Mus Musculus Skin , at Fibroblast Proliferation and Collagen Biosynthesis *Res. J. Pharm. Biol. Chem. Sci.* **3** 5 pp55–61

[4] Taufikurohmah T, Sanjaya I G M, Baktir A, and Syahrani A 2016 HISTOCHEMICAL CHANGES LIVER AND KIDNEY OF MICE EXPOSED TO MERCURY AND RECOVERY WITH NANO GOLD *Jurnal Molekul* **11** 1 pp. 80–91

[5] Tsai C Y, Shiau A L, Chen S Y, Chen Y H, Cheng P C, Chang M Y, Chen D H, Chou C H, Wang C R, and Wu C L 2007 Amelioration of collagen-induced arthritis in rats by Nano gold *Arthritis Rheum.* **56** 2 pp544–554

[6] MacNeil S 2008 Biomaterials for tissue engineering of skin *Materials Today* **11** 5 pp26–35

[7] Tulaihi B Al and Alhabib S 2015 Uncertainties around incretin-based therapies: A literature review *Saudi Pharm. J* **25** 1 pp1–7

[8] Michael S, Sorg H, Peck C T, Koch L, Deiwick A, Chichkov B, Vogt P M, and Reimers K 2013 Tissue Engineered Skin Substitutes Created by Laser-Assisted Bioprinting Form Skin-Like Structures in the Dorsal Skin Fold Chamber in Mice," *PLoS One*, vol. 8, no. 3, 2013.

[9] T. Taufikurohmah, A. Soegianto, I. G. M. Sanjaya, A. Baktir, and A. Syahrani,

"Mercury exposure effects to skin tissue of mus muscullus at fibroblasts cell proliferation and collagen quantity," *Res. J. Pharm. Biol. Chem. Sci.*, vol. 4, no. 4, 2013.

[10] Bourdineaud J, Laclau M, Maury-brachet R, and Gonzalez P 2012 Effects of Methylmercury Contained in a Diet Mimicking the Wayana Amerindians Contamination through Fish Consumption : Mercury Accumulation , Metallothionein Induction, Gene Expression Variations, and Role of the Chemokine CCL2 no. June, pp. 7710–7738

[11] Murai T, Sato M, Nishiyama H, Suga M, and Sato C 2013 Ultrastructural analysis of Nano gold-labeled cell surface microvilli in liquid by atmospheric scanning electron microscopy and their relevance in cell adhesion *Int. J. Mol. Sci* **14** 10 pp. 20809–20819

[12] Martinez-finley E J and Aschner *M Recent Advances in Mercury Research* pp163–171, 2014.

IOP Conf. Series: Journal of Physics: Conf. Series **947** (2017) 012057 doi:10.1088/1742-6596/947/1/012057

Profile of Metacognition of Mathematics Pre-Service Teachers in Understanding the Concept of Integral Calculus with Regard Gender Differences

L Misu[1,3], I K Budayasa[2,4], and A Lukito[2,5]

[1]Department of Mathematics Education Universitas Halu Oleo, Kendari-Indonesia
[2]Department of Mathematics Universitas Negeri Surabaya-Indonesia

Email: [3)]lamisuhamid@yahoo.co.id, [4)]ketutbudayasa@unesa.ac.id, [5)]agunglukito@unesa.ac.id

Abstract. This research is to describe metacognition profile of female and male mathematics' pre-service teachers in understanding the concept of integral calculus. The subjects of this study are one female and 1 male mathematics' pre-service teachers who have studied integral calculus. This research type is an explorative study with the qualitative approach. The main data collection of this research was obtained by using Interview technique. In addition, there are supporting data which is the result of the written work of research subjects (SP) in understanding the question of integral calculus. The results of this study are as follows: There is a difference in metacognition profiles between male and female mathematics' pre-service teachers in the understanding concept of integral calculus in the interpreting category, especially the definite integral concept. While in the category of exemplifying, there is no difference in metacognition profile between male and female mathematics' pre-service teachers either the definite integral concept and the indefinite integral concept.

1. Introduction

The metacognition profile is a natural and an intact description of a person's cognition that involves his own thinking in terms of using his knowledge, and the ability to plan and monitor his thinking process, and evaluate the process and outcome of one's thinking when understands a concept. According to Koriat, (in Goh [7] that Metacognition refers to what people know about cognition in general and about their own cognitive processes, in particular, as well as how they use this knowledge to adjust their informational processes and behavior. While the cognitive notion as described by Gagne [6] is a process that occurs internally within the central nervous system at a time when people are thinking. Then Neisser [12] says that cognition is the acquisition, arrangement, and use of knowledge.

The concept examined in this paper is the concept of integral calculus, consisting of the definite integral concept and indefinite integral concept. To understand the concept of integral calculus using the Bloom theory developed by Anderson et al. [1], and Mayer [11]. There are 7 categories to understanding the concept of integrals based on the theory, namely: Interpreting, Exemplifying, Classifying, Summarizing, Inferring, and Explaining. This paper discusses only two categories of conceptual understanding, i.e: Interpreting and Exemplifying. The interpreting category is defined as changing from

IOP Conf. Series: Journal of Physics: Conf. Series **947** (2017) 012057 doi:10.1088/1742-6596/947/1/012057

one form of representation to another, while the exemplifying category is defined as finding a specific example or illustration of a concept or principle [11].

The metacognition profile in understanding the concept of integral calculus, especially the category of interpreting and exemplifying can be seen from 2 components of metacognition, i.e. cognitive knowledge consists of declarative knowledge, procedural knowledge, and conditional knowledge (Flavell [5]), and metacognition skills consisting of planning, monitoring and evaluating (Dawson [4] and Livingston [9]).

The students' metacognition profile review distinguishes between male and female mathematics' pre-service teachers. Thus, the problems in this research are (1) how the metacognition profile of female mathematics' pre-service teachers in understanding the concept of integral calculus, and (2) how the metacognition profile of male mathematics' pre-service teachers in understanding the concept of integral calculus.

2. Method

This type of research is an explorative study with the qualitative approach. The subjects of this study are mathematics' pre-service teachers who are studying integral calculus, consisting of 1 female and 1 male. Selection of the subject of this study based on the highest score on the Mathematics Ability Test (a score of at least 70), and consider the value of the Student Achievement Index. The main data collection of this research was obtained by using interview technique. In addition, there are supporting data which are the result of written work of research subjects in understanding the question of integral calculus task.

3. Results and Discussion

3.1. The Results of the male mathematics' pre-service teachers interviews about understanding the concept of integral calculus in the category of interpreting

3.1.1. The Indefinite Integral Concepts

The result of male mathematics' pre-service teachers interview in understanding the concept of indefinite integral by interpreting the form of narration to the form of indefinite integral notation as follows: (1) The male mathematics' pre-service teachers can interpret the form "The first derivative function *F*(*x*) to *x* is f(*x*)" into the indefinite integral notation"$\int f(x)\,dx = F(x)+C$", (2) The male mathematics' pre-service teachers can argue that "an indefinite integral is the opposite of a derivative, so it is called anti-derivative, (3) The male mathematics' pre-service teachers may use the procedure to obtain indefinite integral notation based on the definition of anti-derivative (the indefinite integral), (4) The male mathematics' pre-service teachers can determine the hook element of the indefinite integral concept of the derivative, since the indefinite integral is the opposite of the derivative, (5) The male mathematics' pre-service teachers can explain easily in interpreting the narrative form to an indefinite integral notation, and (6) The male mathematics' pre-service teachers can confirm the interpretation of the narrative form to the indefinite integral notation "$\int f(x)\,dx = F(x)+C$"is true.

Based on the interview result of male mathematics' pre-service teachers, the male mathematics' pre-service teachers have used metacognition knowledge and metacognition skills in understanding the concept of the indefinite integral in the category of interpreting.

3.1.2. The Definite Integral Concepts

Based on the results of male mathematics' pre-service teachersinterviews in understanding the concept of the definite integral by interpreting the image form to the form of the definite integral notation as follows: (1) The male mathematics' pre-service teachers can interpret the image form "The region bounded by the curve *y*=*f*(*x*), the lines $x = a$, $x = b$ and the *x*-axis" to the definite integral notation

IOP Conf. Series: Journal of Physics: Conf. Series **947** (2017) 012057 doi:10.1088/1742-6596/947/1/012057

$L=\int_a^b f(x)dx$, (2) The male mathematics' pre-service teachers can not give a reason that the definite integral notation from "The region bounded by the curve $y=f(x)$, the lines $x = a$, $x = b$ and the x-axis" is $L=\int_a^b f(x)dx$, (3) The male mathematics' pre-service teacherswas not able to use the procedure to obtain an the definite integral notation based on the definition of Riemann sums, (4) The male mathematics' pre-service teachers can not determine the element of the hook of the definte integral concept of $\lim_{|p|\to 0}\sum_{i=1}^{n} f(\bar{x}_i)\Delta x_i$, (5) The male mathematics' pre-service teachers find it difficult to explain the steps to calculate the region bounded by the curve $y=f(x)$, the lines $x = a$, $x = b$ and the x-axis" becomes an the definite integral notation $L=\int_a^b f(x)dx$, and (6) The male mathematics' pre-service teachers do not have the procedures to obtain an the definite integral notation based on the definition of Riemann sums.

Based on the interview result of male mathematics' pre-service teachers, then the male mathematics' pre-service teachers can only use knowledge of declarative metacognition to understand the definite integral concept of the category of interpreting.

3.2. The Results of the female mathematics' pre-service teachers interview about understanding the concept of integral calculus in the category of interpreting

3.2.1. The Indefinite Integral Concepts

The results of female mathematics' pre-service teachers interviews in understanding the concept of indefinite integrals by interpreting the form of narration to the form of the indefinite integral notation as follows: (1) The female mathematics' pre-service teachers can interpret the form "The first derivative function $F(x)$ to x is f(x)" into the indefinite integral notation"$\int f(x)dx = F(x)+C$", (2) The female mathematics' pre-service teachers can give the reason that "$D_x[F(x)+C] = f(x)$means an indefinite integral is the opposite of a derivative, (3) The female mathematics' pre-service teachers may use the procedure for obtaining indefinite integral notation based on the definition of anti-derivative (the indefinite integral), (4) The female mathematics' pre-service teachers can define the relationship between the concept of indefinite integral and the derivative of the indefinite integral as the opposite of the derivative, (5) The female mathematics' pre-service teachers can explain easily in interpreting the narrative form to an indefinite integral notation, and (6) The female mathematics' pre-service teachers can confirm that the interpretation of the narrative form to the indefinite integral notation is " $\int f(x)dx = F(x)+C$"is true.

Based on the interview result of female mathematics' pre-service teachers, the female mathematics' pre-service teachers have used metacognition knowledge and metacognition skills in understand the indefinite integral concepts in the interpreting category.

3.2.2. The Definite Integral Concepts

Based on the results of female mathematics' pre-service teachersinterviews in understanding the concept of the definite integral by interpreting the image form to the form of the definite integral notation as follows: (1)The female mathematics' pre-service teachers can interpret the image form "the region bounded by the curve $y=f(x)$, the lines $x = a$, $x = b$ and the x-axis" to the definite integral notation $L=\int_a^b f(x)dx= \lim_{|p|\to 0}\sum_{i=1}^{n} f(\bar{x}_i)\Delta x_i$, (2) The Female mathematics' pre-service teachers can give the reason that the definite integral notation from " the region bounded by the curve $y=f(x)$, the lines $x = a$, $x = b$ and the x-axis" is $L=\int_a^b f(x)dx$,(3) The female mathematics' pre-service teachers are not perfect using procedures to obtain the definite integral notation based on the definition of Riemann sums, (4) The female mathematics' pre-service teachers can determine the hook element of the definite integral

IOP Conf. Series: Journal of Physics: Conf. Series **947** (2017) 012057 doi:10.1088/1742-6596/947/1/012057

concept that is $\lim_{|P|\to 0}\sum_{i=1}^{n} f(\bar{x}_i)\Delta x_i$, (5) The female mathematics' pre-service teachers explains the steps of calculating the area of "the region bounded by the curve $y=f(x)$, the lines $x = a$, $x = b$ and the x-axis" being the definite integral notation $L = \int_a^b f(x)dx$, and (6) The female mathematics' pre-service teachers master the procedure for obtaining efinite integral notation based on the definition of Riemann sums.

Based on the interview result of female mathematics' pre-service teachers, then the female mathematics' pre-service teachers can use metacognition knowledge and metacognition skills in understandingthe definite integral concept of the category of interpreting, except the knowledge of Conditional.

3.3. The Results of the male mathematics' pre-service teachers interview about understanding the concept of integral calculus in the category of exemplifying

3.3.1. The Indefinite Integral Concepts

The result of interviews of male mathematics' pre-service teachers in giving examples of the indefinite integral as follows: (1) The male mathematics' pre-service teachers can exemplify the indefinite integrals based on indefinite integral definitions, is "$\int(2x+3)dx = x^2+3x+C$", (2) The male mathematics' pre-service teachers may argue that "the indefinite integral is the inverse of the derivative" so that the derivative$x^2 + 3x + C$ is $2x + 3$, (3) The male mathematics' pre-service teachers can use the procedure to obtain $x^2 + 3x + C$ ie $\int(2x+3)dx = \frac{2}{1+1}x^{1+1} + \frac{3}{0+1}x^{0+1} + C$, (4) The male mathematics' pre-service teachers can determine the characteristics of the indefinite integral example ie if it is derived to produce the original function in the added number of constants, (5) The male mathematics' pre-service teachers can provide other examples in addition to polynomial functions namely trigonometric functions, exponential functions, and logarithmic functions, and (6) The male mathematics' pre-service teachers are only able to give an example of a simple indefinite integral. While the examples are complicated, for example: $\int \ln x\,dx$ and $\int \log x\,dx$ cannot be completed.

Based on the interviews of male mathematics' pre-service teachers, then the male mathematics' pre-service teachers can only use metacognition knowledge and metacognition skills in giving an example of a simple indefinite integral.

3.3.2. The Definite Integral Concepts

The result of interviews of male mathematics' pre-service teachers in giving examples of the definite integral as follows: (1) The male mathematics' pre-service teachers can provide an definite integral example based on the fundamental theorem of calculus, i.e. $\int_0^4 (2x-1)dx = 12$, (2) The male mathematics' pre-service teachers may argue that "the definite integral is necessarily the region bounded by the function$f(x) = 2x$-1, the lines $x = 0$, $x = 4$ and the x-axis" So the breadth is 12 unit area, (3) The male mathematics' pre-service teachers can use the procedure to obtain the region bounded by the functions of $f(x) = 2x - 1$, the lines $x = 0$, $x = 4$ and the x-axis, ie $\int_0^4 (2x-1)dx = (4^2-4)-(0^2-0) = 12$, (4) The male mathematics' pre-service teachers have not been able to determine the sample characteristics of a definite integral, (5) The male mathematics' pre-service teachers can provide other examples in addition to polynomial functions namely trigonometric functions, exponential functions, and logarithmic functions, and (6) The male mathematics' pre-service teachers are only able to give an example of a simple definite integration. While the examples are complicated, for example: $\int_0^1 \ln x\,dx$ and $\int_0^1 \log x\,dx$ cannot be completed.

Based on the interviews of male mathematics' pre-service teachers, then the male mathematics' pre-service teachers can only use metacognition knowledge and metacognition skills in giving an example of a simple definite integral.

3.4. The Results of the female mathematics' pre-service teachers interview about understanding the concept of integral calculus in the category of exemplifying

3.4.1. The Indefinite Integral Concepts

The result of interviews of female mathematics' pre-service teachers in giving examples of the indefinite integral as follows: (1) The female mathematics' pre-service teachers can instantiate indefinite integrals based on the definition of indefinite integrals, ie $\int (x^{4/3})dx = \frac{3}{7}x^{7/3} + C$, (2) The female mathematics' pre-service teachers may argue that "the indefinite integral is the inverse of the derivative" so that the derivative $\frac{3}{7}x^{7/3} + C$ is $x^{4/3}$, (3) The female mathematics' pre-service teachers can use the procedure to obtain $\frac{3}{7}x^{7/3} + C$ i.e. $\int (x^{4/3})dx = \frac{1}{3/4+1}x^{4/3+1} + C$, (4) The female mathematics' pre-service teachers can determine the characteristics of the indefinite integral example ie if it is derived to produce the original function in the added number of constants, (5) The female mathematics' pre-service teachers can only give examples of indefinite integrals of polynomial functions, root functions, and fractional functions, and (6) The female mathematics' pre-service teachers are only able to give an example of a simple indefinite integral, while a complex example can not yet.

Based on the interview result of female mathematics' pre-service teachers, the female mathematics' pre-service teachers can only use the knowledge and skills of metacognition in giving an example of the simple indefinite integral.

3.4.2. The Definite Integral Concepts

The result of interviews of female mathematics' pre-service teachers in giving examples of the definite integral as follows: (1) The female mathematics' pre-service teachers can provide an definite integral example of course based on the fundamental theorem of calculus, ie $\int_2^5 x^2 dx = 39$, (2) The female mathematics' pre-service teachers may argue that "the definite integral is necessarily the region bounded by the function $f(x) = x^2$, the lines $x = 2$, $x = 5$ and the x-axis" So the breadth is 39 unit area, (3) The female mathematics' pre-service teachers can use the procedure to obtain the region bounded by the functions of $f(x) = x^2$, the lines $x = 2$, $x = 5$ and the x-axis, ie $\int_2^5 x^2 dx = \frac{1}{3}(5)^3 - \frac{1}{3}(2)^2 = 39$, (4) The female mathematics' pre-service teachers have not been able to determine the sample characteristics of an definite integral, (5) The femalesmathematics' pre-service teachershave not been able to provide other examples other than polynomial functions, and (6) The female mathematics' pre-service teachers are only able to give an integral example of a simple one.

Based on the interview result of female mathematics' pre-service teachers, the female mathematics' pre-service teachers can only use the knowledge and skills of metacognition in giving an example of simple definite integral.

The results above show that there are differences in metacognition profiles between male and female mathematics' pre-service teachers in understanding the concept of integral calculus, especially the integral concept of the category of interpreting. The female mathematics' pre-service teachers are able to explain the reason or procedure for obtaining the definite integral notation through the Riemann sum approach. While the male mathematics' pre-service teachers are only able to write the definite integral notation, but cannot explain the reason or procedure to obtain of about the definite integral notation. Means the female mathematics' pre-service teachers are able to analyze an image of the curve region so as to obtain an integral notation. This is not in line with the opinion of Casey [2], Maccoby & Jacklyn

[10] that boys as a group rely on spatial strategies when completing mental-rotation tasks, while girls as groups tend to use verbal strategies to accomplish this task. However, this difference is influenced by motivation and craft factors in learning mathematics. This is as explained by Mitsos and Browne (in Haralambos and Holborn [8]) that female have better learning achievement rates than male, as females are more motivated and work more diligently than male in doing school work.

Then, in the exemplifying category, there is no difference in metacognition profile among male and female mathematics' pre-service teachers. However, the examples given for both the concept of indefinite integral and definite integrals are of course only polynomial functions. While other functions, such as trigonometric functions, root functions, and exponential functions give a simple example. This shows that the knowledge of mathematics' pre-service teachers about integral calculus, especially the Exemplifying category is still low.

4. Conclusion

Based on the results of research and discussion above, it can be concluded that:

In the category of interpreting, the male mathematics' pre-service teachers can use metacognition knowledge and metacognition skills in understanding the concept of indefinite integrals, while the concept of the definite integral only uses declarative knowledge. While in the exemplifying category, the male mathematics' pre-service teachers can use metacognition knowledge and metacognition skills to give examples of the indefinite integral and the definite integrals, but only a simple example;

In the category of interpreting, the female mathematics' pre-service teachers have used metacognition knowledge and metacognition skills in understanding the concept of the indefinite integral and the definite integral, except the conditional knowledge of the concept of the definite integral. While in the exemplifying category, the female mathematics' pre-service teachers can use metacognition knowledge and metacognition skills to give examples of the indefinite integral and the definite integral, but only a simple example;

5. References

[1] Anderson L W, Krathwohl D R, Airasian P W, Cruikshank K A, Mayer R E,PintrichP R, et al 2001 *A Taxonomy for Learning, Teaching, andAssissing: A Revision of Bloom's Taxonomy of Educational Objectives* (NewYork: Longman)

[2] Casey M B, Nuttall R L, and Pezaris E 2001 Spatial Mechanical Reasoning Skils Versus Mathematics Self-Confidence as Mediators of Gender Differences on Mathematics Subtests Using Cross-National Gender-Based Items *Journal for Research in Mathematics Education* **32** pp29-56

[3] Center for Curriculum Development (CCD) 2001 *Learning Module & Learning Method of Curriculum Development CenterMinistry of Education Malaysia* (http://myschoolnet.ppk.kpm.my/bcb8.pdf)

[4] Dawson T L 2008 Metacognition and Learning in Adulthood *Paper presented in ODNI/CHCO/IC Leadership Development Office.*

[5] Flavell J H 1979 Metacognition and Cognitive Monitoringa New Area of Cognitive—Developmental Inquiry *J. American Psychologist* **906**

[6] Gagne R M 1983 Some Issues in the Psychology of Mathematics Instruction *Journal for Research in Mathematics Education* **14** 1.

[7] Goh C 2008 Metacognitive Instruction for Second Language Listening Development: Theory, Practice and Research Implications *RELC Journal* **39** 2 pp188-213

[8] Haralambos and Holborn 2004 *Sociology: Themes and Perspectives Sixth Edition* (London: Harper)

[9] Livingstone J A 1997 *Metacognition: An Overview*Available (http://www.gse.buffalo.edu/fas/shuell/CEP564/Metacog.html)

[10] Maccoby E E and Jacklin C N 1974 *The Psychology of Sex Differences* (Stanford: Stanford University)

IOP Conf. Series: Journal of Physics: Conf. Series **947** (2017) 012057 doi:10.1088/1742-6596/947/1/012057

[11] Mayer RE 2002 *Teaching for Meaningful Learning* (Upper Saddle River, NJ: Prentice-Hall)
[12] Neisser U1976 *Cognition and Reality* (S. Francisco, CA: Freeman)

IOP Conf. Series: Journal of Physics: Conf. Series **947** (2017) 012058 doi:10.1088/1742-6596/947/1/012058

Understanding fraction concepts of Indonesian junior high school students: A case of field independent and field dependent students

K Mamonto[1], D Juniati[2], T Y E Siswono[2]

[1] Posgraduate in Mathematics Education, Universitas Negeri Surabaya, Ketintang, Surabaya 60231, Indonesia
[2] Mathematics Department, Universitas Negeri Surabaya, Ketintang, Surabaya 60231, Indonesia
Corresponding Author. E-mail: [1]karnainmamonto@mhs.unesa.ac.id

Abstract. This study aims to describe the students' understanding of fraction concepts. This research is descriptive with the qualitative approach. The subjects consisted of 1 Field Independent (FI) and 1 Field Dependent (FD) students. Setting subjects based on *Group Embedded Figure Test* (GEFT) results and meeting equivalent mathematical ability. Data obtained techniques through Task-based interview. The results show that FI students were represented a complete fractional concept and express the meaning of fractional notation with its own words.The students use of fractional concept such as the idea of dividing equally, fractional equivalents, and the relative size of the fraction to classify fractions based on certain properties, giving fractional examples and non-fractional examples, and comparing fractions. However, FD students represented declarative farctional concepts, had difficulty in giving a good meaning notation, relation between components, classifying fractions and comparing them. FD students are fixated on fraction notation and the visualization of concrete objects. Overall, these results provide a detailed picture of the understanding of fractional concepts that, FI students tend to use coherent analysis, while FD students tended to be fixated on fractional notation and visualization of concrete objects.

1. Introduction

Understanding concepts is a fundamental ability for students. In coping with the understnding, someone can construct the meaning of instructional messages, whether oral, written or graphic [1]. All students must have the opportunity and support necessary to study in-depth math with understanding [2]. Understanding is a cognitive activity or a person's mental experience in connecting an object to another object [3], Mousley refers to comprehension as the ability of a person to make some connection in the mind and make it possible to perform further abstractions [4]. In relation to the understanding of a mathematical concept, Hiebert and Carpenter suggests that the idea (concept) of mathematics is understood if it is part of the internal network. Mathematics is specially understood if the mental representation is an integral part of an existing network framework. Understanding is determined by stronger networks or more links in the network [5]. Without a strong understanding of concepts, students will learn the rules of operation-using a set of rules-without logic [6]. One of the concepts in mathematics is fractional. Fractions are one of the most challenging cognitive lessons, difficult to teach [7], and is a complex mathematical topics in the primary school curriculum [8]. Even Lamon argues that for most people, their relationship with mathematics began to decline early in elementary school, right after it was introduced to the fraction. Nevertheless, fractions are one of the many concepts in mathematics applied in everyday life [9]. Fractional concepts are abstract ideas about the part-whole relation [6]. In line with that view Charalombus suggests that the part-whole relation as a situation in which a continuous quantity or set of discrete objects is partitioned into equal parts [10]. Lack of understanding of fractional concepts leads to difficulties in terms of calculations with fractions, decimals and concepts, the use of fractions in measurement, and the concept of ratio and proportion [6]. Students who have weaknesses in fractional learning will face considerable challenges and difficulties in other areas of mathematics such as the concept of algebra, measurement, ratio and proportion [11].

Previous research reveals that students tend to use procedural knowledge when answering the issue of a fractional worth that is presented in symbolic form [12]. This study reveals that students demonstrate the use of procedural knowledge when answering the issue of denominations are presented in symbolic form. In some cases, all numerical reasoning is indicated in the procedure they are using. Many students cannot use area diagrams representing fractional symbols worth. Students who successfully connect parts / entire areas with symbolic and pictorial interpretations for one whole and three-quarters show their knowledge is more common and they are better able to apply the notion to the pictorial representation using a number line [12]. However, this study is limited to fractions equivalent to involving students in 3-5 elementary schools. In addition, the study has not considered the different characteristics of each individual in response to stimulate processing information including how to receive, remember and solve problems. These characteristics, attitudes, tendencies tend to be stable and unchanged are related to the cognitive style [13]. Thus the understanding of the individual as a cognitive activity depends heavily on the different cognitive styles. One of the widely studied cognitive styles is the cognitive style that is psychologically divided into two: Field Dependent cognitive (FD) and Field Independent cognitive (FI) styles. Individuals who have cognitive style FD tend to have difficulties in separating incoming information from their contextual environment and are more likely to be influenced by external cues and they become less selective in information absorption. While FI individuals have little difficulty in separating the most important information from their context, are more likely to be influenced by internal cues than external cues, and they are selective in inputting information [14].

This study aims to describe the understanding of the concept of fractional junior high school students in cognitive style Field Independent (FI) and Field Dependent (FD). The fractional concept is the sub-construction of relations from the same parts of a whole (part-whole relationship). This research is a descriptive research qualitative approach with data collection techniques through job-based interviews. The results of this study are expected to contribute to the development of learning in procedural understanding, especially in the operation of fractions

2. Method

The descriptive research with qualititative approaches conducted to explore students' understanding. Subjects in this study consisted of 2 students of grade VII of junior high school in Central Sulawesi consisting of 1 Cognitive Style Field Independent students (FI) and 1 students of cognitive style Field Dependent (FD). Setting the subject begins by giving the Group Embedded Figure Test (GEFT) and Mathematics Capability Test (MCT). GEFT is used to identify students' cognitive styles, and MCT is used to look at mathematical ability of the party. GEFT is given to the group of students in grade 7th. Based on the results of this GEFT test, the subjects who obtained the test score 0-9 were grouped in cognitive-style FD students and who scored 10-18 were categorized into FI cognitive-style students. From the results of the test, selected each 1 FI student hereinafter called the subject of FI, and 1 FD students hereinafter referred to the subject of FD by considering the criteria of equivalent mathematical ability that is with the differences of maximum MCT score 5. Collecting in this research use Task-based interview technique, where subject given Task of Understanding Concept of Fraction (TUCF) then conducted an interview. For example, to determine the fractions and meanings of part relations and overall, the task questions are analysed as shown in Table 1.

IOP Conf. Series: Journal of Physics: Conf. Series **947** (2017) 012058 doi:10.1088/1742-6596/947/1/012058

Table 1. Task questions to determine the fractions and meanings of part and whole relations.

Pictorial to Symbolic	Symbolic to Pictorial
10. Determine the fractions that show the shaded area in the following model! Please provide an explanation! 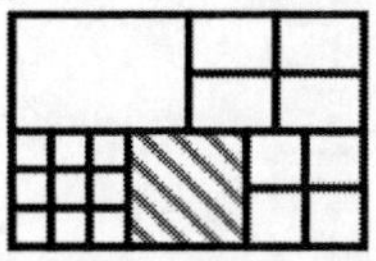	11. Suppose a rectangle represents a whole area. Using the subdivision of the given image, shade the area indicating the fraction $\frac{3}{4}$ 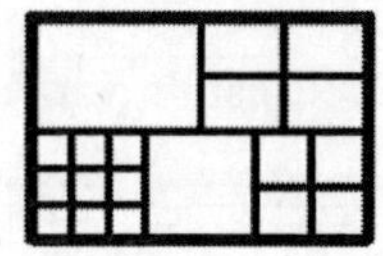
7. What fraction value is indicated by the point p in the following model? Explain! 	

For data validity, used time triangulation. Time triangulation is a way of testing the validity of data by comparing the data collected in two different times. For example, data on TUCF-1 results and interview 1 compared with data from TUCF-2 and interview 2. The result of this triangulation shows that the data is consistent, so obtained valid data, and data TUCF results and interviews used as a reference to analyse data to answer research questions. The process of data analysis using the flow of activities: data condensation, data display, and drawing and verifying Conclusions [15]. During the research processed, selection, focusing on simplification, extracting, and transforming raw field data. Furthermore, the researchers make a summary, encode, develop the theme, generate categories and write memos or analysis notes to be presented. Then, attempts are made to give meaning to the presentation of data as a conclusion.

3. Results and Discussion

3.1. Represents the concept of fractions

Based on the data obtained, the FI subject mentions the definition of fractions by expressing the meaning of the fraction notation, mentioning the fractional components as well as the meaning of the numerator and denominator in its own language by not fixating on the fraction notation. Whereas the subject of FD mentions the definition of a declarative fraction, but in expressing the meaning of the numerator and denominator, the subject of the FD depends on the notation and visualization of the concrete object which he had previously received. These findings show what Witkin et.all, points out, that the characteristics of individuals who have cognitive style of Field Dependent tend to accept existing structures [14].

In the category of relation meaning between the part and the whole of the model area, the measurement (the number line) and the discrete set, when the picture is presented in the shape of the subject area the FI divides into equal parts as the shaded area before giving the fraction symbol represented by the image. Even when presented a more complex picture and the subject of the FI is required to determine the fractional symbol on the model representing the fragment, the subject is able to show the appropriate symbol (see Figure 1). The subject's ability to apply the same divide idea also applies to the opposite case, in which case the subject is able to interpret the fraction symbol before representing it into the fractional model by first dividing the presented image into four equal parts before making the shading representing a fraction (see Figure 2 & 3).

IOP Conf. Series: Journal of Physics: Conf. Series **947** (2017) 012058 doi:10.1088/1742-6596/947/1/012058

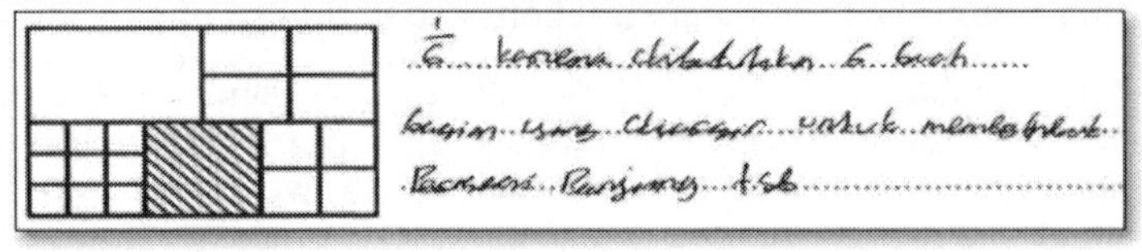

Figure 1. Results of the cognitive style subject field independent work

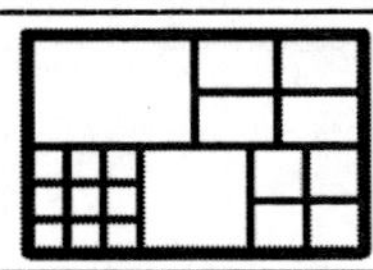

Figure 2. Draw on the problem before it is done

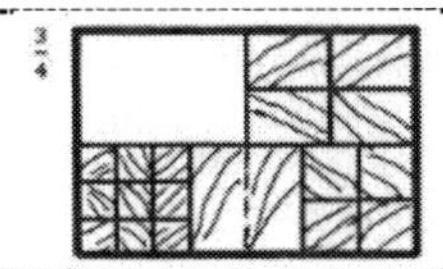

Figure 3. Results of FI student work in making a $\frac{3}{4}$ representation

Similarly, in the measurement model (line number) and discrete set, in the measurement model, the subject is able to determine the fractions which shown by a particular point on the number line (Figure 4). The subject matter determines one point on the number line as a $\frac{1}{4}$fraction by suggesting its meaning that the point is "3rd scale of 4 scales". Then in the discrete set of subjects, the subject suggests that $\frac{4}{16}$because of the number of circles, there are 4 dark circles whereas the total circle is 16.

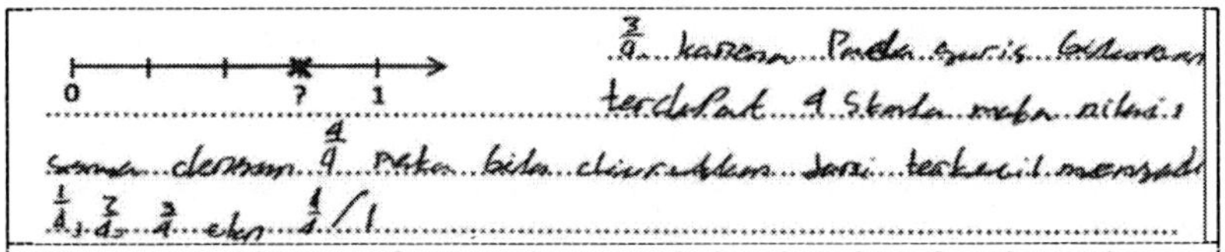

Figure 4. Photos of FI student work in showing fractions that are represented by drawing a number line.

The same ability is also shown by the subject of FD. By utilizing his knowledge of the idea of dividing equally on fractions, the subject determines that the image of the presented area model represents the $\frac{1}{6}$ (See Figure 5). However, the subject can not find the pattern and is affected by the existing sub- sections so as to fail to find the shading area representing $\frac{3}{4}$ for the entire rectangular area presented (See Figure 6 and figure 7). The findings are in line with the theory that FD individuals tend to see things as a whole pattern and find it difficult to separate the whole pattern into parts. Individual FD has a tendency to be easily distracted and easily confused so that lack the ability to solve the problem. This means that FD individuals tend to be easily influenced by the deception elements [14].

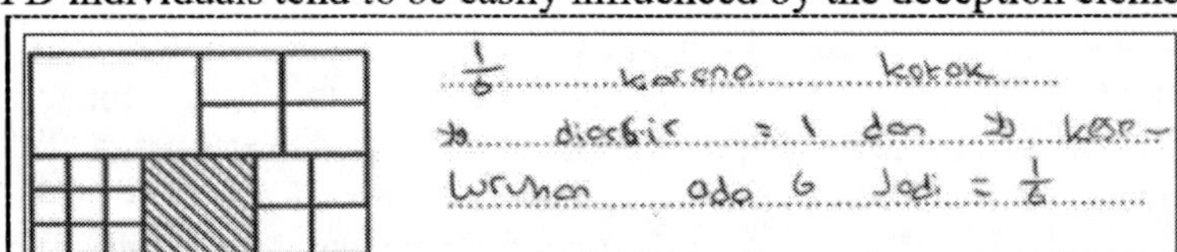

Figure 5. Photo Results of cognitive-style student Field Dependent work

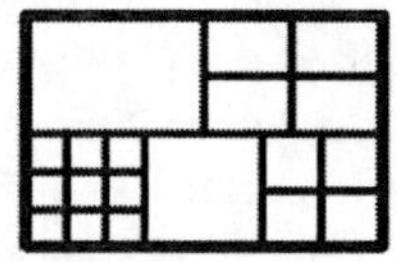

Figure 6. Draw on the problem before it is done

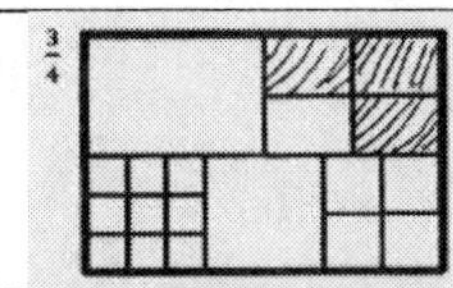

Figure 7. Results of FI student work in making a $\frac{3}{4}$ representation

Similarly, in the measurement model (number line) the subject of FD only suggests its general meaning as in the regional model by saying that $\frac{3}{4}$ on the asterisks means that 3 is part of 4. The findings are in line with the theory that, the individual FD Have a tendency to perceive or understand something globally [14]. The findings also point to differences from previous studies that revealed that students were better able to apply the notion to the pictorial representation using a number line [12].

3.2. Classify fractions by certain properties (in accordance with the concept)

The subject of FI groups the fraction representations presented into fractional groups based on the relation of similarities and differences in fractions. Using the knowledge of the equivalent fractions, the FI subject identifies all the images representing a fraction and classifies the images representing a fraction based on the relationship of similarities and differences. The subject of FI in finding the equation by tracing down to the simple value of a fraction. This finding reinforces Mayer's (2002) statement that classifying occurs when the subject knows that something (for example, an example) falls into a specific category (eg, concept or principle). This finding also points to what Hiebert and Carpenter, suggest that understanding is determined by stronger networks or more links in the network [5].

The subject of the FD classifies fraction representations in simple form, but can not identify similarities in images representing more complex or unusual fragments. This is in line with existing theories that FD individuals tend to have difficulties in separating incoming information from their contextual environment, and more likely to be influenced by external cues [13].

3.3. Provide examples and non examples

In giving examples and non-examples, both the FI and FD subjects provide fractional examples in the regional model, discrete sets and number lines by drawing and giving the fractional value represented by the image. The subject of FI and FD also find fractional shapes/frames representing certain fractions such as images representing fractions of $\frac{2}{3}$and finding images that do not represent $\frac{2}{3}$ fractions, taking into account the general concept or principle of fractions ie fair parts, And whole sections. With reference to the same general concept or principle, the subject provides a model/image form that does not represent a fraction or that belongs not to a fractional instance.

3.4. Leverage fractional concepts to compare fractions.

When presented with problems related to everyday life, in which subjects are asked to make choices on pizza slices of a certain size, the subject of FI selects a pizza slice that has a relative size of $\frac{2}{3}$ compared to pieces of relative $\frac{1}{4}$ and $\frac{1}{3}$ size on the grounds That because the piece is the greatest. The subject of FI says that "to compare the fractions, the note is the numerator and denominator of the fraction. If the denominator is different than seen from the denominator by means of the smaller the denominator, the greater the size of the fraction. If among the fractions of the same denominator it is seen from the numerator, the bigger the numerator the greater the size of the fraction"

In contrast to the subject of FD. The subject of the FD difficulties in determining the larger or smaller size of a fraction of the other fractions, arguing that $\frac{1}{4}$is greater than $\frac{2}{3}$and $\frac{1}{3}$. The subject is influenced by the visualization of the concrete object he has ever received so that it is difficult to compare the fractional size with other fractional sizes. The subject gives the reason why the $\frac{1}{4}$is bigger because if it is made the circle will be bigger in shape. The subject of FD suggests that if pizza is to be formed into four parts, the parts will be larger than the $\frac{2}{3}$part. The subject suggests that larger fractions are fractions that are slightly in part. The findings are in line with previous research that, in general, students are better able to apply fractions in the form of symbolic interpretation relationships with images, and they are better able to apply the notion to representational representations [12]. These findings also suggest that the subject of FD tends to be affected by visualizing the concrete object he or she once received. The subject of FD has dependence on external cues and structures, easily disturbed and easily confused so as to lack the ability to solve problems [14].

4. Conclusion

Based on the findings presented on the results of the study, it is shown that Cognitive Field Independent students represented the concept of fractions well and express the meaning of fractional notation with their own words. Students mention fractional components, express the relation meaning between fractional components by using coherent analysis, and not affected by the visualization of concrete objects. FI students also utilize various ideas/concepts related to fractions (shared ideas), and create a network of coherent fractional concept schemes in interpreting fractional concepts, including presenting fractional concepts into various mathematical representations. While the cognitive-oriented students of Field Dependent represented declarative fractional concepts, difficulty in giving meaning notation well, and still glued to fraction notation and visualization of concrete objects. FD students also tended to be affected by sub-sections contained in the model/image/diagram representing fractions. These findings suggest that, cognitive style differences are one of the decisive factors that can influence students' understanding of the concept of fractions.

5. Acknowledgements

The author would like to thank the Directorate General of Teachers and Education Personnel of the Ministry of Education and Culture of the Republic of Indonesia as a provided scholarship.

References

[1] R. E. Mayer, Rote Versus Meaningful Learning, vol. 41, Ohio: College of Education The Ohio State University, 2002, pp. 226-232.

[2] NCTM, Principles and Standards for School Mathematics, Reston VA: The National Council of Teachers of Mathematics, Inc., 2000.

[3] A. Sierpinska, Understanding in Mathematics, D. P. Ernest, Ed., London: The Falmer Press, 2005.

[4] J. Mousley, "What Does Mathematics Understanding Look Like?," *https://dro.deakin.edu.au/.../mousley-whatdoesmathematics-20...,* pp. 553-560, 2005.

[5] J. Hiebert and T. P. Carpenter, Learning and Teaching With Understanding, D. A. Grouws, Ed., Reston: National Council of Teachers of Mathematics, 1992.

[6] J. A. V. d. Walle, Elementary and Middle School Mathematics, London: Pearson Education, 2007.

[7] S. J. Lamon, Teaching Fractions and Ratios for Understanding: Essential Content Knowledge and Instructional Strategies for Teachers, New York: Routledge Taylor & Francis Group., 2012.

[8] J. L. Cortina, J. Višňovská and C. Zúñiga, "An Alternative Starting Point for Fraction Instruction," *International Journal for Mathematics Teaching and Learning,* pp. 1-28, 2015.

[9] D. Pitta-Pantazi, E. M. Gray and C. Constantinos, "Elementary school students' Mental representations of fractions," in *Proceedings of the 28th Conference of the International Group for the Psychology of Mathematics Education*, Melboune, 2004.

[10] C. Y. Charalambous, "Developing And Testing A Scale For Measuring Students' Understanding Of Fractions," in *Proccedings of the 31st conference of the international group for the psychology of mathematics education*, Seoul, 2007.

[11] F. A. Razak, N. Noordin, R. Alias and R. Dollah, "How Do 13-year Olds in Malaysia Compare Fractions," *Procedia - Social and Behavioral Sciences,* pp. 100-105, 2012.

[12] M. Wong and D. Evans, "Students' Conceptual Understanding of Equivalent Fractions," in *Proceedings of the 30th annual conference of the Mathematics Education Research Group of Australasia*, 2007.

[13] M. A. Guisande, M. F. Páramo, C. Tinajero and S. A. Leandro, "Field dependence-independence (FDI) cognitive style: An analysis of attentional functioning," *Psicothema,* vol. 19, pp. 572-577, 2007.

[14] H. A. Witkin, C. A. Moore, D. R. Goodenough and P. W. Co, "Field-Dependent and Field-Independent Cognitive Styles and Their Educational Implications," *Review of Educational Research,* vol. 47, pp. 1-64, 1977.

[15] M. B. Miles, A. M. Huberman and J. Saldaña., Qualitative Data Analysis: a methods sourcebook, Washington DC: SAGE, 2014.

[16] B. Wooldrige and B. Melanie Haimes, "The field Dependence/Field Independence learning Style: implication for Adult Student Diversity, outcomes assessment and accountability," in *Learning Styles and Learning: A Key to Meeting The Accountability Demands in Educations*, S. J. S. Ronald R. Syims, Ed., New York, Nova Science Publishers, 2006, pp. 237-257.

Students' Representation in Mathematical Word Problem-Solving: Exploring Students' Self-efficacy

A Sahendra[1,3], M T Budiarto[2], and Y Fuad[2]
[1]Postgraduate Program in Mathematics Education, Universitas Negeri Surabaya, Surabaya 60231, East Java, Indonesia
[2]Faculty of Mathematics and Natural Science, Universitas Negeri Surabaya, Surabaya 60231, East Java, Indonesia
[3]MTs Negeri 1 Tulang Bawang, Tulang Bawang 34611, Lampung, Indonesia

Abstract. This descriptive qualitative research aims at investigating student represented in mathematical word problem solving based on self-efficacy. The research subjects are two eighth graders at a school in Surabaya with equal mathematical ability consisting of two female students with high and low self-efficacy. The subjects were chosen based on the results of test of mathematical ability, documentation of the result of middle test in even semester of 2016/2017 academic year, and results of questionnaire of mathematics word problem in terms of self-efficacy scale. The selected students were asked to do mathematical word problem solving and be interviewed. The result of this study shows that students with high self-efficacy tend to use multiple representations of sketches and mathematical models, whereas students with low self-efficacy tend to use single representation of sketches or mathematical models only in mathematical word problem-solving. This study emphasizes that teachers should pay attention of student's representation as a consideration of designing innovative learning in order to increase the self-efficacy of each student to achieve maximum mathematical achievement although it still requires adjustment to the school situation and condition.

1. Introduction

Representation is a description an object or process i.e. words, diagrams, graphics, computer simulations, mathematical equations and others. It can also be used to describe cognitive processes to the understanding of an idea in mathematics [1]. Representation of mathematical requires students to discover and develop a tool or way of thinking in communicating mathematical ideas from the abstract to the concrete, making it easier to understand. It takes place in two stages: internal and external representation. Internal representation is knowledge and structure in memory that is part of the cognitive process i.e. personal notation systems, nature language, visual images and problem-solving strategies while external representations are knowledge and structures in environments such as physical symbols, objects, or dimensions [2]. The process of internal representation is certainly not observable visually and cannot be assessed directly because it is a person's mental activity in mind. In other words, a person who performs an internal representation process in learning mathematics will think about the ideas, thought, or mathematical concepts he is studying in order to clearly understand the problem, connect the problem with the student's knowledge, and make a plan.

External representation is used to describe what the student, teacher, and mathematician. It is grouped into three forms, namely: visual form, symbol, and verbal. The visual form includes diagrams, tables, or graphs, geometry drawings and sketch drawings. The symbols include mathematical models and symbols or mathematical expressions. The verbal form includes both written and spoken words or

IOP Conf. Series: Journal of Physics: Conf. Series **947** (2017) 012059 doi:10.1088/1742-6596/947/1/012059

sentences [3]. When faced with new and complex ideas, it is possible for students to try different representations as a manifestation of their strategies in mathematics problem-solving. The use of various representations in solving a problem is called multiple representation. To be able to identify and represent the same concept in different representational models, conceptual understanding in mathematics it is important [4,5]. Generally, math problems can be a word problem. The word problem is presented in the form of a meaningful narration that can be understood and answered mathematically based on prior learning experience, as well as related to the situation experienced by students in everyday life [6,7]. In solving word problem, the students must understand the text to identify the missing information, obtain a plan to solve for it, and perform calculations to search for it. Solving the word problem in mathematics is an important aspect in mathematic learning and mathematical understanding. The steps to mathematical word problem solving according to Polya includes: (1) understanding the problem, (2) devising a plan, (3) carrying out the plan, (4) looking back [8].

Mathematical word problem give students challenges in applying mathematical thinking in various situations. The solving requires integration of some cognitive processes in which the students need to understand the language and factual information in the problem, translate the problem by using relevant information to create the appropriate mental representation, compile and monitor problem-solving plans, and carrying out appropriate procedural calculations [9]. In addition, knowledge in mathematical word problem can give students confidence to accomplish the task. The lack of strategy in mathematical word problem can decrease the confidence and belief of being success. The success of students in mathematical word problem is supported by psychological factors, not only about the method of learning but also about the beliefs about the ability possessed by students. The beliefs that the students have in performing a particular task or action is called self-efficacy [10].

Math self-efficacy affects mathematical achievement of the students [11]. High self-efficacy created tranquility ahead of the task and confidence in facing the difficult activities. Conversely, a person who doubts about his ability, thinks things in more difficult way than the real truth. [12]. Trust in one's academic ability is an essential component of school success. When individuals are familiar with task demands or activities, they tend to bring out self-efficacy that has been developed as a result of previous experience with similar tasks. Students' self-confidence in the success of problem-solving or mathematical word problems is called mathematical word problem self-efficacy [13]. There is a relationship between self-efficacy and academic achievement. The higher the self-efficacy, the higher the academic ability is. Self-efficacy is also a determinant that influences individual choices in an escape effort, persistence in the face of adversity and the mindset and emotional reactions they experience [14,15]. Thus, representation and self-efficacy are very important parts in the mathematical word problem-solving process, so the solution of mathematical word problem is more structured and in accordance with the goals which are going to be achieved. Therefore, this study aims at examining students' representation in mathematical word problem based on their self-efficacy.

2. Method

This research is descriptive qualitative research, researcher use Polya's problem-solving step. The research subjects are two students of class VIII MTs Negeri 2 Surabaya with equal mathematic ability consisting of two female students, one of them with high self-efficacy and the other with low self-efficacy. The selection of study subjects was according to the results of mathematical ability test, documentation of the result of middle test even semester of 2016/2017 academic year, and results of questionnaire of mathematical word problem self-efficacy scale. The instrument of mathematical word problem self-efficacy scale is an adaptation of the original instrument with an adjustment to the psychological and cultural students of MTs Negeri 2 Surabaya. The data obtained were the provision of mathematical word problem-solving task and interview which were done twice. Giving the second mathematical word problem-solving task and the second interview led to a triangulation of data.

IOP Conf. Series: Journal of Physics: Conf. Series **947** (2017) 012059 doi:10.1088/1742-6596/947/1/012059

3. Results and Discussion

Subjects with high self-efficacy and low self-efficacy exercise all steps of Polya's problem-solving concept. The subject with high self-efficacy, at a stage, in understanding the problem, tend to use symbols in presenting information and presenting what is asked on the question, but not yet able to explain the meaning of symbols used correctly. In addition, they tend to choose representations in the form of visuals and symbols. The next one is manipulating the visual form and symbols according to the plan, then the results are interpreted in the form of a written text. In devising a plan and carrying out the plan, high self-efficacy subject dominantly employed more than one form of representation or multiple representation. Meanwhile, the subject with high self-efficacy looked back problem-solving strategies and recalculated to prove that the answers obtained were true through sketches and formulas that were made and able to solve mathematical word problems appropriately and consistently along had positive perceptions in mathematics. In terms of behavior, the subject with high self-efficacy in mathematical word problem-solving, dominantly used multiple representation to ensure that the answers obtained is correct. In solving the linear equations problem, the subject with high self-efficacy analyzed the probability in answering without using formal operations taught like substitution and elimination methods. In this case, mathematical word problem-solving is considered more effective and efficient. The subject with high self-efficacy assumed that mathematics have an important and useful role in everyday life [16]. This assumption was made as mathematics is one of the favorite subjects for her.

The subject with low self-efficacy tend to use written words in presenting information and presenting what is asked on the matter, choosing representations in visual form or symbols as a solving strategy. The subject with low self-efficacy have not been able to explain the meaning of the symbols used correctly. Furthermore, at the stage of carrying out the plan, the subject manipulated the visual form or symbol according to the plan, then the results were interpreted into the form of written words. At the stage of devising a plan and carrying out the plan, the subject dominantly planned using visual representation. At the looking back stage, the subject rechecked the problem-solving strategy and recalculated to prove that the answer obtained is correct through drawing the sketch or the formula being made. At each stage, the subject tends to use only one form of representation only. Subjects with low self-efficacy have not showed consistency in mathematical word problem-solving [13]. In terms of behavior, the subject with low self-efficacy in mathematical word problem-solving dominantly used only one form of representation based on learning experience. In solving the linear equations problem, the subject with low self-efficacy used formal operations namely the method of elimination and substitution according to the learning experience gained.

The results show that there are several similarities of representational in mathematical word problem-solving between subjects with high and low self-efficacy as shown in Table 1.

Table 1. Similarities of representational in mathematics word problem-solving between subjects with high and low self-efficacy

Problem Solving Step	Similarities between high and low self-efficacy subjects
Devising a plan	Using sketch image based on the subject's understanding of the problem.
Carrying out the plan	- Manipulating sketch drawings that are created to get answers from the given problem. - Interpreting the answers obtained into the form of written words.
Looking back	Examining the sketch drawing.

IOP Conf. Series: Journal of Physics: Conf. Series **947** (2017) 012059 doi:10.1088/1742-6596/947/1/012059

The results show that there are some differences of representational in mathematics word problem-solving between subjects with high and low self-efficacy as shown in Table 2.

Table 2. The results showing the differences of representational in mathematics word problem-solving between subjects with high and low self-efficacy

Problem-Solving Step	Subject with High Self-efficacy	Subject with Low Self-efficacy
Understanding the problem	Presenting data or information that is known to the problem in the form of symbols. Presenting what is asked on symbol form issues	Presenting data or information that is known to the problem in the form of written words matching to the information of the given question. Presenting what is asked on verbal form issues.
Devising a plan	Choosing a problem-solving strategy from a given problem which tends to be more than one form of representation or multiple representation that is the visual form and symbol based on the learning experience. In addition, it is to assure the results obtained on the visual representation.	Choosing a resolution strategy from a given problem which tends to be one of the representations or single representation.
Carrying out the plan	- Manipulating the drawing of sketches and formulas that have been selected to determine the answers. - In the linear equations problem, the subject tends to analyze the possibility of the values of x and y in order to obtain the answer.	- Manipulating the drawing of the selected sketches or formulas to get the expected answers. - In the linear equations problem, the subject tends to do the elimination and substitution method in order to determine the value of the variable which is then used to determine the final answer.
Looking back	The subject tends to use multiple representations of visuals and symbols when looking back the problem-solving strategies and recalculating to prove that the answers obtained are correct.	The subject tends to use only one form of representation of the visuals or symbols when rechecking the problem-solving strategy and recalculating to prove that the answers obtained are correct.

4. Conclusion

Based on the results of data analysis and discussion it can be concluded that students with high self-efficacy in mathematical word problem-solving, presents data or information that is known and asked in relation to the issue into the form of symbols. The strategy of completion of the problem given tend to involve requirement of multiple representation in the visual form of sketches and symbols of formulas or mathematical models, and manipulation of sketch drawings and formulas that have been made. Particularly, for mathematical models on linear equations learning materials, the students analyzed the possible answers that meet the mathematical model that has been made based on the students' experience of working on problems related to linear equations. The results obtained are interpreted in the form of written words. The next one is looking back the problem-solving strategy through drawing sketches and recalculating the conformity of the results on the mathematical models that have been made and also assuring that the steps taken are the best ones. Meanwhile, the student with low self-efficacy presented information that is known and what is asked on the issue in the form of written words. The results are adapted to the problem by choosing a strategy of completion of the given problem which tends to lead on the use of one representation formula in the form of drawing sketches or symbols which are considered as formulas mathematical models only. In addition, the manipulation of the drawing sketch has been made after that. While in the mathematical model of linear equations learning material, the students used the elimination and substitution methods to find a solution. The results obtained were then interpreted into the verbal form of written text. Next, in looking back the problem-solving strategy through sketch drawing or recalculating the conformity of the results on the mathematical model has been made to ensure that the steps taken are the best ones.

References

[1] Hwang W Y, Chen N S, Dung J J, and Yang Y L 2007 Multiple representation skills and creativity effects on mathematical problem solving using a multimedia whiteboard system *Educational Technology and Society* **10** 191–212
[2] Zhang J 1997 The Nature of External Representations in Problem Solving *Cognitive Science* **21**(2) 179–217
[3] Goldin G and Steingold N 2001 Systems of Representations and the Development of Mathematical Concepts *The Roles of Representation in School Mathematics* 1–23
[4] Owens K D, Clements MA 1998 Ken. Representations in spatial problem solving in the classroom *The Journal of Mathematical Behavior* **17** 197–218
[5] Even R 1998 Factors involved in linking representations of functions *The Journal of Mathematical Behavior* **17**(1) 105–21
[6] Bates E T and Wiest L R 2003 Impact of Personalization of Mathematical Word Problems on Student Performance *Mathematics Educator* **14**(2) 17–26
[7] Boonen A J H, de Koning B B, Jolles J, and van der Schoot M 2016 Word Problem Solving in Contemporary Math Education: A Plea for Reading Comprehension Skills Training *Frontiers in psychology* 7th February p 191
[8] Polya G 1945 How to Solve It *The Mathematical Gazette* **30** 181
[9] Desoete A, Roeyers H, and De Clercq A 2003 Can offline metacognition enhance mathematical problem solving? *Journal of Educational Psychology* **95**(1) 188–200
[10] Bandura A 1995 *Self-efficacy in changing societies* (Cambridge: Cambridge University Press) p 334
[11] Bonne L, Lawes E, Bonne, and Lawes 2016 Assessing students' maths selfefficacy and achievement 60–3
[12] Pajares F and Schunk D H 2001 Self-beliefs and School Success: Self-efficacy, Self-concept, and School Achievement 239–66
[13] Akinsola M K and Awofala A O 2009 Effect of personalization of instruction on students' achievement and self-efficacy in mathematics word problems *International Journal of Mathematical Education in Science and Technology* **40**(3) 389–404
[14] Usher E L and Pajares F 2009 Sources of self-efficacy in mathematics: A validation study *Contemporary Educational Psychology* **34**(1) 89–101
[15] Pajares F 1996 Self-efficacy Beliefs and Mathematical Problem-Solving of Gifted Students *Contemporary educational psychology* **21**(4) 325–44
[16] Montague M, Enders C, Dietz S 2011 Effects of cognitive strategy instruction on math problem solving of middle school students with learning disabilities **34**(4) 262–72

IOP Conf. Series: Journal of Physics: Conf. Series **947** (2017) 012060 doi:10.1088/1742-6596/947/1/012060

Students' Self-Monitoring on Mathematics Ability: Cube and Cuboid Problem Solving

N T Lusiana[1,2], A Lukito[1] and S Khabibah[1]
[1]Mathematics Department, Universitas Negeri Surabaya
[2]1st Junior High School Pagentan Banjarnegara

Abstract. This study aims at describing students' activity to understand the behaviors processes called self-monitoring in a cube and cuboid problem solving viewed from mathematics ability. The subjects were eight graders of junior high school who studied surface area and volume of cube and cuboid clussified into high, average and low mathematics abilities. Mathematics ability test to select the subjects the study. Data were collected through self-monitoring task and interviews. Data triangulation was used to verify the credibillity findings. Data analysis was done by data condensation, data display and conclusion drawing and verification. Results showed that students' self-monitoring with high math ability is more fullfilled self-monitoring components. Students with average and low math abilities not fullfilled the component that covers verifying the results during solving the problem. It is expected that teachers must provide different learning treatments to improve students' self-monitoring for better learning outcomes.

1. Introduction

Self-monitoring is the students' activity in systematic way to understand the behavior and the thinking process that did in problem solving. Students have difference expressions and behavioral settings in learning. So, important for students to be done self-monitoring [1]. Each student must be able to determine of the personally problem solving during the learning process by reducing the teacher's intervention. Students should be cognitively aware of what they will do, what they should do, why they do and what outcomes will be made if done. Self-monitoring is needed to improve the function of independence learning in order to achieve maximum learning outcomes. Personal behavior monitoring of activity in a professional and systematic way will improve learning outcomes [2]. Learning outcomes is a description of student mastery in mathematics through the mathematics learning process. That called ability [3]. Self-monitoring of students problems solving consist of the components reading and understanding information, analyzing and planning strategies, implementing and monitoring strategies used, verifying and ensuring that they have used every step of the activity [4].

There were a significant difference between the results of the students' mathematics skills that reinforced by giving a similar sample to those using self-monitoring from preparation to implementation and similar problem solving investigations [5]. Self-monitoring will help students acquire knowledge, improve generalization and transfer knowledge and skills to achieve students learning success [6]. But the two studies did not reveal how the complete description of self-monitoring students in solving problems in cubes and cuboid viewed on differences in students' math abilities. There are three categories mathematical abilities namely high, average and low [7].

Based on the problems, researchers interesting in knowing the profile of students self-monitoring in a cube and cuoid problem solving viewed on mathematical ability. Research subjects are junior high school students who have been able to solve cube and cuboid that can study abstract material such as mathematics [8]. In this study, interviews was used to find out students self-monitoring performed after completing the self-monitoring task that contains a cube and cuboid problem. Interview results show that students with high math ability through all components of self-monitoring. Student with average math ability are not performing verification activities and students with low math ability do not verifying and do not making sure they have used every step of the activity. This means that students with high math ability perform self-monitoring activities more complex than students with other abilities. The impact of a more complex self-monitoring implementation makes students with high math ability able to solve problems more precisely. Therefore, teachers can give extra attention to students with low and average math ability to be able to train their self-monitoring so as to obtain better problem solving results.

2. Method

The study is a qualitative research that aims to describe students self-monitoring in a cube and cuboid problems solving on differences student abilities. The instrument is a mathematical ability test consisting of 5 essay questions, self-monitoring tasks of students used to obtain student's self-monitoring data, and interview used to collect the corresponding data of students self-monitoring more deeply. Interviews were conducted after the students completed the self-monitoring cube and cuboid task.

Mathematical ability test was given to 32 eight graders students who had studied previous cubes and cuboid. Data on mathematical ability test results are used to select subjects three subjects with high, average and low math abilities. Each subject was given self-monitoring tasks and then interviewed. Data triangulation was done to verify the credibility of the findings. Data analysis was done include data condensation, data display and conclusions drawing and verification [9]. The whole research procedures include preparation phase, implementation phase, data analysis and report writing phase.

3. Results and Discussion

3.1. Students Self-Monitoring on High Mathematics Ability

Subjects self-monitoring on high mathematical ability in reading and understanding information components appears reading the problem five times to understand the problem. The subject expresses her understanding of the problem by writing the known and questions in the answer sheet using his own language which is different from the language of the question. The subject emphasizes her understanding by telling about the complete problem and command the problem to be resolved. Subjects use information in a known matter that includes length, width, height and sides, then formulates it to answer questions.

Self-monitoring subjects in analyzing and planning strategies components seen in detecting experiences that have been previously owned when elementary about cube and cuboid. Subjects use other information relating to the problem that is the formula of the cube and cuboid, and trying to change the unit as expected by the problem. The subject tries to solve the problem with a different strategy by first changing the unit for self-monitoring task 1 and by using the cube and cuboid formula for self-monitoring task 2.

Subjects self-monitoring in implementing and monitoring the strategy components emerged when planning the solution step that will be used is to change the unit first, but there are obstacles, then proceed with the preparation of a new plan that is by change the way used. The subject will get a solution if it is at a certain step. Subjects have also made mistakes in problems solving in the form of mistakes to change units and calculations but then fix the error. Subjects make corrections to each step of work with the goal of denying to themselves that there is no mistake her made. The subject is

IOP Conf. Series: Journal of Physics: Conf. Series **947** (2017) 012060 doi:10.1088/1742-6596/947/1/012060

rethinking the steps that have been taken and the results that have been obtained to ensuring the correct answers obtained.

Subjects self-monitoring in the verifies component when finding and changing answers that are considered unreasonable. Subjects try different ways that have been written with the aim of supporting or comparing with the solution that has been obtained.

Subject self-monitoring in ensuring it has used every lagkah of activity components appears on the activity of correcting the calculations until repeating them to ensuring that the results obtained are the correct result. The subject also shows the suitability between the problem and the results obtained through the examination of the solution steps.

Based on the activities that have been done, then the data of subjects self-monitoring on high math ability in cube and cuboid problem solving can be presented in figure 1.

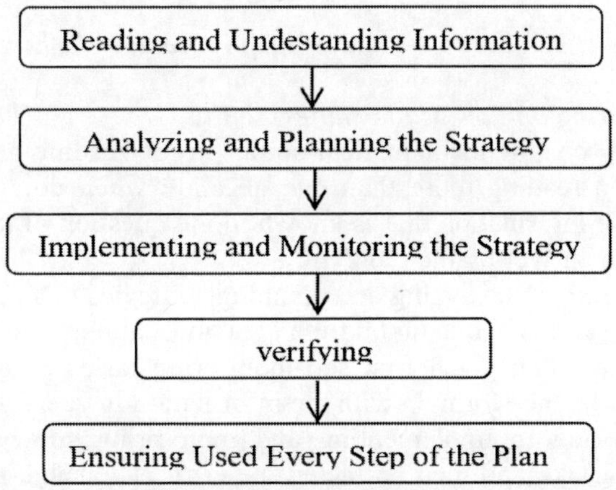

Figure 1. Subject Self-Monitoring on High Mathematics Ability

3.2. Subject Self-Monitoring on Average Mathematics Ability

Subjects self-monitoring on average mathematics ability in reading and understanding informations components of information visible when reading the problem more than once or even five times to understand the problem and to solve it. Subjects express problems using their own words based on comprehension include known and asked. The known stringing subjects such as length, width, height and sides to answer are asked by searching the volume of cubes and cuboid.

Subjects self-monitoring in analyzing and planning strategies components are seen in the activity of detecting previous experiences about finding the area and volume of cubes and cuboid. The subject utilizes the known on the problem and computes it with the formula of the area and volume of cubes and cuboid. The subject tries two different ways and chooses one answer that is considered to have a more appropriate way.

Subjects self-monitoring in implementing and monitoring strategies components emerged when planning, executing plans and then proceeding with completion in accordance with the initial solution plan. The subject is aware made quite a few mistakes so that the subject has to correct the error. The subject believes the answers that have been obtained because having made corrections to the solution steps.

Subject self-monitoring in ensuring used every step of the plan component appears on the activity of examination of the calculation to believe the answers have been obtained. The subject examining completion steps and conformity between the questions and the results obtained.

Based on the activities that have been done, then the data of subject self-monitoring on average math ability in cubes and cuboid problem solving can be presented in Figure 2.

IOP Conf. Series: Journal of Physics: Conf. Series **947** (2017) 012060 doi:10.1088/1742-6596/947/1/012060

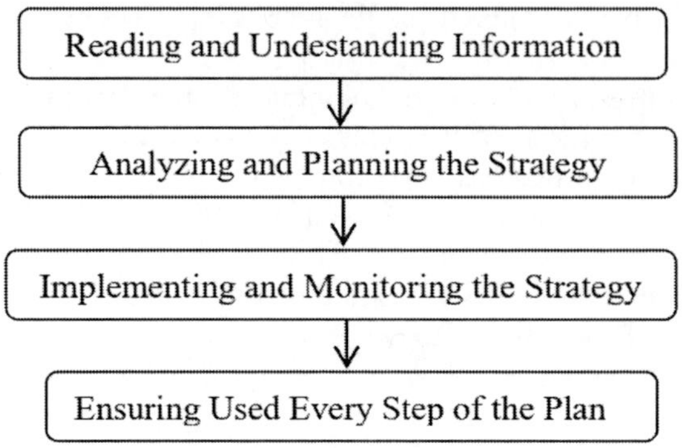

Figure 2. Subject Self-Monitoring on Average Mathematics Ability

3.3. Subject Self-Monitoring on Low Mathematics Ability

Subjects self-monitoring on low mathematical ability in the reading and understanding information component appear when reading more than one question when doing a problem, while and after completion. Subjects use information that is known in the question of length, width, height and cube sides to be formulated in answering the problem.

Subjects self-monitoring on analyzing and planning strategies component is seen in the activity of detecting previous experiences about finding the area and volume of cubes and cuboid that is at the time of the primary and when given first self-monitoring tasks. Subjects using and associate the volume, area and circumference formula with the information in question.

Subjects self-monitoring in implementing and monitoring strategies component arises when knowing has made a mistake and then corrected the error. The subject then proceeds by correcting steps that have been taken during the completion.

Based on activities that have been done, then the data of the subjects self-monitoring on low mathematical ability in cubes and cuboid problem solving can be presented in Figure 3.

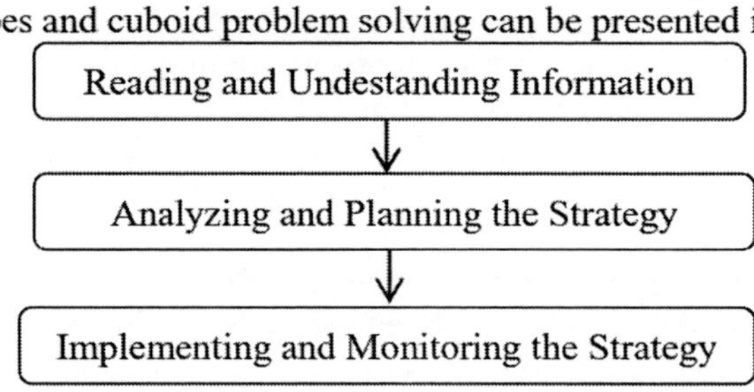

Figure 3. Subject Self-Monitoring on Low Mathematics Ability

4. Conclusion

Based on the research findings and analyzing, concluded that students with high mathematical ability perform all self-monitoring activities components include reading and understanding, analyzing and planning strategies, implementing and monitoring strategies used, verifying and ensuring used every step of the activity. Students with average math ability showed self-monitoring activities components include reading and understanding, analyzing and planning strategies, implementing and monitoring strategies used and ensuring used every step of the activity but not verifying the problem. Whereas students self-monitoring on low math ability perform activities components include reading and understanding, analyzing and planning strategies, implementing and monitoring strategies used, but does not perform verification and ensuring used every step of the activity.

5. Anknowlegement

Indonesian Ministry of Education and Culture. Directorate General of Teachers and Education Personnel, Directorate of Basic Education Teacher Development.

References

[1] Wolf H, Spinath F M, Riemann R, and Angleitner A 2009 Self-monitoring and personality : A behavioural-genetic study *Pers. Individ. Dif* **47** 1 pp25–29

[2] Oliver R M, Wehby J H, and Nelson J R 2015 Helping teachers maintain classroom management practices using a self-monitoring checklist *Teach. Teach. Educ* **51**, pp. 113–120

[3] Uno H B and Mohamad N 2015 *Belajar dengan Pendekatan PAILKEM* (Jakarta: Bumi Aksara)

[4] Goos M and Galbraith P 2000 *A Money Problem : A Source of Insight into Problem Solving Action* pp1–21

[5] Ning L 2011 *An Experimental Research on the Transfer of Mathematics Skills Based On Self-Monitoring Strategy* **4** 1 pp 67–74

[6] Kuan H and Liu J 2016 *Correlation Research on the Application of E-Learning to Students' Self -Regulated Learning Ability , Motivational Beliefs , and Academic Performance* **12** 1 pp.1091–1100

[7] Amala M A and Ekawati R 2016 Profil Proses Matematisasi Horizontal Dan Vertikal Siswa SMP Dalam Menyelesaikan Masalah Kontekstual Pecahan Ditinjau Dari Kemampuan Matematika *Mathedunesa Jurnal Ilmiah Pendidikan Matematika* **3** 5 (ISSN: 2301-9085)

[8] Uno H B and Kuadrat M 2014 *Mengelola Kecerdasan dalam Pembelajaran: Sebuah Konsep Pembelajaran Berbasis Kecerdasan* (Jakarta: Bumi Aksara)

[9] Miles M B, Huberman A M, & Saldana J 2014 *Qualitative Data Analysis A Methods Sourcebook Edition 3* (Washington : SAGE)

Algebraic Reasoning in Solving Mathematical Problem Based on Learning Style

N F Indraswari[1], I K Budayasa[2], R Ekawati[2]

[1] Postgraduate in Mathematics Education, Universitas Negeri Surabaya, Ketintang, Surabaya, 60231, Indonesia
[2] Mathematics Education Department, Universitas Negeri Surabaya, Ketintang, Surabaya, 60231, Indonesia

E-mail: [1]nurfitriyah1804@gmail.com

Abstract. This study aimed to describe algebraic reasoning of secondary school's pupils with different learning styles in solving mathematical problem. This study begins by giving the questionnaire to find out the learning styles and followed by mathematical ability test to get three subjects of 8th-grade whereas the learning styles of each pupil is visual, auditory, kinesthetic and had similar mathematical abilities. Then it continued with given algebraic problems and interviews. The data is validated using triangulation of time. The result showed that in the pattern of seeking indicator, subjects identified the things that were known and asked based on them observations. The visual and kinesthetic learners represented the known information in a chart, whereas the auditory learner in a table. In addition, they found the elements which makes the pattern and made a relationship between two quantities. In the pattern recognition indicator, they created conjectures on the relationship between two quantities and proved it. In the generalization indicator, they were determining the general rule of pattern found on each element of pattern using algebraic symbols and created a mathematical model. Visual and kinesthetic learners determined the general rule of equations which was used to solve problems using algebraic symbols, but auditory learner in a sentence.

1. Introduction

Algebraic reasoning is important because it is the basic of all mathematical contexts, including arithmetic which allows someone to explore mathematical structures [1]. Moreover, algebra is one of the topics in mathematics that is closely related to the use of symbols and numbers to solve equations, analyze functional relations, and determine the structure of representational systems which consist of expressions and relationships [2]. But in reality, many pupils experience difficulty in learning algebra. Tall [3] and Toshiakaira [4] state that the concept of variables and functions becomes a serious obstacle for many pupils and even teachers in Japan. The difficulties in learning algebra can be minimized by identifying the pupils' algebraic reasoning so that teachers can design a learning which develops pupils' algebraic reasoning. Algebraic reasoning is the generalization of the mathematical idea of a particular thing through argumentation, and states formally according to the age of the pupils [5]. Algebraic reasoning is a type of reasoning used in solving algebra problems [6] and problem solving can also be used to develop pupils' algebraic reasoning [7]. The problem solving strategy among pupils is different, this is influenced by the learning style that each pupil has [8]. Learning styles are a complex way in which pupils perceive and feel most effective and efficient in processing, storing, and recalling what they have learned [9]. Learning styles are classified into 3 types; visual, auditory, and kinesthetic [10]. Based on these descriptions, it was concluded that learning styles can influence pupils' algebraic reasoning in solving algebraic problems.

Some studies related to algebraic reasoning include relation on cognitive style and problem solving with algebraic reasoning. Chrystotomu [11] which mentioned the cognitive style of visualizer and verbalizer affect the number sense and algebraic reasoning of pupils in solving the problem. Nathan and Koedinger [12] found pupils' problem solving strategies were versatile, so it indicated their algebraic reasoning is different. However, this study differs from previous research which had been explained. This study was not only limited to identifying the influence of cognitive style on the number sense and pupils' algebraic reasoning as well as the difference of pupils' completion strategy in solving the problems related to algebraic reasoning. This research was more focused to describe algebraic reasoning of pupils with visual, aauditory, and kinesthetic learning style in solving algebraic problems deeply. The indicator of algebraic reasoning pupils was adapted from three stages of algebraic reasoning in solving the problems proposed by Herbert and Brown [13] that consist of seeking patterns, recognizing patterns, and generalizations. Based on the above explanation, the purpose of this study was to describe the algebraic reasoning secondary school's pupils with visual, auditory, and kinesthetic learning style in solving mathematics problems.

2. Method

This study was an explorative study with qualitative descriptive approach. Explorative research means extract the necessary information in a study deeply. The approach used in this study is qualitative descriptive because the setting of study is natural background and the main instrument of research is the researcher. First, the researcher gave a questionnaire of learning style, after that a mathematics ability test which consists of 10 mathematical problems in secondary school was given. After getting one visual, one auditory, and one kinesthetic learning style pupils with equivalent mathematics ability, then the researcher gave the written test which consist of mathematical problem related to algebra. After each category had one subject, they were interviewed based on their answer in solving algebra related mathematics problem. The pupils were given a written test for twice. Then the consistency of the data gotten from the pupils' solution and interview were checked by time triangulation. When the data was said valid, it was ready to be analyzed. Data analysis phase in this research consists of 5 steps; data categorization, data reduction, data presentation, data interpretation, and conclusion. After analysis the data, this study was eliciting the algebraic reasoning of secondary school's pupils with visual, auditory, and kinesthetic learning style in solving mathematical problems.

3. Results & Discusion

Algebra is one of the branches of mathematics focusing on problem solving using symbols instead of constants and variables. Part of algebra used in this study is system of two linear equations. The two variable linear equation system is a system/unity of several similar linear two variables equations. linear two variables equations is an equation that has two variables and the rank of each variable is one. The similar linear two variables equations here is two linear equations which contain the same variable. The algebraic reasoning data of each subject was valid, so that it could be analyzed based on the algebraic reasoning indicator which was adapted from three stages of algebraic reasoning in solving the problems proposed by Herbert and Brown [13]. Here is the problem given to each pupil.

Mr. Toni has a garden which is divided into some plots and each plot consists of apple trees and evergreen trees. Mr. Toni spends different maintenance cost for each plot in every month. Both kind of trees were planted in a kind of pattern as shown in Figure 1.

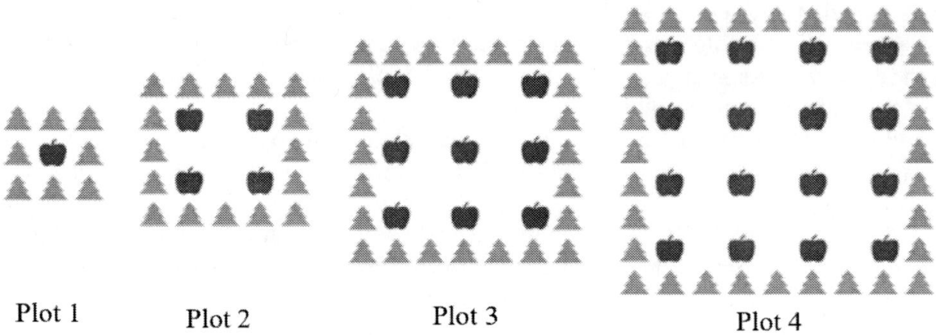

Figure 1. Pattern of Trees and Fruits

- How many apple trees and evergreen trees are there in the 6th-garden plot? Explain your answer!
- Write down the role that helps you identify the number of trees in each garden plot! Explain the role that you have made!
- If the maintenance cost for apple trees and evergreen trees is Rp19.500,00 and Rp46.000,00, respectively in the 1st plot and the 2nd plot, then how much does the maintenance cost for the kth-garden plot? Explain it! (The maintenance cost for each garden plot depends on the number of apple and evergreen trees).

3.1 The Algebraic Reasoning of Visual Learning Style's pupil in Solving Mathematical Problem

When solving the problem, pupil observed the problem in several times and focused to understand the patterns. Then she identified the known information and the asked information. The pupil identified the known and asked information based on her observations on the given problem. Furthermore, in order to find the number of apple and evergreen trees in the sixth plot, pupil represented the amount of both kinds of trees between plots with logical reason. Representing the given problems was like Chrysostomou's opinion [11] which stated algebraic reasoning as the use of various representations related to quantitative situations in a relational way. Furthermore, pupil stated that the number of apples and evergreen trees affected the pattern of each plot since both kinds of trees arranged each plot in the garden. Subject claimed that there was no relationship between the apple trees and evergreen trees regardless the fact that there was a relationship between the number of apple trees and evergreen trees with the order of plots. This indicated that the pupil made a relationship between two quantities.

Pupil declared rules for determining the number of apple and evergreen trees in symbol form. The rule had actually been found by pupil when she worked on the problem (point a). At point a, she had actually stated that to find out the number of apple trees was by using squaring, while to find out the number of evergreen trees was by multiplying the order of the plot by eight. In other words, when deciding on both kinds of trees in the sixth plot he had already recognised the pattern. It meant that the pupil had already made a conjecture related to the rule for determining the number of both kinds of trees in the next plot. To prove the conjecture, pupil used drawing to justify the reason. Pupil drew by observing the pattern of both kinds of trees on each plot. After a relation is established between the illustrations drawn and the rules which had already been found, pupil got same result which meant that the truth of conjecture was proven.

Pupil proved the truth of conjecture, then it could be stated that pupil determined the general rules of the constituent elements of the pattern directly. Pupil said that by using SPLDV, the maintenance cost required in the kth-plot could be determined. Pupils use x and y variables which represented the maintenance costs of each apple and evergreen tree, respectively. Then, pupil created a mathematical model for the first and second plots and look for the values of x and y. *In order to* To determine the general rule of maintenance cost equation required for the kth-plot, pupil directly expressed it in algebraic equation $k^2(3500) + 8k(2000)$.

3.2 The Algebraic Reasoning of Auditory Learning Style's pupil in Solving Mathematical Problem

When solving the problem, pupil read the problem in several times and it sounds like talked to himself about what he understood. Then she identified the known information and the asked information. Furthermore, in order to find the number of apple and evergreen trees in the sixth plot, pupil represented the amount of both kinds of trees between plots using a table, accompanied by logical reason. Representing the given problem was like Chrysostomou's opinion [11] which stated the algebraic reasoning as the use of various representations related to quantitative situations in a relational way. Pupil stated that the amount of apples and evergreen trees affected the pattern of each plot, since both kinds of trees arranged each plot in the garden. Then, pupil identified the relationship between both kinds of trees with the order of plots. Pupil claimed there were no relationship between the number of apples and evergreen trees, whereas there was a relationship between the number of apple and evergreen trees with the order of plots.

Pupil declared rules for determining the number of apple and evergreen trees in symbol form. The rule had actually been found by pupil when she worked on the problem (point a). At point a, she had actually stated that to find out the number of apple trees was by using squaring, while to find out the number of evergreen trees was by multiplying the order of the plot by eight. In other words, when deciding on both kinds of trees in the sixth plot he had already recognised the pattern. It meant that the pupil had already made a conjecture related to the rule for determining the number of both kinds of trees in the next plot. To prove the truth of her conjecture, pupil proved it by drawing. Pupil drew by observing the pattern of both kinds of trees on each plot. After being matched between the illustrations drawn with the rules which had already found, pupil got same result which meant that the truth of conjecture was proven.

Pupil proved the truth of conjecture, then it could be stated that pupil determined the general rules of the constituent elements of the pattern directly. Pupil said that by using SPLDV, the maintenance cost required in the kth-plot could be determined. Pupils use x and y variables which represented the maintenance costs of each apple and evergreen tree, respectively. Then she made a mathematical model for the first and second plots, then found out the values of x and y. She added the maintenance cost required for each kind of tree to determine the general rule of the maintenance cost required equation in the kth-plot. However, pupil declared it in a word (verbally) but she had not expressed it directly by using algebraic symbols. This confirmed that the pupil has been determine the general rule of equation used to solve the problem even though the pupil still wrote it in a sentence rather than in algebraic symbols.

3.3 The Algebraic Reasoning of Kinesthetic Learning Style's pupil in Solving Mathematical Problem

When solving the problem, pupil read the problem in several times and it sounds like talked to himself about what he understood. Then she identified the known information and the asked information. Furthermore, in order to find the number of apple and evergreen trees in the sixth plot, pupil represented the amount of both kinds of trees between plots using a table, accompanied by logical reason. Representing the given problem was like Chrysostomou's opinion [11] which stated the algebraic reasoning as the use of various representations related to quantitative situations in a relational way. Pupil stated that the amount of apples and evergreen trees affected the pattern of each plot, since both kinds of trees arranged each plot in the garden. Then, pupil identified the relationship between both kinds of trees with the order of plots. Pupil claimed there were no relationship between the number of apples and evergreen trees, whereas there was a relationship between the number of apple and evergreen trees with the order of plots.

Pupil expressed rules for determining the number of apple and evergreen trees in algebraic symbol form. The rule was obtained from the difference between the number of apple trees and evergreen trees between plots, thus the pattern of both kinds of trees was gotten. This implied that the pupil made a rule-related to conjecture in determining the number of both kinds of trees in the next plot. Pupil drew the pattern of the 6th-plot which consist of 13 evergreen trees surrounding and apple trees inside for confirming the truth of the conjecture. After being matched between the illustrations drawn with the

IOP Conf. Series: Journal of Physics: Conf. Series **947** (2017) 012061 doi:10.1088/1742-6596/947/1/012061

rules which had already found, pupil got same result which meant that the truth of conjecture was proven. It meant the pupil had proven the truth of the conjecture she had made accompanied by logical reason.

Pupil proved the truth of conjecture, then it could be stated that pupil determined the general rules of the constituent elements of the pattern directly. Pupil said that by using SPLDV, the maintenance cost required in the kth-plot could be determined. Pupils use x and y variables which represented the maintenance costs of each apple and evergreen tree, respectively. Then, pupil created a mathematical model for the first and second plots and look for the values of x and y. She added the maintenance cost required for each kind of tree to determine the general rule of the maintenance cost required equation in the kth-plot. However, the pupil declared it in separated algebraic symbol form, she had not yet written the equation in algebraic symbol form completely. She said that she forgot to write the equation in the complete form which was desired in the given problem. This confirmed that the pupil had been determine the general rule of equation used to solve the problem.

4. Conclusion

As kinesthetic pupils sought the pattern, visual and auditory pupils had already recognized the patterns formed based on the representations they had already made. In addition, when pupils proved the truth of conjecture, the three pupils proved it by drawing the patterns. But there was a difference in finding the number of both kinds of trees. Visual and auditory pupils looked for the number of evergreen trees by counting, whereas kinesthetic pupil used arithmetic series. The visual and kinesthetic pupil determined the general rule of equation used in solving problems using algebraic symbols, while the auditory pupil expressed the general rule that she made verbally (in a word). Based on this it could be said that the visual pupil, auditory pupil and kinesthetic pupil had already satisfied all three algebraic reasoning indicators which were adapted from Herbert and Brown [13].

References

[1] Aljaberi N M 2015 University Pupils' Learning Styles and Their Ability to Solve Mathematical Problems *Int J. of Buss. and Soc. Sci.* **6** pp 152-65

[2] Chrysostomou M, Tsingi C, Cleanthous E, and Pantazi P D 2013 Examining Number Sense and Algebraic Reasoning Through Cognitive Styles *Educ. Stud. Math.* **83** pp 205-23

[3] DePorter B and Hernacki M 2003 *Quantum Learning: Membiasakan Belajar Nyaman & Menyenangkan* Suyono (Bandung: PT.Mizah Pustaka)

[4] Herbert K and Brown H 2000 Pattern as Tools for Algebraic Reasoning (B. Moses, Ed.) Algebraic Thinking Grade K-12 Reading from *NCTMs School Based Journals and Other Publication* pp 123-28

[5] James B W and Gardner L D 1995 Learning Styles: Implications for Distance Learning *Version of Record Online*

[6] Kaput J and Blanton M 2005 *Algebrafying the elementary mathematics experience in a teacher-centered, systemic way* (Mahwah, NJ: Lawrence Erlbaum Associates).

[7] Lew H C 2004 Developing Algebraic Thinking in Early Grades: Case Study of Korean Elementary School Mathematics *The Math Edu* **8** pp 89-106

[8] Ontario 2013 *Paying Attention to Algebraic Reasoning K-12* (Toronto: Queen's printer for ontario)

[9] Richardson K, Berenson S and Staley K 2009 *The J. of Math. Behav.* **28** pp 188–99

[10] Tall D 1991 *Advanced mathematical thinking* (The Netherlands: Kluwer Academic Publisher)

[11] Toshiakaira F 2003 *Probing pupils' understanding of variables through cognitive conflict: Is the concept of a variable so difficult for pupils to understand* N A Pateman B J, Dougherty and J T Zilliox

[12] Windsor W 2010 Algebraic Thinking: A Problem Solving Approach *Shapping the Future of Mathematics Education: Proceesing of the 33th annual conference of the Mathematics Education Research Group of Australia* (Fremantle: MERGA)

[13] M J Nathan and K R Koedinger 2000 Teachers' and researchers' beliefs about the development of algebraic reasoning *J. Res. Math. Educ* **31** pp 168–90

Development of Problem-Based Learning Oriented Teaching Learning Materials to Facilitate Students' Mastery of Concept and Critical Thinking Skill

M Reza[1], M Ibrahim[2], Y S Rahayu[3]

[1]Universitas Negeri Surabaya, Indonesia

Email: [1]muhamadreza@mhs.unesa.ac.id, [2]musliminibrahim@unesa.ac.id, [3]yunirahayu@unesa.ac.id

Abstract. This research aims to develop problem-based learning oriented teaching materials to improve students' mastery of concept and critical thinking skill. Its procedure was divided into two phases; developmental phase and experimental phase. This developmental research used *Four-D Model*. However, within this research, the process of development would not involve the last stages, which is disseminate. The teaching learning materials which were developed consist of lesson plan, student handbook, student worksheet, achievement test and critical thinking skill test. The experimental phase employs a research design called *one group pretest-posttest design*. Results show that the validity of the teaching materials which were developed was good and revealed the enhancement of students' activities with positive response to the teaching learning process. Furthermore, the learning materials improve the students' mastery of concept and critical thinking skill.

1. Introduction

As one of educational aspects, learning science plays a vital role in upgrading education quality, particularly in forming qualified human resources, who are able to think critically and creatively to overcome problems and apply the knowledge in daily life. The fundamental dilemmatic problem often faced by the primary school teachers is the fact that mostly the students cannot correlate between what they have learned and how to use it in real life [1]. This happens on account of their comprehension of the material, which is merely in abstract forms, i.e. unlike things they need in the society. The conceptual understanding is important for every student because by extensively and profoundly knowing it, they would be able to apply it for various purposes [2]. One of mental processes in understanding concept is critical thinking skill which is defined as cognitive process of gaining knowledge.

According to Kirmizi [3], critical thinking is briefly a mental process for perceiving the world using our existing knowledge. Akyuz [4] stated that critical thinking is a thinking skill which consists of mental processes of discernment, analysis and evaluation. Critical thinking comprises of such processes as featuring intelligence about a matter, deciding and judging accurately [5]. In detail, critical thinking is a way of approaching and solving problems based on persuasive, logical and rational arguments, which involve verifying, evaluating and choosing the right answer to a given task and reasoned rejection of other alternatives solutions [6].

To create and escalate students' mastery of concept and critical thinking skill, the researcher uses Problem-Based Learning (PBL). This model is one of an instructional model in which a problem is designed as the center of the learning [7]. The problem is designed such that the students can solve in small groups and reflect on their experiences [8]. The model may improve the teaching and learning process because it focuses more on the developing students as self-directed learners compared to traditional models that encourage students to be spoonfed by teachers [9]. Thus, this model can be considered as a student-centered method of teaching that involves learning through solving unclear but genuine problems. In detail, students analyse the problem and context and apply deductive and inductive processes to understand the problem in order to find a possible solution or solutions [10]. This model turns the students from passive information recipients to active ones, free self-learners and problem solvers, and slides the emphasis of educational programs from teaching to learning [11].

In order to manifest a significant learning pattern, the teacher should maintain the teaching in such a way, so that the students can get many experiences. Teachers as the learning management at educational units are expected to prepare and develop qualified learning media, including lesson plans, student worksheet, assessment sheet, and student teaching materials to reach the learning goals. Therefore, it is necessary to conduct developmental research to evolve the science learning oriented to the learning model based on the problem to train the students.

2. Method

This research is a developmental research, as, within it, the science teaching learning material oriented to problem-based learning model is developed. The program is designed to help the students' mastery of concept and critical thinking skill. The teaching and learning material comprise lesson plans, student teaching materials, student worksheet, and assessment sheet which were tested in the classroom. The developed learning materials implemented in a class with 25 pupils of the fourth grade. The research procedure is divided into two phases, i.e. developmental phase and experimental phase.

2.1 Developmental phase

The developmental phase aims to produce better learning through revision and validation from experts. This developmental research refers to 4D (*Four-D Model*) by Thiagarajan [12]. The procedure of the development consists of four stages namely, define, design, develop, and disseminate. However, within this research, the process would not involve the last stage, i.e. disseminate. The researcher chooses the model because it is detailed in its stages, systematical, and suitable for the development of learning oriented to PBL model.

2.2 Experimental phase

This phase employs a research design called "*One group pretest-posttest design*".The observed variable consists of 7 elements. They are validation of learning media, implementation of the lesson plans, students' activities, students' responses, concept comprehension, critical thinking skill, and obstacles in learning. Therefore, to determine the quality of the developed teaching and learning material, it is required to obtain three kinds of data, i.e. validity, practicality and effectiveness.

3. Results and Discussion

The results of developing the science teaching and learning material oriented to problem-based learning model to train the students' mastery of concept and critical thinking skill are described as follows:

IOP Conf. Series: Journal of Physics: Conf. Series **947** (2017) 012062 doi:10.1088/1742-6596/947/1/012062

3.1 Validation of Learning Media

The result of the validation of learning media showed in table 1.

Table 1. The result of learning media validation

No	Learning Media	Average Score	Assessment Category
1.	Lesson plans	3.83	Good/Valid
2.	Student teaching materials	3.76	Good/Valid
3.	Student worksheet	3.73	Good/Valid
4.	Concept comprehension test	3.75	Good/Valid
5.	Critical thinking skill test	3.65	Good/Valid

3.2 Readability of Student Teaching Materials and Student Worksheet

The percentage of students' reading material readability and student worksheet for the content aspects got the interesting response percentage of 84% which means 84% of the students were very interested in the teaching material and the student worksheet, while 16% of the students were not. For the aspect of performance for teaching materials and the student worksheet, the researcher obtained percentage of students who were interested in the performance of students teaching materials which was 92% and 88% for student worksheet, while 8% for those who were not interested in the performance of students learning materials and 12% for student worksheet.

In terms of difficulty aspect of the explanation of activities in the teaching materials and worksheet, 20% of the students argued that it was difficult, while 80% of them argued that it was not difficult. Meanwhile, for the image aspect, 88% of students thought that the drawings on the teaching materials and 84% for the worksheet were easy to understand. 12% of them thought that the drawings on the student teaching materials and 16% of them for the student worksheet were difficult to understand. Hereinafter, on the aspect of the question, 88% of students said that the questions in student teaching materials were understandable, and 84% of them said so for the student worksheet. On the other hand, 12% and 16% of the students argued that the questions on the teaching materials and worksheet were hard to be understood.

3.3 Lesson Plan Implementation

All learning activities in Lesson Plan were implemented, with the average value obtained by each aspect in the amount of 3.75 was a good category.This indicates that the lesson plan using the overall problem-based learning model has been implemented in the classroom learning process.

3.4 Student Activity

During the learning, the researcher conducted an observation of the student activities. The observation results can be presented as in Table 2.

IOP Conf. Series: Journal of Physics: Conf. Series **947** (2017) 012062 doi:10.1088/1742-6596/947/1/012062

Table 2. The result of student activities observation

No	Students' Activity	Meeting I (%)	Meeting II (%)	Average Percentage of Students' Activity (%)
1	Listening/paying attention to the teacher	12	9	10.5
2	Taking note and discussing with student/teacher	13	8	10.5
3	Reading and doing the student worksheet and teaching materials	14	14	14
4	Preparing the equipment and materials for the experiment/observation	13	13	13
5	Working in group	12	13	12.5
6	Presenting the group work	6	8	7
7	Discussing with the other student and the teacher and getting any response from the teacher concerning about the presentation	10	13	11.5
8	Asking/responding the question from the other student or group, or even the teacher in class discussion	9	11	10
9	Giving conclusion related to the learning	7	9	8
10	Behaving irrelevantly	4	2	3
	Total	100	100	100

3.5 Students' response

To elicit the students' response towards the learning, a questionnaire is given to the subjects. The analysis in the percentage of subjects' response is described as follows: for the material/content, teaching materials (learning media), student worksheet, learning atmosphere and the way teachers teach, 80% of students were very interested and the rest were interested enough. For the learning materials/content components, teaching materials (learning media), student worksheet, learning environment and how teachers teach, 83.2% of the students stated that it was very new and 16.8% of others stated that those aspects were clear enough. Regarding the students' responses to the language component of learning media, material/content of teaching materials, examples of problems, worksheet, experiment/observation in the student worksheet, 80.8% of students stated that it was very easy, 14.4% students stated that it was quite easy and 4.8% of students stated that it was quite difficult. For the response to the component of the teacher's explanation and guidance during the learning of concepts and the completion of the worksheet, 84% of students stated that it was very clear, while the other 16% stated quite clear.

Regarding the components of the teacher's way of using the problem-based learning model, which are orienting problem, organizing students, guiding investigation, developing and presenting the work, analyzing and completing it, 82.4% of the students said it was very good, 11.2% stated it was good enough, 4% stated it was less good. The students responded to the learning component using the problem-based learning model, which are orienting problem, organizing students, guiding the investigation, developing and presenting the work, analyzing and evaluating, 80% of them stated it was very easy, 16% of them stated that it was quite easy, and 4% of them stated it was quite difficult. The students' response to the critical thinking skill component, using a credible source, writing reasons based on existing data or arguments, accepting or rejecting arguments, ideas, or decisions accompanied by clear and logical reasons, developing and maintaining an opinion, logical reasoning, planning experiments and planning, defining terms using their own words, being able to think openly, giving information well, and making conclusions was 81.2% of students stated it was very easy, 13.6% of them stated it was easy enough, and 5.2% of them stated it was quite difficult. As the response to critical thinking test items and concept comprehension test, 84% of the students stated it was very easy, 16% of the students stated it was quite easy, 86% stated it was very new and 14% of them stated it was quite new. Thus, in general, the students give a very positive response to learning science with PBL.

3.6 Concept comprehension test

The test consisted of 10 items of multiple choice questions and was given twice, i.e. pretest and final test (post-test). The average value of the students' concept comprehension test at the pretest time was in the amount of 5.20 and 8.52 for the posttest. In the pretest, the number of students was six, while in post-test, all students got the complete value on the mastery criterion (value ≥ 7). The average sensitivity of concept comprehension test was 0.33 (item of sensitivity to learning process). The average score of gain-score in concept comprehension test was 0.67 with a g-moderate category.

3.7 Critical thinking skill test

The test was measured using the form of a description test. It consisted of ten items that have been adapted to the indicators of critical thinking skill. In the pretest, all students did not gain the average score of 4.08, while at the post test there was only one student who did not gain the average score of 7.49. Average sensitivity was about 0.34 with category "sensitive". Then, the average gain-score value of critical thinking skill was 0.58 with the g-moderate category.

4. Conclusion

Based on the results of media experiment, analysis, discussions, and findings in the learning process, it can generally be concluded that the media of science learning model based on developed problems have been valid, practical, and effective to improve the understanding of concepts and train the critical thinking skill of students.

References

[1] Muslich M 2008 *KTSP Pembelajaran Berbasis Kompetensi dan Kontekstual* (Jakarta: Bumi Aksara) p 40

[2] Ibrahim M 2005 *Asesmen Berkelanjutan. Konsep Dasar, Tahapan Pengembangan dan Contoh Surabaya* (Surabaya: Unesa University Press)

[3] Kirmizi F S, Saygi C, and Yurdakal, I H Determine 2015 The relationship between the disposition of critical thinking and the perception about problem-solving skills *Procedia - Social and Behavioral Sciences* **191** pp 657- 61

[4] Akyuz H I and Samsa S 2009 The effects of blended learning environment on the critical thinking skills of students *Procedia Social and Behavioral Sciences* **1** pp 1744–1748

[5] Sarigoz O 2012 Assessment of the high school students' critical thinking skills *Procedia - Social and Behavioral Sciences* **46** pp 5315 – 5319

[6] Florea N M and Hurjui E 2015 Critical thinking in elementary school children *Procedia - Social and Behavioral Sciences* **180** pp 565 – 572

[7] Jonassen D H and Woei H 2008 All problems are not equel: implication for problem based leraning *Interdiciplinary Journal of Problem-based Learning* **2**(2) pp 6-28

[8] Selcuk G S and Caliskan S 2010 A small-scale study comparing the impacts of problem-based learning and traditional methods on student satisfaction in the introductory physics course *Procedia Social and Behavioral Sciences* **2** pp 809–813

IOP Conf. Series: Journal of Physics: Conf. Series **947** (2017) 012062 doi:10.1088/1742-6596/947/1/012062

[9] Harun N F Yusof K M Jamaludin M Z and Hassan S A H S 2012 Motivation in problem-based learning implementation *Procedia - Social and Behavioral Sciences* **56** pp 233 – 242

[10] Etherington M B 2011 Investigative primary science: aproblem-based learning Approach *Australian Journal of Teacher Education* **36** (9) pp 53-74

[11] Akınoglu O and Tandogan R O 2007 The Effects of Problem-Based Active Learning in Science Education on Students' Academic Achievement, Attitude and Concept Learning *Eurasia Journal of Mathematics, Science & Technology Education* **3**(1), pp 71-81

[12] Ibrahim M 2005 *Asesmen Berkelanjutan. Konsep Dasar, Tahapan Pengembangan dan Contoh*. (Surabaya: Unesa University Press)

IOP Conf. Series: Journal of Physics: Conf. Series **947** (2017) 012063 doi:10.1088/1742-6596/947/1/012063

Training of Students' Critical Thinking Skills through the implementation of a Modified Free Inquiry Model

S A Hadi[1], E Susantini[1], and R Agustini[1]

[1]Faculty of Mathematics and Natural Science, Universitas Negeri Surabaya, Surabaya 60231, East Java, Indonesia

Email: endangsusantini@unesa.ac.id

Abstract. This research aimed at training students' critical thinking skills through the implementation of a modified free inquiry learning model. The subjects of this research were 21 students of Mathematics Semester II. Using *One-Group Pretest-Posttest Design,* the data were analyzed descriptively using N-gain indicator. The results indicate that the modified free inquiry learning model was effective to train students' critical thinking skills. The increase in the students' critical thinking skills viewed from the value of *N-Gain* has a range of values with the categories of medium and high with a score between 0,25-0,95. Overall, the change in *N-Gain* score of each student and each indicator of critical thinking skills is as increasing with a moderate category. The increase of *N-Gain* value is resulted from the fact that the students were directly involved in organizing their learning process. These criteria indicate that the modified free inquiry learning model can be used to train students' critical thinking skills on photosynthesis and cellular respiration materials. The results of this research are expected to be nationally implemented to familiarize students with andragogy learning style which places the students as the subjects of learning.

1. Introduction

Critical thinking is a sensible idea of what is believed and what needs to be done [1]. Critical thinking skills can also be interpreted as the ability to think logically, reflectively, and productively to assess the situation in order to make the right decision [2]. The students' thinking skills are important to be improved, in line with the discourse of improving the quality of education in accordance with learning objectives. The purpose or the success of a learning process is achieved when it meets three important criteria, namely the aspects of knowledge, attitude and skills. These three criteria are formed from the habits and reinforcement that become the character which relies on someone's mindset [3].

Learning activities should be designed to direct the students in developing their attitudes, knowledge, and critical thinking skills in varying combinations and emphases [4]. Practicing critical thinking skills in varying emphases can be accessed in some indicators achieved in a learning process. The indicators selected in this research included: making decisions to prepare the steps of problem-solving experiments, analyzing, making inferences, evaluating, and drawing conclusions. These indicators were adapted from Ennis [1]. The efforts to direct students in practicing critical thinking skills can be accomplished using student-centered learning models. The students' which is already at the formal operational stage (adults) tend to perform *independent* learning styles that allow them to gain a deep understanding of an issue in accordance with the purpose of critical thinking skills [5].

Critical thinking skills which are identical to *independent* learning style can be applied by using constructivist learning models, such as inquiry-based learning. Learning models with investigation activities, such as inquiry, involve students fully in the teaching and learning activities either in groups or individual, which is one way to develop students' critical thinking skills [6]. The inquiry learning model is used to train students' critical thinking skills because the syntax of inquiry has a correlation with the indicators of critical thinking skills developed in this research. The selected five indicators had been adjusted to the cognitive level of the students who were in the formal operational phase. The abilities at the formal operational stage consist of thinking abstractly, testing hypothesis, and forming concepts which do not depend on physical reality [7]. The students' cognitive level became the reference for the researcher to determine the appropriate level of inquiry for the sample of the experiment, by applying the modified free inquiry model.

The results of the observations conducted in July 2016 showed that the students' critical thinking skills, especially the subjects of this research, were still relatively low. The students tended to listen to the explanations from the lecturer without being actively involved in the learning process. The material that should have been implemented in investigation activities which involved students actively was not optimal because it was delivered by using lecturing method. Such learning activities tend to make the students memorize the material without understanding the content of the learning material presented. The topic in this research, photosynthesis and cellular respiration, is chosen regarding the fact that this material required the students to understand, apply, analyze factual, conceptual and procedural knowledge. The students' curiosity related to the process of photosynthesis and cellular respiration may influence their learning positively so that the students could process, reason and present the information related to the learning material.

There have been several studies investigating the training of critical thinking skills through the implementation of inquiry learning model by previous researchers, such as Qing *et al.* The study suggests that the critical thinking skills of prospective teachers increased when inquiry-based learning was implemented [8]. Greenwald *et al* state that the inquiry-based learning is an alternative solution substituting conventional learning to improve the students' critical thinking skills and clinical reasoning [9]. Sutarma *et al* state that there is an increasing trend in the students' critical thinking skills with the implementation of inquiry learning model compared with the direct learning model [10]. Nevertheless, the previous research conducted by researchers did not explain the level of inquiry of the sample used, so that it did not give an illustration of the types of inquiry applied to the research sample. Therefore, in this research, a modified free inquiry model was implemented to train students' critical thinking skills with various considerations that had been previously analyzed.

2. Research methods

The type of this research is pre-experimental research with *one group pretest-posttest design.* The subjects of this research were 21 mathematics students in the mathematics and natural science faculty of IKIP Mataram who were joining Basic Biology course. This research was conducted for 3 weeks starting from 17th to 29th of May 2017. The learning materials which were applied to train the students' critical thinking skills were developed by the researcher. The modification of free inquiry-based learning material on photosynthesis and cellular respiration material is developed as learning material. The learning materials implemented in this research had passed the validity test with valid results.

The research instrument used to measure the students' critical thinking skills was an essay test instrument consist of 5 items. The test was said to be sensitive to differentiate the students' ability before and after receiving the learning process because the test instrument had a positive sensitivity index with scores between 0,52 – 0,68 [11]. This test was given in two stages, the *pretest* and the *posttest* to determine the students' critical thinking skills before and after the learning process.

The teaching and learning process was carried out for 3 meetings supported with a student worksheet on each learning material to be discussed. The student worksheet was developed with the orientation of the modified free inquiry model, so that the percentage of students' involvement in the learning process became more dominant. The steps provided in the student worksheet involve problem orientation, formulating problems, formulating hypotheses, identifying and defining variables, developing work procedures, analyzing data, compiling inferences, evaluating, and drawing conclusions. Students could develop their critical thinking skills through the steps provided in the student worksheet.

The score of students' critical thinking skills derived from subsequent tests is categorized according to the level of critical thinking skills of each student. Critical thinking skill level consists of 4 levels, that are very less critical level with test score t satisfy $25.00 < t \leq 43.75$, less critical level with test score t satisfy $43.75 < t \leq 62.50$, critical level with test score t satisfy $62.50 < t \leq 81.25$ and very critical level with test score t satisfy $81.00 < t \leq 100.00$ [12].

A improvement of the students' score of critical thinking skills was analyzed by using *N-Gain* equation [13]. The *N-Gain* value of the students' critical thinking skills can be defined by the criteria of score as the following: high> 0.70; Moderate 0.3 - 0.70; Low <0.30 [14]. The results of the students ' critical thinking skill level analysis and the increase of N-Gain score become the benchmark in determining whether this developed learning tool is effectively used to train students' critical thinking skills.

3. Results and Discussion

3.1. The Scores of Students' Critical Thinking Skills

The score of the students' critical thinking skill and the increase of *N-Gain* score are presented in Table1.

Table 1. Results of N-Gain analysis of students' critical thinking skills

Students'	Scores		Category	*N-Gain*	Category
	Pretest	*Posttest*			
A1	10	65	Critical	0,61	Middle
A2	10	60	Less Critical	0,55	Middle
A3	0	25	Very Less Critical	0,25	Low
A4	20	60	Less Critical	0,5	Middle
A5	15	50	Very Less Critical	0,41	Middle
A6	35	90	Very Critical	0,85	High
A7	0	70	Critical	0,7	High
A8	0	60	Less Critical	0,6	Middle
A9	0	70	Critical	0,7	High
A10	10	60	Less Critical	0,55	Middle
A11	10	60	Less Critical	0,55	Middle
A12	10	70	Critical	0,67	Middle
A13	0	70	Critical	0,7	High
A14	0	80	Critical	0,8	High
A15	20	85	Very Critical	0,81	High
A16	0	95	Very Critical	0,95	High
A17	15	60	Less Critical	0,65	Middle
A18	0	60	Less Critical	0,6	Middle
A19	5	70	Critical	0,68	Middle
A20	20	95	Very Critical	0,94	High
A21	0	75	Critical	0,75	High

Table 1 shows that, in *the pretest,* none of the students achieved the minimum accepted score that had been set, which was 60. While, in the *posttest,* 19 students achieved the minimum accepted score and 2 students did not pass the test because of several factors, such as the students' activities were not productive, the inability of the students in understanding the concepts, and several internal factors faced by the students themselves. Unproductive activities, such as do not paying attention to lecturer's explanation, became a crucial cause of student's failure in passing the test. The intensity of the student's attention toward learning had an effect on the level of memorization/absorption of information [14].

IOP Conf. Series: Journal of Physics: Conf. Series **947** (2017) 012063 doi:10.1088/1742-6596/947/1/012063

Students were said to complete the teaching and learning classically with the percentage of 90.48%. Thaiposri and Wannapiroon in their research state that the learning activities with an inquiry-based learning model can help students develop their critical thinking skills [15]. The samples used in the research of Thaiposri and Wannapiroon were junior high school students, while the samples of this research were the first-semester university students.

The students' incompleteness of the critical thinking skill test could also be caused by the concept understanding on the photosynthesis and respiration material. The level of the students' concept understanding on the learning material had an effect on their ability in solving the critical thinking skill test questions. The students with the code A3 and A5 who did not pass the test or not complete showed that their concept understanding was still low. This finding is evidenced with answers of students A3 and A5 on critical thinking skills tests that deviate far from the concept of the answer that should be. The A3 and A5 student answer sheets are shown in Figure 2. The failure in understanding concepts would results on the inability to complete the critical thinking skill tests. Concept understanding has a positive correlation with students' critical thinking skills. Asmawati in her research states that critical thinking skills are the basic requirement for the students in understanding the concepts in the discipline [16]. The research by Asmawati did not explain the importance of concept understanding in order to be critical in a problem solving as presented in this research.

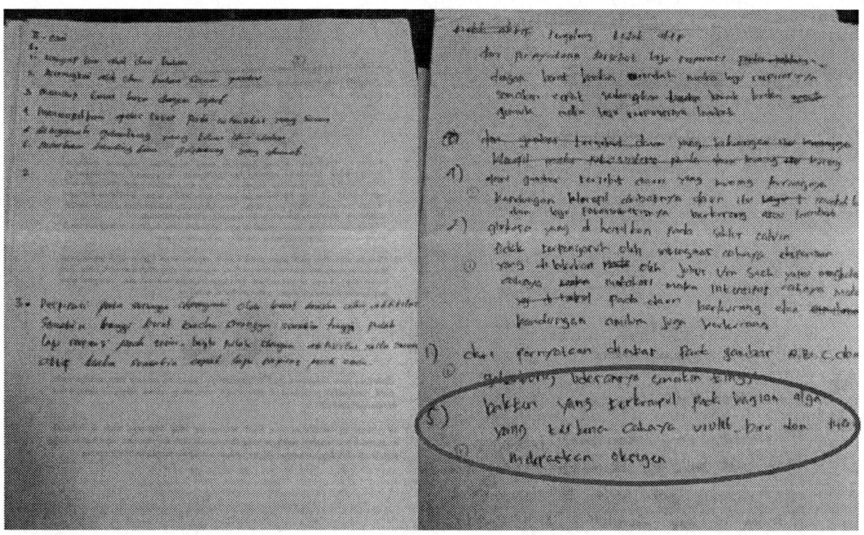

Figure 1. The A3 & A5 Student Response Sheet indicating the concept of incomprehension

Figure 1 shows that, students A3 and A5 have not mastered the concept of photosynthesis and cellular respiration well. The answer given by the students are not based on the correct concept. For example, in number 5 answer, students stated "bacteria collected in algae that gterkena violet light, blue and red release oxygen". The correct concept is "all algae release oxygen as a by-product of the process of photosynthesis carried out by algae. Algae exposed to violet, blue, and red light release the most amount of oxygen compared to light-induced algae with other color spectra. The abundance of oxygen released by algae, invites aerobic bacteria to gather more around the algae releasing more oxygen, which is exposed to violet, blue and red lights". The student's answer in Figure 1 shows that A3 and A5 students have not mastered the concept well so that it becomes one of the causes of his unresponsiveness in answering critical thinking skills test.

3.2. The completeness of the Indicators of Critical Thinking Skills

The indicators compiled in this research are important to be analyzed whether they were stated to be completed or not, because the indicators are the teachers' references in determining whether the learning process in the class is successful. The completeness of the critical thinking skill indicators is presented in Table 2.

Table 2. Results of N-Gain analysis and completeness of each critical thinking indicator

Indikator	Proportion Scores		Category	*N-Gain*	Category
	Pretest	*Posttest*			
Develop an experimental procedure	0,07	0,7	Complete	0,68	Middle
Analyze	0,12	0,64	Complete	0,6	Middle
Arrange inferences	0,1	0,77	Complete	0,75	High
Evaluate	0,1	0,67	Complete	0,63	Middle
Compile conclusions	0,05	0,62	Complete	0,6	Middle

The results of the analysis of the increase in the score of critical thinking skills *(N-Gain),* which has been presented in Table 2 show that none of the indicators were completed in the *pretest*, while in the *posttest* all indicators were completed. The *N-Gain* value of each indicator ranges from 0.6 to 0.75 with medium and high categories. These criteria indicate that the learning process using a modified free inquiry model is effective to train students' critical thinking skills on photosynthesis and respiration materials. The level of critical thinking skills of students after the application of learning tools developed, put the students at critical level and very critical although there are 2 students who are still at a less critical level. These results are supported by Kitot *et al* their research, stating that the inquiry-based learning model is effective to be used to enhance students' critical thinking skills [17]. The research by Kitot *et al* only compares the inquiry learning model with conventional models without specifying the type of inquiry used. The indicator with the lowest score increase was analyzing and making conclusions. Students still had difficulties in analyzing and making conclusions because the students were still not maximal in preparing editorial sentences, but the students undersood the concepts well. This was proven that the students were abe to answer the questions although their sentences were not correctly organized.

4. Conclusion

Based on the result of the research, it can be concluded that the modified free inquiry learning model is effective to be used to train students' critical thinking skills on photosynthesis and respiration materials. This is evidenced by the increase in scores of the *N-Gain,* both the *N-Gain* of each student and the *N-Gain* of each indicator. The findings of this research indicate that critical thinking skills have a possitive correlation with concept understanding.

Acknowledgement

I would like to express my gratitude to advisor 1 and 2 who have participated in the writing of this paper until it can be published and useful for many people. I would also like to thank IKIP Mataram institute which has given permission to conduct this research.

References

[1] Ennis R H 1996 *Critical Thingking* (New York: Prentice Hall Inc)
[2] Desmita 2009 *Psychology Development of Learners* (Bandung: Remaja Rosdakarya)
[3] Kuswana W S 2013 *Thinking Taxonomy* (Bandung: Remaja Rosdakarya)
[4] Ibrahim M 2008 *Innovative Learning Model of IPA Through Meaning* (Surabaya: Departemen Pendidikan Nasional Balitbang)
[5] Jhonson E B 2002 *Contextual Teaching and Learning, What It Is and Why it's here to Stay* (California: Corwin Press)
[6] Kardi S 2008 Direct Inquiry *Learning Model, Technology Science, and Society* (Surabaya: UNESA)

[7] Slavin R E 2009 *Educational Psychology Theory Into Practice 9th ed* (New Jersey: Pearson Education, Inc)
[8] Qing Z, Jing G, and Yan W 2010 Promoting reservice Teachers Critical Thinking Skills By Inqury Based Chemical Experiment *Elsevier ltd.* (Doi: 10.1016/j.sbpro.2010.03.737)
[9] Greenwald R R and Quitadamo I J 2014 A Maind of Their Own: Using Inquiry Based Teaching to Build Critical Thinking Skills and Intelectual Engagement in an Undergraduate Neuroanatomy Course *The Journal of Undergraduate Neurosciense Education JUNE*
[10] Sutarma I N, Arnyana I B P, and Swasta I B J 2014 Effect of Inquiry Learning Model on Critical Thinking Skills and Scientific Performance on Biology Lesson of Class XI IPA SMA Negeri 2 Amlapura *e-Journal Program Pascasarjana Ubiversitas Pendidikan Ganesha* Vol **4**
[11] Okonkow and Osuji 2006 *Measurement and Evaluation* (Nigeria: National Open University of Nigeria)
[12] Setyowati S and Mosaik 2011 Implementation of Cognitive Conflict Approach in Physics Learning to Grow Students' Critical Thinking Ability *Jurnal Pendidikan Fisika Indonesia*
[13] Hake R R 1999 *Analyzing Change/Gain Scores* (Hatteras Street, Woodland Hills, CA, 91367 USA. Indiana University)
[14] Ratumanan T G 2015 *Learning and Learning As Well As The Factors That Affect The 3rd Edition* (Surabaya: Unesa University Press)
[15] Thaiposri P and Wannapiroon 2014 Enhancing Students Crtical Thinking Skills Through Teaching and Learning by Inqury Based Learning Activities Using Special Network and Cloud Cpmputing *Elsevier* (Doi: 10.1016/j.sbpro.2015.02.013)
[16] Asmawati and Eka Y S 2016 Student Worksheet (LKS) Using Guided Inquiry Model to Improve Critical Thinking Skill and Mastery of Student Concept *Jurnal Pendidikan Fisika Universitas Muhammadiyah Metro* Vol **3** No 1
[17] Kitot Awg K, Ahmad A R, and Seman A A 2010 The Effectiveness of Inquiry Teaching in Enhancing Students Critical Thinking *Elsevier* (Doi: 10.1016/j.sbpro.2010.10.037)

IOP Conf. Series: Journal of Physics: Conf. Series **947** (2017) 012064 doi:10.1088/1742-6596/947/1/012064

The Profile of Student Misconceptions on The Human and Plant Transport Systems

M Ainiyah[1], M Ibrahim[2], M T Hidayat[3]

[1,2,3]Science Education Department, Postgraduate, Universitas Negeri Surabaya, Campus Ketintang Surabaya 60231, Indonesia.

Email : mujizatul.ainiyah23@gmail.com

Abstract. This research aims to identify misconceptions on the humans and plants transportation systems. The research was done in the 8th grade in Indonesia. Data were collected to use a three-tier test. This type of research was used survey design. Content analysis was used to analyze the misconception data. The results of this research were the location of misconception of each student is different. The highest misconceptions identified in this research, namely: a) arteries that drain blood to the heart (73.3%); b) veins that drain blood from the heart (70.0%); c) place of oxygen and carbon dioxide exchange occurs in the veins (66.7%); d) blood pressure in veins greater than in capillaries (63.3%); e) absorption of water occurs diffusion and absorption of minerals occurs osmosis (76.7%); f) transport of photosynthesis process occurs by diffusion (66.7%); g) photosynthesis process occurs during the day (63.3%); and h) process of evaporation of water through the leaves are guttation (56.7%). The results of this research show that the level of students misconceptions on the of human and plant transport systems is still high so that it can serve as a reference to improve the learning process and the reduction of student misconceptions.

1. Introduction

The learning process in several areas in Indonesia still not takes place effectively because there is still a lot of students who have not yet reached the optimal mastery learning. The learning process is said to be effective if it can find out the condition of the students for achieving learning progress optimally. Failure in learning can be caused by students learning process difficulties resulting from misconception within the students. Learning will be more effective if teachers know students learning difficulties, one of which was caused by the misconception on students, because by knowing the learning difficulties experienced by the student teachers can find solutions to the issues facing students so that learning can be effective and can reduce misconceptions experienced students.

The research that has been done researchers in the journal "Science Education" has shown that students have their own student conceptions about the phenomenon of nature [1]. Siswa came to school with early knowledge about the phenomena that will be taught in learning, but many students develop an understanding of science concepts only to a limited extent only after the learning process in school [2]. If the initial construction conflicts with scientific concepts that are accepted by the experts then students are having a misconception, it acts as a barrier and prevents success in learning [3].

The level of understanding of the ability of students in receiving the subject matter varies between one student with other students. Some students have not yet understood the concept well and have a different concept with a concept embraced by the experts so that students have difficulty in understanding the concept to explain various symptoms of nature. students bring the initial idea about the natural phenomenon during learning in class, but there are times when the idea is inconsistent with the idea of generally accepted by experts [4]. The discrepancy of understanding the concept of beginning students with the concept of experts is often referred to as a misconception.

The misconception is the idea or view of a concept belonging to someone not in accordance with the agreed concept and is considered correct by the experts. The views are not appropriate (wrong) it is resistant and persistent [5]. The conception of the truth is the conception that fits the scientific concept based on the consensus of the experts [6]. The misconception that occurs in students caused by several sources, including names of students, teachers, textbooks used, context, and how to teach the teacher [7].

Identification of the misconception of students in learning science is the most important part [8; 9; 10; and 11]. Identification the misconception is the first step in teaching science. The role of science teachers is to know the misconception of students and guide students towards a better understanding through direct activities, investigation, and interaction with peers and adults [12]. Investigation on children's early knowledge is essential for students and teachers [13].

Students carry misconception themselves to school when learning science and the misconception that owned these students be a hindrance in developing new concepts. This is what led to the misconception that most students due to the initial concept (preconceptions) which is then brought to a formal education. When students have a misconception on any level, the role of teachers is important because it is necessary to remove the misconception that occurs in students. If the misconception is not removed from the students, then the student will feel unfamiliar with his new knowledge because students can't afford to assimilate new knowledge to the level that exists [14].

Research on understanding the concept of diffusion and osmosis obtained the results that the degree of misconception about the movement of the particles on the concept of diffusion and osmosis is quite high [15]. Research on the misconception on photosynthesis and respiration in plants obtained the results that the process of photosynthesis as energy providers process [16].

Concepts in science are the concepts that are interrelated and cannot be separated so that the learning of science should be taught in an integrated to obtain the understand students are intact don't break back. The material transport system of humans and plants is a material that has a connectedness with other concepts which include the circulatory system, diffusion and osmosis, and transportation systems in plants. This material has a high level of understanding of one concept with another concept that is interconnected. An act of diagnosis by using three-tier tests need to be done to detect the misconception of students on the material transportation system of humans and plants.

Three tiered test is the most valid form of test instruments, reliability, and correct the misconception of students rather than to identify instruments that shaped one tier and two tier test [17]. The three-tiered test is a three-tier test that is composed of three levels. The first level (one-tier) in the form of multiple choice to know the content, the second level (two-tier) is the choice of reason based on answers on the first level, and the third level (three-tier) in the form of a question the assertion of belief of the answer which has been selected in the first and second level [18; 19; and 20]

From the explanation above, this research aims to identify the misconception of students on the material transportation system of humans and plants. This research is expected to know the layout of the misconception of students that cause learning difficulties and low levels of student mastery learning outcomes.

2. Research method

Type of this research is descriptive research i.e. describe or illustrate misconceptions level students on the material transportation system of humans and plants. The research was done on the 8th grade Junior High School in Indonesia amounted to 30 students. The research design used, namely the research survey. Data collection method using three-tier-based tests. The research instrument used by researchers is the sheet three-tier test. Data analysis using quantitative descriptive. Content analysis was used to analyze the misconception data. Analysis of the level of understanding and misconceptions students can be done by looking at the results answer the student after working on the tests given. The categories that classify student answers written in Table 1.

Table 1. Guidelines for Assessment of Student Answers

First-tier	**Second-tier**	**Third-tier**	**Category**
True	True	sure	Understand concept (TK)
True	Wrong	sure	Misconception (MK)
Wrong	True	sure	Misconception (MK)
Wrong	Wrong	sure	Misconception (MK)
True	True	Not sure	Understand the concept but not sure (PKTY)
True	Wrong	Not sure	Not understand the concept (TTK)
Wrong	True	Not sure	Not understand the concept (TTK)
Wrong	Wrong	Not sure	Not understand the concept (TTK)

[21]

3. Result and Discussion

Investigation of knowledge and misconceptions of students is very important for students and teachers in learning science [8; 9; 10; and 11]. The Three-tiered test can use identify student misconceptions on the matter of human and plant transport system [17]. The profile and location of the misconception that students vary. The profile of each student's misconceptions in the transportation system of humans and plants can be presented in Figure 1.

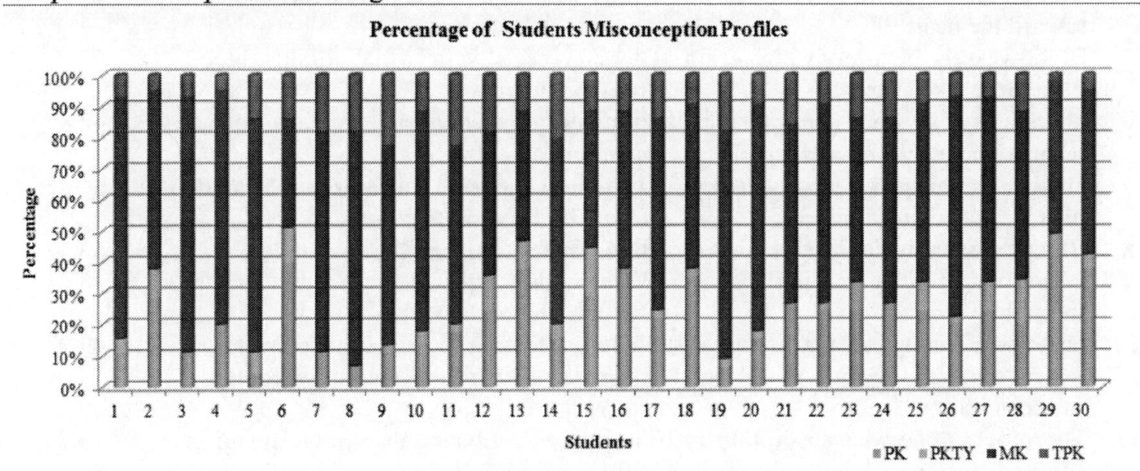

Figure 1. Diagram of the percentage of student misconceptions

Description:

PK: Understand the concept
MK: Misconception
PKTY: Understand the concept but not sure
TPK: don't understand the concept

The results of each student misconceptions on the material system of transportation along with humans and plants can be known is still quite high. The layout of the misconceptions and the level of understanding of the concept that every student is different, because of the ability of each student's knowledge in a different construct. Some students have not yet understood the concept well and have a different concept with a concept embraced by the experts so that students have difficulty in understanding the concept to explain various symptoms of nature.

The misconceptions students can be caused by a few things, namely learning resources are used well students, the misconception that happens to teachers, students in the condition of incomplete reasoning, intuition is wrong, use of learning methods that teachers are less precise, and the experience of interacting with students in daily life [7]. Misconceptions owned these students be a hindrance in developing new concepts. This is what led to the misconception that most students due to the initial concept (preconception) which is then brought to formal education [14].

The knowledge beginning of the phenomenon of students will be taught in learning, but many students develop an understanding of science concepts only to a limited extent only after the learning process in school as revealed by [2; 4]. Students who are having a misconception to occur because of imperfections in the process of constructive knowledge students with preconception, so that the students can understand the concept only partly intact and not be a barrier to students to understand the material in learning [3].

Analysis of student answers every question on the three-tier test based on the guidelines for assessment of student answers in Table 1 [20]. The students responded with a confident in providing confidence profile high i.e CRI > 2.5. The calculation results of the percentage of the misconception that occurs on the item by using three-tiered test results that can be obtained are presented in Table 2.

Table 2. The percentage of students Misconceptions on human and plant transport system

No.	Scientific Concepts	The concept of Students	Percentage of MK (%) (N=30)
1	Booth left heart thick-walled	The porch right heart thick-walled	50.4
2	A form of white blood cells irregular (ameboid) and has a core	A form of white blood cells concave discs in the middle and has no core	46.9
3	Characteristics of vessels thick-walled, elastic, and is located in the body	Characteristics of vessels thick-walled, not rubbery, and located near the surface of the body	47.8
4	Blood vessels vein the flow of blood toward the heart	Blood vessels the artery the flow of blood toward the heart	73.3
5	Blood vessels the artery which drain blood from the heart	Blood vessels vein which drains blood from the heart	70.0
6	Circulatory system in humans occurs in an enclosed and double	Circulatory system in humans occur openly and double	42.4
7	Oxygen changed and carbon dioxide occurs in capillaries	Oxygen changed and carbon dioxide occurs in the venous vessels	66.7
8	Blood pressure in capillaries greater	Blood pressure in capillaries smaller	63.3
9	Water and nutrient transport occurs through xylem vessels	Water and nutrient transport occurs through phloem vessels	48.6
10	Factors affecting the transport of water to xylem the form of the pull of transpiration	Factors affecting the transport of water to xylem the form of the absorption roots	49.2
11	The process of water absorption in soil in diffusion	The process of water absorption in soil in diffusion	76.7
12	The process of absorption of minerals from the soil in osmosis	The process of absorption of minerals from the soil in osmosis	76.7
13	The time span of the process of photosynthesis occurs when there is light	The time span of the process of photosynthesis during the day	63.3
14	Results of the Transport process of photosynthesis in diffusion	Results of the Transport process of photosynthesis in diffusion	66.7
15	The process of evaporation of water through the leaves going through the pull of transpiration	The process of evaporation of water through the leaves occurs in guttation	56.7

The layout of the misconceptions students on the transportation system of humans and plants can be seen in Table 2. The question of the test used in this study amounted to 15 grains of matter. Misconceptions occur due to a concept belonging to someone not in accordance with the agreed concept and thought by experts [5].

The results of students misconception analysis in Table 2 that the highest misconception profiles occurred in the diffusion and osmosis process of 76.7%. The Students have difficulty in understanding the process diffusion and osmosis. Students having difficulty between dissolved substances or solvents that moved in the process of diffusion and osmosis. The results of the previous research also State that students experience the misconception is quite high on the concept of movement of particles in diffusion and osmosis [15].

The high misconception profiles can be caused by several things. The learning ethics teachers only explain the material in fragment-fragment, incomplete, and it is difficult to understand. The teachers are less familiar with the human and plants transport system because the teacher in question is having trouble physics teacher so to teach the material in integrated and teachers less understand a concept on biological material, so that in explaining the material only fragment-line. Teacher not being able to in providing integrated science materials description. This condition can cause the ability of students in construct preconception with new knowledge to form concepts whole cannot happen perfectly, so the misconception of students is high enough on the material transportation system of humans and plants.

The role of teachers very important necessary in eliminating the misconception that occurs in students. If the misconception is not removed from the students, then the student will feel unfamiliar with his new knowledge because students can't afford to assimilate new knowledge into existing levels [14]. Misconception students can be known and guide students towards a better understanding can be done through direct activities, investigation, and interaction with peers and adults as revealed by [12]. Teachers should know the knowledge preconception and the misconception that experienced by students because both of these factors is an important factor for improving student mastery learning.

4. Conclusion

Identification of the misconception that occurs students on the transportation system of humans and plants shows that the overall level of student's misconception is quite high that is on average about 60%. The location of misconception of each student is different. The highest misconception profiles were the absorption of water occurs diffusion and absorption of minerals occur osmosis i.e 76.7%. The other high misconceptions related to blood vessels, blood pressure, photosynthesis process, and guttation. The misconception that the students caused by the initial knowledge of students, students ability in the constructed knowledge, learning methods used by teachers, and understanding teacher.

5. References

[1] Treagust D F 1995 *Diagnostic assessment of students ' science knowledge. In Learning science in the schools* Glynn S M & Duit R (Eds.) New Jersey Lawrence Erlbaum Associates Publishers

[2] Duit R and Treaguat D F 2003 Conceptual change: a powerful framework for improving science teaching and learning *International Journal of Science Education* **25** (6), 671-agricultural 688

[3] Allen M 2010 *Misconceptions in Primary Science* (England Open: University Press)

[4] Fach M, de Boer Tanja, and Parchmann Ilka 2007 Results of an Interview Study of u.s. Bases for The Development of Tools for Supporting Stepped Stoichiometric Problems *Chemistry Education Research and Practice* **8** (1) 13-31

[5] Ibrahim M 2012 *A Series of innovative Learning concepts, Misconceptions, and How his education* (Surabaya: Unesa University Press) p 13

[6] Novak J d and Gowin D B 1984 *Learning How to Learn* (Cambridge: Cambridge University Press)

[7] Paul S 2013 *The misconception and the change of concept in physics education* (Jakarta: PT Grasindo) p 82.
[8] Osborne R 1985 *The children's own concepts. In Primary Science: Taking the plunge* D Harlen (ed) (London: Heinemann Educational) pp 75-agricultural 91
[9] Staver J R 2007 Teaching science France *International Bureau of Education*
[10] Selley N J 1999 *The art of constructivist teaching in the primary school* (Great Britain: David Fulton Publishers)
[11] Ozmen H 2004 Some Student Misconceptions in Chemistry: A Literature Review of Chemical Bonding *Journal of Science Education and Technology (JRST)* **13** (2)
[12] Yip D 1998 Identification of misconceptions in novice biology teachers and remedial strategies for improving biology teaching *International Journal of Science Education* **20** (4) 461- 477
[13] Ross K, Lakin L, and Callaghan P 2000 *Teaching secondary science* (London: David Fulton Publishers)
[14] Puspa V 2015 Student Misconceptions About Plant-A Sri Lankan Transport Example *European Journal Of Science And Mathematics Educations* **3** pp 275-288
[15] Louis O A, Lioyd B H 1993 Freshman Biology Majors Misconceptions about Diffusion and Osmosis *Paper Presented at the Annual Meeting of the National Association for Research in Science Teaching* (Atlants, GA, April 15-19, 1993) SE **053 672** DE 362 399
[16] Treagust D F and Haslam F 1986 Evaluating secondary students ' misconceptions of photosynthesis and respiration in plants using a two -agricultural tiered diagnostic instrument *Paper presented at the 59th annual meeting of the National Association for Research in Science Teaching* San Francisco California March 28 the use of **31**
[17] Haki P and Ali E 2010 Development of a Three-tiered Test to Assess Misconceptions About Simple Electric Circuit *The Journal Of Educational Research* **103** pp 208-222
[18] Demet K Z 2014 Using Three-Tier Diagnostic Test to Assess Students ' Misconceptions of States of Matter *Eurasia Journal of Mathematics, Science, & Technology Education* **10** (5) pp 509-521
[19] Yasin K 2005 *The Diagnosis of Eleventh Grade Students ' Misconceptions about Geometric Optics by a Three-tiered Test* (Thesis: Middle East Technical University)
[20] Fatma T 2005 *Developing a Three-tiered Test to Assess High School Students ' Misconceptions Concerning Force and Motion* (Thesis: Middle East Technical University)
[21] Arslan H, Ozge, Ceyhan Cigdemoglu, and Christine Moseley 2012 A Three-Tier Diagnostic Test to Assess Pre-Service Teachers ' Misconceptions about Global Warming, Greenhouse Effect, Ozone Layer Depletion, and Acid Rain *Internation Journal of Science Education* **34** (11) p 1667-1686

IOP Conf. Series: Journal of Physics: Conf. Series **947** (2017) 012065 doi:10.1088/1742-6596/947/1/012065

Spatial Skill Profile of Mathematics Pre-Service Teachers

R O E Putri [1]

[1]Universitas Islam Majapahit, Mojokerto, Indonesia
*Corresponding author: rizky.prims@gmail.com

Abstract. This study is aimed to investigate the spatial intelligence of mathematics pre-service teachers and find the best instructional strategy that facilitates this aspect. Data were collected from 35 mathematics pre-service teachers. The Purdue Spatial Visualization Test (PSVT) was used to identify the spatial skill of mathematics pre-service teachers. Statistical analysis indicate that more than 50% of the participants possessed spatial skill in intermediate level, whereas the other were in high and low level of spatial skill. The result also shows that there is a positive correlation between spatial skill and mathematics ability, especially in geometrical problem solving. High spatial skill students tend to have better mathematical performance compare to those in two other levels. Furthermore, qualitative analysis reveals that most students have difficulty in manipulating geometrical objects mentally. This problem mostly appears in intermediate and low-level spatial skill students. The observation revealed that 3-D geometrical figures is the best method that can overcome the mentally manipulation problem and develop the spatial visualization. Computer application can also be used to improve students' spatial skill.

1. Introduction

Problem solving ability is one of the main goals in learning [1]. Someone will do the mental process using all the knowledge possessed and determine the right strategy to solve the problem. Polya [2] discussed two kinds of problems, namely (a) problems to find, can be theoretical or practical, abstract or concrete, including puzzles, and (b) problems to prove, to show that a statement is true or false). In mathematics, problems usually take the form of math problems, but not all math problems considered as a problem. Based on the results of preliminary observation and interviews with several mathematics teachers in junior high school, many students have difficulty in solving mathematical problems related to geometry. This is because students find it difficult to imagine an object in their minds. The ability to visualize objects into the student's mind, is often referred as spatial ability. Sorby [3] stated that spatial ability is a mental process involving the ability to move things mentally and change the point of view when looking at an object.

Spatial ability of a person can affect mathematical ability, this is supported by the results of research showing that spatial ability has a positive relationship with students' ability to solve problems ([4], [5], [6], [7]). The field of science in mathematics that most often requires spatial ability is geometry. To overcome such problems, the teacher has done various ways during the learning process. However, scaffolding given by teachers sometimes lacks the spatial aspect, so that students with low spatial skills will tend to have trouble in understanding the material provided ([8], [9], [10]). To train and improve students' spatial skills, teachers must be able to bridge the real world with the minds or imaginations of students [3].

The goal of this research is describing the spatial ability of mathematics pre-service teachers. So it is expected to see the extent to which spatial ability affects a person's ability to solve mathematical problems related to geometry. Knowing this will create appropriate scaffolding for geometric material with respect to the spatial ability aspect. The main reason for conducting this research to junior high school students is that students are in the early stages of formal operations, meaning students have started thinking abstractly [11]. Considering that there is need for learning that takes into account every aspect of the student's ability, in this case is spatial ability.

2. Method

Quantitative research method was used to investigate pre-service teachers' spatial ability. This research was conducted at Universitas Islam Majapahit, Mojokerto. The participants of this research were 25 mathematics pre-service teachers that already took Geometry course. Three instruments were used to collect data in this study. First instrument was spatial ability test, The Purdue Spatial Visualization Test (PSVT) that known as The Purdue Visualization of Rotations (ROT). This test is adapted from spatial visualization test that was developed by Dr. Roland Guay in 1976 from Purdue University. The main purpose of this test is to assess the ability to visualize rotated objects. This test consist of 30 problems in spatial rotation visualization, shown below as an example

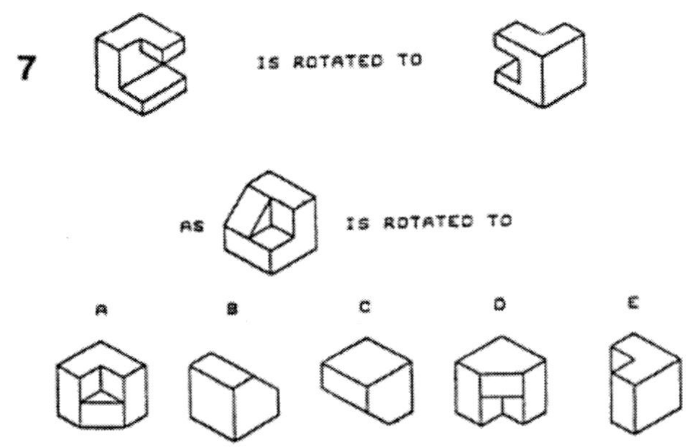

Figure 1. Problem in PSVT

On this test, there are three directions for participants to solve the problems [12]

1. Observing how the three-dimensional object in the first row is rotated.
2. Examining what if the object in the second row is rotated in the same direction.
3. Selecting among the five possible answers (A, B, C, D, and E) in the third row.

The participants were divided into three categories based on the spatial ability test result. The three categories were high, intermediate, and low level of spatial skills. Scoring level of the spatial ability test based on this table:

Table 1. spatial ability categories

	Score (x)	Category
1.	$80 \le x \le 100$	High
2.	$60 < x < 80$	Intermediate
3.	$0 \le x \le 60$	Low

The second instruments were mathematical ability test that consist of five problems related to geometry subject. This instrument was used to measure the participants' ability in solving geometry problems so that the correlation between spatial ability and geometrical problem solving ability can be known.

The third instrument was questioner that consists of 20 questions about the participants' perspective about learning Geometry. The questioner was written in Bahasa Indonesia as the participants' nationality language.

3. Results

IOP Conf. Series: Journal of Physics: Conf. Series **947** (2017) 012065 doi:10.1088/1742-6596/947/1/012065

Data of spatial ability were collected in the second week of the semester, so that initial ability of the 25 participants can be measured. Sixty minutes was given to the participants to solve the Purdue Spatial Visualization Test (PSVT) and each number must be done in 2 minutes. The result shows that mostly participants had intermediate level of spatial ability, and the rest were in high and low level of spatial skill. The table below shows the percentage of spatial skills possessed by the participants

Table 2. Spatial Ability Result

Number of Participants	Percentage (%)	Category
5	20	High Spatial Ability
17	68	Intermediate Spatial Ability
3	12	Low Spatial Ability

Most participants got wrong answers for several same numbers, such as: 16, 24, 28, 30. The three-dimensional object in those numbers are rotated around more than one axis as the picture shown below

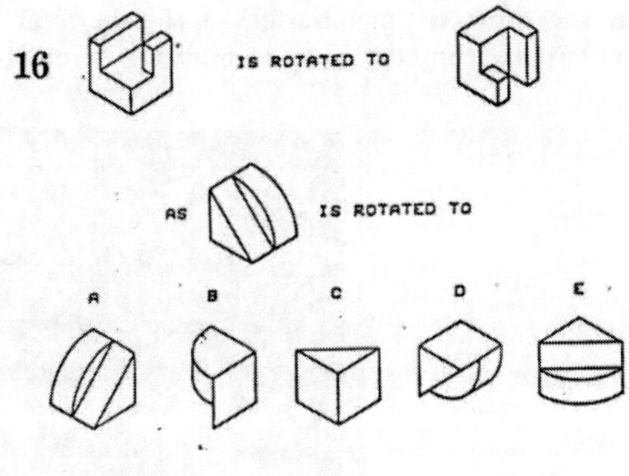

Figure 2. Two times rotated object

The participants in the high level of spatial ability were the only group who could answer those numbers correctly. In other hand, none of three participants in the low level of spatial ability had the right answer.

The mathematical ability test was done in two times, first time was in the middle of the semester and the last time was in end of the semester as the final test. The result shows that there is a positive correlation between spatial ability and geometrical problem solving ability, in other words participants in the high level of spatial ability tend to have better score than the other group of participants. It can be seen in the picture below.

The first picture (a) was done by high spatial ability pre-service teacher as the first participant. We can see from the picture, the first participant was able to draw two planes intersect correctly. In this state we can say that the high spatial ability participant was able to imagine the position of the two intersect planes and noticed that the intersection of two planes was a line. The second picture (b) was done by intermediate spatial ability participant. The second participant was not able to draw correctly. Accroding to the interview, the intermediate spatial level participant had difficulty in imagining the intersection, alhough the intermediate participant understand that the intersection of planes is line. The last picture (c) was done by the low spatial level participant. As predicted, the third participant was not able to solve the drawing task.

IOP Conf. Series: Journal of Physics: Conf. Series **947** (2017) 012065 doi:10.1088/1742-6596/947/1/012065

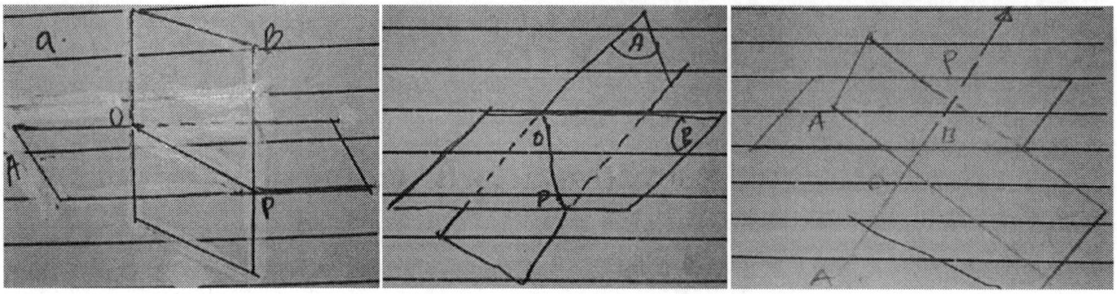

(a) (b) (c)

Figure 3. The Result of Mathematical Ability Test

The questioner was given to the participants in the end of the semester, so that they could respond and analyze the instructional geometry course for the whole semester. The result shows that most participants could understand the geometry material easier if using real 3-D or graphical object to represents geometrical object than just using the pictures in the text book.[13]

(a) (b)

Figure 4. Real 3-D and graphical object to represents geometrical object

4. Discussion and Conclusion

The purpose of this study is investigate the spatial ability of mathematics pre-service teachers and find the best instructional strategy that facilitates this aspect. According to the result of the Purdue Spatial Visualization Test (PSVT), 68% of the participants possessed intermediate level of spatial ability, 20% in the high level of spatial ability and the rest was in the low level of spatial ability as 12%. Participants in intermediate and low spatial ability levels had difficulty in solving spatial rotation problems that the three-dimensional object are rotated around more than one direction. The results support the findings of Caissie, Vigneau and Bord [14], Bordner and Guay [12]who stated that most students have difficulty in solving rotation spatial problems with more than one directions.

The result also shows that there was a positive relation between spatial ability and mathematical ability in geometrical problems solving. Participants in high level of spatial ability tend to had better mathematical ability than the other levels. It was also found in the study was carried out by Turğut and Yilmaz, [4], Rabab'h and Arsaythamby [5], Verdine, et. al. [6], Hannafin et al. [7], and Yenilmez & Kakmaci [15]. According to the questioner results, during the geometry lesson activities, almost all the participants stated that it was easier to understand the geometrical concepts if using real geometrical objects. This result is similar to the earlier study, which was carried by Akayuure, Asiedu-Addo, and Alebna [16] who stated that real geometrical objects would help students understand about geometrical concepts. Geometry computer application was also used during this study to assist the low spatial abilty group understand easily. This method supports the findings of Sorby [3] and Roca-González, et al. [17] who also used virtual technologies to help their students understand the geometry concepts.

In conclusion, the present study primarily confirms previous findings that spatial ability is one of very important aspect in geometry. Learning activities should support students in all levels of spatial ability.

References

[1] NCTM 2000 *Learning and Teaching Geometry, K-12* (Rseton, Virginia: National Council of Teachers of mathematics)
[2] Polya G 2004 *How to Solve it: A New Aspect of Mathematical Method. Second Edition* (Princeton: Princeton University Press)
[3] Sorby S A 2009 Developing 3-D spatial visualization skills *Engineering Design Graphics Journal*, **2** 63.
[4] Turğut M and Süha Y 2012 Relationships Among Preservice Primary Mathematics Teachers' Gender, Academic Success and Spatial Ability *International Journal of Instruction.* **5** 2 (e-ISSN: 1308-1470 *www.e-iji.net)*
[5] Rabab'h B and Arsaythamby V 2015 Spatial Visualization as Mediating between Mathematics Learning Strategy and Mathematics Achievement among 8th Grade Students *International Education Studies;* **8** 5 (doi:10.5539/ies.v8n5p1)
[6] Verdine B N *et al* 2013 Deconstructing Building Blocks: Preschoolers' Spatial Assembly Performance 1 Relates to Early Mathematical Skills *CDEV journal* pp 1–14 (DOI: 10.1111/cdev.12165)
[7] Hannafin R D, Truxaw M P, Vermillion J R, and Liu Y 2010 Effects of Spatial Ability and Instructional Program on Geometry Achievement *The Journal of Educational Research* **101** 3 (DOI: 10.3200/JOER.101.3.148-157)
[8] Bakker A, Smit J, and Wegerif R 2015 *Scaffolding and dialogic teaching in mathematics education: Introduction and review* (ZDM Mathematics Education 47: 1047)
[9] Battista M T 1990 Spatial Visualization and Gender Differences in High School Geometry *Journal for Research in Mathematics Education* **3** 21 pp47-60.
[10] Clements D H 1998 Geometric and Spatial Thinking in Young Children *Opinion paper National Science Foundation.*
[11] Piaget J 2000 *Piaget's theory of cognitive development* *Childhood cognitive development The essential readings* pp33-47.
[12] Bodner G M and Guay R B 1997 The Purdue visualization of rotations test *The Chemical Educator* **4** 2 pp1-17.
[13] Yılmaz H B 2009 On the Development and Measurement of Spatial Ability *International Electronic Journal of Elementary Education* **2** 1 pp83-96.
[14] Caissie A F, Vigneau F, and Bors D A 2009 What does the Mental Rotation Test measure? An analysis of item difficulty and item characteristics. *Open Psychology Journal* **2** 1 pp94-102.
[15] Yenilmez K and Ozlem K 2015 Investigation of the Relationship between the Spatial Visualization Success and Visual/Spatial Intelligence Capabilities of Sixth Grade Students *International Journal of Instruction* **8** 1 pp 189-204
[16] Akayuure P, Asiedu-Addo S K, and Alebna V 2016 Investigating the Effect of Origami Instruction on Preservice Teachers' Spatial Ability and Geometric Knowledge for Teaching *International Journal of Education in Mathematics, Science and Technology* **4** 3 pp198-209.
[17] Roca-González C, *et al* 2017 Virtual Technologies to Develop Visual-Spatial Ability in Engineering Students *Eurasia Journal of Mathematics, Science & Technology Education* **2** 13.

IOP Conf. Series: Journal of Physics: Conf. Series **947** (2017) 012066 doi:10.1088/1742-6596/947/1/012066

A Metacognitive Profile of Vocational High School Student's Field Independent in Mathematical Problem Solving

L Nugraheni[1,*], I K Budayasa[1] and S T Suwarsono[2]
[1]Universitas Negeri Surabaya, Jalan Ketintang Baru XII no.34, Ketintang, Gayungan, Surabaya 60231, Jawa Timur Indonesia.
[2]Universitas Sanata Dharma, Mrican, Tromol Pos 29, Yogyakarta 55002, Indonesia.

[*]Email: likninnugraheni@mhs.unesa.ac.id

Abstract. The study was designed to discover examine the profile of metacognition of vocational high school student of the Machine Technology program that had high ability and field independent cognitive style in mathematical problem solving. The design of this study was exploratory research with a qualitative approach. This research was conducted at the Machine Technology program of the vocational senior high school. The result revealed that the high-ability student with field independent cognitive style conducted metacognition practices well. That involved the three types of metacognition activities, consisting of planning, monitoring, and evaluating at metacognition level 2 or aware use, 3 or strategic use, 4 or reflective use in mathematical problem solving. The applicability of the metacognition practices conducted by the subject was never at metacognition level 1 or tacit use. This indicated that the participant were already aware, capable of choosing strategies, and able to reflect on their own thinking before, after, or during the process at the time of solving mathematical problems.That was very necessary for the vocational high school student of Machine Technology program.

1. Introduction

The study results of the Trends in International Mathematics and Science (TIMSS) revealed that the questions used to measure the ability of learners fall into four categories: (1) low to measure ability up to the knowing level, (2) intermediate to measure ability up to the applying level, (3) high to measure ability up to the reasoning level, (4) advanced to measure ability up to the reasoning level with incomplete information. While the achievements of Indonesian children as revealed in several TIMSS reports were less encouraging, 95% of the Indonesian children were only up to the intermediate level, while almost 40% of the Taiwanese learners are able to reach high and advanced levels by Kemendikbud [1]. This shows that the mathematical ability of the Indonesian children was very low, especially in terms of mathematical problem solving. The low level of mathematical problem solving ability of the Indonesian in general, particularly the vocational high school student of Machine Technology Program (known as SMK Pemesinan) in East Java should be improved.

The problem-solving stage that this study refers is one proposed by Polya [2]. Polya in [2] suggested the detailed steps that should be taken by students so that problem solving can be done effectively and efficiently and obtain the right solution. Polya's problem-solving steps direct student to always be aware of their potential abilities and able to manage those abilities to be used in problem solving. Polya problem-solving steps is understand the problem, devise a plan, carry out the plan, look back.

Some researchers have pointed out that metacognition is an important issue in problem solving as well as in the acquisition and application of learning skills in various fields of discovery by Flavell in [3], Panoura and Philippou, 2005 in [4]. One advantage of involving metacognition processes in solving trigonometric math problems is to establish a strong, thorough and profound understanding of trigonometric math problems. This enables the student of vocational high school student of Machine

IOP Conf. Series: Journal of Physics: Conf. Series **947** (2017) 012066 doi:10.1088/1742-6596/947/1/012066

Technology Program (SMK Pemesinan) to not only use the formulas, but also are aware of the process and the results of their thinking about the meaning and usefulness of the trigonometric formulas. It is very necessary by the students of to solve mathematic problems, in particular trigonometry serving as a support for other participant. So it is expected to improve the quality of education in vocational high school student of Machine Technology Program (SMK Permesinan) in East Java Indonesia.

Acharya in [5] cited that if the cognitive style of student can be accommodated in learning, it will result in improving learning attitudes and increasing thinking skills, academic achievement, and creativity. Information about cognitive styles can help teachers at school become more sensitive to the differences the student have in the classroom. Therefore, by knowing the cognitive styles of learners, the teacher knows the right way to teach in the classroom.

Many scholars argue [3,4,6,7,8,9] about the notion and components of metacognition, but essentially they put an almost equal emphasis on the components. This study therefore only focused on metacognition experience component which includes planning, monitoring, and evaluating. If one can evaluate the process and result of thinking, it will allow increasing awareness about process and result of one's thinking and ultimately will increase his or her learning results.

Swartz and Pepkins [8] distinguish the four levels of metacognition: tacit use, aware use, strategic use, and reflective use which is the development of metacognition ability. The tacit use or level 1 is the level of one's metacognition with the indicator: not thinking about the decision taken or the thinking process it does. The aware use or level 2 is the level of metacognition of a person with the indicator: aware of what and why doing a particular thinking. The strategic use or level 3 is the level of one's metacognition with the indicator: having the ability to regulate their own thinking, consciously using specific strategies that improve the thinking accuracy. The reflective use or level 4 is the level of one's metacognition with the indicator: having the ability to reflect on his own thinking before, after or during the process and thinking about how to proceed and how to improve.

Witkin in [10] stated that cognitive style is a characteristic that is reflected from each individual and these characteristics can be influenced by two factors, namely: the factors related to the influence of external stimuli and the factors related to the influence of an individual person. Witkin in [11] described the characteristics of the cognitive style of the Field Independent are (a) analytical, competitive, independent, and individualistic, (b) having its own goals, objectives, strategies and reinforcement, (c) intrinsically motivated, (d) less social skills and preferring individual tasks, (e) structured and well-organized in learning.

Based on this above-mentioned problem, it is important for the student of SMK Permesinan who are highly capable and cognitive field independent to realize knowledge of their metacognitive ability. The goal is to be able to solve mathematical problems either in the field of mathematics itself or its application in the subject matter specialization in line with each area of expertise.

2. Research Method

2.1 Research Type

The type of this research is explorative with qualitative approach. To get in-depth descriptive data about the metacognition profile of the SMK Pemesinan student with high-cognitive and field-independent cognitive techniques in solving trigonometric mathematical problems, the student are given the tasks of solving mathematical problems (TPM-1) followed by interviews. At the time of the interviews, the student are asked to explain the problem solving done in accordance with the steps of solving the problem according to Polya. The data of the problem solving tasks and the interviews were combined, then described qualitatively and the results were in the forms of written or oral words or descriptions of the research participant and then analyzed. Therefore this research used a qualitative descriptive approach.

2.2 Research Instruments
In addition to the researcher as the main instrument in this research, other instruments such as mathematical ability test (known as TKM), Group Embedded Figures Test (GEFT) test, problem solving test (known as TPM-1), and interview guidelines were needed.

The sample problem solving test is presented as follows: note Figure 1

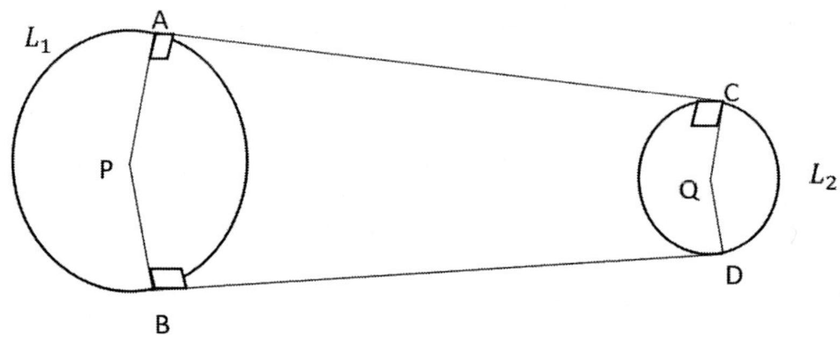

Figure 1. Figure mathematical ability test (TPM).

A large circle L_1 with a diameter of 14 cm and a small circle L_2 has a diameter of 10 cm. If it is known that the length of the tangent line of the outer alliance of the two circles is AC = BD = 50 cm, then specify the magnitude of $\angle$ APB!

2.3 Participant
The participant of this study was a tenth-grade student of SMK Permesinan. The basic consideration is that mathematics is the foundation of support for most other participant taught in SMK Permesinan in the tenth, eleventh, or twelfth grades. Based on the 2013 curriculum in [12] the trigonometric materials were taught in SMK Permesinan. Some criteria of the participant of this study were a high-ability student with field independent cognitive-styles. Theme was only one student participating in this study.

The criteria for the selection of research participant include: (1) the mathematical ability test (TKM) scores by taking participant from high-ability groups. The participant is regarded as highly capable, if the scores > 80, (2) GEFT scores, the student' abilities are divided into two cognitive style groups in [10]: (a) high-ability students, if their GEFT scores > 9 and the students belong to a group of highly capable cognitive field-independent participant, (b) for highly-skilled students, if their GEFT scores are between 0-9 then the students include high field-dependent cognitive skills, (3) if from 2 categories in (2) .a) there is at least one subject that meets, then the next step is to ask the teacher's judgment about whether the selected participant has a score that corresponds to the day-to-day ability of his class and can express his oral and written opinions. The selected participant is known as TFI.

3. Results and Discussion

Based on the results of TFI subject analysis in solving mathematical problems starting from the stage of understanding mathematical problems of TPM-1, creating a mathematical problem solving plan of TPM-1, implementing mathematical problem solving of TPM-1, and reviewing the results of mathematical problem solving of TPM- 1, it can be concluded that the metacognition profile conducted by TFI subjects in solving the problems of trigonometry of TPM-1 by considering the frequency of metacognition activity implementation at each stage and level is presented in table 1.

IOP Conf. Series: Journal of Physics: Conf. Series **947** (2017) 012066 doi:10.1088/1742-6596/947/1/012066

Table 1. Implementation of TFI Metacognition Activities Implementation In Trigonometry Problem Solving (TPM-1)

Problem Solving Stages	Metacognition Stages and Levels											
	Planning				*Monitoring*				*Evaluating*			
	1	2	3	4	1	2	3	4	1	2	3	4
Understanding problems	0	2	0	0	0	1	1	5	0	1	1	1
Problem solving planning	0	0	1	2	0	0	1	5	0	2	1	1
Implementing the problem solving planning	0	2	0	0	0	1	5	5	0	2	2	4
Reviewing the problem solving results	0	0	0	0	0	0	1	0	0	2	1	0
Total	0	4	1	2	0	2	8	15	0	7	5	6

From Table 1, the results of metacognition process analysis conducted by TFI in the TPM-1 trigonometric problem solving was in the TFI planning stage at the metacognition level 2 or aware use, 3 or strategic use, and 4 or reflective use. Specifically, with four items at level 2 or aware use, this indicated that TFI was aware of what and why doing certain thoughts at the understanding problem stage, and executing the problem-solving plan. There was one item at the metacognition level 3 or strategic use, it indicated that TFI could regulate its own thinking and consciously use specific strategies that improved the accuracy of thinking at the planning stage of making problem-solving plans. There were two items at the metacognition level 4 or reflective use, it indicated that TFI had the ability to reflect on its own thinking before, after or during the process and thinking about how to proceed and how to improve in the planning phase of making a problem-solving plan. From table 1 above, TFI did not do planning at the reviewing problem solving result but directly did monitoring and evaluating.

At the monitoring stage TFI was at the metacognition level 2 or aware use, 3 or strategic use, and 4 or reflective use. There were two items s at the metacognition level 2 or aware use, this indicated that TFI was aware of what and why doing certain thoughts scattered at the monitoring stage when understanding the problem, and executing the TPM-1 trigonometric problem-solving plan. There were a total of 8 items at the metacognition level 3 or strategic use, this indicated that TFI could regulate its own thinking and consciously use specific strategies that improved the accuracy of its thinking at the monitoring stage when understanding the problem, making problem-solving plans, implementing problem-solving plans, and reviewing TPM-1 problem solving results. There were a total of 15 items at the metacognition level 4 or reflective use, it indicated that TFI had the ability to reflect on its own thinking either before, after or during the process and thinking about how to proceed and how to improve during the monitoring phase of problem solving, making problem-solving plans, and implementing a TPM-1 trigonometric problem-solving plan.

At the evaluation stage TFI is at metacognition level 2 or aware use, 3 or strategic use, and 4 or reflective use. With 7 items at the metacognition level 2 or aware use, this suggested that TFI was aware of what and why do certain thoughts scattered at the stages of understanding problems, creating problem-solving plans, implementing problem-solving plans, and reviewing the results of TPM-1 trigonometric problem solving. As there were 5 items at the metacognition level 3 or strategic use, it revealed that TFI could regulate its own thinking and consciously use specific strategies that improved the accuracy of its thinking at the evaluating stages when understanding problems, making problem-solving plans, implementing problem-solving plans, and reviewing problem-solving results. As there were 6 items at the metacognition level 4 or reflective use, it showed that TFI had the ability to reflect on its own thinking before, after or during the process and thinking about how to proceed and how to improve at the evaluating stages of understanding the problem, making problem-solving plans, and implementing a TPM-1 trigonometric problem-solving plan. Table 1 showed that TFI was never at the metacognition level 1 or tacit use, it pointed out that TFI had not never thought of decisions made on the thinking process either in the contextual or formal problem solving.

These results showed that high-ability subject with independent field cognitive style performed the metacognition activities well involving the three types of metacognition activities, i. e. planning, monitoring, and evaluating at the metacognition level 2 or aware use, 3 or strategic use, and 4 or reflective use in mathematical problem solving. This is in contrast to the research of Mania Nosratinia, and ShirinAdibifar [13]. The results showed that both experimental and control groups were both given the pretest of writing. The experimental group received the metacognitive strategy training, while the control group received a regular teaching program at a language school. At the end of the training, both groups were given a post-test and the results showed that the experimental group was statistically better in their post-tests. In addition, the independent-field learners outperformed the dependent-field groups in their post-tests. The results of Maria Nosratinia and Shirin Adibifar showed that the metacognition training on writing skills while solving mathematical problems had not been explored, the metacognition profile thus needed to be revealed in solving mathematical problems.

In addition, the research results by Agoestanto, Sukestiyarno and Rochmad in [14] showed that the students' mathematical critical thinking skills with FI cognitive style were better than the ones of FD cognitive style to conclude, assumptions, deductions, and interpretations. While in the aspect of evaluation of argument, the students' critical thinking abilities with FD cognitive style were better than the one of students with FI cognitive styles. The results of Agoestanto's study show only the critical thinking skills of FI and FD students only, but had not shown the metacognition profile of the subjects. Thus the metacognition profile needed to be revealed.

The research by Effendi in [15] concerning "Implementation of Creative Problem Solving Model to Improve The High School of Student's Metacognitive indicated that (1) improvement of metacognitive ability of students who applied the Creative Problem Solving model was significantly better than the one of the students who only employed the conventional learning; (2) There were significant differences in metacognitive enhancement abilities among the students who obtained the creative problem-solving models and one of the students who received the conventional learning in terms of previous, high, middle and low student mathematics levels. There was an increase in metacognitive ability of the experimental class students who previously had high and moderate mathematical skills, it was significantly better than students' metacognitive abilities in the control class. However, the increased metacognitive ability of the lower students and the level of prior mathematical knowledge in the experimental and control classes did not differ significantly. In this study the students' metacognition abilities given the creative problem solving were increased, but not yet explained on the level of metacognition to what extent the improvement was. In addition, the cognitive styles that considerably affected the learning outcomes were also not discussed in the research by Effendi. It therefore needs to be further investigated concerning the levels of metacognition and the cognitive styles.

4. Conclusion

The result of this research revealed that the subjects of TFI carried out the metacognition activities with very high frequency and dynamic. TFI's metacognition activities involved the three types of activities: planning, monitoring, and evaluating already at the metacognition level 2 or aware use, 3 or strategic use, and 4 or reflective use in solving mathematical problems. The frequency of metacognition activities done by the subjects of TFI, never at the metacognition level 1 or tacid use. This suggested that the subjects of TFI already had consciousness, was able to choose a strategy, and was able to reflect on their own thinking either before, after, or during the process of solving trigonometric problems. The implementation of good metacognition activities when solving trigonometric math problems is necessary for machining vocational students.

5. References

[1] Kemendikbud 2014 *Panduan Pelatihan Guru Matematika SMA/SMK dalam Penerapan Kurikulum 2013* (Jakarta: Kemendikbud)

[2] Polya G 1973 *How to Solve It*. Second Edition. (Princeton: University Press Princeton)

[3] Flavell J H 1979 Metacognition and cognitive Monitoring, A New area of Cognitive-Developmental Inquiry *American Psychologist*.

[4] Panaoura A and Philippou G 2005 *The Measurement of Young Pupils' Metacognition Ability in Mathematics* (http://citeseerx.ist.psu.edu/viewdoc/download?doi=10.1.1.120.3575&rep=rep1&type=pdf)

[5] Acharya and Ms Chandrama. 2002 Studens Learning Style and Their Implication for Teachers *Center for Development of Teaching and Learning* September 2002 Vol. **5** No. 6.

[6] Brown H D 1993 *Principles of language learning and teaching* (Indiana: Prentice Hall) (https://s3.amazonaws.com/academia.edu)

[7] Fisher R 2007 *Thinking about Thinking: Developing Metacognition in Children* (http://www.teachingthinking.net)

[8] NCREL 2007 *Metacognition* (http://info@ncrel.org)

[9] Livingston and Jennifer A 1997 *Metacognition: An Overview* (http://gse.buffalo.edu/fas/shuell/cep564/metacog.htm)

[10] Witkin H A, Oltman P K, Raskin E and Karp S 1971 *A Manual for the Group Embeded Figures Test* (Palo Alto, CA: Consulting PsikologyPrees)

[11] Witkin H A, Moore C A, Goodenough D R and Cox P W 1977 Field Dependent & Field Independent Cognitive Style and their Educational Implicationc. *Review of Education Researcch Winter*. Vol.**47** No.1 Page 1-64 (http://dx.doi.org/10.21070/pedagogia.v2il.48)

[12] Depdikbud 2016 *Permendikbud Nomor 24 Tahun 2016 Tentang Kompetensi Inti dan Kompetensi Dasar Pelajaran Pada Kurikulum 2013* (Jakarta :Depdikbud)

[13] Adibifar S and Nosratinia M 2014 The Effect of Teaching Metacognitive Strategies on Field Dependent and Independent Learners' Writing Published by Elsevier Ltd. Procedia - *Social and Behavioral Sciences* **98** (2014) 1390 – 1399.

[14] Effendi A 2017 Implementation of creative Problem Solving Model to Improve The High School Student's Metacognitive (http://iopscience.iop.org/article/10.1088/1742-6596/812/1/012065/pdf)

[15] Agoestanto A, Sukestiyarno Y L and Rochmad 2016 Analysis of Mathematics Critical Thinking Students in Junior High School Based on Cognitive Style (http://iopscience.iop.org/article/10.1088/1742-6596/824/1/012052/pdf)

Acknowledgements
The author would like to thank the Ministry of Research and Technology who have granted the scholarship to the author.

IOP Conf. Series: Journal of Physics: Conf. Series **947** (2017) 012067 doi:10.1088/1742-6596/947/1/012067

Developing Learning Model P3E to Improve Students' Critical Thinking Skills of Islamic Senior High School

Bahtiar[1], **Y S Rahayu**[2], **Wasis**[3]

[1,2,3]Universitas Negeri Surabaya, Indonesia

Email : Yunirahayu@unesa.ac.id

Abstract. This research aims to produce P3E learning model to improve students' critical thinking skills. The developed model is named P3E, consisting of 4 (four) stages namely; organization, inquiry, presentation, and evaluation. This development research refers to the development stage by *Kemp*. The design of the wide scale try-out used *pretest-posttest group design*. The wide scale try-out was conducted in grade X of 2016/2017 academic year. The analysis of the results of this development research inludes three aspects, namely: validity, practicality, and effectiveness of the model developed. The research results showed; (1) the P3E learning model was valid, according to experts with an average value of 3.7; (2) The completion of the syntax of the learning model developed obtained 98.09% and 94.39% for two schools based on the assessment of the observers. This shows that the developed model is practical to be implemented; (3) the developed model is effective for improving students' critical thinking skills, although the n-gain of the students' critical thinking skills was 0.54 with moderate category. Based on the results of the research above, it can be concluded that the developed P3E learning model is suitable to be used to improve students' critical thinking skills.

1. Introduction

Natural science is clasically divided into two parts, namely; (1) the physical sciences whose objects are matter, energy, and the transformation of matter and energy, (2) while the biological sciences whose objects are the living things and their environment [1]. As one branch of natural sciences, physics is the study of behavior and its control of nature in various forms of symptoms. Students' knowledge of science still need to be improved due to the fact that is still far from expectation.

The activeness in science learning lies in two things, namely physically active actions or hands-on and active thought or minds-on[2]. Both of these must be balanced, because being active physically is not enough, students also need to gain experience. Science learning is done by using various ways (techniques) such as observation, exploration, and experimentation. The standard competence of physics graduate in Senior High Schools requires the students to do experimentation, such as formulating problems, proposing and testing hypotheses, defining variables, designing and assembling instrument, collecting, processing and interpreting data, drawing conclusions, and communicating the results of the experiment in oral and in written forms. To meet these competencies, teachers are required to facilitate, motivate, direct, and guide students in practical activities and discovery.

Critical thinking skill is a skill that must be developed to face the 21st century. To support critical thinking, students must develop their process skills [4]. The development of science process skills

IOP Conf. Series: Journal of Physics: Conf. Series **947** (2017) 012067 doi:10.1088/1742-6596/947/1/012067

allows students to construct and to solve problems and to do critical thinking [6]. This possibility can occur because most of the components of critical thinking are also the components of science process skills, such as; *observing, testing hypotheses, predicting, classifying, designing experiments, measuring,analysis, and synthesis*[7]. Critical thinking skills are considered to be the same as scientific thinking skills in science [8-10]. Therefore, critical thinking skills and science process skills need to be taught in science learning.

One of the learning models that can gain critical thinking skills is *inquiry* model [11], one of *inquiry* models that is able to arouse critical thinking skill is *guided inquiry*. According to [12]*guided inquiry* model can help students gain a better understanding, and students are more interested when they are actively involved. The acquisition of knowledge or concepts of physics in senior high schools should be fostered and developed through *inquiry* especially *guided inquiry* based on the students' curiosity [12]. The students are given the opportunity to solve problems through practical activities, and this model is suitable with the developmental level of students.

2. Research method

This research is a developmental research, where the subjects of this research were ten grader students in the 2016/2017 academic year. The topic that is taught were temperature and heat. The products of this research were syllabus, lesson plan, students' textbook, student worksheet and assessment sheet *(LP)* of the temperature and heat material integrated with P3E learning model. The development of this model was carried out through several stages, as follows.

2.1 Analysis Stage

This stage is to accommodate the needs *(needs assessment)* in the field. The needs analysis stage was reviewed through field study and literature study. Analysis of literature study consisted of; theoretical analysis supporting the learning model, syntax analysis, goal analysis, and learning environment analysis.

2.2 Design Stage

This stage was carried out in several steps, such as: the initial design of the learning model and the learning material by determining the focus on improving the critical thinking skills. The next was the validation of the model and the learning material the -i ($i \geq 1$) by experts. The revised results of the feedback from the experts were presented in *Focus groups discuss* (FGD). The objective of the FGD was to obtain input from three (3) validators of campus and ten (10) participants from various universities. The results of FGD were analyzed quantitatively and descriptively. The draft of the product that had been revised was then tried out (the-i, $i \geq 1$). The try out was only conducted on small populations. The technique sampling was done through *purposive random sampling*. The results of the try out analysis produced practical and effective learning model and learning material. If the results of the analysis were not practical and effective, they would be revised, and the learning model and the learning material were back to draft II-i, then tried out again the i-th ($i = 1,2,3, ..., n$) and analyzed, otherwise if the results of the analysis were practical and effective, it produced P3E learning model and learning material which were valid, practical, and effective.

2.3 Development Stage

The purpose of this stage was to know the improvement of students' critical thinking skills after using the products developed. The activity at this stage is called wide try-out. The samples were selected using *clustur random sampling* technique. The wide try-out was carried out for 6 meetings at the two *madrasah* (Islamic schools), namely MAN 1 grade XB and MAN 2 Model grade XC Mataram. Before the P3E model was applied, the researcher gave a pretest of critical thinking skill. After the pretest was carried out, the P3E learning model was implemented using valid and practical learning material. The posttest of critical thinking skill was given after the the learning process. The design of the try out and the wide try out of P3E learning model used *one-group pretest-posttest design,* as shown in Table 1.

IOP Conf. Series: Journal of Physics: Conf. Series **947** (2017) 012067 doi:10.1088/1742-6596/947/1/012067

Table 1. One group pretest posttest design

Pretest	*Treatment*	*Posttest*
O_1	X	O_2

Information:
O_1 : Pretest of critical thinking skills.
X : Treatment given to the students in the form of P3E learning model and the learning material developed.
O_2: Posttest of critical thinking skills.

[13].

The data about the students' responses were obtained through a questionnaire which was given at the end of the learning activity using P3E learning model. The validity of the data of the learning model and learning material were obtained by using validation sheet which was filled/assessed by the experts. The data of the implementation of learning model were obtained using observation technique by four observers. The data on the legibility and difficulty level of the textbooks and worksheets were obtained through *close procedure* technique while the difficulty level of the student textbook and student worksheets were done by highlighting the sentence which were not understood. The data of the students' responses were obtained through questionnaire method. The data of the obstacles encountered in the learning process were obtained through observation methods done by the observers. The data of science process skills were obtained through investigation activity *(LKS and LP)* performed by students which included eight indicators of science process skills. The data of critical thinking skills *(KBK)* were obtained through an essay test consisting of 5 questions. These questions were organized based on three (3) criterion of critical thinking skill indicators, namely: (1) analyzing, (2) synthesizing, and (3) concluding.

3. Result and Discussion

3.1 Result of Validity of P3E Learning Model

Table 2. Recapitulation of validation results

No	Component	Validation Score			Item Average	Category
		V1	V2	V3		
1	P3E Model	3.7	3.6	3.8	3.7	Valid
2	Lesson plan	4.41	4.76	4.90	4.70	Very Valid
3	Student worksheet	4.05	4.40	4.64	4.37	Very Valid
4	Textbooks	4.20	4.47	4.47	4.38	Very Valid
5	The assessment sheet	3.8	3.8	3.4	3.67	Valid
Average Validator Rating		4.03	4.21	4.24	4.16	Very Valid

Table 2 above shows that the result of the validation of the three validators for all components of the P3E model was 3.7 with valid category [14]. The table shows that the results of the expert study on the *items* of organization, investigation, presentation and evaluation have an average of 4.53 with a very valid category [14]. The achievement of this quality was because the development of the learning material had undergone through several stages, namely needs analysis, concept analysis, task analysis, discussion with MA physics teachers, and had been validated by experts (validator). The investigation component appeared on the content feasibility with an average value of 4.57 (very valid), the details are as follows; (1) fostering the ability to formulate problems (4.33 = very valid); (2) fostering the ability to formulate hypotheses (4.33 = very valid); (3) fostering the ability to identify variables (4.33 = very valid); (4) fostering the ability to define operational variable (DOV) (4.33 = very valid); (5) fostering the experimental activity (4.67 = very valid), (6) fostering the ability to collect experimental data (4.67 = very valid), (7) fostering the ability to analyzed data (4.67 = very valid) (8) fostering the ability to make conclusions on the results of the experiment (4,67 = very valid), thus the average component of student worksheet was 4.37 with very valid catagory [14]. The results of the validation of the student textbook from the three validators show that the average of the assessments from the three validators for all components was 4.38 and the average value of each component was 4.0 <SV≤5,0. Thus, the category of each component as well as all the components were very valid [14]. The results of the validation of the assessment sheet of the students' critical thinking from the three validators were 1 question item which was determined as valid with revision, which was number 1, and 4 question items were determined as valid without revision. The assessment average was 3.67. The reliability of the assessment sheet based on the assessment from the validators was 94.44%, so that the assessment sheet was said to be reliable which was ≥0,75 (75%) [3]. The assessment sheet of critical thinking skills developed by researcher can be said to be valid to improve the students' critical thinking skills.

3.2 Results of the Effectiveness P3E learning model

The effectiveness of the P3E learning model could be seen from the achievement of the learning process after the implementation of P3E learning model, in which the achievement of the students of MAN 1 Mataram reached 93.10% with complete category, while the achievement of the students of MAN 2 Model Mataram reached 96, 77% with complete catagory. The average score of MAN 1 Mataram was 0.51 and the average score of MAN 2 Model Mataram was 0.47. Thus, the science process skills can be classically catagorized as complete. The research conducted by Nurex suggested that learning using science process skill approach can improve students' critical thinking skills by 71, 25% [15]. The research conducted by Haryono suggest that science process skilll based learning model is effective in improving students' science process skill from from 46.08% to 67.27% [15] . The result development students' post test science process skills in this research was greater than that of the pretest, and the students' learning outcomes were classically completed because they had the completeness of ≥85%. These results are presented in Table 3 and Table 4.

Table 3. The results of students' process skills

MAN 1 Mataram			MAN Model 2 Mataram		
Score		N-Gain	Score		N-Gain
Pretets	Postest		Pretest	Postest	
50.54	75.75	0.51	5 4.94	76.01	0.47
Completeness Percentage		93.10%			96.77%
Category		T			T

Table 4. The results of students' critical thinking skills

MAN 1 Mataram				MAN 2 Model Mataram			
Pre	**Post**	N-Gain	Category	Pre	Post	N-Gain	Category
61.21	**82,41**	0.50	S	62.10	82.42	0.53	S
Completeness Percentage		93.07%		Completeness Percentage		92.63%	

The N-gain of the critical thinking skills of the students of MAN 1 Mataram was 0.50 with medium category, and the N-gain of the critical thinking skills of the students of MAN 2 Model Mataram was 0,53 with medium category. Thus, it then can be said that the developed learning model was effective to improve critical thinking skills, although the N-Gain of the students' critical thinking skills in total was 0.54 moderate categoy.

Based on the results of the research and the theoretical study above, the students' understanding related to the critical thinking skill and science process skill can be improved. This is supported by the results of the research by Liliasari that science concept understanding can be imporoved through developing students' thinking skills [16]. In line with this, teachers feel easier to teach science to students whose critical thinking skills have been developed.

IOP Conf. Series: Journal of Physics: Conf. Series **947** (2017) 012067 doi:10.1088/1742-6596/947/1/012067

4. Conclusion

Based on the results and the discussion of the research,it can be concluded that: (1) the sophisticated P3E learning model is valid based on assessments from the validators. The validity score was viewed in terms of supporting theory, syntax, social systems, supporting systems, and instructional impact with an average of 3.7 with valid category. Therefore, the developed learning model can be implemented in schools, (2) The P3E learning model is practical to improve students' critical thinking skills. Based on the results of the completeness of the P3E learning model, each syntax has increased from the begining of the learning until the end of the learning with very good category. This is supported by the students' response toward the learning, with the accumulation of the students' interest reaching 100%. In addition, there is no any obstacle which cornered to the impementation of the developed P3E learning model, which means that the P3E learning model can be easily applied (practical), (3) the P3E learning model is effective to improve critical thinking skills, although the n-gain of the students' critical thinking skills with 0.54 with the moderate category.

Some suggestions are displayed as: (1) the implementation of the P3E learning model was only on the material of temperature and heat in grade X to improve students' critical thinking skills. It needs to be implemented more extensively on other physics materials and other science subjects, (2) he samples of the wide try out were only carried out in 2 schools. It is expected that other researchers increase the number of the research samples to strengthen the generalization of the research results, (3) ased on the results of n-gain, where it is still in the moderate category. Further in-depth study is expected to improve students' critical thinking skills through a structured exercise and the availability of adequate reference sources.

Further research is necessary to be conducted to analyze the effect of P3E learning model on the other variables, such as; thinking skills and problem solving skills

References

[1] Kemp 1994 *Instructional Design: A plan for unit and course development* (Belmont: Fearon-Pitman Pub)

[2] National Research Council (NRC) 2000 *Inquiry and the National Science Education Standards: A Guide for Teaching and Learning* [Online] http://books.nap.edu/html/inquiry_addendum/notice.html accessed 9 February 2013

[3] Depdiknas 2006 *Peraturan Menteri Pendidikan Nasional No. 23 tahun 2006* tentang Standar Kompetensi Lulusan untuk Satuan Pendidikan Dasar dan Menengah. Jakarta

[4] BSNP T 2006 *Standar Isi* (Jakarta: BadanStandar Nasional Pendidikan)

[5] Kalelioglu F & Gilbahar Y 2013 The Effect of Instructional Techniques on Critical Thinking and Critical Thinking Disposition in Online Discussion *Education Technology & Society* **17**(1) 248-258

[6] Kemble 1999 *National Academy of Science* (Washington DC: National Academies Press)

[7] Hassard J 2005 *The Art Teaching Science* (New York: Oxford University Press)

[8] Ergazaki M 2000 Biotechnology/Genetic Engineering *Research on Teaching and Learning in Critical Thinking Context* http://www.iubs.org/patras.biotech-criticalthinking.pdf access pada 18 Desember 2015.

[9] Pomahac G, Thelma M Gunn, Lance M Grigget 2007 Bioethics and Critical Thinking in a Science Classroom *Conference Prisiding* http://www.umanitoba.ca/confer-guy.1438.pdf accessed 18 Desember 2015

[10] Schafersman S D 1991 *Introduction to Critical Thinking* [Online] http://www.freeinquiry.com/critical-thinking.html accessed 12 November 2012

[11] Eggen Paul and Kaucahak Don 2012 *Strategi dan Model Pembelajaran: Mengajar Konten dan Keterampilan Berpikir, Edisi 6* (Jakarta: PT. Indeks)

[12] Nur M 2011 *Teori-Teori Belajar* (Surabaya: Universitas Negeri Surabaya)

[13] Arikunto Suharsimi 2006 *Dasar-Dasar Evaluasi Pendidikan* (Jakarta: Bumi Aksara)

[14] Ratumanan T G dan Laurens 2006 *Evaluasi Hasil Belajar Yang Relevan Dengan Kurikulum Berbasis Kompetensi* (Surabaya: Unesa University Press)

[15] Nur M 2011 Hasil-Hasil Penelitian dan Pengembangan Perangkat Pembelajaran bermuatan Keterampilan Berpikir dan Perilaku Berkarakter Pendukung Pendidikan IPA Bertaraf Internasional *paper presented in Seminar Nasional Pendidikan MIPA Universitas Negeri Sriwijaya* 17 September 2011

[16] Liliasari 2009 *Berpikir Kritis dalam Pembelajaran Sains Kimia Menuju Profesionalitas Guru* (Bandung: Program Studi Pendidikan IPA Sekolah Pascasarjana UPI)

IOP Conf. Series: Journal of Physics: Conf. Series **947** (2017) 012068 doi:10.1088/1742-6596/947/1/012068

Students' Thinking Procesess on Quadrilateral Concept: A Case of Providing Non-Examples

E B Rahaju[1]

[1]Mathematics Departement, Universitas Negeri Surabaya

E-mail: endahrahaju@unesa.ac.id

Abstract. Characteristic of someone understand the concepts is the ability to give the examples and non-examples and the reasons. The thinking process in determining the example and non-examples of a concept will be different because of its cognitive style.
This is an explorative study which qualitatively analyze the thinking processes of field independent (FI) and field dependent (FD) students in providing non examples of quadrilateral concepts. The results of the study indicate that: 1) The essential characteristics of each shape in the quadrilateral as the basis for the FI participant to determine the non-examples, so that the shape which does not meet those characteristics are selected as non-examples. The selected first non-example tends to be similar with the grouped shapes. 2) The essential characteristics of each shape in the quadrilateral as the basis for the FD participant to determine the non-examples, so the shapes have striking differences are chosen as non-examples to make her be easier to give the reasons. The determination of the essential characteristics of a quadrilateral is important in determining non-example of quadrilateral. Based on the selection of examples and non-examples of quadrilateral will facilitate to determine the relationship of many concepts in the quadrilateral.

1. Introduction

The main task of teaching mathematics in school is to know the thinking process of students in learning mathematics and how mathematical knowledge is interpreted in the mind. Suryabrata [1] argues that in the process of thinking appear mental activities to form understanding, form opinions and make conclusions. One of the goals given a math lesson at the school level is that students understand the concept of mathematics. One indicator of understanding the concept of mathematics is being able to provide an example and a non-example of a concept. Sjogren [2] argues that the concepts can be obtained through three stages: (a) categorization stage, (b) eliminating inappropriate categories and combining appropriate categories to form a concept, and (c) concluding a concept so as to give examples and non-examples of concepts with its categorization.

In understanding a concept, students have different ways. This is appropriate with Tayler [3] who states that: "*dimensions of individual differences, among others such as intelligence, reasoning ability, creativity, cognitive style and gender*". Where as Slameto [4] is of the opinion that the cognitive style is the way a person create perceptions and compiles information (stimulus) that comes from the surrounding environment that is influenced by the experiences that have been owned. Based on psychological aspects, Witkin [5] grouped cognitive styles into two, are field independent (FI) and field dependent (FD) types.

IOP Conf. Series: Journal of Physics: Conf. Series **947** (2017) 012068 doi:10.1088/1742-6596/947/1/012068

Furthermore Adelina [6] and Witkin [5] stated that: *"cognitive styles are the characteristic, self consistent modes of functioning which individuals show in their perceptual and intelectual activities", "cognitive style represent a person's typical modes of perceiving, remembering, thinking and problem solving".*

Based on the Rahaju's research [7], showed that there are differences in the thinking process of the FI and FD in understanding concept of quadrilateral. While one of indicator to understand a concept is *capability* to give examples and non-examples of a concept. In general in the process of learning mathematics, teachers only provide examples or ask students to give examples of a concept. Therefore researcher was interested in conducting a research that aims to explore thinking process of students in determining non-examples of the quadrilateral concept.

2. Method

This research aims to explore the thinking processes of students in determining non-examples of quadrilateral based on cognitive styles. The instrument used to categorize students' cognitive styles is GEFT (group embedded figure test), an instrument developed by Witkin [8]. At GEFT, participants were asked to thicken a simple figure found on complex figure. Tracing the thinking process of participants in determining non-examples of quadrilateral was done through the task given to the participants. In this task, participants were asked to express how they were gave non-examples of a concept in quadrilateral. Their way can be expressed with think aloud. To obtain more in-depth and accurate data about participants' thinking processes, this study conducted task-based interviews. The participants in this research was two students grade VII public junior schools' in Surabaya, Indonesia with cognitive different style, namely field independent (FI) and field dependent (FD). The control variables chosen in this research were equivalent mathematical abilities and had the same gender.

3. Result and Discussion

The search for the thought processes done by both the participants in determining non-examples of quadrilateral, served according to the order of the concept of selected by both the participants.

3.1. Research Results FI Subject

S: (*In giving of the reasons to select non-example of square, subject saw that figure*

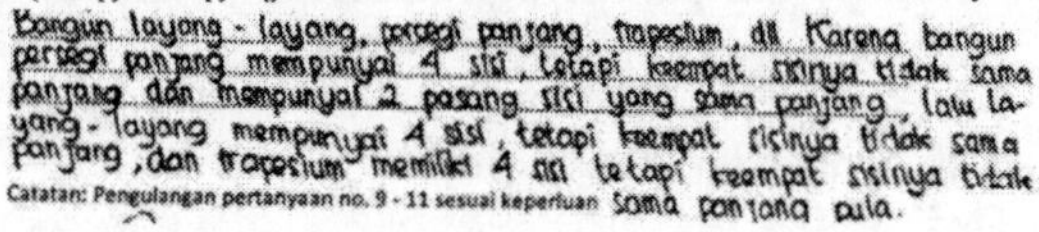
Bangun layang - layang, persegi panjang, trapesium, dll Karena bangun persegi panjang mempunyai 4 sisi, tetapi keempat sisinya tidak sama panjang dan mempunyai 2 pasang sisi yang sama panjang, lalu layang - layang mempunyai 4 sisi, tetapi keempat sisinya tidak sama panjang, dan trapesium memiliki 4 sisi tetapi keempat sisinya tidak sama panjang pula.
Catatan: Pengulangan pertanyaan no. 9 - 11 sesuai keperluan

The kite has 4 sides of different length, the rectangle has 4 sides of different length and 2 pairs of the same length, the trapezoid has 4 sides of different length

S: The first time, I think is a shape that have no four sides of equal length. Rectangel and square easily seen difference does it make which is to long different sides

Figure 3.1 Task based interview non-example of square

Based on the results in figure 3.1, it shows that the FI in selecting non-examples square focus on long edge attributes and suggests that the FI concludes the characteristic "four sides of equal length" is an essential characteristic for a square. So the non-examples selected are shapes that does not have these characteristics.

S: (*When determine reasoning non-example of rectangle (parallelogram and trapezoid), subject look at the figures in the previous question while making some scratch*

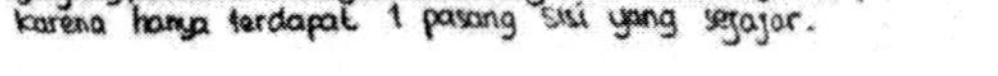
Bangun persegi. Karena semua sisinya sama panjang. Bangun jajar genjang, karena sisinya tidak tegak lurus. Bangun trapesium, karena hanya terdapat 1 pasang sisi yang sejajar.

The square because the four sides are the same length, parallelogram because the sides are not perpendicular, the trapezium because it has only one pair of parallel sides

IOP Conf. Series: Journal of Physics: Conf. Series **947** (2017) 012068 doi:10.1088/1742-6596/947/1/012068

S: The first time I think to find non-example of rectangle is quadrilateral who have no characteristic rectangle. The essential characteristic of the rectangle is to have 2 pairs of sides of different lengths, all angles are right and 2 pairs of parallel sides. I choose square, because most easy the difference with rectangle, where the long sides. Square had four sides of equal length while rectangle have two pairs sides of the unequal length.

Figure 3.2 Task based interview non-example of rectangle

Based on the results in figure 3.2, it shows that FI which is the essential characteristic of the rectangle is "having 2 pairs of sides of different lengths, right angles and 2 pairs of parallel sides". So the non-examples of the rectangle selected are square, parallelogram, and trapezoid. The non-example selected are the shape that does not have one of the essential characteristics of a rectangle, but a square is an incorrect non-example because FI is still affected by the rectangular shape obtained during the math learning in elementary school, which has two pairs of sides of different lengths.

S: (*In giving of the reasons to select non-example of parallelogram (trapezoid and kite), subject look at the figures in the previous question*)

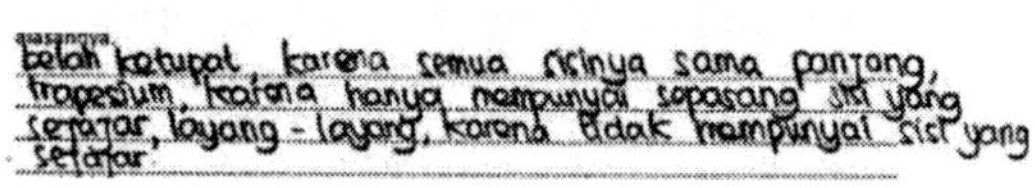

The rhombus because it has four sides of equal length, the trapezium because it has only one pair of parallel sides, the kite has no parallel sides

S: The first time I think to find non-example of parallelogram is quadrilateral who have no characteristic parallelogram. The essential characteristic of the parallelogram is to have 2 pairs of sides of different lengths, and 2 pairs of parallel sides. I choose rhombus, because most easy the difference with parallelogram, where the long sides. Rhombus had four sides of equal length while parallelogram have two pairs of side of the unequal length.

Figure 3.3 Task based interview non-example of parallelogram

Based on the results in Figure 3.3, it shows that the essential characteristics of the parallelogram are "having 2 pairs of sides of different lengths, and 2 pairs of parallel sides". So the FI chooses non-examples of the parallelogram are rhombus, trapezium and kite. The rhombus is an incorrect non-example of the parallelogram, because FI is still affected by the parallelogram shape obtained during the math learning in elementary school, which has 2 pairs of sides of different lengths.

S: (*In giving of the reasons to select non-example of kite (rhombus and parallelogram), subject look at the figures in the previous question*)

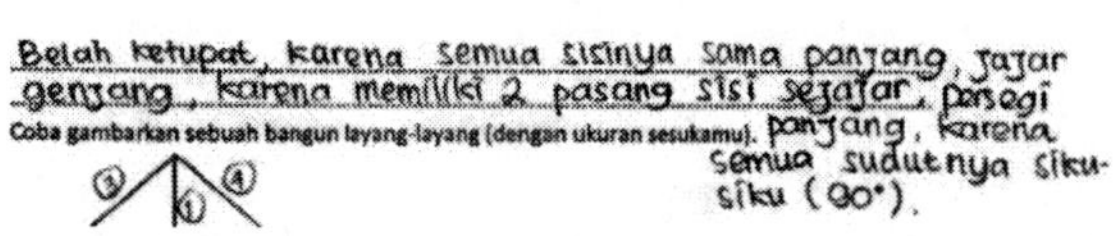

The rhombus because it has four sides of equal length, the parallelogram because has 2 pair parallel sides, the rectangle has four right angles

S: the essential characteristic of the kite are "having 2 pairs of sides adjacent to different lengths and pairs of opposite corners equal.

Figure 3.4 Task based interview non-example of kite

Based on the results in Figure 3.4, it shows that the FI chose non-example of kite is a shape that does not have one of the essential characteristics are rhombus, parallelogram and square. The non-examples of kite selection are an incorrect, because the essential characteristics of the kites concluded by FI are less precise. The fault is also caused by the FI is still affected by the form of kite that he studied in elementary school

S: The essential characteristic of rhombus are "having four sides the same length, each other the angles is opposite and the diagonals perpendicular to each other

IOP Conf. Series: Journal of Physics: Conf. Series **947** (2017) 012068 doi:10.1088/1742-6596/947/1/012068

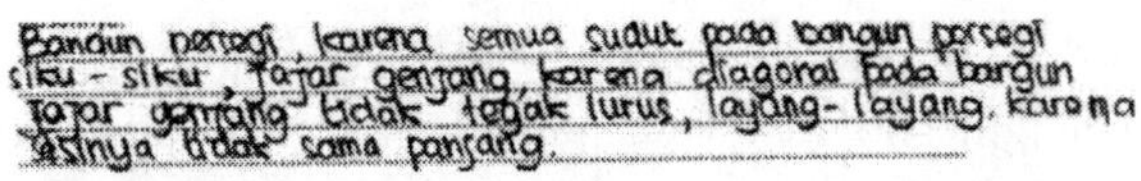

The square has four right angles, the parallelogram because the diagonal are not perpendicular, the kite because the sides are not the same length

S: (*In giving of the reasons to select non-example of rhombus, subject look at the figures in the previous question*)

Figure 3.5 Task based interview non-example of

Based on the results in Figure 3.5, it shows that the FI selects non-examples of rhombus are shapes that do not have any of the essential characteristics, are rectangles, parallelograms and kites. The square is an incorrect non-example because FI participant is still affected by the rhombus shape obtained during the math learning in elementary school, which has acute angle and obtuse angle.

S: (*In giving of the reasons to select non-example of trapezoid, subject look at the figure in the previous question*)

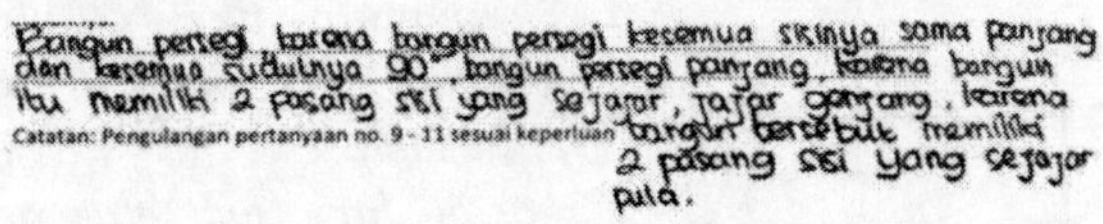

The square because the four sides are the same length and the angle is right, rectangle and parallelogram have 2 pairs of parallel sides

S: Characteristic easy on the trapezoid are most have two sides of equal length and a pair of parallel sides. So non-example trapezoid, who do not have the characteristics of a trapezoid

Figure 3.6 Task based interview non-example of trapezoid

Based on the results in Figure 3.6, it shows that the FI selects non-examples of trapezoid are shapes that have two pairs of parallel sides, since the trapezoid essential characteristics is to have exactly a pair of parallel sides.

3.2. Research Results FD Subject

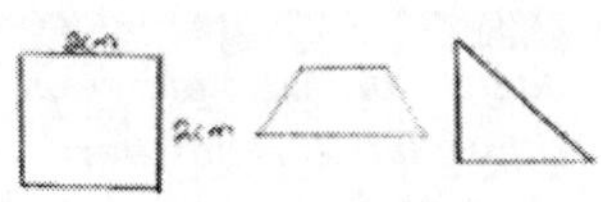

(*The participant was drawing non-example rectangle. When will gave second non-example, she was looking previous figure*)

S: First what I think to find non-example is find differences easy. Square most easily the difference with rectangle, because all sides of equal length .Trapezoid not had four right angle. Triangle only

Figure 3.7 Task based interview non-example of rectangle

Based on the results in Figure 3.7, it shows that according to FD which is the essential characteristic of the rectangle are "having 2 pairs of sides of different lengths, and four right angles". She drew the first non-examples, to make it easier to give the reason for the non-example selection. Non-examples selected are shapes that has a striking difference with the rectangle. Square is an incorrect non-example because she is still affected by the rectangular is a shape always having different side lengths as she were studied in elementary school.

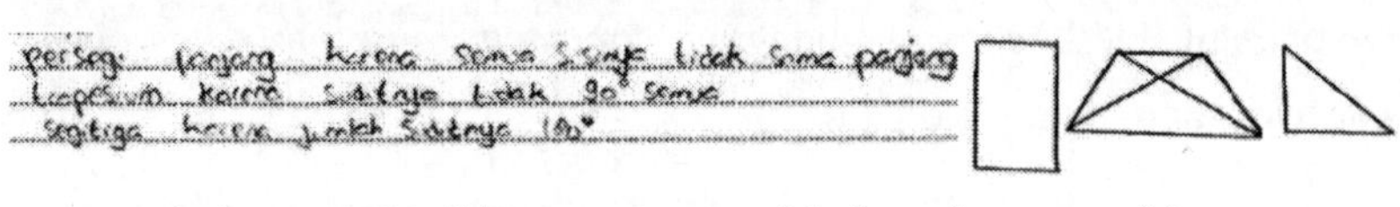

(*When participant will give non-example and the reason, she was looking previous figure before and then drawing non-example of rectangle*)

rectangle because its sides are not equal in length, trapezoid has no right angle, triangle of angle number 180°

S: I think to find non-example of square is find differences easy, who have no four sides of equal length and has no four right angle.

Figure 3.8 Task based interview non-example of square

IOP Conf. Series: Journal of Physics: Conf. Series **947** (2017) 012068 doi:10.1088/1742-6596/947/1/012068

Based on the results in Figure 3.8, it shows that according to the FD which is the essential characteristic of the square are "having four sides of equal length and four right angles". She drew the first non-examples, to make it easier to give the reason for the non-examples selection. The non-example selected is a shape has a striking difference with a square, is that does not have any of the essential characteristics.

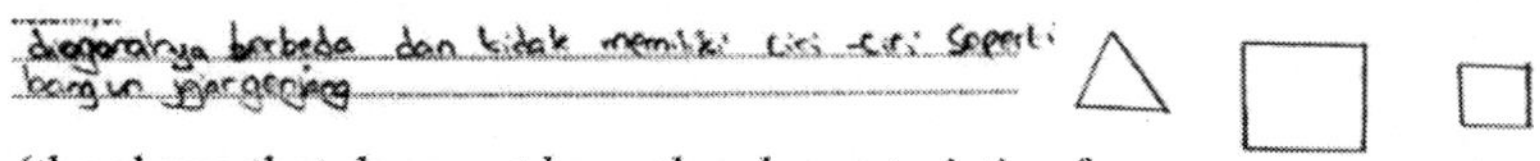

(the shape that does *not have the characteristic of parallelogram and diagonally different*

S: I think to find non-example is find differences easy with parallelogram. Triangle only have three sides. Square have diagonal that intersect perpendicular, while a parallelogram not

(She draw non-example of parallelogram without saw the previous figure. She was giving the reasons for the selection with each non-example look at an figure has already been prepared earlier)

Figure 3.9 Task based interview non-example of

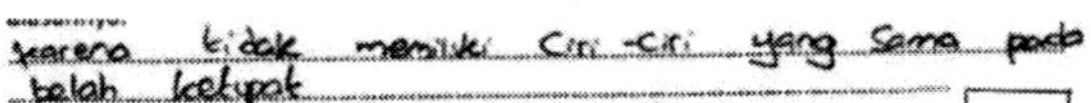

(The shape that does not have the characteristics of rhombus)

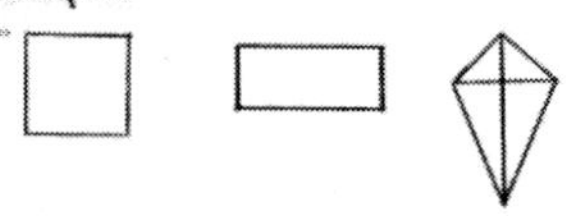

S: I think to find non-example is find differences easy with rhombus, has no right angle. Square and rectangle have right angle and kite has no four sides of equal length

(She wrote the reasons in general first, then drew the non-examples. She gives the reason for each non-sample has chosen by looking at the figure has created)

Figure 3.10 Task based interview non-example of rhombus

S: First what I think to find non-example is find differences easy with kite, has no right angle and have two pairs sides of the

(The subject of drawing non-example of kite about previous figure. The subject of giving the reasons for the selection with each non-example look at an figure has already been prepared earlier)

Figure 3.11 Task based interview non-example of

Based on the results in Figure 3.10, it shows that according to the FD which is the essential characteristic of the rhombus are "do not have right angle and have four sides of equal length". She drew the first non-examples, to make it easier to give the reason for the non-examples selection. The non-example selected is a shape has right angle or has no four sides of equal length. The square is an incorrect non-example because she is still affected by the rhombus is a shape always having acute angle and obtuse angle as she were studied in elementary school.

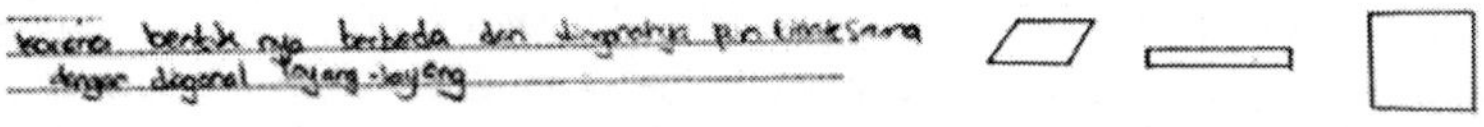

S: I think to find non-example is find differences easy with kite. It has no right angle and have two pairs of sides are unequal length.

(She wrote the reasons in general first, then drew the non-examples. She gives the reason for each non-sample has chosen by looking at the figure has created)

Figure 3.11 Task based interview non-example of

Based on the results in Figure 3.11, it shows that the FD which is the essential characteristic of the kite are "do not have right angle and have two pairs of sides are unequal length". She drew the first non-examples, to make it easier to give the reason for the non-examples selection. The non-example selected is a shape have right angle or have diagonals do not intersect perpendicular. The reason given for the parallelogram

IOP Conf. Series: Journal of Physics: Conf. Series **947** (2017) 012068 doi:10.1088/1742-6596/947/1/012068

chosen as a non- example of kite does not correspond to the essential characteristics, but rather she sees the diagonal intersection. This shows that she is more focused on the figure of the kite.

karena bentuk dan sisinya berbeda

because the shape and sides are different

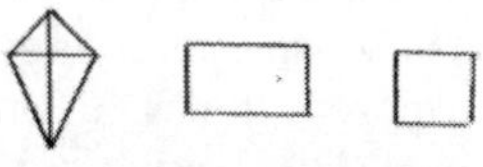

S: To make it easier to give reasons for the selection non-examples of trapezoid, I drew the example first. I think to find non-example is find differences easy with trapezoid, it has a pair of parallel sides. Square and rectangle have two pair of parallel sides and kite has no pair of parallel sides

(She wrote the reasons in general first, then drew the non-examples. She gives the reason for each non-sample has chosen by looking at the figure has created)

Figure 3.12 Task based interview non-example of

Based on the results in Figure 3.12, it shows that according to the FD which is the essential characteristic of the trapezoid is "has a pair of parallel sides". She drew the first non-examples, to make it easier to give the reason for the non-examples selection. The non-example selected is a shape have no parallel sides or have two pairs of parallel side

3.3. Discussion

Based on the results (Picture 3.1 – 3.6) showed the FI thinking process in determining the non-example of a shape in a quadrilateral was by looking to essential characteristics of each shape. So to determine a non-example of a shape in a quadrilateral, she chose a shape has no one of the essential caracteristics of the shape. To ensure the truth of non-example selection reasons, she considered the quadrilateral figures was contained in her previous task. Based on her reasons of the non-example selection performed, it showed that attributes of side length and the measure of angle are very important. She concerned with the parallelism attribute when determining non-examples of trapezoid, since an essential characteristic of the trapezoid is it has a pair of parallel sides. She chose the first non-example, it is the similar shapes of the ones has been grouped, whereas in the second and third of non-examples she tent to choose shapes that have striking differences. There are mistakes that she conducted in selecting non-examples of rectangles and parallelogram. This happened because she was affected by the rectangular and the parallelogram shapes always have two pairs of sides with different lengths. The mistake indicated that she was still influenced by familiar shapes at the time of math learning in elementary school, even though she has already known the essential characteristics of each shape.

Based on the results (Picture 3.7 – 3.12) showed the FD thinking process in determining the non-examples of a quadrilateral was by looking to essential characteristics of each shape. So to determine a non-example of a quadrilateral shape, she chose a shape has a very different to others. She gave non-examples for each shape in quadrilateral in the form of figures, then she gave a general reason for the non-example selection. Through the figures she can create easily providing the reasons for choosing any non-examples. She tent to chose triangles as non-examples in quadrilateral because they have many different sides. Based on the reason of the given non-example selection, it showed that side length, measure of angle and number of sides are the important attributes. She took parallelism attribute when selecting a non-example of trapezoid, since it has a pair of parallel sides is an essential characteristic of trapezoid. She chose the first non-example was square or rectangle because their characteristic is very familiar. While the second and third non-examples, she tend to choose shape that have striking differences with the grouped shape. There is a mistake in selecting non-examples rectangle and parallelogram, because she still assume that rectangle or parallelogram have two pairs of sides of different lengths.

By observing the thought processes of the two participants in determining non-examples of quadrilateral, it appears that they do the thinking steps through the formation of understanding by looking at the esential characteristics of a shape, so they can make the conclusion that the non-example of a shape

IOP Conf. Series: Journal of Physics: Conf. Series **947** (2017) 012068 doi:10.1088/1742-6596/947/1/012068

in a quadrilateral, it is a shape that does not have one or more esential characteristics. Once confirmed there are one or more essential characteristics that are not met, they can conclude that a shape is a non-example[1]. Looking at the way of both participants to determine the non-example of quadrilateral, there are striking differences. This can be seen when specifying non-examples, the FI can directly provide the reason for the non-examples selection, whereas the FD to provide the reason for the non-example selection must be through drawing the non-example. By looking at the non-example figure her creates, the FD can easily see that the figure non-example does not have esential characteristic a shape in quadrilateral. The way of the FI in giving the non-examples of quadrilateral show that she tends to be consistently not easily influenced by changing the situation. While the FD shows tend to be consistently easily influenced by changes in the situation. The conditions of both participants are in accordance with Adelina's [6] and Witkin's [5].

To learn the quadrilateral comprehensively, it is necessary to consider the relationship between the shapes in it. The non-example selection of a quadrilateral has an important to determining the relationship between the shapes in the quadrilateral. If the non-example selection of each shape in a quadrilateral is done correctly, then there will be no mistake in determining the relationship in quadrilateral. For example an FD made a mistake in selecting a square as a non-example of rectangular, she would be able to result in no relationship between square and rectangle. This indicates that the wrong non-example selection will result in an error in determining the relationship of two shapes.

4. Conclussion

The thinking process of the FI and FD in determining non-examples of each shape in quadrilateral is done an understanding based on the essential characteristics of quadrilateral. Based on these characteristics, the participants holds that a non-example selection of a quadrilateral is a shape which does not have any of these essential characteristics of each shape in quadrilateral. So they make the conclusion that the non-examples of quadrilateral which they chooses has one or more characteristic difference. The FI participant give two kinds of the non-examples of a quadrilateral: the first non-example of quadrilateral which a shape has some similarities and the next non-example is a shape which has a striking difference. The FI provide a direct reason for the non-example selected. While the FD participant in giving non-examples of a quadrilateral, she tends to choose triangle or square or rectangle as the first non-example. She chose a shape has a striking difference with a shape has been grouped as the next non-example. The non-example selection begins with a generally accepted reason (for all non-examples selected) and continues with non-example drawing. The FD give the reasons for each non-example selection after viewing the figure it has created.

5. References

[1] Suryabrata Sumadi 2006 *Educational Psychology* (Jakarta: PT Raja Grafindo)
[2] Sjogren J 2011 *Concept Formation in Mathematic* (Sweden: Acta Universitatis Gothoburgensis)
[3] Taylor, Lloyd J. Ortega, R David 2003 The Application Of Goldratt's Thinking Process To Problem Solving Las Vegas: *Proceedings of the Academy of Strategic Management* **2**(3) 89–104.
[4] Slameto 2003 *Learning and Its Affecting Factor* (Jakarta: Rineka Cipta)
[5] Witkin H A & Goodenough DR 1981 *Cognitive Styles: Essence and origin* (New York: International University Press)
[6] Guisande MA, et all 2007 Field Dependence-Independence (FDI) Cognitive Style: an analysis of attentional functioning *Psicothema* **19**(4). 78-109
[7] Rahaju EB 2016 Thinking Process of Junior High School Students Based on Cognitive Style In Understanding Concepts of Quadrilateral (Dissertation. Surabaya: Postgraduate Unesa)
[8] Witkin HA, Oltman, P. K, Raskin E, Karp S A 1971 *A Manual For Embedded Figures Tests* (California: Consulting Psychologists Press Inc)

Technological Pedagogical Content Knowledge of Prospective Mathematics Teacher in Three Dimensional Material Based on Sex Differences

M A Aqib[1,*], M T Budiarto[1], P Wijayanti[1]

[1]Departmen of Mathematics-Universitas Negeri Surabaya, Indonesia Jl. Ketintang, Surabaya 60231
[*]*Corresponding Author*. Email: mohammadaqib@mhs.unesa.ac.id

Abstract: The effectiveness of learning in this era can be seen from 3 factors such as: technology, content, and pedagogy that covered in Technological Pedagogical Content Knowledge (TPCK). This research was a qualitative research which aimed to describe each domain from TPCK include Content Knowledge, Pedagogical Knowledge, Pedagogical Content Knowledge, Technological Knowledge, Technological Content Knowledge, Technological Pedagogical Knowledge and Technological, Pedagogical, and Content Knowledge. The subjects of this research were male and female mathematics college students at least 5th semester who has almost the same ability for some course like innovative learning, innovative learning II, school mathematics I, school mathematics II, computer applications and instructional media. Research began by spreading the questionnaire of subject then continued with the assignment and interview. The obtained data was validated by time triangulation.This research has result that male and female prospective teacher was relatively same for Content Knowledge and Pedagogical Knowledge domain. While it was difference in the Technological Knowledge domain. The difference in this domain certainly has an impact on other domains that has technology components on it. Although it can be minimized by familiarizing the technology.

1. Introduction

The age of globalization has an impact on everything such as communications, informations and tools that more advanced from the past [1]. This also affects the learning. Learning that was originally using traditional learning system must became a modern learning system or in other words learning must using technology that develops today. In learning teacher should have extensive knowledge of the technology that to be taught, pedagogy skills in classroom management based on technology, and technical knowledge about technology that to be used [2]. In general there are three main factors that affectthe effectiveness of learning with technology such as:technology, pedagogy, and materials that to be taught. A knowledge framework that links these three factors in complexly called Technological Pedagogical Content Knowledge (TPCK) [3].

There are seven knowledge domains from TPCK which is a combination of the three bodies components such as: content, pedagogy, and knowledge. The seven knowledge domain are Content Knowledge (CK), Pedagogical Knowledge (PK), Technological Knowledge (TK), Pedagogical Content Knowledge (PCK), Technological Content Knowledge (TCK), Technological Pedagogical Knowledge (TPK), and Technology, Pedagogy and Content Knowledge (TPaCK) [3]. CK is a knowledge about learning materials that to be taught, in this research is math lessons include facts, concepts, principles, and rules [3,4]. PK is knowledge of methods and processes in learning include knowledge about classroom management, assessment, and lesson plans [3,4]. PCK is the knowledge of the learning process in terms of the material that to be taught [3,4]. TK is knowledge of a variety in technologies, from low –tech technology to digital technology [3,4]. TCK is the knowledge of how technology can provide a new representation from a particular subject, in this research is math lesson [3,4]. TPK Is knowledge about how various technologies can be used in the learning process [3,4]. TPaCK is the knowledge needed by teachers to integrate technology in the learning process for each subject matter [3,4]. Description of each domains knowledge from TPCK and relationship among them can be seen in Figure1.

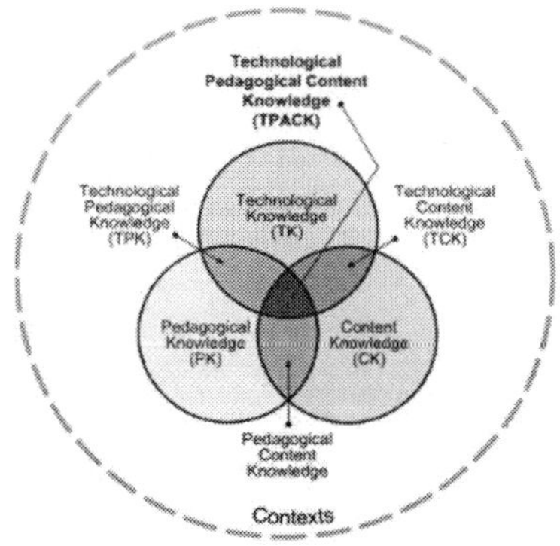

Figure 1. Description of TPCK components
Figure was adopted from Koehler and Mishra

In learning, teacher must know the various domains of knowledge contained in TPCK and know the relationship between these domains. Many teachers experience a lack of preparation in using today's technology because they earn a bachelor's degree when technology is more simple than today [3]. This leads prospective teacher who will become teacher have a big role in learning with technology. That because technology will continue to evolve over time. Mathematics is one example of learning that technology's necessary needed. Technology can become the bridge of abstract thinking in learning mathematics. The three dimensional material is one of the mathematical content area which it called space & shape [5].

One of relevant study is from Jordan entitled "The influence of gender on beginning teachers' perceptions of their Technological Pedagogical Content Knowledge" which results there are significant differences between male and female beginner teachers. Females rated their knowledge higher than males in one domain that is Pedagogy Knowledge and man rated their knowledge higher than female in six other domains [6]. Data retrieval is done by using closed questionnaire only, so the data obtained only in form of perception and can't describe completely. From that matter, research about TPCK based on sex differences that use qualitative approach in order to get the whole description is urgently needed.

2. Research methods

This research is qualitative descriptive research. Qualitative approach is used so that it produces descriptive data in the form of written or oral words from people and observed behavior. The subjects of this research were mathematics college students at least 5th semester that already thought some course like innovative learning, innovative learning II, school mathematics I, school mathematics II, computer applications and instructional media.The subjects consist of two college students selected based on the criteria that the two students have nearly equal ability and have different sex. Research began by spreading the questionnaire of subject then continued with the assignment and interview.

The instrument in this research is task-based interview guide. The task-based interview guide is used as a directive in the interview to obtain the information that is needed. Task in this research is to create lesson plans with help of technology on three dimensional materials. Meanwhile, the type of interview used in this research is semi-structured that it contains about extracting more information about the task undertaken by college student. The interview was conducted after the college students made the lesson plansbased on technology. After the data obtained, validation of the data is using time triangulation with breaks time about three weeks.

IOP Conf. Series: Journal of Physics: Conf. Series **947** (2017) 012069 doi:10.1088/1742-6596/947/1/012069

Data analysis in this research are data reduction, data presentation, and conclusion [7]. In data reduction, data based on interview is coded according to the indicators, selecting and disposing of unnecessary information. Then the data is classified according to each TPCK domains. In the data presentation, there are a collection of information that arranged systematically so it can be drawn a conclusion. In conclusion, the researcher gives an overview for each domain in TPCK.

3. Results and Discussion

The description of each TPCK domains for prospective teacher based on sex differences are presented in Table 1.

Table 1. Description of TPCK prospective teacher

Domain	**Description**	
	Male Prospective Teacher	Female Prospective Teacher
CK	Give an examples of facts, concepts, principles and rules but memorized in definition so that the definition appeared as a opinion of subject	Give an examples of facts and concepts and explaining its definition in detail but memorized it
	Explain the detail of three dimensional concepts	Explain the detail of three dimensional concepts
PK	Give an example of various techniques and methods of both cognitively and psychomotorically assessment and its functions	Give an examples of various techniques and methods of both cognitively and psychomotorically assessment and its functions
	Knew examples of models, approaches, methods and the outline of its definitions	Give an examples of models, approaches, and methods in general
	Explain theoretically about classroom management and learning management	Explain theoretically about classroom management and learning management
TK	Explain the technical about using hardware	Explain the technical about hardware
	Explain the latest of technology and interested in using it	Not explain about the latest of technological development
	Give an examples of various types applications but limited when using it	Give an examples of various applications but felt difficult when using it
PCK	Adjusted the selection of models, methods, and approaches to each other by focusing on the material that being taught	Adjusted the selection of models, methods, and approaches to each other by focusing on the curricula
TCK	Represented the three dimensions concepts according to the technology that he used	Can't applied the representation of facts, concepts, principles and rules with technology that she used in the three dimensions and limited on gave an images
TPK	Choosed technology that effectively and efficiently based on the design of learning	Choosed technology that is practically and easy to use
	Selected technologies that can enhanced student activity	Selected technologies that can enhanced student activeness
TPaCK	Maked technology as a reinforcement of concepts in learning of three dimensional learning	Maked technology as a conceptual presenter in three dimensional learning

IOP Conf. Series: Journal of Physics: Conf. Series **947** (2017) 012069 doi:10.1088/1742-6596/947/1/012069

Based on Table 1,it describe seven domains of TPCK include CK, PK, TK, PCK, TCK, TPK and TPaCK. It shows that there are similarities and differences between male and female prospective teacher. Based on obtained data, show that for CK and PK domains almost has no differences. For the PCK domain, both male and female prospective teacher is much alike in selection of models, methods, approaches, and related to each other. Slightly differentiate that male students are more concerned with the suitability of the material that to be taught while the female prospective teacher focus more on the suitability of the applicable curriculum.

The big difference between male and female prospective teacher was found in TK domain. This is due to significant differences in terms of gender-related attitudes to ICT, ICT skills, and use of ICT [8]. Male prospective teacher knew various types of applications but limited when using it where female prospective teacher knew the various applications but felt difficult when using it. Here it is seen that male prosepctive teacher are indirectly more able to use computer applications than female prosepctive teacher. Female prosepctive teacher have difficulty in using computer applications, unlike male prosepctive teacher although their use is still limited in basics. This is because both men and female have significant differences in ICT skills with male have higherscores than female [9]. Differences in TK domain certainly have an impact on other domains that have technological components such as TCK, TPK, and TPaCK.

In TCK domain, male prospective teacher can represent concept of the three dimensional material according to the technology that used while the female prospective teacher are less able to apply the representation of facts, concepts, principles and rules on the three dimensional according to the technology that used. This is certainly not separated from the technology that they used, geogebra selected by male prospective teacher is easy in represent the concept compared to technology namely power point that used by female prospective teacher [2].

In TPK domain, male prospective teacher choose technology that is effectively and efficiently in terms of learning design while female prospective teacher choose technology that is practical and easy to use. In contrast to male prospective teacher, female prospective teacher feel unable to use geogebra so that she used power point which in fact it is the application that she mastered. There are an indication that female teachers have low self-confidence in using ICT in learning compared to male teachers [10]. The same is also on TPaCK domain where male prospective teacher utilize technology as a reinforcement concept in the learning of the three dimensional material while the female prospective teacher utilize technology only as the presentation of the material in learning three dimension material.

The big difference between male and female students in terms of technology can actually be minimized. As it can be seen that male and female prospective teacher can overcome technical problems related to computers. This happens because the use of laptops and LCD have become common place in education. It can be concluded that the difference of knowledge about technology between male and female prospective teacher can be minimized if the technology has been frequently used. The more and more computers are used in schools it will reduce the effect of differences in use between male and female [11].

4. Conclusion

Based on the result of the research, male and female prospective teacher candidate students are the same for the CK and PK domain. In the TK domain, male prospective teacher knew various types of applications but limited when using it where female prospective teacher knew the various applications but felt difficult when using it. In the PCK domain, male students are more concerned with the suitability of the material that to be taught while the female prospective teacher focus more on the suitability of the applicable curriculum. In the TCK domain, male prospective teacher knew various types of applications but limited when using it where female prospective teacher knew the various applications but felt difficult when using it. In the TPK domain, male prospective teacher choose technology that is effectively and efficiently in terms of learning design while female prospective teacher choose technology that is practical and easy to use. Last in the TPaCK domain, male prospective teacher utilize technology as a reinforcement concept in the learning of the three dimensional material while the female prospective teacher utilize technology only as the presentation of the material in learning three dimension material.

IOP Conf. Series: Journal of Physics: Conf. Series **947** (2017) 012069 doi:10.1088/1742-6596/947/1/012069

5. Acknowledgement
I would like to express my gratitude to advisor 1 and 2 who have participated in the writing of this paper until it can be published and useful for many people. I would also like to thank UNESA which has given permission to conduct this research.

References
[1] Niess M L 2005 Preparing teachers to teach science and mathematics with technology: Developing a technology pedagogical content knowledge *Journal of teaching and Teacher Education* ,509 – 523
[2] Mainali B R 2008 Using dynamic geometry software Geogebra in developing countries: A case study of impressions of mathematics teachers in Nepal *International Journal for Mathematics Teaching & Learning* p1-16. 16p
[3] Koehler M J & Mishra P 2009 *What is technological pedagogical content knowledge?* (Contemporary Issues in Technology and Teacher Education.60-70)
[4] Schmidt DA 2010 Technological Pedagogical Content Knowledge (TPACK): The Develoment and Validation of an Assesment Instrumen for Preservice Teacher *JRTE*.**42**(2) 123-149
[5] Shiel G, Perkins R, Close S & E Oldham 2007 *PISA Mathematics: A Teacher's Guide* (Dublin: Department of Education and Science)
[6] Jordan K 2013 *The influence of gender on beginning teachers' perceptions of their Technological Pedagogical Content Knowledge (TPACK)*(Australian Educational Computing, 28(2))
[7] Miles M B, Huberman A M & Saldana J 2014 *Qualitative Data Analysis*(Thousand Oaks, CA: Sage)
[8] Kay R 2006 Addressing gender differences in computer ability, attitudes and use: The laptop effect *Journal of Educational Computing Research* **34**(2), 187-21
[9] Markauskaite L 2006 Gender issues in preservice teachers' training: ICT literacy and online learning.Australasian *Journal of Educational Technology* **22**(1), 1-20
[10] Jamieson-Proctor R, Burnett P, Finger G, & Watson G 2006 ICT integration and teachers confidence in using ICT for teaching and learning in Queensland state schools *Australasian Journal of Educational Technology* **22**(4), 511-530)
[11] Koh J H L, & Sing C C 2011 Modeling pre-service teachers' technological pedagogical content knowledge (TPACK) perceptions: The influence of demographic factors and TPACK constructs *Australian Society for Computers in Learning in Tertiary Education (ASCILITE) conference*, Hobart, 4-7 December, 2011

IOP Conf. Series: Journal of Physics: Conf. Series **947** (2017) 012070 doi:10.1088/1742-6596/947/1/012070

The written mathematical communication profile of prospective math teacher in mathematical proving

K V Pantaleon[1,5,6], D Juniati[2], A Lukito[3] and K Mandur[4]

[1,2,3]Faculty of Mathematics and Science Universitas Negeri Surabaya, Indonesia
[4,5] STKIP St. Paulus Ruteng-Indonesia

E-mail: [6]kristianuspantaleon@mhs.unesa.ac.id

Abstract. Written mathematical communication is the process of expressing mathematical ideas and understanding in writing. It is one of the important aspects that must be mastered by the prospective math teacher as tool of knowledge transfer. This research was a qualitative research that aimed to describe the mathematical communication profile of the prospective mathematics teacher in mathematical proving. This research involved 48 students of Mathematics Education Study Program; one of them with moderate math skills was chosen as the main subject. Data were collected through tests, assignments, and task-based interviews. The results of this study point out that in the proof of geometry, the subject explains what is understood, presents the idea in the form of drawing and symbols, and explains the content/meaning of a representation accurately and clearly, but the subject can not convey the argument systematically and logically. Whereas in the proof of algebra, the subject describes what is understood, explains the method used, and describes the content/meaning of a symbolic representation accurately, systematically, logically, but the argument presented is not clear because it is insufficient detailed and complete.

1. Introduction

Written mathematical communication is an intellectual activity that requires students to express their ideas or mathematical thoughts in writing. Written communication is one of the important aspects [1] that must be mastered by teachers/prospective math teacher as tool of knowledge transfer, for at least three major reasons. First, when students communicate their mathematical thinking and reasoning they involve during math instruction, to think, share ideas, and clarify their understanding [2,3]. Second, communication helps teacher to recognize the student's ability, understand their mathematical understanding, and then make the appropriate decisions to help them [4]. Third, classroom communication is powerful tool for teacher to assess students' learning and can create a safe environment, exploring ideas, and genuine dialogue [4]. Written communication is commonly used in education to measure how much students absorb the material presented by asking students to write down what they think or understand on the answer sheet. Writing is tool used to describe ideas, relationships, situations, or mathematical arguments, and present abstract ideas in a readable form that can be consumed by others. Thus, written mathematical communication is the process of realizing what is to be conveyed in the form of written. This communication covers all aspects of mathematics including proving as the essence of mathematics [5] because one of the purposes of proving is to communicate statements into deductive systems [6]. Therefore, the mathematical proving must be communicated precisely, coherently and clearly through logical reasoning so that it can be accepted.

IOP Conf. Series: Journal of Physics: Conf. Series **947** (2017) 012070 doi:10.1088/1742-6596/947/1/012070

The research about mathematical communication has been investigated in many studies. Dewi [7] studied about the differences in mathematical communication profiles of prospective teachers in solving mathematical problems based on gender and level of math ability using three aspects as the criteria, namely: accuracy, completeness, and fluency. Dewi put more emphasis on the results of mathematical communication; those are the truth of ideas based on mathematical rules, the adequacy of information, and speaking skills, than the process of mathematical communication itself. Furthermore, Wichelt and Kearney [8] conducted research about communication in mathematics learning using open-ended questions. They claimed that students' mathematical vocabulary comprehension was better when using open-ended questions. This study also put more emphasis on the results of mathematical communication, namely mathematical vocabulary comprehension. Kabael [9] studied about the communication skills of math teachers. Kabael found out that junior high school math teachers had no communication skills in the mathematical language as expected of a mathematics teacher. Kabael put more highlighted on the skill of mathematical communication than process. Related to the mathematical proving, many studies have shown that written communication in mathematical proving is a problem to students. Yackel and Hanna [10] found out that developing an understanding of mathematics proof and proving remains a challenge for many students. This is understandable because by its very nature, the mathematical proof is very sophisticated and seems to be much more challenging intellectually than many other parts of the school mathematics curriculum. Mathematical proof is one of the most difficult topics for students to learn [11,12]. Moreover, the problem of proof seems to be more complicated because the aim of a problem to prove is to show conclusively that a certain clearly stated assertions is true, or else to show that it is false [13]. Proving processes is an endeavour that demands high-order thinking skills to construct ideas and express them logically and systematically. Therefore, to communicate the proof requires a broader and more comprehensive understanding of knowledge such as conceptual understanding, procedural knowledge, methods of proof, and reasoning abilities. Ovez and Ozdemir [14] through the literature review concluded that proving skills of prospective mathematics teachers was not at a sufficient level. The study also showed that prospective mathematics teachers had some difficulties in understanding proof and proof writing. This fact is also supported by the results of Moore's research [10] which concluded that students probably were not ready to use proof and deductive reasoning as tool for solving mathematical problems and developing mathematical knowledge. These disadvantageous situations indicate that prospective mathematics teachers are not ready to communicate their mathematical ideas through proving. Prospective teachers should be able to communicate because teacher is one of the key aspects of students successful in learning achievement [15].

The focus of this study was on the process of written mathematical communication when subject solved the mathematical proving. In describing this research, the researchers analyze the subject's written assignment result and task-based interview result. Communication activities were analyzed by reference to the written mathematical communication theory framework. Based on the results of the analysis, it is concluded that the mathematical communication activities in proving are (1) explaining what is understood, (2) describing the method used, (3) presenting the idea in the form of drawing and symbols, (4) explaining the content/meaning of a representation, and 5) conveying the argument. In general the subject could convey her ideas appropriately and coherently, but less of clearly because the description was insufficient detailed and complete.

2. Method

This research was a qualitative research. Data collection started by giving math ability test to 48 students of mathematics education program, STKIP St. Paulus-Ruteng. The results of this test were used to categorize students based on their level of mathematical ability. In the next stage, the researchers gave the proving task to the 48 students. Based on the results of the analysis, the researchers selected one of 48 students with moderate math skills as the research subject. This determination was based on the consideration that the mathematical communication profile of students with moderate math skills can at least provide an overview for the other two groups. In the next stage, the subject of research was given Task I which contained two questions of proof, each on geometry and algebra. In addition, to describe the ability of prospective teachers in proof, the researchers also assigned the same task to other 47 students in the same class. The work results of the subject on Task I was then analyzed based on the theoretical framework of mathematical communication that had been prepared. In addition, the researchers also conducted interviews based on the work of the subject. While the work of other 47 students analyzed to get a general description of the proof ability. Furthermore, the researchers performed triangulation data by giving Task II which was similar to Task I to get valid data about the written mathematical communication profile of prospective mathematics teacher. The results of this work were then analyzed and used as the basis for interviewing the subject. The results of the analysis in the first and second stages were then compared to obtain the final data of mathematical communication profile of prospective mathematics teacher.

3. Results and discussions

The results of the mathematics test showed that of 48 students who took the test, only one student (2%) had high math skills, fourteen students (29%) had moderate math skills, and sixty-nine students (69%) had low math skills. Based on the results of the analysis, the researchers selected one of 48 students with moderate math skills as the research subject to whom the proving tasks were given.

The following description is the result of the study consisting of the general description of students' ability in mathematical proving, and the data description about the written mathematical communication profile in the proof of geometry and algebra.

3.1. An overview of student ability in mathematical proving

Task I was done by 48 students. The task consists of two questions, each measures proving skills in geometry and algebra. The two intended questions are: (1) Prove that if ABCDEF is a regular hexagonal with side length is a unit then the area of BCEF is $a^2\sqrt{3}$ unit; and (2) Prove that for every integer n, if n^2 is odd then n is odd. Based on the results of the analysis, it was found that for the first question, about geometry, only three students (6.25%) could do it correctly; three students (6.25%) could do it thoroughly but made various procedural and conceptual errors, while 42 students (87.5%) could not solve it completely. All of the students could draw the regular hexagonal correctly, but they could not find the right strategy to determine the area of BCEF. Most of them, divided the regular hexagonal into triangles and rectangle and then tried to find the length of EC by using knowledge of trigonometry and Pythagoras formulas but they failed. For the second number, about algebra, 20 students (41.7%) could solve it correctly, and the remaining 28 students (58.3 %) could do it thoroughly but made various mistakes both conceptual and procedural. In this question, most of the students chose indirect proof method using contraposition but some of them were wrong in determining it. The most common errors were algebraic manipulation and misconceptions about hypotheses and conclusions based on statements to be proved.

3.2. The written mathematical communication profile on the proving of geometry

Based on in-depth analysis, it was found out some mathematical communication activities conducted by the subject when did the proving on the geometry tasks. On the Task I, the question given was: "Prove that if ABCDEF is a regular hexagonal with side length is a unit then the area of BCEF is $a^2\sqrt{3}$ unit". The first communication activity conducted by the subject when doing this question was to explain what was understood from the problem. The subject began the proof by presenting correctly and clearly what was known and what to be proven from the question. Furthermore, the subject presented the steps of proof systematically and logically that began by presenting the idea in the form of images and symbols. The subject drew regular hexagonal correctly and then gave the names of the regular hexagonal by writing the composing letters at each vertex (see Figure 1 (a)). The subject then explained the content/meaning of the image. The subject argued that the area of BCEF is equal to the area of the rectangle. To simplify the subsequent completion, the subject drew again the area to be proved by referring to the first image that has been created (see Figure 1 (b)). The subject could draw this idea correctly but in less detail so that it was rather difficult to understand. The mathematical symbols used were also less consistent. And then, based on the new picture, the subject constructed and presented argument to show that the area of BCEF is $a^2\sqrt{3}$ unit but the subject could not convey the argument correctly and logically. The argument presented was not clear because it was not well-detailed. The subject said that the length of $x = \sqrt{2}a$ but she did not give an explanation of why it was so.

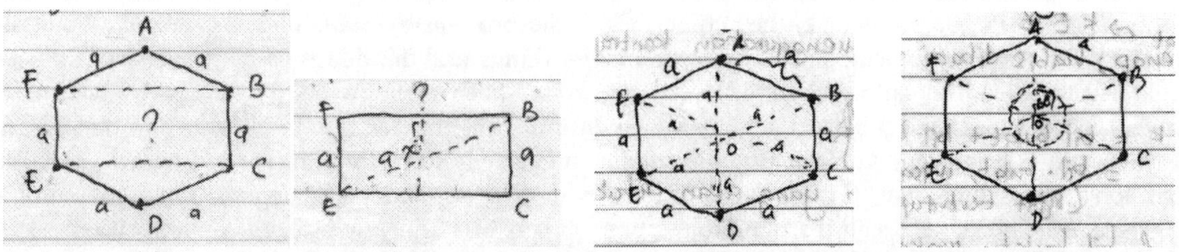

(a) (b) (c) (d)

Figure 1: Student's answer on the proof of geometry

Furthermore on the Task II, the question given was: "Prove that the regular hexagonal area with the side length is a is $\frac{3}{2}a^2\sqrt{3}$ unit". The subject began the proof by explaining what she understood, related to the known element and the conclusion to be proven. The subject could explain the idea correctly, systematically and completely. Then the subject presented systematically the steps of proof that began by presenting the idea in the form of pictures and symbols (see Figure 1 (c)) and followed by an explanation regarding to the content/meaning of the drawing. The explanation given could be well understood because it was quite simple. To facilitate the subsequent completion, the subject presented again the idea in the form of drawing of the regular hexagonal, including a circle in it (see Figure 1(*d*)), and followed by describing the contents of the drawing. The subject explained her idea systematically and logically by using simple language so that it could be well understood. Based on the new drawing the subject conveyed argument to show that the area of a regular hexagonal with side length is a is $\frac{3}{2}a^2\sqrt{3}$ unit. The subject conveyed the argument accurately but not systematically.

Based on the description above, it appears that the sequence of activities of the written mathematical communication on the proving of geometry is: (1) explaining what is understood, (2) presenting the idea in the form of drawing and symbols, (3) explaining the content/meaning of a representation, (4) presenting the idea in the form of drawing and symbols, and then closed by (5) conveying an argument. The subject explained what is understood correctly and then presented the idea in the form of drawing and symbols in detail and complete. The subject can explain the content/meaning of a representation accurately and clearly, but the subject cannot convey the argument systematically. The mathematical symbols used are also less consistent.

3.3. The written mathematical communication profile on the proving of algebra

The results of the analysis showed that there were some mathematical communication activities conducted by the subject when doing the proving tasks on algebra. In task I, the question given was: "Prove that for every integer n, if n^2 is odd then n is odd". The first mathematical communication activity raised was explaining what was understood namely the elements that were known and to be proven. The subject could explain what is understood correctly and completely. According to the subject, the known elements are n is integer and n^2 is odd, and the element to be proved is n is odd. Furthermore, the subject explained the method to be used namely indirect proof by using contraposition. This second step was not done by the subject on the proof of geometry. The subject could not explain this idea in detail and complete. The ideas presented were also unclear because they were not well ordered and were not accompanied by adequate explanations. Based on the prescribed method, the subject presented the proving steps systematically and logically, beginning with putting forward the argument, and continued by explaining the symbolic representation contained in the argument. The subject described these ideas correctly but not in detail and incomplete. The subject ended the proof by conveying argument systematically and logically and using simple and understandable language. She said: "So, it is proven that if n is even then n^2 is even, so it is true that if n^2 is odd then n is odd". The subject could show conclusively that the statement was true.

Further, in the Task II, the question given was: "Prove that for every integer n, if n^2 is even then n is even". As in the previous task, the subject began the proving by explaining the elements that she understood from the problem. Those were the known things and the things to be proved. The subject explained the idea in detail and complete, starting from what was known and followed by what to be proved. According to the subject, the known elements are n is integer and n^2 is even, and the element to be proved is n is even. Additionally, the subject suggested the method to be used namely proof by contraposition, which was preceded and ended with a brief and clear explanation about atomic propositions. The subject said that the proposition: $p \rightarrow q$ was equivalent to its contraposition: $\sim q \rightarrow \sim p$. By referring to the contraposition, the subject re-explained the elements that are known and which will be proved. According to the subject, the known element is n is odd, and the element to be proved is n^2 is odd. The subjects could explain the ideas correctly, detailed, and complete. Based on this chosen method, the subject presented the proving steps. The subjects could present these steps in a systematic way, accompanied by explanations for each step taken. The subject also explained the meaning of any symbolic representations used in the argument using simple and understandable language. The subject ended the proving by arguing that the given statement was true. She said: "Proven, if n is odd then n^2 is odd, with $n \in \mathbb{Z}$, so it is true that if n^2 is even then n is even".

Based on the description above, it appears that the sequence of activities of the written mathematical communication on the proving of algebra is: (1) describing what is understood, (2) describing the method used, (3) conveying the arguments, (4) explaining the content/meaning of a symbolic representation, and then closed by (5) conveying the argument. Generally, the subject could convey all of the ideas appropriately and coherently, but less of clearly because the description was insufficient detailed and incomplete.

3.4. The written mathematical communication profile on the proving of algebra

The results description above showed that, there is quite difference in the pattern of mathematical communication on the two questions given, both in Task I and Task II. To prove the truth of statements in the field of geometry, the subject tends to use images as mental representations. These images are used to reduce the concept abstraction, and make it more concrete so that all the important elements involved in reaching the solution will become apparent. The images are helpful to the subject in explaining her ideas in detail and completely, logically and systematically. Moreover, these images can help the reader to navigate the subject's thinking flowing well. An explanation that directly refers to a concrete image is more easily understood so that the truth of the proposed ideas can be easily accepted. Furthermore, when working on the proof of geometry, the subject do not convey the method or strategy used, but directly use the known facts to show the truth of the statement to be proved. It means that the subject uses direct proof method. While on the proof of algebra, the subject explicitly explains the method that will be used that is indirect proof with contraposition. The problem given on the proving of algebra cannot be solved by direct proof. If using direct proof, the subject will face difficulty in performing algebraic manipulations to show conclusively that the statement is true. This condition forces the subject to search for other suitable methods. Therefore, the subject always begins her proof by altering the initial statement into another equivalent form called the contraposition. By using contraposition, the subject can easily show conclusively that the statement is true.

4. Conclusions

Based on the results of this study, it can be concluded that in the proof of geometry, the subject explains what is understood, presents the idea in the form of drawing and symbols, and explains the content/meaning of a representation accurately and clearly, but the subject cannot convey the argument systematically and logically. Whereas in the proof of algebra, the subject describes what is understood, explains the method used, and describes the content/meaning of a symbolic representation accurately, systematically, logically, but an argument presented is not clear because it is insufficient detailed and incomplete.

References

[1] Wichelt L and Kearney NE 2009 Communication: a vital skill of mathematics *Science and Mathematics* Education *Commons* (DigitalCommons@University of Nebraska–Lincoln) p. 6
[2] Silver E A and Smit M S 1996 Building discourse communities in mathematics classrooms: a worthwhile but challenging journey *Communication in Mathematics, K-12 and Beyond* eds P C Elliot and M J Kenney (Reston, Virginia: NCTM) pp. 20-28
[3] NCTM 2000 *Principles and Standards for School Mathematics*. Reston (The National Council of Teacher of Mathematics, Inc) pp 60-345
[4] Pourdavood R G and Wachira P 2015 Importance of mathematical communication and Discourse in Secondary Classrooms *Global Journal of Science Frontier Research: F Mathematics and Decision Sciences* (Global Journals Inc. (USA)). **15** 17
[5] Knuth J E 2002 Secondary school mathematics teachers' conceptions of proof *Journal for Research in Mathematics Education*. **33** 379
[6] Stavrou S G 2014 Common Errors and misconceptions in mathematical proving by education undergraduates. *IUMPST: The Journal*. **1**. 1
[7] Dewi I 2009 Profil komunikasi matematika mahasiswa calon guru ditinjau dari perbedaan jenis kelamin *Disertasi* (Surabaya: Unesa)
[8] Wichelt L and Kearney N E 2009 Communication: A vital skill of mathematics. *DigitalCommons@University of Nebraska – Lincoln*
[9] Kabael T 2012 Graduate student middle school mathematics teachers' communication abilities in the language of mathematics. *Procedia-Social and Behavioral Sciences* 55, 809 – 815
[10] Yackel E and Hanna G 2003 Reasoning and Proof. *A Research Companion to Principles and Standards for School Mathematics* eds. J Kilpatrick, W G Martin and D Schifter (Reston: NCTM) pp. 231-345
[11] Heinze A 2004 The proving process in mathematics classroom – method and results of a video study *Proc. of the 28th Conf. of the Int. Group for the Psychology of Mathematics Education,* **3** p. 41
[12] Mejia-Ramos J P and Inglis M 2009 Argumentative and proving activities in mathematics education research *Proc. of the ICMI Study 19 Conf.: Proof and Proving in Mathematics Education* eds. F. L. Lin, et al, **2** p. 88
[13] Polya G 1973 *How To Solve It; A New Aspect of Mathematical Method* (New Jersey: Princeton University Press) p. 154
[14] Ovez F T D and Ozdemir E 2014 The investigation of prospective mathematics teachers' proof writing skills and proof self-efficacy *Procedia - Social and Behavioural Sciences* **116** 4075
[15] Ma'rufi, I Ketut B and Dwi J 2016 Pedagogical content knowledge: knowledge of pedagogy novice teachers in mathematics learning on limit algebraic function *The 1st International Conference on Mathematics, Science, and Computer Science (ICMSC) 2016* (AIP Publishing) p. 1

IOP Conf. Series: Journal of Physics: Conf. Series **947** (2017) 012071 doi:10.1088/1742-6596/947/1/012071

Drilling Students' Communication Skill through Science, Environment, Technology, and Society (SETS)-Based Learning

B L Al-Farisi[1], Tjandrakirana[2] and R Agustini[3]
[1,2,3]Science education, Post-graduate of Universitas Negeri Surabaya

Email:[1] rudianaagustini@unesa.ac.id

Abstract. Student's communication skill paid less attention in learning activity at school, even though communication skill is needed by students in the 21st century based on the demands of new curriculum in Indonesia (K13). This study focuses on drilling students' communication skill through science, environment, technology, and society (SETS)-based learning. The research is a pre-experimental design with a one-shot case study model involving 10 students of ninth-grader of SMPN 2 Manyar, Gresik. The research data were collected through observation method using communication observation sheet. The data were analyzed using the descriptive qualitative method. The result showed that students' communication skill reached the completeness of skills decided both individually and classically in the curriculum. The fundamental result of this research that SETS-based learning can be used to drill students' communication skill in K13 context.

1. Introduction

Communication is defined as a process of conveying information containing the meaning from one party to others in order to have a mutual understanding [1]. Communication is a process of delivering a statement from a person to another as a consequence of social relationship [2]. Communication skills is the skill in conveying ideas or opinions, and thoughts to others in clear, logical, purposeful and systematic way so that the meaning can be understood by others [3]. Students require communication skills in term of knowledge dissemination. Learning science is basically aimed to provide learners with abilities to become scientists/researchers. In academic tradition, a researcher as a manufacturer of information will generate research results, ideas, and theories that are then communicated to the public through various forms of media [4].

The fact that happened so far is learners hesitated when asked to put forward their opinion or to do a presentation. Whereas, ideas the students gained from learning activities need to be communicated to other people as implementation of knowledge dissemination. Based on the pre-research data and the results of interview with a teacher of science education (IPA) subject in one of the secondary schools (junior high school) in Gresik, it was clear that the students' communication skills need to be paid more attention. In accordance with the written in *Permendikbud RI* No. 104 Year 2014 that the students' individual completeness on the skill aspect (in this case the communication skill) is set based on the optimum performance namely 2.67 (B-). The students are considered classically complete if the percentage of classical completeness is >75%.

The learning process requires an approach that can help students to understand the concept or matter can be communicate them with others effectively. Science, Environment, Technology and

Society (SETS) constitutes one of the recommended approaches in the teaching-learning process at the secondary level [5]. Science-Technology-Society (STS) is one approach for curriculum designed to make concepts and simple processes existing on the subject of social science more precise and relevant to the students' lives [6]. According to Poedjiadi, the SETS-based learning basically gives an understanding on the relevance of science, technology and society, and is a vehicle to train learners' assessment sensitivity towards the impact of environment as a result of the development of science and technology [7]. The SETS-based learning has three fundamental principles, namely: (1) close relation between science, environment and society; (2) learning process based on constructivism; (3) covering five domains consisting of the domain of knowledge, attitude, process, creativity and application [8].

A research conducted by Dass suggested that the application of the SETS approach shows the improvement of learners' achievement in terms of mastering the concepts and processes; ability to apply concepts and the process of science in a new situation, most notably in real life [9]. Other studies on the SETS approach also showed positive results; the increase of learners' attention towards their lessons. Learners are able to connect concepts they learned with experience they gained in real life through the SETS learning [10]. Another study by Rusilowati and team demonstrated that the SETS-based learning was effective in developing students' understanding towards natural science (IPA) [11]. Another research result also described that learning process through science-technology-society (STS) had more significant impact on making students understand the science concept than textbook oriented learning [12].

The purpose of this research is to promote students' communication skills through SETS-based learning. Ten students of ninth-grader in junior high school were involved in this research. The result showed that 80% of the student achieved completeness that decided in communication skills. This was because SETS-based learning conditioned students in many activities to interact and communicate one another as well as help train their communication.

2. Method

This research is a pre-experimental design with one-shot case study model in which there was a group treated and then observed [13]. Subjects of the research were 10 students of Class IX of Even Semester Year 2016/2017 located in junior high school (SMPN) 2 Manyar, Gresik, East Java. The treatment given was a learning activity through SETS-based learning conducted during 3 meetings. A learning materials based on SETS which was aimed to train students' communication skills was applied. The learning materials was previously validated by two experts and was claimed valid.

The research instrument used a communication observation sheet that was developed by researchers. Moreover, aspects of observed communication skills consist of four components: skill in asking questions, skill in answering them, skill in coming up with ideas or opinions, and skill in oral presentation.

Analysis of students' communication skill was obtained from observation conducted by two observers. Data obtained were analyzed for understanding the completeness of students' communication skills both individually and classically. Individual completeness was calculated by dividing students' maximum scores and then timing them with 4. Student scores obtained from observation were converted to *Permendikbud RI* No. 104 Year 2014 [14] as provided on Table 1.

Table 1. Score Range of Competence Skills

Score	Predicate
3,85 – 4,00	A
3,51 – 3,84	A-
3,18 – 3,50	B+
2,85 – 3,17	B
2,51 – 2,84	**B-**

Score	Predicate
2,18 – 2,50	C+
1,85 – 2,17	C
1,51 – 1,84	C-
1,18 – 1,50	D+
1,00 – 1,17	D

Individual completeness for competence of communication skills was decided based on optimum performance namely 2,67 (B-). However, classical completeness was calculated by dividing all complete students over students in total and then timing them with 100%. The students are considered classically complete if the percentage of classical completeness is >75%.

3. Result and discussion

Analysis of students' communication skills was obtained from data of observation results conducted by two observers observing students by using the observation instrument sheet of communication skills. The analysis results of the observation score on communication skills can be seen in Table 2.

Table 2. Analysis Result of Students' Communication Skills

Student	Students' Communication Skill Aspect				$\bar{x}$	P	C
	K1	K2	K3	K4			
SW1	3.0	2.7	3.3	3.0	3.0	B	T
SW2	3.0	1.7	3.0	2.3	2.5	C+	TT
SW3	1.7	3.0	3.3	3.3	2.8	B-	T
SW4	2.3	3.0	2.7	3.3	2.8	B-	T
SW5	2.0	3.7	3.0	2.7	2.8	B-	T
SW6	2.7	3.0	2.7	3.0	2.8	B-	T
SW7	3.3	2.7	3.0	3.3	3.1	B	T
SW8	2.7	2.7	2.3	2.7	2.6	B-	T
SW9	3.0	2.3	3.0	3.3	2.9	B	T
SW10	2.7	2.3	2.3	2.3	2.4	C+	TT

Explanation
K1: Asking a question — $\bar{x}$: average
K2: Answering a question — P : predicate
K3: Giving opinion — C : completeness
K4: Oral presentation — T : completed; TT : incomplete

Based on the observation results presented in Table 2, it is known that 8 of 10 students are declared to have reached the completeness. The results automatically show that 80% of students have classically completed. A learning process should ideally not only customize the students to receive the information, to remember, and to memorize. Moreover, students are expected to convey what they have received to others, including give and respond ideas. Communication skills which are trained in this study focuses on the students' communication skills both verbally and non-vocal vocal, covering several aspects, namely: ask a question, answer a question, put forward ideas, and oral presentation. Analysis of students' communication skill was also served in graph as shown on **Figure 1**. Students' achievement in gaining the completeness of both individual and classical communication skill was because the SETS-based learning is able to create a situation which enables students to actively interact with one another and with teachers. This will help students train their communication skill.

IOP Conf. Series: Journal of Physics: Conf. Series **947** (2017) 012071 doi:10.1088/1742-6596/947/1/012071

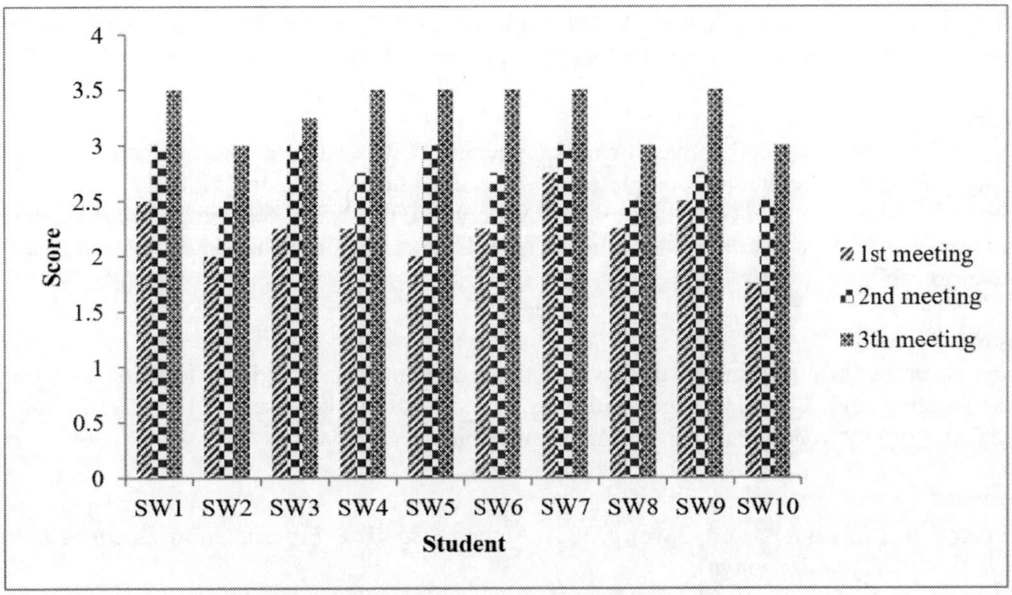

Figure 1. Analysis Result of Students' Communication Skills

The learning activity is adjusted with stages of the SETS-based learning style consisting of some stages: (1) invitation, (2) concept development, (3) concept application, (4) concept establishment, (5) assessment [7]. The first stage known as introduction constitutes invitation which is designed to make students focused on learning. In this research a teacher starts learning activity with appreciating and motivating students. The introduction process was done well according to needs, materials, methods, environment, and teachers' skills as well, learning process would be optimal [15]. The second stage of SETS model is the formation or development of concept. The formation process of the concept in this research was conducted through discussion, in which students were divided into some small groups based on teachers' decision. Each group was asked to discuss the topics provided in LKS. The students had opportunity to engage in dialogue and to share opinion on a concept that encouraged them to ask and answer a question or give opinion within the group during discussion. Discussion held by small groups also helped students develop the concept well. This is suitable for socio-constructivism learning method theory arguing that if a teacher can provide a supporting social environment, students can develop their knowledge and understanding well [16].

The third stage is the application of the concept in the form of analysis of the issue or problem solving. At this stage, the teacher displayed an issue closely related to daily life. Students were then asked to discuss the issue with their respective groups. The problems presented in learning can foster students' spirit to find solutions to a problem [17; 11]. Presentation activities took place after class discussion of the stage 2 and stage 3 completed, each group was asked to convey the discussion results through oral presentations in front of the classroom and then continue with the interview session. The main achievement of learning can be seen from the meaning and the understanding. That can be achieved if students are able to explain ideas derived from facts, ask a relevant question, and apply it in daily life. [18].

During the process of the formation of concepts and the analysis of issues/problems (stage 2 and stage 3) teachers need to straighten out when there is a misconception over the activities in progress. The fourth stage in learning SETS is known as the stage of the establishment of the concept. If there is no misconception during the process of concept formation, as well as after the end of the analysis of issues and problem solving, teachers still need to develop the concept as syntax of SETS learning (4th stage) through an emphasis on the key concepts that students need to know. The fifth stage of the SETS learning model is assessment. At this stage, checking students' understanding through questions

IOP Conf. Series: Journal of Physics: Conf. Series **947** (2017) 012071 doi:10.1088/1742-6596/947/1/012071

about biotech material was conducted. Assessment is aimed to find out toward extent the students understand the material through understanding check either literally or orally.

4. Conclusion

Based on the research described above, it can be concluded that student's communication skill trained by looking into the achieving of completeness of each student both individually and classically. Therefore, science environment technology and society (SETS)-based learning can be considered a learning model which is suitable for the demand of curriculum in order to train students' communication skill.

5. Acknowledgements

Thanks to all parties contributing in this research, to the expert for validated learning materials and giving constuctive advice. I am also grateful to principal of junior high school (SMPN) 2 Manyar for giving mc opportunity to do the research in his institution.

6. Reference

[1] Endang S, Mulyani S, and Suyetty 2011 *Mengaplikasikan Keterampilan Dasar Komunikasi* (Jakarta: Erlangga)
[2] Effendi O I 2004 *Dinamika Komunikasi* (Bandung: Remaja Rosdakarya)
[3] Cangara H 2006 *Pengantar Ilmu Komunikasi* (Jakarta: Raja Grafindo Persada)
[4] Kurniawan A T 2011 *J. Komunikasi Massa.* **4** 12-17
[5] Binadja A, Wardani S, and Nugroho S 2008 *J. Inovasi Pendidikan Kimia.* **1** 256- 262
[6] Mansour N 2009 *Bulletin of science, technology and society.* **4** 287-297
[7] Poedjiadi A 2010 *Sains Teknologi Masyarakat* (Bandung: Remaja Rosdakarya)
[8] Rusmansyah and Irhasyuarna Y 2003 *J. Pendidikan dan Kebudayaan* **9** 95-109
[9] Dass P M 2005 *J. Issues in Teacher Education* 95-109
[10] Yoruk N Morgil I and Secken N 2010 *J. Natural Science* **12** 1417-1424
[11] Rusilowati A, Binadja A, Supriyadi and Widyatmoko A 2014 *International Conference on Mathematics, Science, and Education* (ICMSE 2014)
[12] Akcay B and Akcay H 2015 *J. Education in Mathematics Science and Technology* **3** 37-45
[13] Sugiyono 2012 *Metode Penelitian Kuantitatif Kualitatif dan R&D* (Bandung: Alfabeta)
[14] Permendikbud 2014 *Tentang Penilaian Hasil Belajar oleh Pendidik pada Pendidikan Dasar dan Pendidikan Menengah* (Jakarta: Depdiknas)
[15] Tim Pengembangan MKDP Universitas Pendidikan Indonesia 2011 *Kurikulum dan Pembelajaran* (Bandung: Rajawali Press)
[16] Santrock W J 2011 *Educational Psychology* (New York: McGraw-Hill Companies Inc.)
[17] Kardi S 2013 *Sains Teknologi Masyarakat* (Surabaya: Universitas Negeri Surabaya Press)
[18] Mctighe J and Seif E 2013 *J. Pennsykvania Educational Leadership* **24** 122-129

IOP Conf. Series: Journal of Physics: Conf. Series **947** (2017) 012072 doi:10.1088/1742-6596/947/1/012072

Students' metacognitive activities in solving the combinatorics problem: the experience of students with holist-serialist cognitive style

B N Trisna[1, 2], I K Budayasa[1] and T Y E Siswono[1]

[1] Surabaya State University, Indonesia
[2] STKIP PGRI Banjarmasin, Indonesia
*corresponding author: bennytrisna@mhs.unesa.ac.id

Abstract. Metacognition is related to improving student learning outcomes. This study describes students' metacognitive activities in solving the combinatorics problem. Two undergraduate students of mathematics education from STKIP PGRI Banjarmasin were selected as the participants of the study, one person has a holist cognitive style and the other a serialist. Data were collected by task-based interviews where the task contains a combinatorial problem. The interviews were conducted twice using equivalent problem at two different times. The study found that the participants showed metacognitive awareness (A), metacognitive evaluation (E), and metacognitive regulation (R) that operated as pathways from one function to another. Both, holist and serialist, have metacognitive activities in different pathway. The path of metacognitive activities of the holist is AERCAE-AAEER-ACRECCECC-AREERCE with the AERAE-AER-ARE-ARERE pattern, while the path of metacognitive activities of the serialist is AERCA-AAER-ACRERCERC-AREEEE with the AERA-AER-ARERER-ARE pattern. As an implication of these findings, teachers/lecturers need to pay attention to metacognitive awareness when they begin a stage in mathematical problem solving. Teachers/lecturers need to emphasize to students that in mathematical problem solving, processes and results are equally important.

1. Introduction

Metacognition is used to refer to the awareness individuals have of their own thinking; their evaluation of that thinking; and their regulation of that thinking [1]. Metacognitive awareness (A) relates to individuals' awareness of where they are in the process of solving a problem. It also includes their knowledge of what needs to be done, what has been done, and what might be done in problem solving situations. Metacognitive evaluation (E) refers to judgements about one's thinking processes, capacities and limitations in problem solving situations. For example, individuals could be making a judgement regarding the effectiveness of their thinking or of their strategy choice. Such an evaluative function assumes some awareness of the individuals' thinking processes and anticipates the possible regulation of those processes. Metacognitive regulation (R) draws upon individuals' knowledge about self and strategies, including how and why they use particular strategies, and uses executive skills (such as planning and setting goals) to optimise the use of their own cognitive resources.

Metacognition plays a role in mathematical problem solving. Metacognition is related to improving student learning outcomes [2][3][4]. Silver argued that failure or success in solving mathematical problems is influenced by the use of metacognition [5]. Similarly, Cardelle-Elawar reported that most students who had difficulties in mathematics less use cognitive or metacognitive strategies [6].

IOP Conf. Series: Journal of Physics: Conf. Series **947** (2017) 012072 doi:10.1088/1742-6596/947/1/012072

An experimental study reported that there were differences in mathematics achievement based on students' cognitive style. The study also reported differences in the use of cognitive and metacognitive strategies based on cognitive styles [7]. Another study showed that good critical thinkers engaged in more metacognitive strategies, especially high-level planning and high-level evaluating strategies [8]. Among various dimensions of cognitive styles, Pask's Holist-Serialist is influential to student learning and problem solving [9]. Therefore, this study aims to describe students' metacognitive activities in solving a combinatorics problem based on Holist-Serialist cognitive style. According to Biryukov, combinatorial problems have non-algorithmic characters. Solving problems of this domain is useful for developing mathematical reasoning, critical thinking abilities, and thus it leads to activating their metacognitive skills [10].

2. Method

This is an explorative research with qualitative approach. The qualitative approach was chosen because the study was natural, no treatment was given to the subjects. In addition, the data are descriptive in the form of a series of words and pictures taken from interviews, field notes, photos, videos, and documents.

Two undergraduate students of mathematics education from STKIP PGRI Banjarmasin were selected as the participants of the study, one person has a holist cognitive style and the other a serialist. Selection of the participants using a questionnaire adapted from the Study Preference Questionnaire (SPQ) developed by Ford as revised by Clarke [11].

Data were collected by task-based interviews where the task contains a combinatorial problem. The problem solving process is divided into 4 stages from Polya, that is understanding the problem, devising a plan, carrying out the plan, and looking back [12]. At each stage, subjects are interviewed to reveal their metacognitive activities during this stage.

The interviews were conducted twice using equivalent problem at two different times. Data from interviews supported by observations, answer sheets, and recordings (audio and video) were analyzed qualitatively, including data review, categorization, reduction, display and interpretation.

The participants were given a task containing a combinatorial problem, number 1 for the first interview and number 2 for the second.

1. How many different 4-digit numbers can be formed from the digits 1, 2, 3, 4, 5, 6, 7, 8, 9 if digits 8 and 9 always appear in every number and repetitions are not allowed?
2. How many different 5-digit numbers can be formed from the digits 1, 2, 3, 4, 5, 6, 7, 8, 9 if digits 2 and 5 always appear in every number and repetitions are not allowed?

The results showed that the participants' responses in both interviews were consistent. Therefore, the following will only be exposed to the results of the first interview.

3. Result and discussion

3.1. Metacognitive activities of the holist (H)

3.1.1. Understanding the problem. H was assigned a task of combinatorial problem. To understand the problem, H said that first he will analyze the problem by reading it. This means H showed metacognitive awareness (A), i.e. he knows what to do to understand the problem. Furthermore, H said that he would read the problem with quick and silent. When asked why using that strategy, H reasoned that he would concentrate on finding important information in the matter. He also believes that by quick reading he would be faster and easier to understand the problem. That way is also his habit. This information indicates that he judged the effectiveness of the strategy he will used in reading to understand the problem (E) before he decided to use the strategy (R).

After he tried to find the information (C = cognition) by reading the given problem, H explained his understanding of the given problem. He showed in which part of the problem he obtained the information (by pointing). He explained the meaning of 'repetitions are not allowed', "If there are 4 boxes, the digits in the boxes should not be the same." His responses indicate that he already has knowledge of what is known and asked, even though it is not mentioned explicit. Thus, it can be said that he knows about the problem (A). Furthermore, he judged the correctness of his understanding of the problem (E) by expressing his belief that his understanding is correct. This is reinforced by an explanation of his understanding as mentioned earlier.

It appears that the pathway of metacognitive activities of subject H is AERCAE. If cognitive activity C is ignored, then the pattern of metacognitive activities of H in understanding the problem is AERAE. This pattern not only complements what Wilson and Clarke have founds, AER, AERE and ARE in overall problem-solving activities [1], but also explains that metacognitive activities that includes awareness, evaluation, and regulation has arisen in the process of understanding the problem.

3.1.2. Devising a plan. After he understood the problem, H stated that he was thinking about how to do it, imagining work processes, and the end result. This means that he knows what to do (A) before solving the given problem. He claimed that he knows some of ways in which he can solve the problem (A), but the way he described as "making boxes" was the first way he thinked and he was not sure to work using the other way. This shows that H judged the effectiveness of the way he would use to solve the problem (E). He also expressed his confidence to apply the way. Again, this indicates that he has judged, not only on the effectiveness of the way to be used, but also himself capacity (E). Furthermore, based on the judgedments, H sets out to use the "make boxes" way as he would use to solve the given problem (R).

The path of metacognitive activities of H in devising a plan is AAEER with an AER pattern (ignoring the same sequential repetition of activities). This pattern is a standard pattern proposed by Wilson and Clarke [1]. He said that metacognitive evaluation is often preceded by metacognitive awareness and anticipating metacognitive regulation. However, these results show that these three metacognitive activities all appear in this stage, which in the study by Wilson and Clarke are still not clearly visible.

3.1.3. Carrying out the plan. After he setted out his plan to solve the problem, H said that he would start working according to the plan. It showed that H knows what to do (A). Next, H started to solve the problem (C), and he worked on his plan (R).

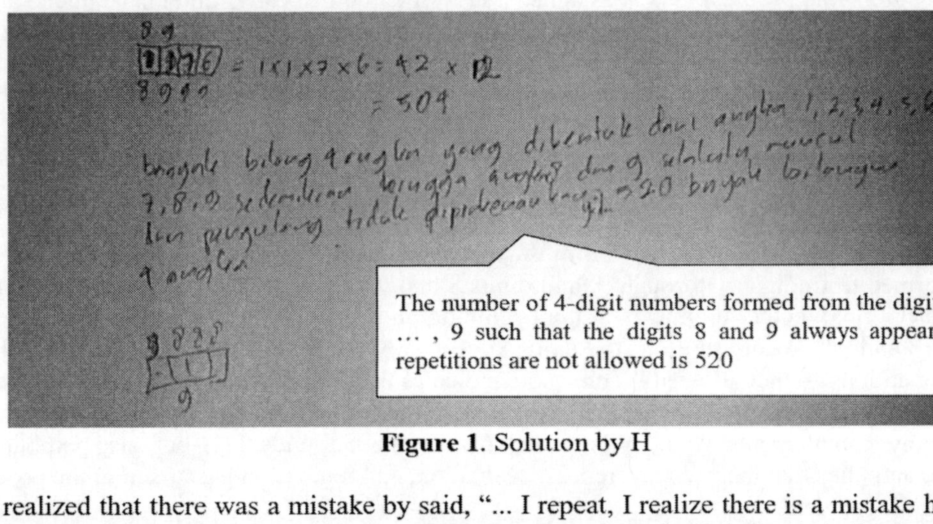

Figure 1. Solution by H

H realized that there was a mistake by said, "... I repeat, I realize there is a mistake here, it should be 8, 9 written above. Here 1, 1, since 8, 9 each represent a single digit." This indicates that he judged the correctness of his thought process (E). The H's statement indicates that the boxes he created should be written with 'the number of digits that can be placed in a particular position', not even with

IOP Conf. Series: Journal of Physics: Conf. Series **947** (2017) 012072 doi:10.1088/1742-6596/947/1/012072

'digits to be placed.' Therefore, he corrected it (C) by writing the digit 8 above the first box and 9 above the second box, and then replacing the contents of the first and second boxes with 1 (Figure 1). H continued his work (C), but he again realized there was a mistake, "... in here there are three possibilities, in here there are three possibilities. So there are... 3, 6, 9. It should be here multiplied by 9 (counting while mumbling). Oh ... it should be here 12." This indicates an evaluation by H on his mistaken thought process (E) followed by correcting the error (C) by replacing 42 x 9 to 42 x 12 (Figure 1). Then H continued the work until it is done (C). In solved the problem, he has used the principle of multiplication.

It appears that the pathway of metacognitive activities of H is ACRECCECC with ARE pattern. This pattern is in line with Wilson and Clarke's finding [1]. This finding also show that metacognitive awareness, metacognitive evaluation, and metacognitive regulation, all three can appear at this stage in combinatorial problem solving.

3.1.4. Looking back. After he solved the problem, H said, "Now I want to check." It showed that he knows what to do (A). Then he planned the check (R). Next, he checked his work (E). This can be seen in his response, "I pay attention to my work from the beginning, how I do it, the process is like that, right. The calculation (he checked the calculation), is correct." This quote also shows that H judged the correctness of his thought process in solving the problem (E) indicated by the words "right" and "correct", supported by his statement, "I think this answer is correct."

To make sure, then he thought of different ways (R) to find answers. After he found the answer in a different way (C), then he compared the result with the previous answer. This shows that he judged the correctness of his thought process in solving the problem (E) by comparing the results of the two solutions he made, even though the latter solution was not written. Based on the checks, he believed that the solution is correct.

It appears that the pathway of metacognitive activities of H is AREERCE with ARERE pattern. This pattern not only complements what Wilson and Clarke have founds, AER, AERE and ARE in overall problem-solving activities [1], but also explains that metacognitive activities that includes awareness, evaluation, and regulation has arisen in the process of looking back.

3.2. *Metacognitive activities of serialist* (S)

3.2.1. Understanding the problem. S was assigned a task of combinatorial problem. To understand the problem, he said that first he would read the problem. This means that he showed metacognitive awareness (A), i.e. he knows what to do to understand the problem. S said that he would read silently from beginning to end. When asked why he should read like that, he reasoned that it was his habit and made it easy for him to understand the problem. This information indicates that S judged the effectiveness of the way he will used in reading to understand the problem (E) before he decided to use the way (R).

After he tried to find the information (C) by reading the given problem, he explained his understanding of the problem. He said, "From this problem, I know ... so there are 4-digits numbers that can be formed from digits 1 through 9, and digits 8 and 9 should always appear in the formation of these numbers. So, whether it appears at the beginning or at the end, they should appear, there are always digits 8 and 9." According to S, the digits appear only once, because on the question there is the phrase "repetitions are not allowed." This quote indicates that S understands the problem (A).

It appears that the pathway of metacognitive activities of S is AERCA with AERA pattern. This pattern not only complements what Wilson and Clarke have founds [1], but also explains that metacognitive activities that includes awareness, evaluation, and regulation has arisen in the process of understanding the problem.

3.2.2. Devising a plan. After he understood the problem, S stated that he tried to find the right formula to do the problem, thought to solve the problem, and devised steps to work. This means that S knows what to do (A) before solving the problem. Furthermore, he recalled materials related to the problem, similar problems, also previous and successful work strategies (A). Before decided to use "make boxes" way, he believed that his strategy can be used to solve the problem. That is, S judged the effectiveness of the strategy he will used (E), and then decided to use it (R), the same way he used to solve similar problems.

The path of metacognitive activities of S in devising a plan is AAER with an AER pattern. This pattern is a standard pattern proposed by Wilson and Clarke [1]. He said that metacognitive evaluation is often preceded by metacognitive awareness and anticipating metacognitive regulation.

3.2.3. Carrying out the plan. After he setted out the plan, S said he would carry out his plan. It shows that he knows what to do (A). Next, he started to solve the problem (C), and he worked on the plan (R).

S changed his strategy from 'writing all possibilities' to 'writing only some parts.' Furthermore, he utilized the patterns he found to calculate the many possible 4-digit numbers if the numbers 8 and 9 in different positions (Figure 2). He said that the reason changed the strategy, "because I cannot write it down anyway, because after I know 'oh it is 6 times', so I immediately write, oh in here 6, definitely" and "I shortened by writting down the codes." This indicates that S judged the effectiveness of the strategy used to solve the problem (E), and then for that reason S changed its strategy (R).

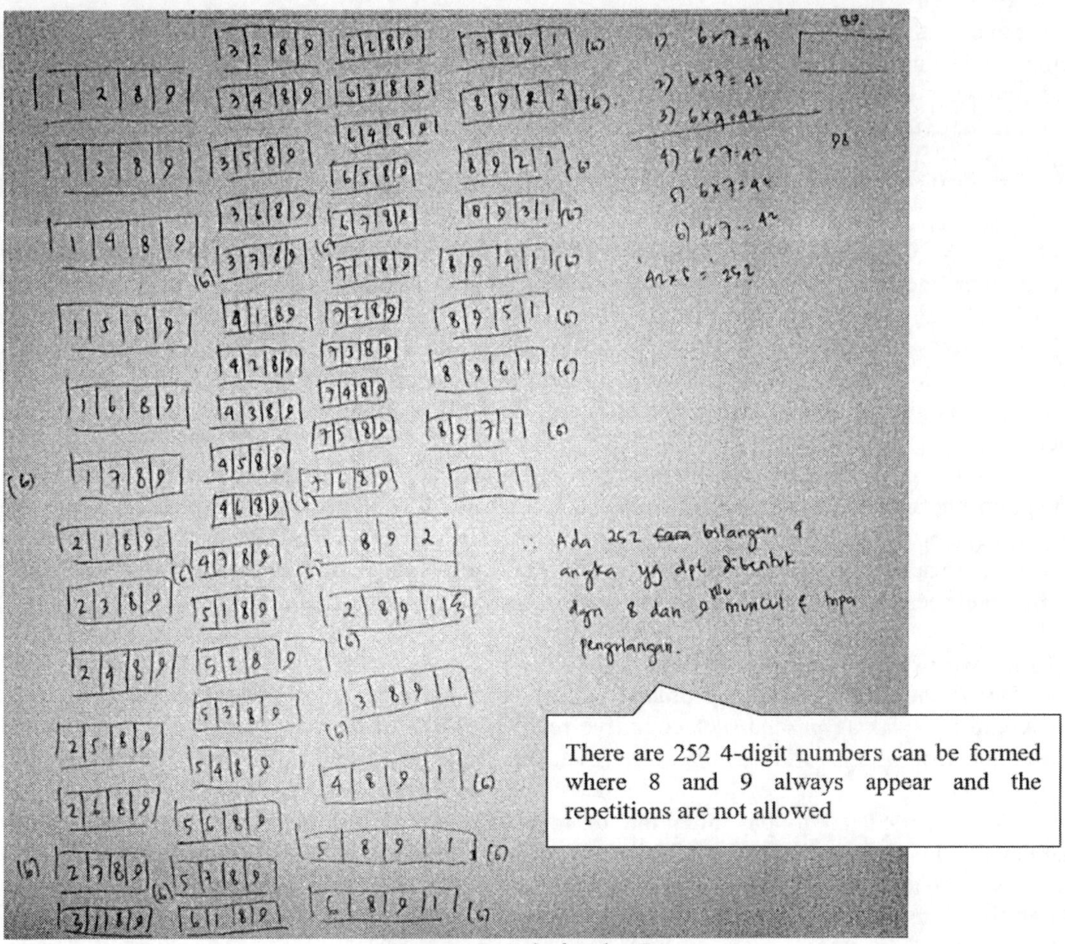

Figure 2. Solution by S

IOP Conf. Series: Journal of Physics: Conf. Series **947** (2017) 012072 doi:10.1088/1742-6596/947/1/012072

Furthermore, S worked with a new strategy (C) until he finds a pattern similarity if positions 8 and 9 are exchanged. He said, "... Then when the position is reversed, it means there are also the same digit pairs." He realized that his work will be shorter (Figure 2). The response of the S indicates that he, once again, judged the effectiveness of his strategy and realized that there was a new strategy that makes his work shorter (E), so he changed his strategy again (R). Then he worked on a new strategy (C) until finally he found the answer, 252.

It appears that the pathway of metacognitive activities of H is ACRERCERC with ARERER pattern. This pattern complements Wilson and Clarke's findings [1]. This finding also show that metacognitive awareness, metacognitive evaluation, and metacognitive regulation, all three can appear at this stage in combinatorial problem solving.

3.2.4. Looking back. After he solved the problem, S claimed to need to check. It shows that he knows what to do (A). Then, he planned the check (R), "... I looked back at the list I made." Next, he checked his work (E).

S believed that the answer is correct, because he has made a list of possibilities in sequence, "I do not think there is any more because at the time of doing this I do it in sequence." S also said, "After I finished checking and I am sure with the answer then I make the conclusion." It can be interpreted that S judged the truth of his thought process in solving the problem (E) and he believed the correctness of the answer (E).

It appears that the pathway of metacognitive activities of H is AREEEE with ARE pattern. This pattern is in line with Wilson and Clarke's finding [1] and also explains that metacognitive activities that includes awareness, evaluation, and regulation has arisen in the process of looking back.

Both participants, holist and serialist, showed metacognitive activities in different pathway. The path of metacognitive activities of the holist is AERCAE-AAEER-ACRECCECC-AREERCE with the AERAE-AER-ARE-ARERE pattern, while the path for the serialist is AERCA-AAER-ACRERCERC-AREEEE with the AERA-AER-ARERER-ARE pattern. Wilson & Clarke [1] and Barbacena & Norina [13] did not elaborate the problem-solving process in the stages, so the results clarify their findings by detailing the paths and patterns at each stage of problem solving.

In each stage, metacognitive activities begins with a metacognitive awareness. This indicates that metacognitive awareness is a prerequisite for other metacognitive activities, i.e. regulation and evaluation. As an implication, teachers/lecturers need to pay attention to metacognitive awareness when they begin to understanding the problem, devising a plan, carrying out the plan, and looking back.

Both show more metacognitive evaluation activities than metacognitive awareness and regulation. This is in line with the findings of Barbacena and Norina [13]. This can be explained as the tendency of participants to result rather than process, whereas in solving a mathematical problem, results and processes must receive balanced attention. Further, both subjects use different strategies to solve the problem, especially at the stages carrying out the plan and looking back.

4. Conclusions and recommendations

In solving the combinatoric problem, holist and serialist showed metacognitive awareness, metacognitive evaluation, and metacognitive regulation that operated as pathways from one function to another. Both, holist and serialist, have metacognitive activities in different pathway. However, in every stage it always begins with metacognitive awareness activities. As an implication, teachers/lecturers need to pay attention to metacognitive awareness when they begin a stage in mathematical problem solving. They show more metacognitive evaluation activities than metacognitive awareness and regulation. This can be explained as the tendency of participants to result rather than process, whereas in solving a mathematical problem, result and process must receive balanced attention. Therefore, teachers/lecturers need to emphasize to students that in mathematical problem solving, processes and results are equally important.

Mathematical problem solving involves the complex interaction between cognitive and metacognitive activities. While this study is more focused on metacognitive activities. Therefore, in the future, studies can be developed to know the interaction, for example by observing the interaction between thinking skills (cognition) and metacognitive function, so more can be utilized to improve student learning outcomes.

5. References

[1] Wilson J and Clarke D 2004 Towards the Modelling of Mathematical Metacognition *Math. Educ. Res. J.* **16** 25–48
[2] Biggs J B and Telfer R 1987 *The Process of Learning* (Prentice-Hall of Australia)
[3] Birenbaum M 1996 Assessment 2000: Towards a Pluralistic Approach to Assessment *Alternatives in Assessment of Achievements, Learning Processes and Prior Knowledge* ed M Birenbaum and F J R C Dochy (Dordrecht: Springer Netherlands) pp 3–29
[4] Wilson J and Jan L W 1993 *Thinking for Themselves: Developing Strategies for Reflective Learning.* (ERIC)
[5] Silver E 1985 The Teaching and Assessing of Mathematical Problem Solving *Rest. NCTM*
[6] Cardelle-Elawar M 1992 Effects of teaching metacognitive skills to students with low mathematics ability *Teach. Teach. Educ.* **8** 109–21
[7] Abdollahpour M A, Kadivar P and Abdollahi M H 2006 Relationship between cognitive style, cognitive and meta-cognitive strategies with academic achievement *Psychol. Res.* **8** 30–44
[8] Ku K Y L and Ho I T 2010 Metacognitive strategies that enhance critical thinking *Metacognition Learn.* **5** 251–67
[9] Bajraktarevic N, Hall W and Fullick P 2003 Incorporating learning styles in hypermedia environment: Empirical evaluation *Proceedings of the workshop on adaptive hypermedia and adaptive web-based systems* pp 41–52
[10] Biryukov P 2004 Metacognitive Aspects of Solving Combinatorics Problems *Int. J. Math. Teach. Learn.* **5**
[11] Clarke J A 1993 Cognitive style and Computer-Assisted Learning: problems and a possible solution *Res. Learn. Technol.* **1** 47–59
[12] Polya G 1973 *How to Solve It* (Princeton, New Jersey: Princeton University Press)
[13] Barbacena L B and Norina R.Sy 2015 Metacognitive Model in Mathematical Problem Solving *BU Fac. e-Journal* **12** 16–22

IOP Conf. Series: Journal of Physics: Conf. Series **947** (2017) 012073 doi:10.1088/1742-6596/947/1/012073

Critical Thinking Skills of an Eighth Grade Male Student with High Mathematical Ability in Solving Problem

Ismail[1]

[1]*Department of Mathematics, Universitas Negeri Surabaya*

[1]Email: ismail@unesa.ac.id

Abstract. This study aims to describe student's critical thinking skill of grade VIII in solving mathematical problem. A qualitative research was conducted to a male student with high mathematical ability. Student's critical thinking skill was obtained from a depth task-based interview. The result show that male student's critical thinking skill of the student as follows. In understanding the problem, the student did categorization, significance decoding, and meaning clarification. In devising a plan he examined his ideas, detected his argument, analyzed his argument and evaluated his argument. During the implementation phase, the skill that appeared were analyzing of the argument and inference skill such as drawing conclusion, deliver alternative thinking, and problem solving skills. At last, in rechecking all the measures, they did self-correcting and self-examination.

1. Introduction

Mathematics needs to be given to all of students ranging from elementary school to equip students with the ability to think logically, analytical, systematic, critical, and creative as well as the ability to cooperate. The competencies are required, so that learners can have the ability to acquire, manage, and use information to survive in ever-changing, uncertain, and competitive state. Critical thinking skills involve the integration of several capabilities, they are: the ability of observation, analysis, reasoning, judgment, and decision making [1].

Critical thinking is a skill that is very important in the 21st century besides the other skills such as creative thinking, communication skills and collaboration skills [2]. Those skills are recognized as an integral part of students who are prepared to stock the more complex life and in the environment work on the 21st century. Critical thinking is a skill that is needed by everyone. Why is critical thinking important that need to be learned? There are several reasons, among others: Critical thinking is a skill that is required in any job, when studying any field of science, to solve any problems, and is a valuable asset to one's career. Critical thinking is very important in the 21st century, because this is an era of information and technology [3]. One must respond to changes quickly and effectively, thus require a flexible intellectual skills, the ability to analyze information, and integrate multiple sources of knowledge to solve problems. Critical thinking is analytic and reflective [4].

Based on the limited test results showed that the students' ability to provide the correct arguments related to the question of critical thinking in schools is still low. This is indicated by the many students who can provide the argument correctly to the question, but the question asked students to give arguments.

IOP Conf. Series: Journal of Physics: Conf. Series **947** (2017) 012073 doi:10.1088/1742-6596/947/1/012073

Critical thinking skills can be developed through the study of mathematics. The problem, as long as teachers in the class did not develop critical thinking skills optimally through learning. Based on observations limited to the teacher in the classroom and review of lesson plans Ismail shows that teachers are more oriented to the contents of the subject matter [5]. Teaching the teachers are still lacking in the development of students' critical thinking skills.

Critical thinking is thinking that involve testing, connecting, and evaluate all aspects of the problem. Creative thinking is the thinking that is the authenticity and reflective and produce something as complex and "new." Facione identified 6 main critical thinking skills, the interpretation, analysis, evaluation, inference, explanation and self-regulation [6].

To be able to develop critical thinking skills of students, teachers need to find the right strategy in learning accomplishments [4]. Questions that can be asked of teachers are critical to the students. There are various examples of critical questions for example, Is there another way? What if ...? What is wrong? Problems that develop critical thinking skills students have different characteristics compared with ordinary matter. A matter of critical thinking is a matter that involves the analysis, synthesis, and evaluation of a concept.

Based on this background the authors formulate the following research questions: How critical thinking skills of male students of eight grade of junior high school ability in solving mathematical problems? Based on the research questions, the general purpose of this study is to describe the critical thinking skills of male students of eight grade junior high math ability in solving mathematical problems. Problems that develop critical thinking skills students have different characteristics compared with ordinary matter. A matter of critical thinking is a matter that involves the analysis, synthesis, and evaluation of a concept.

Based on this background the authors formulate the following research questions: How is the critical thinking skills of male students of eight grade of junior high school ability in solving mathematical problems?

Based on the research questions, the general purpose of this study is to describe the critical thinking skills of male students of eight grade junior high math ability in solving mathematical problems.

2. Review of literature

2.1 Thinking

Thinking is an internal mental processes that build and operate the mental representation of information. This is in line with the statement Solso, thinking is defined as the process of generating a new mental representation, through the transformation of information that involve complex interactions between mental attributes such as ratings, abstraction, reasoning, imagination, and problem resolution. Based on some of the above opinion, thinking is required in decision making and problem solving[7]. Thus, in this study is meant to think is all mental activity can be observed from the behavior that appears in the form of statements and results of problem resolution.

2.2 Critical Thinking

Critical thinking is no cynical skepticism [6]. Critical thinking is an open-minded without vacillating. Critical thinking is analytical thinking. Critical thinking is judging without judgment and strong without dogmatic. Critical thinking skills is the main interpretation (interpretation), analysis (analysis), evaluation (evaluation), inference (inference), explanation (explanation), and self-regulation (self-regulation). Interpretation means to understand and express the meaning of a variety of experiences, situations, data, events, assessments, conventions, beliefs, rules, procedures, or criteria. Analysis of means to identify the purpose and the true relationship between statements, questions, concepts, description, or other forms of representation of the other, is intended to express the conviction, judgment, experience, reason, information, or opinions. Including checking ideas, detecting arguments, and analyzing arguments as a sub-analysis skills. Evaluation means assessing the credibility of statements or other representations, intended to reveal the perception, experience, situation, judgment, belief, or opinion of someone, to assess the strength of logic, purpose and the true relationship between the statements, descriptions, questions or other forms of representation. Inference means to identify and secure elements needed to withdrawal reasonable conclusion; to form conjectures and hypotheses; to consider the relevant information and the consequent data, statements, principles, evidence, judgments, beliefs, opinions, concepts, descriptions, questions, or other forms of representation to another. Explanation (explanation) means capable of reassuring things and the way that is coherent with the results of reasoning. This means to express and justify reasoning in terms of evidence, the conceptual, methodological, criteria, and contextual based on the consideration, and to present one's reasoning in the form of a convincing argument. Self-regulation means self-awareness to monitor the activities of a person's cognitive, elements that are used in these activities and the results, particularly by applying skills in analysis and evaluation of the self-assessment with a view to asking, confirm, validate, or correct any of the results or reasoning someone.

2.3 Problem Solving and Critical Thinking Skills

Problem solving is an attempt to find a way out of the difficulties in order to achieve a goal that is not so easy to be reached soon. Polya (1985) proposed four steps to resolve the problem, which is to understand the problem, problem-solving strategies to plan, the implementation of problem-solving strategies, and check back all the steps that have been undertaken [8]. Phase understand the problem, without any understanding of the given problem, students may not solve the problem correctly, then the students should be able to plan problem-solving strategies. Completion of the problem, in this phase is very dependent on the student experience more creative in preparing the resolution of a problem, if the plan for settlement of the problems has been made either in writing or not. The next step is the student is able to solve the problem, in accordance with the plans that have been prepared and considered appropriate. The final step of the process of resolving the problem by Polya is checked back all the steps that have been done starting from the first phase up to the third phase. With this model it is unnecessary errors occur can be corrected so that students can find answers that really correspond to a given problem.

Relationship between problem solving and critical thinking skills can be presented with the table 1.

Table 1. Relationship between problem solving and critical thinking skills

No.	problem-solving steps	Critical Thinking Skills
1	Understanding the problem	Interpretation: - categorization, - decode - clarification of meaning
2	Devising a plan	Analysis: - Examination of ideas - Identification of the argument - Analysis of arguments Evaluation: – Assessment of the claim (statement) , – Evaluation arguments
3	Carrying out the plan	Inference: - Questioning the claims, alternative thinking, - drawing conclusions, - problem solving, - decision making,

IOP Conf. Series: Journal of Physics: Conf. Series **947** (2017) 012073 doi:10.1088/1742-6596/947/1/012073

		Explanation: - presentation of the problem, - declare the result, - justification procedure, - the articulation of the argument
4	Looking back	Regulation yourself: - research himself - correcting himself

Relations troubleshooting steps with critical thinking skills in Table is a reference to see the critical thinking skills of students in solving problems of critical thinking math problems. The components contained in the table is used to assess students' critical thinking skills in mathematics.

3. Research Methods

The purpose of this study was to describe the critical thinking skills of male students in class VIII SMP in solving mathematical problems. Based on the object of obtaining a description of the critical thinking skills of male students of class VIII junior high math ability in solving mathematical problems. To accomplish this use of qualitative research. Qualitative research approach used in order to understand the phenomenon of what is experienced research subjects holistically at a special natural context and look at what is behind the symptoms that occur.

This research was conducted in the Luqman Al Hakim Junior High School of Surabaya. Selected junior high school students of eight grade to this consideration, the first based on the age of students is currently on a formal level, so being able to think more abstractly to produce answers that are critical, both students at this level have sufficient knowledge and experience in the materials basic math, because it has been past the primary school level in which there are materials, such as numbers, shapes geometry or other forms of algebra, and three students at this level is still the basic education level, so that the results of the exploration level of critical thinking this could be the foundation or guidelines for the next education level.

Based on the opinions expressed previously, that the cognitive abilities influential in the way of thinking to solve mathematical problems, researchers found critical thinking skills in mathematical problem solving was also influenced by the mathematical skills of students. With this study, researchers wanted to know how critical thinking skills in male students with high math skills. Thus, in this study the researchers classify subjects into three groups: subjects with mathematical ability is high, medium and low.

Techniques used in the selection of subjects is sought subjects who meet the criteria based on the results of mathematical ability Mathematical Ability Test (SMEs) and subjects that can communicate the idea clearly and allow the (potential) explore critical thinking skills. Selection of subjects starting from the highest level and then the lower level. Subjects selected based on the ability of mathematics math skills test results compiled by the researchers themselves. The questions were given to the test, the matters relating some of the material has been obtained prior to the study testing. For your consideration, note also the result of the average daily in learning mathematics in the classroom according to information (information / data) math teachers concerned. Based on test results obtained mathematical skills and a few other considerations that have been mentioned above, students were grouped into three groups: high, medium, and low.

Subjects were eight grade junior high school. Selection of research subjects is done by selecting a male subject with high math skills by comprehension mathematical abilities test.

Data collection instrument in this study is the researchers themselves, while supporting instruments are (1) Tests of mathematical ability (2) issues critical thinking math problem and (3) Guidelines for interviews: interviews conducted in this research is interview-based problems, which do by means of subjects given the task of writing and given time to finish it. Afterwards, subjects were interviewed

IOP Conf. Series: Journal of Physics: Conf. Series **947** (2017) 012073 doi:10.1088/1742-6596/947/1/012073

based on the work already done. During the interview conducted with the assistance handycam observation. Purpose of the interview is to investigate the process of critical thinking of students who do research subjects in solving critical thinking mathematical problems. Through interviews are expected to explore and reveal the students' critical thinking skills.
Methods of data collection in this study is Critical Thinking Problems Math (CTPM) and in-depth interviews (in-depth interviews) based CTPM. After the data collected is then performed checking the validity of the data with (1) a test of credibility, that is the truth test data obtained or known as the internal validity, (2) the reliability test, the test of consistency or constancy of data obtained using triangulation of time, (3) transferability test or external validity testing is not done, since the research findings are not generalizable, and (4) confirmability test or test of objectivity will be fulfilled by itself, when the internal validity of the study are met.

Furthermore, after the data is valid, then the data analysis that includes categorization / classification of data, data reduction, data interpretation, and conclusion.

4. Conclusion and Discussion

4.1 Conclusion

The results showed that the critical thinking skills of eighth grade students of junior high math ability in solving mathematical problems are as follows:
Phase 1: Understanding the problem

At this stage used interpretation skills characterized by the presence of sub-sub-indicators: categorization, decode, and clarify the meaning.
Categorization: disclose information, description, statement, or the required data related to the problem posed.
Decode: read the statement, information or data are required and can express what they know and ask related problems presented.
Clarification of meaning: expressing or clarifying statements, information or data related to the problem posed.
Phase 2: Planning Problem Solving Strategies

At this stage of Strategic Planning Problem Solving use skills of analysis: examination of ideas, arguments identification and analysis of the arguments and skills evaluation: assessing arguments. Examination of ideas: expressing arguments and descriptions that will be used to make a claim concerning the issues raised, only on the problem with a critical question: "What's wrong?", And the problem with a critical question: "Assess whether the statement ... What do you think about ... ? ", while the problems with critical questions" that meet certain conditions "does not appear.
Identification arguments: declare the arguments or the basics need making a conclusion concerning the issues raised, only on the problem with a critical question: "What's wrong?", And the problem with a critical question: "Assess whether the statement ... What do you think about ...? ", while the problems with critical questions" that meet certain conditions "does not appear.
Analysis arguments: declare the arguments or the basics of calculations related to the problems posed needed in making a conclusion on the next step is implementation-related problem-solving strategies.
Assessing argument: states convinced of the veracity arguments concerning the issues raised and expressed confidence once with the conclusions made, after some time the calculation process.

Phase 3: Implementation Strategies Problem Solving

In this stage used the skills of inference: inference, and problem solving skills and an explanation: the presentation of the problem, justification of procedures, and the articulation of the argument.
Conclusion: propose drawing conclusions based on the analysis of available information on the problem and based on the results of calculations related to the problems posed.
Problem Solving: mention the additional information needed to resolve the problems.

IOP Conf. Series: Journal of Physics: Conf. Series **947** (2017) 012073 doi:10.1088/1742-6596/947/1/012073

Presentation of the problem: explain the relationship between the information contained on the issue and the interpretation on the matter referred to convincing arguments used in the conclusion.
Justification procedure: explain the steps taken to find solutions to the problems posed answer.
Articulation argument: explain arguments or reasons answers used in drawing conclusions related to the problems posed.
Phase 4: Looking Back All Steps Have Done

In this stage the use of self-regulation skills characterized by the presence of sub-sub-indicators: self-correcting and self-study.
Proofreading yourself: revisiting the calculation and the results of analysis that has been done before taking a final decision. At problems with a critical question: "What's wrong?" The students aware of miscalculation, and can fix it up properly.
Research myself: On the issue of "who meet certain conditions with a solution of more than one" describes how well the method used, but on three other issues critical thinking skills that do not appear.

4.2 Discussion

Based on the conclusions of this research there are several suggestions submitted:

1. Based on the results of critical thinking skills "alternative thinking" appears on male students with high math skills. Based on this, the teacher in the classroom needs to develop in solving mathematical problems allow students to use the skills "alternative thinking". Thus the critical thinking skills "alternative thinking" can develop better in all students.
2. Based on the research skills of analysis with sub indicators argument analysis and identification of optimal arguments appear in male students with high math skills. Based on the researchers gave suggestions for teachers in the classroom in order to provide more opportunities for students to put forward an argument either orally or in writing. For example in learning more discussions about mathematical problems in it requires a lot of arguments in solving the problem.
3. Problem of critical thinking with critical questions that "meets certain conditions" and likely to bring a lot more than one correct answer, provide opportunities for students raises critical thinking skills optimally. Based on the research of male students with high math skills can be optimally achieve it. With that researchers gave suggestions for teachers in the classroom to give students the chance to get more variations of critical thinking math problems, so that all critical thinking skills can be achieved optimally.

5. References

[1] Ennis R 1992 Critical thinking: What is it? *Proceedings of the Forty-Eighth Annual Meeting of the Philosophy of Education Society Denver*, Colorado, March 27-30. Retrieved from http://www.ed.uiuc.edu/PES/92_docs/Ennis.HTM.

[2]The Partnership for 21st Century Skills 2009 *P21 Framework Definitions* Retrieved from http://www.21stcenturyskills.org.

[3] Huitt W 1998 Critical thinking: An overview. *Edu Psy Int*. Valdosta, GA: Valdosta State University. Retrieved from, http://www.edpsycinteractive.org/topics/cogsys/critthnk.html

[4] Krulik S and Rudnick J A 1999 Innovative Tasks to Improve Critical and Creative Thinking Skills. In Stiff L V and Curcio F R (eds) *Developing Mathematical reasoning in Grades K-12. 1999 Year book* (Reston: The National Council of teachers of Mathematics, Inc.) pp.138-145

[5] Ismail 2012 *Laporan Pelaksanaan Kegiatan In-House Training Peningkatan Mutu Pembelajaran SMP Mata Pelajaran Matematika Di Kab. Ponorogo 2012*. Surabaya: Jurusan Matematika FMIPA Universitas Negeri Surabaya

[6] Facione PA 2011 *Critical thinking: What It Is and Why It Counts.* Retrieved from http://www.insightassessment.com/padaf files/what & why98.padaf.

[7] Solso R L 1995 *Cognitive Psychology* (Boston: Allyn and Bacon)

[8] Polya G 1985 *How To Solve It* 2nd ed (New Jersey: Princeton University Press)

IOP Conf. Series: Journal of Physics: Conf. Series **947** (2017) 012074 doi:10.1088/1742-6596/947/1/012074

Quadcopter Path Following Control Design Using Output Feedback with Command Generator Tracker LOS Based At Square Path

A T Nugraha[1] dan T Agustinah[2]

[1,2]Department Of Electrical Engineering, Institut Teknologi Sepuluh Nopember, Surabaya, Indonesia
[1]E-mail: [1]anggaratnugraha@yahoo.co.id, [2]trihastuti@ee.its.ac.id

Abstract. Quadcopter an unstable system, underactuated and nonlinear in quadcopter control research developments become an important focus of attention. In this study, following the path control method for position on the X and Y axis, used structure-Generator Tracker Command (CGT) is tested. Attitude control and position feedback quadcopter is compared using the optimal output. The addition of the H∞ performance optimal output feedback control is used to maintain the stability and robustness of quadcopter. Iterative numerical techniques Linear Matrix Inequality (LMI) is used to find the gain controller. The following path control problems is solved using the method of LQ regulators with output feedback. Simulations show that the control system can follow the paths that have been defined in the form of a reference signal square shape. The result of the simulation suggest that the method which used can bring the yaw angle at the expected value algorithm. Quadcopter can do automatically following path with cross track error mean X=0.5 m and Y=0.2 m.

1. Introduction

As the development of methods used to deal with control problems in quadcopter [1], nonlinear control techniques is used for controlling hover. More back stepping nonlinear method can approach the characteristics of nonlinear systems. Thus, the calculation steps used more complicated iteration. Even with such complicated calculations, quadcopter capable of following the path corresponding references from the condition values of x, y and z. But when quadcopter by disruption of the constant wind then the response generated ripple and robust experience.

In addition, the use of a linear approach as one of the other control strategies in solving quadcopter control described in [2] - [5]. One way to obtain the optimal conditions quadcopter tracking controller is using Linear Quadratic Tracking (LQT). Furthermore, adding the models that quadcopter is able to follow the reference signal and minimize overshoot is applied in which a simple linear model approach is able to keep representing the nonlinear models of quadcopter. At the time quadcopter tracking, the controller is also able to reduce their interference. When tracking the z axis by interference, then the system cannot handle distractions well when tracking [2].

Other conditions which are also described in tracking problems using step signal treated with the linear approach is the method of Linear Quadratic Regulator (LQR) by adding the integral effect / compensator [3]. Systems that are linearized first and added concept integrator/compensator. This system capable of making quadcopter tracking by using a reference signal form step without any overshoot although the time needed to reach steady state for a long time.

In [4], approached through linearization of nonlinear systems with state feedback gain is obtained by using the LMI pole placement. Furthermore, the use Takagi-Sugeno fuzzy approach was taken into account to implement a linear control algorithm on each linear subsystem. The simulation results of fuzzy control system Takagi-Sugeno show that the response system can perform exact trajectory tracking system. However, the response on the condition of roll and pitch angle cannot follow the reference signal and the oscillation occurs.

An advance improvement to be made to improve the system response to the quadcopter in tracking issue is using the optimal method that is optimal output feedback by adding the structure of the Command-Generator Tracker (CGT) [5], this improvement is capable of tracking in the form of a circle, square or helix / spiral with, The method which is used to obtain the reference signal tracking conformity will result the tracking error as small as possible. To maintain a stable attitude, use of output feedback controller that performs more reliable H∞ so quadcopter in dealing with interference from outside.

In addition, the use of Static Output Feedback (SOF) [6] is used to overcome the limitations of a measurable state where SOF design has the advantage of requiring only measurable signals from the plant to be controlled. SOF controller used on the helicopter are H∞ performance as controls for tracking the position and attitude. Loop shaping techniques used in tracking control design position with the Gain controller Ricatti searched using the solution equation. The control algorithm is presented produces a robust control against disturbance.

Representation of reference input in Command Generator Tracker (CGT) structure is discuss in [5], [7], and [8]. Reference signal that used in trajectory or following path can be from all kind of signal. this structure can guarantee zero steady-state tracking error. The arrangement of the heading of a plant to a path or reference signal is performed with Line of Sight (LOS) algorithm [9]-[10].

The aim of this paper is to design a control system using output feedback with a Command-Generator Tracker (CGT) based on Line of Sight (LOS) to do following path by arranged the heading. The path in this systems are circle and square. The quadcopter moves in following given path in hover condition.

2. Quadcopter Model

Physical modeling will be complex if there are no assumptions used to simplify the equation on a quadcopter. Quadcopter has 6 degrees of freedom (DOF) with 12 outputs, which is 6 out of 12 outputs determine attitude of quadcopter. The kinematic and dynamic quadcopter models are delivered base on Newton-Euler with some simplification assumptions. quadcopter structure which is assumed rigid and symmetry. Propeller structure which is assumed rigid also thrust and drag force is proportional to the squared of propeller velocity.

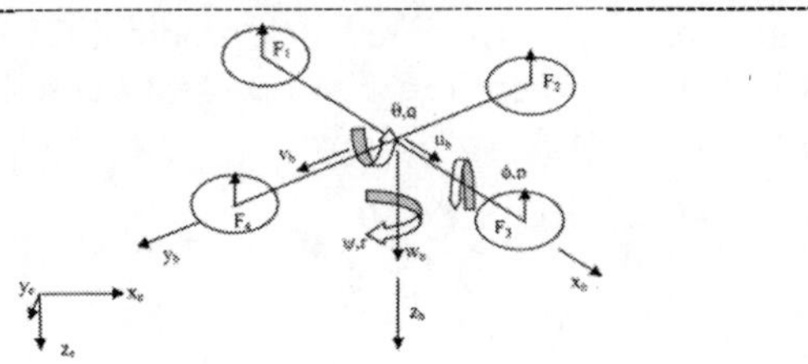

Figure 1. All Figure for State of Quadcopter (*e* for Earth Frame and *b* for *Body Frame Quadcopter*)

The modeling of translational axis lies on the earth coordinate. The transformation from body frame coordinate to earth frame coordinate is using rotation matrix . This matrix is obtained from multiplying rotational matrix times x, y, z axis, expressed by:

$$ {}_{B}^{E}R=\begin{bmatrix} c\theta c\psi & c\phi s\psi + s\phi s\theta s\psi & s\phi s\psi \\ c\theta s\psi & c\phi c\psi + s\phi s\theta s\psi & s\phi c\psi + c\phi s\theta s\psi \\ s\theta & s\phi c\theta & c\phi c\theta \end{bmatrix} \tag{1} $$

With s is the notation of sin, c is the notation of cos

IOP Conf. Series: Journal of Physics: Conf. Series **947** (2017) 012074 doi:10.1088/1742-6596/947/1/012074

2.1. LQR Output Feedback with Command-Generator Tracker (CGT)

Command-Generator Tracker (CGT) is a control design structure that provided compensator according to system requirements to get small tracking errors for the desired reference input $r(t)$. Linear time invariant is written in differential equation which The plant equation and output performance $z(t)$ are:

$$\dot{X}(t) = Ax(t) + Bu(t) + Dw(t) \tag{2}$$

$$y(t) = Cx(t) \tag{3}$$

$$z(t) = Hx(t) \tag{4}$$

The initial conditions of the reference signal r(t) is satisfied with the following differential equation:

$$r^{(d)} + a_1 r^{(d-1)} + \ldots + a_d r = 0 \tag{5}$$

with d as the order and ai as coefficient.

In [7], CGT has made based on dynamic signal reference. However, to fulfill the desired design specification, designing CGT is not using all derivatives from reference signal.

In this control technique, the path following problem will be converted into regulator problem in which error is regulated to 0. To create the whole result of path following system, error tracking modified is necessary:

$$\Delta(s)e = \Delta(s)r - \Delta(s)Hx = -H\xi \tag{6}$$

ξ is modified state vector,

To express dynamic from $\xi(t)$, (5) is operated in Δ(p) to get (6), which p is d/dt in time domain. Define the dynamics by multiplying augmented that operated in Δ(p) so that the modified system is as follows:

$$\dot{\tilde{x}} = \tilde{A}\tilde{x} + \tilde{B}\tilde{u}$$

$$\frac{d}{dt}\begin{bmatrix} \varepsilon \\ \xi \end{bmatrix} = \begin{bmatrix} G & \vdots & 0 \\ & \vdots & -H \\ \cdots & \vdots & \cdots \\ 0 & \vdots & A \end{bmatrix} \begin{bmatrix} \varepsilon \\ \xi \end{bmatrix} + \begin{bmatrix} 0 \\ B \end{bmatrix} \mu \tag{7}$$

LQ regulator design implemented on this system (7). If the state goes to zero, then the tracking error e(t) will be vanished. So, state feedback output of the system is [7]- [8]:

$$\tilde{u} = \begin{bmatrix} K_\varepsilon & K_p \end{bmatrix} \begin{bmatrix} \varepsilon \\ C\xi \end{bmatrix} \tag{8}$$

The control law [9] has solved using LQ regulator with output feedback problem. The solution of this problem adopted the concept as described in [6]-[7]. Therefore, the necessary conditions of the augmented system (7) for LQ regulator with output feedback are as follows:

$$\frac{\partial H}{\partial S} = \tilde{A}_C{}^T P + P\tilde{A}_C + \tilde{C}^T \tilde{K}^T R\tilde{K}\tilde{C} + Q = 0 \tag{9}$$

$$\frac{\partial H}{\partial P} = \tilde{A}_C S + S\tilde{A}_C{}^T + X = 0 \tag{10}$$

$$\frac{1}{2}\frac{\partial H}{\partial K} = R\tilde{K}\tilde{C}S\tilde{C}^T - \tilde{B}^T PS\tilde{C}^T = 0 \tag{11}$$

Which $\tilde{A}_C = \tilde{A} - \tilde{B}\tilde{K}\tilde{C}$ and $X = E\{x(0)x^T(0)\}$. If the weighting matrix $R > 0$ and $\tilde{C}S\tilde{C}^T$ is a nonsingular, the output feedback gain $\tilde{K}$ has the following form:

$$K = R^{-1}\tilde{B}^T PS\tilde{C}^T(\tilde{C}S\tilde{C}^T)^{-1} \tag{12}$$

By using LQ regulator method with output feedback on X and Y, the gain control value is obtained.

2.2. LOS (Line of Sight) Law on Steering Equations

IOP Conf. Series: Journal of Physics: Conf. Series **947** (2017) 012074 doi:10.1088/1742-6596/947/1/012074

The enclosure based strategy is used to drive $e(t)$ toward zero, then direct the velocity vector to the point of intersection $P_{los}^n = [X_{los}, Y_{los}]^T$ to match the heading of the path, this path is implicitly defined by the order in which the waypoints has been defined. The path directly involved the determination, direct assignment $\chi^{(t)}$.

$$\tan(\chi_d(t)) = \frac{\Delta y(t)}{\Delta x(t)} = \frac{y_{Los} - y(t)}{x_{Los} - x(t)} \tag{13}$$

The centre coordinate of a vehicle $\{b\}$ is defined Pn=[X,Y]T and used a circle with a radius R>0, then drawn from the centre point of the vehicle $\{b\}$. That circle will cut a path into two points, which one of the dots is (X_{los}, Y_{los}). The calculation of two unknown LOS points $P_{los}^n = [X_{los}, Y_{los}]^T$ can be found by solving two equations below:

$$[x_{los} - x(t)]^2 + [y_{los} - y(t)]^2 = R^2$$
$$\tan(\alpha_k) = \frac{y_{k+1} - y_k}{x_{k+1} - x_k} = \frac{y_{los} - y_k}{x_{los} - x_k} = constant \tag{14}$$

Where the equation is the Pythagoras equation, whereas the equation shows the angle of the path to the axis of the earth x is a_k. The slope between the two waypoints is constant and also applies for the magnitude of the slope to the points between them. The process of calculating the enclosure analytically based on [9]-[10] can be explained by:
Argument 1: for $|\Delta x| > 0$, So the equation

$$y_{los} = \frac{\Delta y}{\Delta x}(x_{los} - x_k) + y_k \tag{15}$$

With $\Delta x := x_{k+1} - x_k$ and $\Delta y := y_{k+1} - y_k$ are difference position x and y between two *waypoints*.

3. Design Control System

Inner loops are used to stabilize roll, pitch, and yaw angles that guarantee performance. The controller is designed based on the Static Output Feedback (SOF) problem, in which the feedback output gain K is obtained by completing the Linear Matrix Inequality (LMI) iterative algorithm. The LMI form that uses Schur Complement is

$$\begin{bmatrix} P_nA + A^TP_n + Q + L_n^TR^{-1}L_n & P_nB & P_nD \\ B^TP_n & R & 0 \\ D^TP_n & 0 & -\gamma^2 I \end{bmatrix} \leq 0 \tag{16}$$

The result of state system linearization is for roll model, pitch, and yaw, then it becomes

$$x_{inner} = [\phi \quad p \quad \upsilon \quad \theta \quad q \quad v \quad \psi \quad r]^T \tag{17}$$

In the design of the rotation controller, the main variables that controlled are the angle of roll and pitch, also three speeds i.e. angular speed of roll, pitch, and yaw. Therefore, the output vector of this controller is

$$x_{inner} = [\phi \quad \theta \quad \psi \quad p \quad q]^T$$
$$\tilde{u} = [[K_\varepsilon \quad K_p]\begin{bmatrix} \varepsilon \\ C\xi_p \end{bmatrix} \tag{18}$$

Inertia moment for roll and pitch have the same value ($J_{xx} = J_{yy} = 0.03\ \mathrm{kg.m^2}$), so the roll dynamic equals with the pitch dynamic. These are the models of state space linier roll/pitch.

The best result is obtained by the parameter value as follows:

$\gamma = 0.7$, R = 10 and Q = diag {755, 3, 0.1}.

with the iterative LMI feasibility method, the result are P, K, and L matrix after the 12th iteration.

$$\begin{bmatrix} \dot\phi \\ \ddot\phi \\ \dot v \end{bmatrix} = \begin{bmatrix} 0 & 1 & 0 \\ 0 & 0 & 800 \\ 0 & 0 & -15 \end{bmatrix}\begin{bmatrix} \phi \\ \dot\phi \\ v \end{bmatrix} + \begin{bmatrix} 0 \\ 0 \\ 15 \end{bmatrix}\Delta u; y_\phi = C_\phi \begin{bmatrix} \phi \\ \dot\phi \\ v \end{bmatrix} = \begin{bmatrix} 1 & 0 & 0 \\ 0 & 0.1 & 0 \end{bmatrix}\begin{bmatrix} \phi \\ \dot\phi \\ v \end{bmatrix} Z_\phi = Hx_\phi = [0.3 \quad 0 \quad 0]\begin{bmatrix} \phi \\ \dot\phi \\ v \end{bmatrix} \tag{19}$$

IOP Conf. Series: Journal of Physics: Conf. Series **947** (2017) 012074 doi:10.1088/1742-6596/947/1/012074

$$p = \begin{bmatrix} 29.8331 & 0.1572 & 0.4897 \\ 0.1572 & 0.0156 & 0.0545 \\ 0.4897 & 0.0545 & -\gamma^2 I \end{bmatrix}; K = \begin{bmatrix} 0.7346 & 0.0817 \end{bmatrix} \quad (20)$$

The objective control position of path following can make the quadcopter follow the desired path when it is in hover condition. Therefore, the reference signal that given is for performing path following x position, which is a sinusoidal signal with a frequency of 0.1047 rad / s or equal to 0.0167 Hz with a phase equal to -0.26 rad. Here is the reference of signal equation:

$$x_d = -\sin(0.05\pi t) \quad (21)$$

If it modified using Laplace transform, then the obtained is:

$$\Delta(p)x_d = p^2 + 0.02 = 0 \quad (22)$$

$$\dot{x}_d = \begin{bmatrix} 0 & 1 \\ = 0.02 & 0 \end{bmatrix} x_d = Gx_d \quad (23)$$

By substituting the reference model and the system model, the result of quadcopter dynamics for the x axis is as follows:

$$\begin{bmatrix} \dot{\varepsilon} \\ \ddot{\varepsilon} \\ \dot{x} \\ \ddot{x} \\ \dot{v} \end{bmatrix} = \begin{bmatrix} 0 & 1 & 0 & 0 & 0 \\ -0.02 & 0 & -1 & 0 & 0 \\ 0 & 0 & 0 & 1 & 0 \\ 0 & 0 & 0 & 0 & 136.2683 \\ 0 & 0 & 0 & 0 & -15 \end{bmatrix} \begin{bmatrix} \varepsilon \\ \dot{\varepsilon} \\ x \\ \dot{x} \\ v \end{bmatrix} + \begin{bmatrix} 0 \\ 0 \\ 0 \\ 0 \\ 15 \end{bmatrix} U_1 \quad (24)$$

With output performance as follows

$$\tilde{y} = \tilde{C}\tilde{x} = \begin{bmatrix} J & 0 \\ 0 & C \end{bmatrix} \begin{bmatrix} \varepsilon \\ x \end{bmatrix} = \begin{bmatrix} 1 & 0 & 0 & 0 \\ 0 & 1 & 0 & 0 \\ 0 & 0 & 1 & 0 \\ 0 & 0 & 0 & 1 \end{bmatrix} \begin{bmatrix} \varepsilon \\ \dot{\varepsilon} \\ x \\ \dot{x} \\ v \end{bmatrix} \quad (25)$$

The path which want to be assigned as a reference signal to the quadcopter is in circle shape (in a 3-dimensional field). Therefore, the reference signal that given for Y position is:

$$Y_d = -1 - \cos(0.05\pi t) \quad (26)$$

The polynomial characteristic equation is

$$\Delta(p)y_d = p^3 + 0.02p = 0 \quad (27)$$

The polynomial characteristic equation of the reference signal with orde d=3 is written in a matrix state space with 3×3 dimension. Since the following model CGT structure is a representation of the reference signal, not all derivatives of the reference signal are needed to be used in the system structure. So that in this design a modification can be written as follows:

$$\Delta(p)y_d = 0.02p = 0 \Rightarrow \dot{y}_d = \begin{bmatrix} 0 & 1 \\ 0 & -0.02 \end{bmatrix} y_d = Gy_d \quad (28)$$

thus, the modification of multiple augmented system for the dynamic position on the Y axis is

$$\begin{bmatrix} \dot{\varepsilon} \\ \ddot{\varepsilon} \\ \dot{y} \\ \ddot{y} \\ \dot{v} \end{bmatrix} = \begin{bmatrix} 0 & 1 & 0 & 0 & 0 \\ 0 & -0.02 & -1 & 0 & 0 \\ 0 & 0 & 0 & 1 & 0 \\ 0 & 0 & 0 & 0 & -136.2683 \\ 0 & 0 & 0 & 0 & -15 \end{bmatrix} \begin{bmatrix} \varepsilon \\ \dot{\varepsilon} \\ y \\ \dot{y} \\ v \end{bmatrix} + \begin{bmatrix} 0 \\ 0 \\ 0 \\ 0 \\ 15 \end{bmatrix} U_1 \qquad \tilde{y} = \begin{bmatrix} 1 & 0 & 0 & 0 \\ 0 & 1 & 0 & 0 \\ 0 & 0 & 1 & 0 \\ 0 & 0 & 0 & 1 \end{bmatrix} \begin{bmatrix} \varepsilon \\ \dot{\varepsilon} \\ y \\ \dot{y} \\ v \end{bmatrix} \quad (29)$$

IOP Conf. Series: Journal of Physics: Conf. Series **947** (2017) 012074 doi:10.1088/1742-6596/947/1/012074

The output feedback gains $\widetilde{K}$ for the X, Y position controllers are obtained as follows:

$$\widetilde{K}_x = [-4.418 \quad -9.6604 \quad 9,4440 \quad 5.3433]$$
$$\widetilde{K}_y = [15.9326 \quad 23.4931 \quad -15.8108 \quad -6.2181] \tag{30}$$

4. Simulation Results

Matlab is a software that used to know the effectiveness of quadcopter path following. Quadcopter is simulated by testing its movement, whether it is capable to doing the following path for square path as a tracking reference for its movement. The heading of quadcopter is considered to ensure the smoothness movement and maintain its heading as shown in Figure 2.

With adding Line of Sight (LOS) algorithm to calculate the heading of quadcopter, therefore yaw angle (ψ) can be calculated too. By observing the yaw angle, quadcopter can keep up the reference signal from the output controller of the outer loop with an average cross track error of 0.0245 rad. The comparison between reference and the actual yaw angle of quadcopter, it can be concluded that the heading is maintained well can be seen in figure 3.

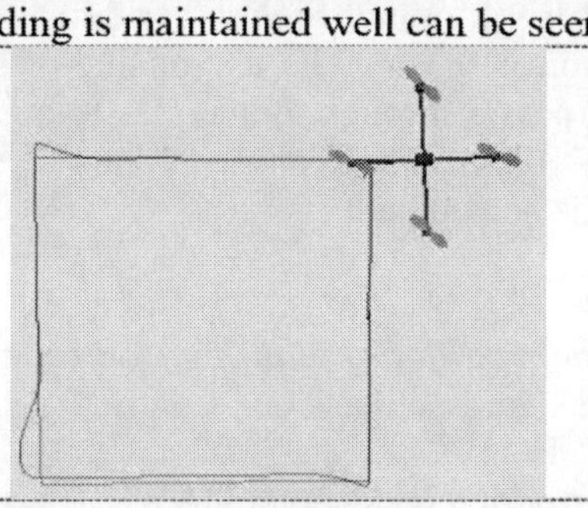

Figure 2. Quadcopter *Path Following* on X and Y-axis

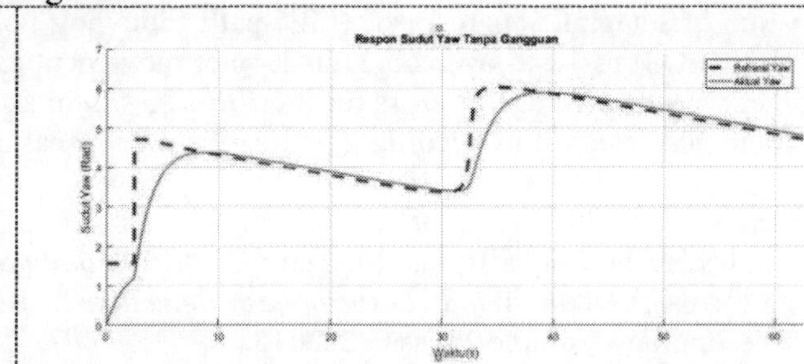

Figure 3. Quadcopter *Path Following* on Yaw Angle.

Quadcopter position on the X-axis which is able to follow the reference can be seen in figure 4. Path following is start at t=15s and the delay is about 2 seconds, that cause error of 0.3 m however the quadcopter still can follow the reference path..

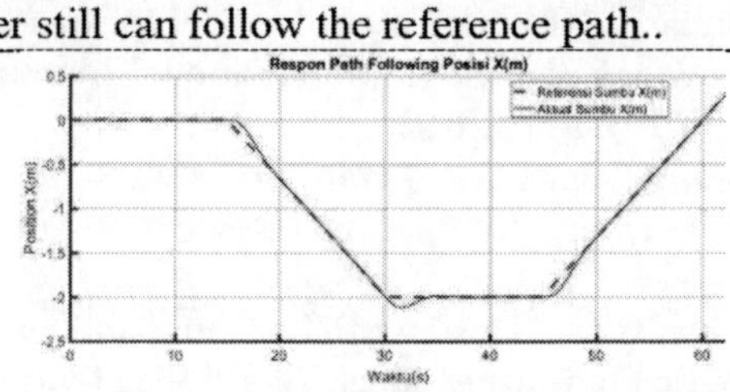

Figure 4. Quadcopter *Path Following* on X-axis.

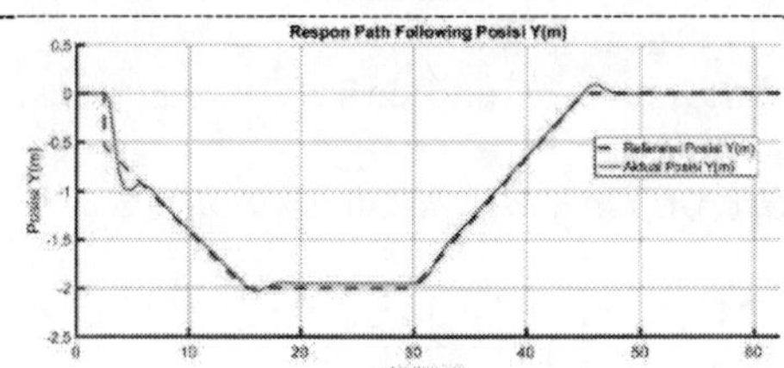

Figure 5. Quadcopter *Path Following* on Y-axis.

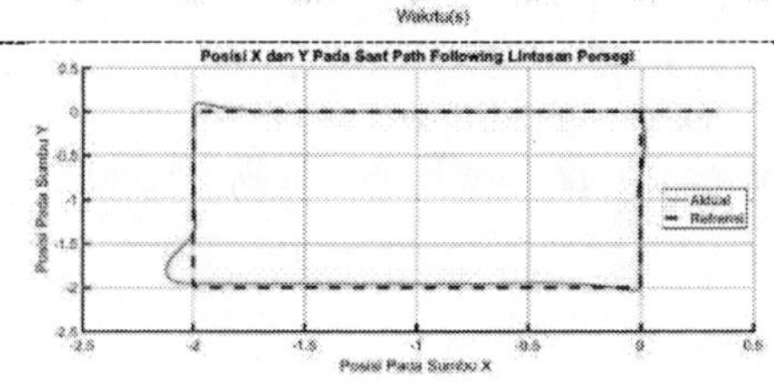

Figure 6. Comparison between Reference and Actual Tour on Square Path.

IOP Conf. Series: Journal of Physics: Conf. Series **947** (2017) 012074 doi:10.1088/1742-6596/947/1/012074

The movement on the Y-axis where the quadcopter has a delayed about 0.4 seconds at the first waypoint and it cause an error path about 0.3 m but the quadcopter can return to the reference path in the Figure 5.

Square shaped reference signal is also added in to the system, as shown in Figure 6. The result is, the quadcopter keeps up follow the path, but there is a deviation occurred at X=0.5 m and Y=0.2 m. LOS algorithm assists the quadcopter controller by giving heading to turn at an angle of 90° smoothly and keeping on track to the reference.

5. Conclusion

The output feedback control method with H∞ performance is used in the inner loop while the Command-Generator Tracker structure controls the model following by adding the LOS algorithm used on the outer loop. The control system which using output feedback with command generator tracker based on LOS on the Path Following Quadcopter problems, it can provide performance as desired.

The quadcopter is able to follow the reference signal. Furthermore, the heading of the quadcopter is also able to follow according to the reference signal obtained from Line of Sight algorithm (LOS). The value of Integral Square Error (ISE) path following position is 0.3 m. The use of line of sight algorithm (LOS) assist to overcome quadcopter movement as it passes through 900 angles to become smoother with the average of cross track error is X=0.5 m and Y=0.2 m. The proposed control system quadcopter also can follow according to the reference signal in the square path.

References

[1] Cabecian D, Cunha R, and Silvestre C., "*A Globally Stabilizing Path Following Controller for Rotorcraft With Wind Disturbance Rejection",* IEEE Transactions on Control *System*s Technology Volume 23 issue 2, 2015.

[2] Suicmez E C and Kutay A T., "*Optimal path tracking control of a* quadcopter UAV", International Conference on Unmanned Aircraft *System*, May 2014

[3] Vivas E C, Pena G M V and Rodriguez F C I, "Simulation of The Quadrotor Controller with LQR with Integral Effect," ABCM Symposium Series in Mechatronics, vol. 5, pp. 390-399, 2012.

[4] Yacef F, Bouhali O, Khebbache H and Boudjema F., "Takagi-Sugeno model for quadcopter modelling and control using *nonlinear state feedback* contro*ll*er", International Journal of Control Theory and Computer Modelling (IJCTCM), May 2012.

[5] Agustinah T, Isdaryani FF, and Nuh M," *Tracking* Control of Quadrotor Using Static Output Feedback with Modified Command-Generator Tracker," International Review of Automatic Control (I.RE.A.CO.), Vol. 9, N. 4, July,2016.

[6] Gadewadikar J, Lewis F L, Subbarao K and Chen B M, "Design of H∞ Command and Control Loops for Unmanned Aerial Vehicles using Static Output-Feedback," in Proceedings of the 46th IEEE Conference on Decision and Control, New Orleans, 2007

[7] Lewis F L and Syrmos V L., "Optimal Control *System* – 2nd ed", John Wiley & Sons, Inc., Canada, 1995.

[8] Lewis F L, Stevens B L dan Johnson E N., "*Aircraft Control and Simulation* – 3nd ed", John Wiley & Sons, Inc., Canada, 2016

[9] Tsourdos A, White B A and Shanmugavel M., "Cooperative path planning of Unmanned Aerial Vehicles", John Wiley & Sons, Ltd, 2011.

[10] Fossen T I., "Handbook of Marine Craft Hydrodynamics and Motion Control", John Wiley & Sons, Inc., Canada, 2011

[11] Quanser, Quanser Q-Ball X-4 User Manual, Quanser Innovate Educate, 2010, Doc. 888 Rev.2

IOP Conf. Series: Journal of Physics: Conf. Series **947** (2017) 012075 doi:10.1088/1742-6596/947/1/012075

Introducing Nine-Point Circle to Junior High School Students

S Fiangga[1,*], M A N Azizah[1], R N K Rini[1], and A N Hidayanti[1]

[1]Department of Mathematics, State University of Surabaya, Ketintang Street, Surabaya 60231, Indonesia

*shofanfiangga@unesa.ac.id

Abstract. The concept of circles is an ancient concept that has appeared since Ancient Egypt from which this concept gives many significant contributions in mathematics' development until now. Nevertheless, the concept of circles hides many uncover mysterious features that are of applications in mathematics. One of the mysterious features is the Nine-Point Circle. This Nine-point circle is also known as Euler's circle, six-point circle, Feuerbach's circle, the twelve-point circle, and many others. Because of these different names, there have been misunderstand among mathematicians about the Nine-Point Circle's history. Besides, the discussion of Nine-Point Circle can be used to be an initial material to explain elementary geometry topic in junior high school's level curriculum of 2013. Therefore, this concept needs to be delivered to the students as a geometry introduction. A possible form of the integration historical aspect of Nine-point circle is suggested in this paper as well as its importance in the curriculum of 2013.

1.Introduction

Based on the government regulation issued by the of ministry educational and cultural of Republic Indonesia number 24 in 2016 about curriculum, the curriculum to be used for junior high school is Curriculum 2013[1]. Curriculum 2013 is a curriculum that covers three important domains namely cognitive domain, psychomotor domain, and affective domain. In the recent Curriculum 2013, the mathematics textbook has implemented content with perspective of history in mathematics. This historical content is implemented in order to make sure that the mathematics concept is taught meaningfully. Nevertheless, the improvement of mathematics curriculum in Curriculum 2013 has omitted several competencies and mathematics materials. This development should be administered properly such that important competencies would not disappear.

In general, the integrating of history of mathematics (HOM) in classroom mathematics has become an issue in mathematics education research [2]–[5]. One of the implementation of history of mathematics in the classroom practice is the integration of HOM in the mathematics text book. In the mathematics textbook, the perspectives of history may be presented into three ways. They are (1) the perspectives of history are attached in the beginning part of chapter before the main part as motivational resources, (2) the perspectives of history are attached in the main part of chapter as conceptual development activities resource, and (3) the perspectives of history are attached in the end part of chapters alternative fun game resources[6]. In addition, the perspective of history can also be used for introducing some mathematics material in school. One of historical content that can be introduced is Nine-Point Circle. Nine-Point Circle is a circle that passing through nine significant concyclic points defined from any triangle. The related material of Nine-Point Circle to junior high school material is the topic of circumcircle of triangle. Unfortunately, this material has been removed from the textbook and curriculum basic competence. However, in trigonometry material of senior high school mathematics, the concept of triangle circumcircle is used for proving sine rule. Therefore, the circumcircle of triangle needs to be taught to student before teaching about the sine rule in senior high school.

As the preparation for the teaching the circumcircle of triangle, student needs to know the construction of Nine-Point Circle. Nine-Point Circle can be constructed by the special line of triangle, altitude line, which has been taught in triangle material of 7th grade. With the fact that the Nine-Point Circle construction can be introduced before the teacher explanation about circumcircle of triangle in 8th grade in circle topic, easier content could be accepted by the students and restore the assumption of Nine-Point Circle historical discovery to student. This paper would discuss the opportunity to improve the material of excircle of triangle using HOM in the recent curriculum 2013.

2. Method

Literature review is used in this study to develop possible implementation of Nine-Poinr Circle. The materials are collected by reviewing and combining some related sources of Nine-Point Circle and school mathematics through books, journals, proceedings, and internet. these collected materials are processed into some steps as shown in diagram bellow (Figure 1):

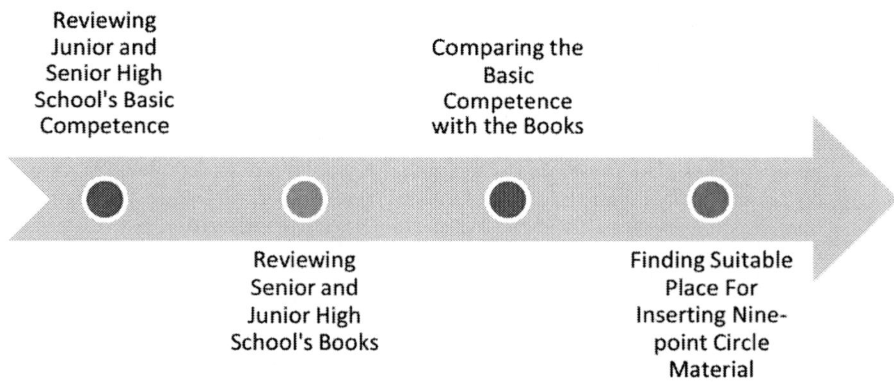

Figure 1. Steps to processed the collected materials

In the first step, the basic competence of senior and junior high school was reviewed to elicit information about suitable material for the topic of Nine-Point Circle. Review is conducted from not only the basic competence stated on the textbook, but also the delivery package of the textbooks. Subsequently, both of basic competence and books are compared to elicit the correlation between mathematics textbooks with the basic competence. In this step, all materials, questions, and quiz are also reviewed. Then, the last step is finding suitable place for inserting Nine-Point Circle as introduction based on review results. In this paper is also added High Order Thinking Skills concept related to curriculum 2013 as the government suggests to exist in every material.

3. Results

3.1 Construction of Nine-Point Circle

As the introduction of circumcircle material, students can construct Nine-Point Circle which is used to explore problems related to altitude line and perpendicular bisector of triangle in 7th grade. The steps of Nine-Point Circle's construction as follows:

a. Draw a triangle ABC. (see Figure 2)
b. Construct the midpoints of the three sides. Label them as L, M, N. (see Figure 3)

IOP Conf. Series: Journal of Physics: Conf. Series **947** (2017) 012075 doi:10.1088/1742-6596/947/1/012075

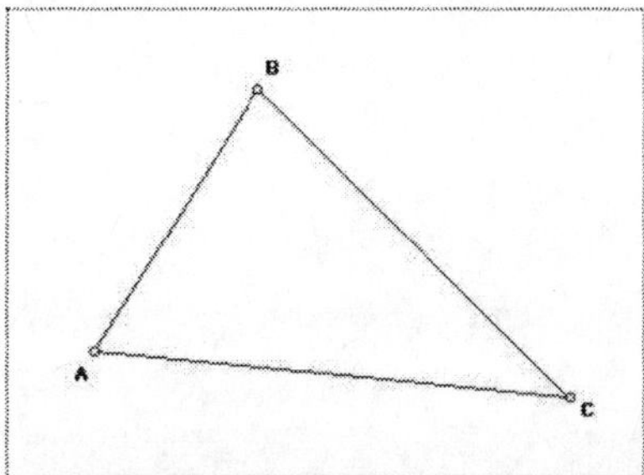

Figure 2. Triangle ABC

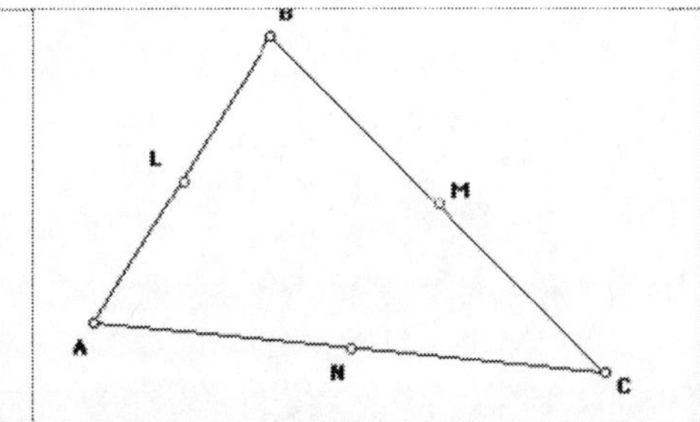

Figure 3. Midpoint L, M and N.

c. Construct the altitudes of the triangle ABC. Label the intersections of them with the triangle as D, E, F. Label the point of intersection of the three altitudes as H. (see Figure 4)

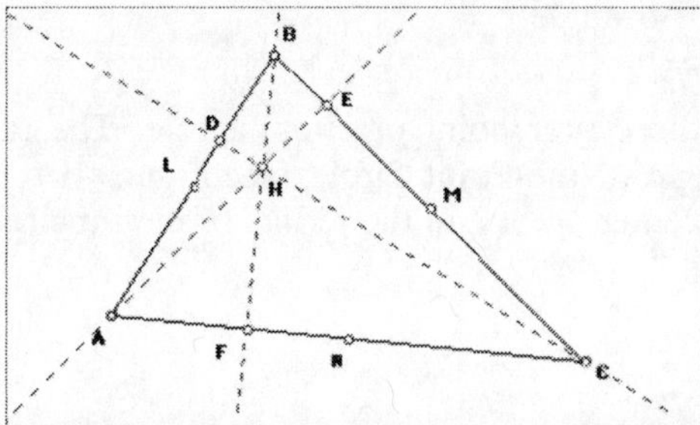

Figure 4. Triangle altitudes and their intersection

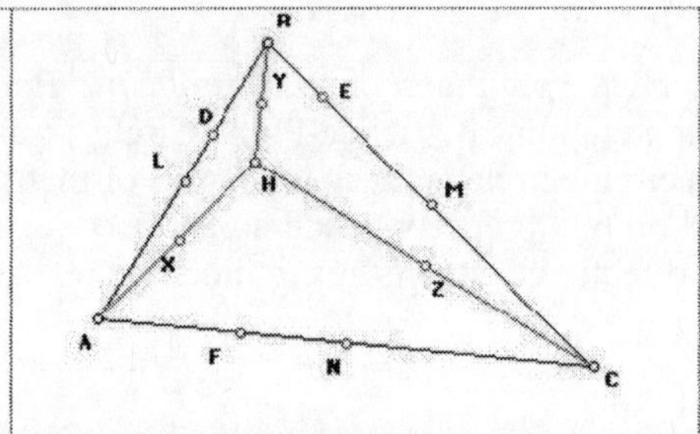

Figure 5. Midpoints of segments AH, HB and CH

d. Construct the midpoints of the segments AH, HB, CH. Label them as X, Y, Z. (see Figure 5)

e. Find the center of the Nine-Point Circle. Construct triangle LMN and find intersection point of three perpendicular bisectors of triangle LMN as the center of Nine-Point Circle. Label the center as U. (see Figure 6)

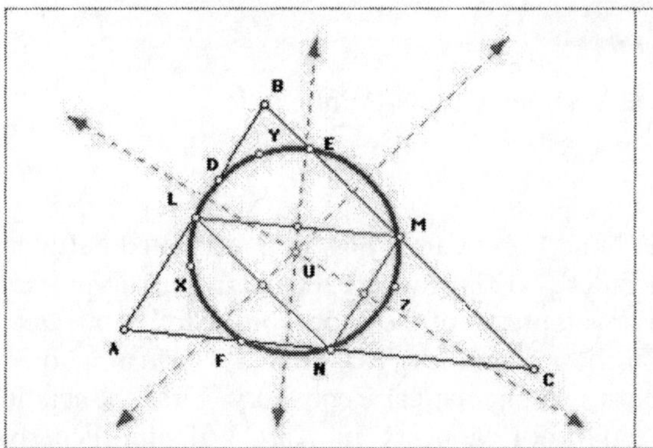

Figure 6. Center of Nine-Point Circle

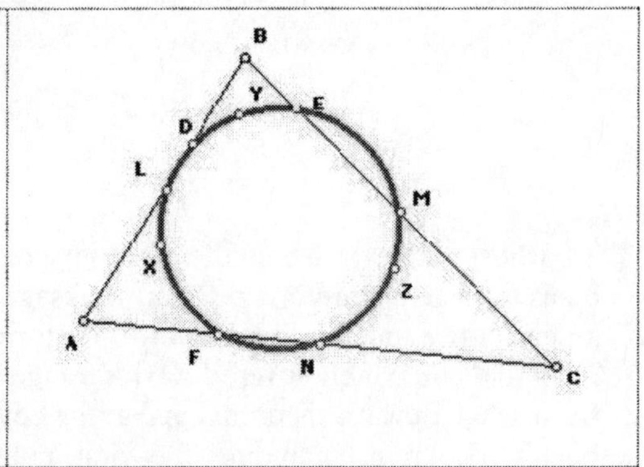

Figure 7. Nine-Point Circle

f. Draw circle passing through point L, M, N, D, E, F, X, Y and Z. (see Figure 7)

3.2 Properties

3.2.1 Medial triangle and orthic triangle

The Nine-Point Circle of a reference triangle is the circumcircle of both the reference triangle's medial triangle (with vertices at the midpoints of the sides of the reference triangle) (see Figure 8) and its orthic triangle (with vertices at the feet of the reference triangle's altitudes) (see Figure 9).

This properties help students to conclude the definition of circumcircle that "circumcircle is circle that is outside the triangle and passing through all of vertices of the triangle"[9].

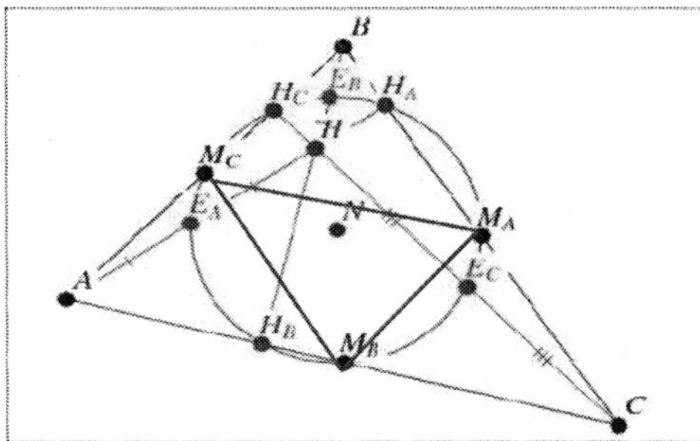

Figure 8. Circumcircle of medial triangle

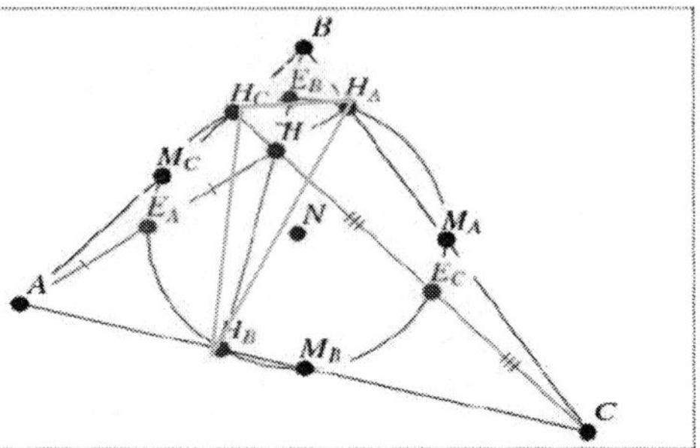

Figure 9. Circumcircle of orthic triangle

3.2.2 The center point and the radius of Nine-Point Circle

The center point of Nine-Point Circle is also the center point of circumcircle. The radius of a triangle's circumcircle is twice the radius of that triangle's Nine-Point Circle. (see Figure 10)

Based on this property, student can discover the center point and the radius of circumcircle. Then, they can directly construct the circumcircle.

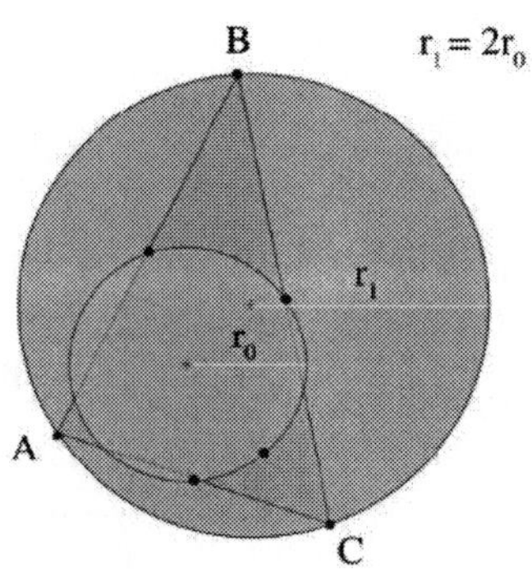

Figure 10. Center point and radius of Nine-Point Circle

3.3 History

The earliest author to whom the discovery of the Nine-Point Circle has been attributed is Euler, but no one has ever given a reference to any passage in Euler's writings where the characteristic property of this circle is either stated or implied [10]. Historical development of the Nine-Point Circle appeared first in 1804 until 1860 which covered a wide range of territory such English, French, German, and Swiss. Start from 1804 Bevan's theorem appears in Leybourn's Mathematical Repository. First enunciation of Feuerbach's Theorem, including the first published proof, appears in Karl Wilhelm Feuerbach's *EigenschafteneinigermerkwiirdigenPunkte des geradlinigenDreiecks*, along with many other interesting proofs relating to the Nine-Point Circle in 1822 discovered as the former Nine-Point Circle. However, Feurerbach only found the six point circle. These points are the midpoints of the three sides of thetriangle and the feet of the triangle perpendicular bisectors (MA, MB, MC, HA, HB, and HC).The circle is officially designated the "Nine-Point Circle" (*le cercle des neufpoints*) by Olry Terquem in 1842. Terquem published the second analytical proof of the theorem that the Nine-Point Circle touches the incircle and the circumcircle of triangle. He was the first person that recognized the added significance of the three midpoints between the triangle's vertices and the orthocenter. And he also called the circle as the Nine-Point Circle.

4. Discussion

Permendikbud Republic Indonesia number 24 in 2016 states that curriculum 2013 (K13) is the official curriculum used in elementary school, junior high school, and senior high school [1]. This government regulation declares the structure and basic design of curriculum 2013. There are core competence (KI) and basic competence (KD) that should be taught to the students. In curriculum 2013, 2 core competences and 60 basic competences for mathematics' subject in junior high school is stated to be taught. Furthermore, students need to improve their High-Order Thinking to satisfy knowledge aspect. Therefore, problems that require student to think creatively and analytically are needed.

Basic competence (KD) 3.11 of curriculum 2013 for 7th grade students of junior highs school is associate the formula of circumferences and areas of various types of rectangles (square, rectangle, rhombus, parallelogram, trapezoid, and kit) and triangle. Understanding special lines of triangle is part of this competence. In mathematics' textbook for junior high school student 7th grade 2nd semester 2016 revise edition, this material is taught in learning activity 8.6 in which the definition of altitude line, median line, perpendicular bisector, angle bisector, and how to construct them are included [7]. The use of historical perspective in mathematics' learning, giving non-routine problems can improve motivation and intellectual wealth of students [8]. Therefore, based on the historical aspect of Nine-Point Circle, the material can be developed as non-routine problem for students after learning about special lines of triangle. Students can construct Nine-Point Circle by applying how to construct altitude line and perpendicular bisector that they have learned. In detail, before students construct it, history of Nine-Point Circle can be inserted to gain the student interest.

The related material of Nine-Point Circle to junior high school material is circumcircle of triangle's material. Unfortunately, this material has been removed from the textbook and curriculum basic competence. Removing incircle and circumcircle in curriculum 2013 is not appropriate because they will be used as important prerequisite topic for the student in senior high school. In 11th grade of senior high school, basic competence (KD) 3.9 state that the students are able to explain sine and cosine rule. In proofing these rules, the students need to use the idea of circumcircle (see the Figure 11 for the example of problem and Figure 12 for the solution of the problem). Therefore, circumcircle material needs to be taught in Junior High School as the prerequisite before sine and cosine is taught. This circumcircle material is suggested to be inserted to curriculum inside circle chapter before basic competence (KD) 3.8 explain and construct interior tangent line and exterior tangent line of two circle. Then, Nine-Point Circle can be inserted as introduction of this new basic competence (KD).

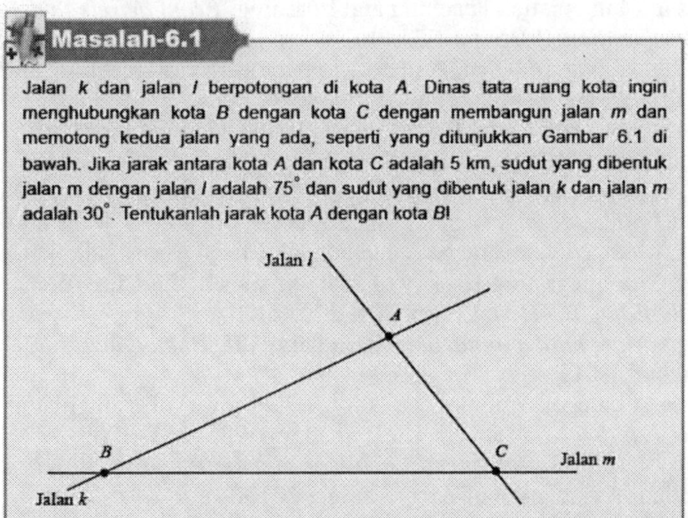

Figure 11. Problem in mathematics textbook 11th grade (obtained from [7])

IOP Conf. Series: Journal of Physics: Conf. Series **947** (2017) 012075 doi:10.1088/1742-6596/947/1/012075

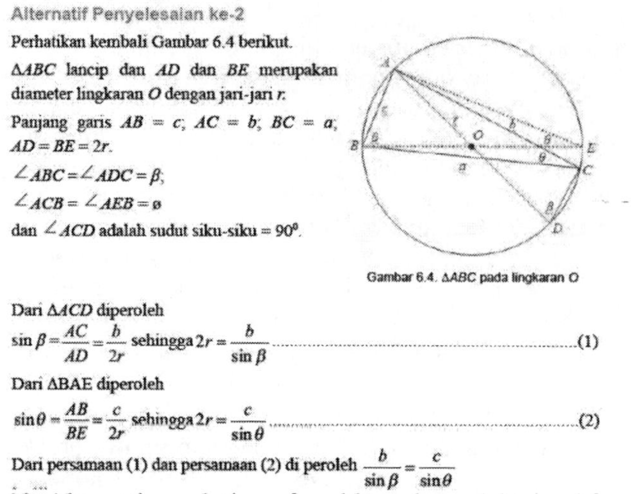
Alternatif Penyelesaian ke-2

Perhatikan kembali Gambar 6.4 berikut.

ΔABC lancip dan AD dan BE merupakan diameter lingkaran O dengan jari-jari r.

Panjang garis $AB = c$; $AC = b$; $BC = a$; $AD = BE = 2r$.

$\angle ABC = \angle ADC = \beta$;

$\angle ACB = \angle AEB = ø$

dan $\angle ACD$ adalah sudut siku-siku = 90^0.

Gambar 6.4. ΔABC pada lingkaran O

Dari ΔACD diperoleh

$\sin\beta = \frac{AC}{AD} = \frac{b}{2r}$ sehingga $2r = \frac{b}{\sin\beta}$ ....(1)

Dari ΔBAE diperoleh

$\sin\theta = \frac{AB}{BE} = \frac{c}{2r}$ sehingga $2r = \frac{c}{\sin\theta}$ ....(2)

Dari persamaan (1) dan persamaan (2) di peroleh $\frac{b}{\sin\beta} = \frac{c}{\sin\theta}$

Figure 12. Alternative solution of problem above (obtained from [7])

5. Conclusion

The perspective of history also can be used to introduce some mathematics material in school, for instance Nine-Point Circle. Nine-Point Circle is a circle that passing through nine significant concyclic points defined from any triangle. The related material of Nine-Point Circle to junior high school material is circumcircle of triangle's material. Unfortunately, this material has been removed from the textbook and curriculum basic competence. Nevertheless, in trigonometry material of senior high school mathematics are found that the excircle of triangle concept is used to prove sine rule. Therefore, this excircle material is suggested to be inserted to curriculum inside circle chapter before basic competence (KD) 3.8. Then, Nine-Point Circle can be inserted as introduction of this new basic competence (KD).

References

[1] Kemendikbud, *Peraturan Pemerintah Kementerian Pendidikan dan Kebudayaan (Permendikbud) Republik Indonesia nomor 24 tahun 2016*. Indonesia, 2016.

[2] L. Radford, Of Love, Frustration, and Mathematics: A Cultural-Historical Approach to Emotions in Mathematics Teaching and Learning, *From beliefs to dynamic affect systems in mathematics education*, 2015, pp. 25–49.

[3] M. Doorman and J. van Maanen, A historical perspective on teaching and learning calculus, *Aust. Sr. Math. Teach.*, vol. 22, no. 2, pp. 4–14, 2008.

[4] A. S. Ozdemir, S. Goktepe, and I. Kepceoglu, Using Mathematics History to Strengthen Geometric Proof Skills,*Procedia - Soc. Behav. Sci.*, vol. 46, pp. 1177–1181, 2012.

[5] B. Savizi, Applicable problems in the history of mathematics: practical examples for the classroom,*Teach. Math. its Appl.* , vol. 26, no. 1, pp. 45–50, Mar. 2007.

[6] C. Tzanakis, A. Arcavi, C. C. de Sa, M. Isoda, C. Lit, M. Niss, J. P. de Carvalho, M. Rodriguez, and M.-K. Siu, Integrating History of Mathematics in the Classroom: An Analytic Survey, *Hist. Math. Educ. ICMI Study*, pp. 201–240, 2000.

[7] Kemendikbud, *Matematika untuk SMP/MTs kelas VII: Buku Guru, seri edisi lengkap*. Jakarta: Kemendikbud, 2013.

[8] U. T. Jankvist, A categorization of the 'whys' and 'hows' of using history in mathematics education, *Educ. Stud. Math.*, vol. 71, no. 3, pp. 235–261, 2009.

[9] G. A. Venema, *Exploring Advanced Euclidean Geometry with GeoGebra USA*. USA: Mathematical Association of America, Inc, 2013.

[10] J. S. MacKay, *History of the Nine Point Circle. Proceedings of the Edinburgh*, 1892. [Online]. Available:http://jwilson.coe.uga.edu/emt668/emt668.folders.f97/anderson/geometry/geometry1project/historyofninepointcircle/history.html.

Institute of Physics
Dirac House, Temple Back
Bristol BS1 6BE UK

ISSN: 1742-6588
ISBN 978-1-5108-5565-6

9 781510 855656

CURRAN ASSOCIATES INC.
proceedings.com